VOLUME SIX HUNDRED AND SIX

METHODS IN ENZYMOLOGY

Radical SAM Enzymes

METHODS IN ENZYMOLOGY

Editors-in-Chief

ANNA MARIE PYLE
Departments of Molecular, Cellular and Developmental Biology and Department of Chemistry
Investigator, Howard Hughes Medical Institute
Yale University

DAVID W. CHRISTIANSON
Roy and Diana Vagelos Laboratories
Department of Chemistry
University of Pennsylvania
Philadelphia, PA

Founding Editors

SIDNEY P. COLOWICK and NATHAN O. KAPLAN

VOLUME SIX HUNDRED AND SIX

METHODS IN ENZYMOLOGY

Radical SAM Enzymes

Edited by

VAHE BANDARIAN
Department of Chemistry,
University of Utah, Salt Lake City,
UT, United States

ELSEVIER

ACADEMIC PRESS
An imprint of Elsevier

Academic Press is an imprint of Elsevier
50 Hampshire Street, 5th Floor, Cambridge, MA 02139, United States
525 B Street, Suite 1650, San Diego, CA 92101, United States
The Boulevard, Langford Lane, Kidlington, Oxford OX5 1GB, United Kingdom
125 London Wall, London, EC2Y 5AS, United Kingdom

First edition 2018

ISBN: 978-0-12-812794-0
ISSN: 0076-6879

For information on all Academic Press publications
visit our website at https://www.elsevier.com/books-and-journals

Publisher: Fiona Pattison
Acquisition Editor: Ashlie Jackman
Editorial Project Manager: Shellie Bryant
Production Project Manager: Divya Krishna Kumar
Cover Designer: Miles Hitchen

Typeset by SPi Global, India

CONTENTS

CONTRIBUTORS

Eyal Akiva
Department of Bioengineering and Therapeutic Sciences, University of California, San Francisco, CA, United States

Kylie D. Allen
Department of Chemistry and Biochemistry, Gonzaga University, Spokane, WA, United States

Patricia C. Babbitt
Department of Bioengineering and Therapeutic Sciences; Department of Pharmaceutical Chemistry; Quantitative Biosciences Institute, University of California, San Francisco, CA, United States

Edward D. Badding
Department of Chemistry, The Pennsylvania State University, University Park, PA, United States

Emily P. Balskus
Department of Chemistry and Chemical Biology, Harvard University, Cambridge, MA, United States

Vahe Bandarian
Department of Chemistry, University of Utah, Salt Lake City, UT, United States

Tadhg P. Begley
Department of Chemistry, Texas A&M University, College Station, TX, United States

Dhananjay M. Bhandari
Department of Chemistry, Texas A&M University, College Station, TX, United States

Smaranda Bodea
Merck & Co., Inc., Boston, MA, United States

Squire J. Booker
Department of Biochemistry and Molecular Biology; Department of Chemistry; The Howard Hughes Medical Institute, The Pennsylvania State University, University Park, PA, United States

Joan B. Broderick
Department of Chemistry and Biochemistry, Montana State University, Bozeman, MT, United States

William E. Broderick
Department of Chemistry and Biochemistry, Montana State University, Bozeman, MT, United States

Shoshana D. Brown
Department of Bioengineering and Therapeutic Sciences, University of California, San Francisco, CA, United States

Nathan A. Bruender
Department of Chemistry and Biochemistry, St. Cloud State University, St. Cloud, MN, United States

Leah B. Bushin
Department of Chemistry, Princeton University, Princeton, NJ, United States

Amanda S. Byer
Department of Chemistry and Biochemistry, Montana State University, Bozeman, MT, United States

Sara Calhoun
Department of Bioengineering and Therapeutic Sciences; Graduate Program in Biophysics, University of California, San Francisco, CA, United States

Julia D. Cramer
Department of Chemistry, University of Hawaii at Manoa, Honolulu, HI, United States

Min Dong
Department of Chemistry and Chemical Biology, Cornell University, Ithaca, NY, United States

Sean J. Elliott
Department of Chemistry, Boston University, Boston, MA, United States

Dmytro Fedoseyenko
Department of Chemistry, Texas A&M University, College Station, TX, United States

Lauren K. Gadsby
Department of Biochemistry and Molecular Biology, The Pennsylvania State University, University Park, PA, United States

Tyler L. Grove
Department of Chemistry, The Pennsylvania State University, University Park, PA, United States

Gemma L. Holliday
Department of Bioengineering and Therapeutic Sciences, University of California, San Francisco, CA, United States

Yilin Hu
Department of Molecular Biology and Biochemistry, University of California, Irvine, CA, United States

Stella Impano
Department of Chemistry and Biochemistry, Montana State University, Bozeman, MT, United States

Joseph T. Jarrett
Department of Chemistry, University of Hawaii at Manoa, Honolulu, HI, United States

Andrew Jasniewski
Department of Molecular Biology and Biochemistry, University of California, Irvine, CA, United States

Sumedh Joshi
Department of Chemistry, Texas A&M University, College Station, TX, United States

Judith P. Klinman
California Institute for Quantitative Biosciences; Department of Chemistry; Department of Molecular and Cell Biology, University of California, Berkeley, CA, United States

Joseph W. LaMattina
Department of Chemistry, The Pennsylvania State University, University Park, PA, United States

Julia K. Lewis
Department of Chemistry, University of Utah, Salt Lake City, UT, United States

Hening Lin
Howard Hughes Medical Institute, Cornell University, Ithaca, NY, United States

Nilkamal Mahanta
Department of Chemistry, Texas A&M University, College Station, TX, United States

Stephanie J. Maiocco
Department of Chemistry, Boston University, Boston, MA, United States

Ana M. Martins
California Institute for Quantitative Biosciences, University of California, Berkeley, CA, United States

Erin L. McCarthy
Department of Biochemistry and Molecular Biology, The Pennsylvania State University, University Park, PA, United States

Elizabeth C. McDaniel
Department of Chemistry and Biochemistry, Montana State University, Bozeman, MT, United States

Elaine C. Meng
Resource for Biocomputing, Visualization, and Informatics, Department of Pharmaceutical Chemistry, School of Pharmacy, University of California, San Francisco, CA, United States

Rolf Müller
Department of Microbial Natural Products, Helmholtz Institute for Pharmaceutical Research Saarland, Helmholtz Center for Infection Research, Saarland University, Saarbrücken, Germany

Haoran Pang
Department of Biochemistry, Duke University Medical Center, Durham, NC, United States

Ursula Pieper
Department of Bioengineering and Therapeutic Sciences, University of California, San Francisco, CA, United States

Lee Rettberg
Department of Molecular Biology and Biochemistry, University of California, Irvine, CA, United States

Markus Walter Ribbe
Department of Molecular Biology and Biochemistry; Department of Chemistry, University of California, Irvine, CA, United States

Andrej Sali
Department of Bioengineering and Therapeutic Sciences; Department of Pharmaceutical Chemistry; Quantitative Biosciences Institute, University of California, San Francisco, CA, United States

Bastien Schnell
Department of Microbial Natural Products, Helmholtz Institute for Pharmaceutical Research Saarland, Helmholtz Center for Infection Research, Saarland University, Saarbrücken, Germany

Mohammad R. Seyedsayamdost
Department of Chemistry; Department of Molecular Biology, Princeton University, Princeton, NJ, United States

Kazuki Tanifuji
Department of Molecular Biology and Biochemistry, University of California, Irvine, CA, United States

Lindsey M. Walker
Department of Chemistry, Boston University, Boston, MA, United States

Bo Wang
Department of Chemistry, The Pennsylvania State University, University Park, PA, United States

Yuanyou Wang
Department of Chemistry, Texas A&M University, College Station, TX, United States

Robert H. White
Virginia Polytechnic Institute and State University, Blacksburg, VA, United States

Kenichi Yokoyama
Department of Biochemistry, Duke University Medical Center, Durham, NC, United States

Anthony P. Young
Department of Chemistry, University of Utah, Salt Lake City, UT, United States

Yugang Zhang
Department of Chemistry and Chemical Biology, Cornell University, Ithaca, NY, United States

Wen Zhu
California Institute for Quantitative Biosciences; Department of Chemistry, University of California, Berkeley, CA, United States

PREFACE

As early as five decades ago, the notion that radicals would be involved in enzymology was a heretical thought, and was rarely invoked, even if a reasonable polar mechanism could not be drawn. However, elucidation of the role of vitamin B_{12} in the reaction catalyzed by adenosylcobalamin dependent enzymes revealed that enzymatic transformations could involve free radicals as discrete intermediates. The intermediacy of the highly reactive oxidant, 5′-deoxyadenosyl radical (5′-dAdo•), which was postulated in the context of these transformations, revealed how unactivated substrates could be poised for a myriad of complex radical mediated transformations initiated by H-atom abstraction.

The pioneering studies of Frey and colleagues on lysine 2,3-aminomutase (LAM) revealed H-atom abstraction events in an enzyme that instead of a cobalamin, appeared to use *S*-adenosyl-L-methionine (SAM) and a 4Fe—4S cluster to catalyze similar transformations. LAM catalyzes reductive cleavage of SAM to generate 5′-dAdo•, which subsequently activates lysine *via* H-atom abstraction to initiate the reaction. Direct observation of many of the proposed reaction intermediates with both native substrate and substrates analogs cemented the paradigm that 4Fe-4S/SAM enzymes catalyze radical-mediated transformations. But, had it not been for bioinformatics, the story may have ended there and this volume of *Methods in Enzymology* would not exist.

In 2001, using the sequence of LAM as a guide, Sofia and colleagues proposed the presence of so-called radical SAM superfamily, which is distributed across all domains of life. The key insight was the realization that a CxxxCxxC motif is almost always present in these proteins. Subsequent studies have revealed that the motif binds the 4Fe—4S cluster, which in turn binds SAM. With decreasing costs and the widespread use of genome sequencing over the last decade, the community has come to realize that the radical SAM superfamily is, in fact, widespread and includes well over 100,000 examples across various genomes.

Despite their ubiquity, biochemical studies of the radical SAM superfamily have been slow because of numerous challenges. Most radical SAM enzymes are oxygen sensitive and all manipulations must be carried out under anoxic conditions. In most cases, the substrates are not known. Furthermore, obtaining these proteins in cofactor replete form can be an additional impediment. Methods for *In vitro* reconstitution of the cluster(s) are often unique to each enzyme and are usually discovered through trial and error. It is fair to say that *in vitro* reconstitution is still an art! Finally, the discovery that some members require additional auxiliary clusters, accessory proteins, or other cofactors, such as B_{12}, further highlight the challenges of working with this superfamily.

Fortunately, these challenges have not limited the field, but instead, have spurred many labs to search for creative solutions. Indeed, in the last decade, numerous radical SAM enzymes have been characterized; many of which are involved in biological transformations, which among others, include metabolic reactions, RNA/DNA modification, and biosynthesis of natural products.

In this *Methods in Enzymology* volume I have tried to highlight the challenges and solutions in the field by inviting chapters that detail various aspects of radical SAM enzymology. Many of the contributions focus on specific methods for a particular enzyme, while others focus on general methods necessary for the discovery and characterization of new members. However, the common goal is to provide a starting point for characterization of unknown radical SAM enzymes.

The increasing pace of discovery over the last decade in the radical SAM field foreshadows an exciting future, though there are still many challenges. The sheer numbers of radical SAM enzymes of unknown function underscores how little we know about this class of enzymes. However, as functional annotations begins to catch up with bioinformatics, predicting an enzyme's biological role may become possible. Unfortunately, obtaining recombinant proteins from a heterologous host in cofactor replete form is likely to remain a challenge for the foreseeable future; but, the solution is likely to come from the nexus of this field and the iron–sulfur cluster biogenesis community. If the last decade is any indication of things to come, we are in for a wild and exciting future. I hope this snapshot of current methods in the field helps accelerate discoveries.

Vahe Bandarian

CHAPTER ONE

Atlas of the Radical SAM Superfamily: Divergent Evolution of Function Using a "Plug and Play" Domain

Gemma L. Holliday[*,1,2], Eyal Akiva[*], Elaine C. Meng[†], Shoshana D. Brown[*], Sara Calhoun[*,‡,3], Ursula Pieper[*,4], Andrej Sali[*,§,¶], Squire J. Booker[||,#,**], Patricia C. Babbitt[*,§,¶,1]

[*]Department of Bioengineering and Therapeutic Sciences, University of California, San Francisco, CA, United States
[†]Resource for Biocomputing, Visualization, and Informatics, Department of Pharmaceutical Chemistry, School of Pharmacy, University of California, San Francisco, CA, United States
[‡]Graduate Program in Biophysics, University of California, San Francisco, CA, United States
[§]Department of Pharmaceutical Chemistry, University of California San Francisco, San Francisco, CA, United States
[¶]Quantitative Biosciences Institute, University of California, San Francisco, CA, United States
[||]Department of Biochemistry and Molecular Biology, The Pennsylvania State University, University Park, PA, United States
[#]Department of Chemistry, The Pennsylvania State University, University Park, PA, United States
[**]The Howard Hughes Medical Institute, The Pennsylvania State University, University Park, PA, United States
[1]Corresponding authors: e-mail address: gemma.l.holliday@gmail.com; babbitt@cgl.ucsf.edu

Contents

[2] Current address: Medicines Discovery Catapult, Mereside, Alderley Park, Alderley Edge, Cheshire, United Kingdom.
[3] Current address: Department of Energy, Joint Genome Institute, Walnut Creek, CA, United States.
[4] Current address: National Agricultural Library, Agricultural Research Service, United States Department of Agriculture, Beltville, MA, United States.

Methods in Enzymology, Volume 606
ISSN 0076-6879
https://doi.org/10.1016/bs.mie.2018.06.004

Abstract

The radical SAM superfamily contains over 100,000 homologous enzymes that catalyze a remarkably broad range of reactions required for life, including metabolism, nucleic acid modification, and biogenesis of cofactors. While the highly conserved SAM-binding motif responsible for formation of the key 5′-deoxyadenosyl radical intermediate is a key structural feature that simplifies identification of superfamily members, our understanding of their structure–function relationships is complicated by the modular nature of their structures, which exhibit varied and complex domain architectures. To gain new insight about these relationships, we classified the entire set of sequences into similarity-based subgroups that could be visualized using sequence similarity networks. This superfamily-wide analysis reveals important features that had not previously been appreciated from studies focused on one or a few members. Functional information mapped to the networks indicates which members have been experimentally or structurally characterized, their known reaction types, and their phylogenetic distribution. Despite the biological importance of radical SAM chemistry, the vast majority of superfamily members have never been experimentally characterized in any way, suggesting that many new reactions remain to be discovered. In addition to 20 subgroups with at least one known function, we identified additional subgroups made up entirely of sequences of unknown function. Importantly, our results indicate that even general reaction types fail to track well with our sequence similarity-based subgroupings, raising major challenges for function prediction for currently identified and new members that continue to be discovered. Interactive similarity networks and other data from this analysis are available from the Structure-Function Linkage Database.

ABBREVIATIONS

BioB biotin synthase
5′-dA 5′-deoxyadenosyl
HMM hidden Markov model
Knowns sequences of known function
LAM L-lysine 2,3-aminomutase
LipA lipoyl synthase
MDA multidomain architecture
Met methionine
MSA multiple sequence alignment
NrdG anaerobic ribonucleotide reductase activase
PFL-AE pyruvate formate lyase activating enzyme

PhnJ	alpha-D-ribose 1-methylphosphonate 5-phosphate C–P lyase
PQQ	pyrroloquinoline quinone
RSS	radical SAM superfamily
SAM	*S*-adenosylmethionine
SFLD	Structure-Function Linkage Database
SPL	spore photoproduct lyase
SSN	sequence similarity network
TIM	triose phosphate isomerase
Unknowns	sequences of unknown function

1. INTRODUCTION: OVERVIEW OF THE RADICAL SAM SUPERFAMILY

The widely studied radical *S*-adenosylmethionine (SAM) superfamily (RSS) was originally defined using bioinformatics techniques to survey the 650 RSS members available at that time. It described a homologous group of enzymes united by their utilization of SAM in a radical mechanism (Sofia, Chen, Hetzler, Reyes-Spindola, & Miller, 2001). The original sequence set came from 126 species representing all Kingdoms of life and included many of the first RSS enzymes to be characterized: L-lysine 2,3-aminomutase (LAM) (Moss & Frey, 1987), biotin synthase (BioB) (Lotierzo, Tse Sum Bui, Florentin, Escalettes, & Marquet, 2005; Reyda, Dippold, Dotson, & Jarrett, 2008), lipoyl synthase (LipA) (Cicchillo, Iwig, et al., 2004; Miller et al., 2000), pyruvate formate-lyase activating enzyme (PFL-AE) (Knappe & Sawers, 1990; Vey et al., 2008), and anaerobic ribonucleoside-triphosphate reductase activase (NrdG) (Padovani, Thomas, Trautwein, Mulliez, & Fontecave, 2001). From a chemical perspective, to be considered a member of the RSS, Sofia defined three characteristics that were minimally required:

- A unique three-cysteine motif that binds the [Fe_4S_4] cluster, leaving the apical iron free to bind the SAM moiety in a bidentate manner. This canonical motif, [CX_3CX_2C], is associated with a single domain and largely conserved in known structures (Fig. 1A). The vast majority of RSS proteins (more than 90%) contain this motif and the rest exhibit several different variations that have either cysteine residues one and two or two and three separated by two or three other residues. The number of residues between alternate cysteine pairs (cysteines 1 and 2 or cysteines 2 and 3) can vary from 3 to 22.

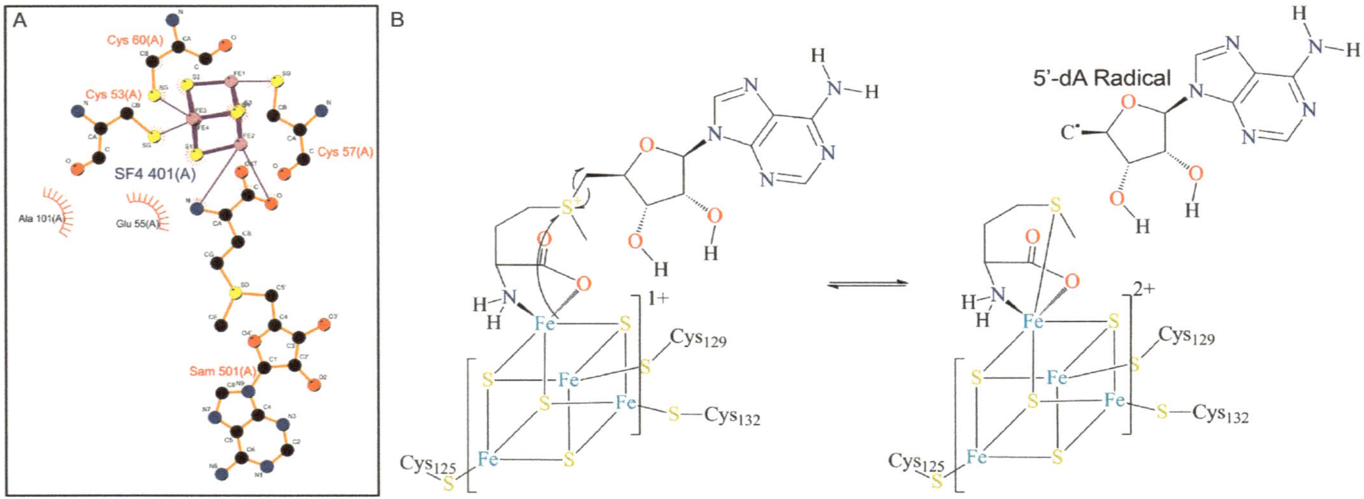

Fig. 1 (A) [Fe_4S_4]-binding motif from biotin synthase (PDB: 1R30). Image created using LigPlot+ (Laskowski & Swindells, 2011). (B) The common activation step associated with the canonical RSS.

- A common activation step involved in formation of the 5′-deoxyadenosyl (5′-dA) radical and methionine (Met) (Fig. 1B).[a]
- A requirement for an external electron donor to catalyze the initial reduction of the $[Fe_4S_4]$ cluster.

In this work, we refer to this superfamily as the "canonical" RSS to distinguish it from several other unrelated superfamilies that use SAM in a radical-like reaction.

The original set of 650 sequences has now grown over 150-fold (not counting sequences from the large compilations of metagenomic data now coming available). Known functions reveal that RSS enzymes catalyze a dizzying array of disparate and essential chemistries that range from the formation of complex metal cofactors (Dinis, Wieckowski, & Roach, 2016) (e.g., the formation of the [FeFe] hydrogenase metallocofactor in HydG (Pilet et al., 2009) and HydE (Nicolet et al., 2008)) to the formation of more than half of the over two dozen known organic cofactors (for example, biotin (Lotierzo et al., 2005; Reyda et al., 2008), lipoic acid (Cicchillo, Lee, et al., 2004; Miller et al., 2000), menaquinone (vitamin K) (Hiratsuka et al., 2008), and pyrroloquinoline quinone (PQQ) (Barr et al., 2016; Puehringer, Metlitzky, & Schwarzenbacher, 2008)). They are also involved in the modification of nucleic acids, often via methylation of aromatic carbon centers, repair of DNA dimers (as in spore photoproduct lyase (SPL) (Benjdia, Heil, Barends, Carell, & Schlichting, 2012; Yang & Li, 2015)), the formation of the wybutosine base on tRNA (Young & Bandarian, 2011), and the formation of complex natural products such as antibiotics (Mahanta, Hudson, & Mitchell, 2017a, 2017b) (for example, nosiheptide (LaMattina et al., 2017; Yu et al., 2009) and bleomycin (Tao et al., 2007)). Despite many excellent reviews (see Broderick, Duffus, Duschene, & Shepard, 2014; Dowling, Vey, Croft, & Drennan, 2012; Grell, Goldman, & Drennan, 2015; Lanz & Booker, 2012; Vey & Drennan, 2011; Wang et al., 2014 for some recent examples) and an ever-growing corpus of work that addresses mechanistic and structural features of individual or small groups of RSS enzymes, relatively little is known about the

[a] Although it has long been assumed that the only chemical feature that all RSS members have in common is the formation of the 5′-dA radical from SAM by the structurally conserved iron–sulfur cluster-binding motif, even this most fundamental and "obligate" descriptor of the canonical RSS no longer holds. At least one exception to this rule has now been identified. Although it is clearly homologous to other canonical RSS sequences, a tryptophan methyltransferase TsrM involved in the biosynthesis of the thiopeptide antibiotic thiostrepton, does not catalyze formation of a 5′-dA radical (Blaszczyk et al., 2016; Pierre et al., 2012). A few other similar as yet uncharacterized cobalamin-dependent methylases that methylate sp^2-hybridized carbon centers may represent additional exceptions.

structure–function relationships across all of the members of this complex and fascinating superfamily because the huge proportion of RSS enzymes remain experimentally uncharacterized.

As with the chemical perspective, the RSS is difficult to define from a sequence and structural perspective due to the varied multidomain architectures (MDAs) in which the core superfamily motif is embedded. For the global analysis described here, we started with the sequence sets that represent the superfamily domain containing the SAM-binding motif available from the widely used Pfam (Finn et al., 2016) and InterPro (Finn et al., 2017) resources, then expanded and curated the set to obtain over 100,000 nonredundant sequences representing the RSS superfamily (see Section 3.1).

Besides the canonical RSS, we note that at least three other homologous sets of other fold types catalyze RSS-like reactions using mechanisms that differ in key ways from those of the RSS described here. We have provisionally defined these groups as superfamilies because the members within each are sufficiently diverse that we expect they may catalyze additional reactions besides those for which they are named. We were unable to find sequence or structural relationships between the canonical RSS and any of these other superfamilies. As a result, we do not include any of the members of these other superfamilies in the canonical RSS.

These superfamilies include the radical SAM phosphonate metabolism superfamily (Kamat, Williams, & Raushel, 2011) that catalyzes the demethylation and cyclization of a phosphine by alpha-D-ribose 1-methylphosphonate 5-phosphate C—P lyase (PhnJ), the radical SAM phosphomethylpyrimidine synthase superfamily (Coquille et al., 2013) that catalyzes the phosphomethylpyrimidine synthesis reaction, and the radical SAM 3-amino-3-carboxypropyl radical-forming superfamily involved in diphthamide biosynthesis (Zhang et al., 2010). For the latter, both the radical formation and structural origin of the SAM-binding motif show differences from the canonical RSS, i.e., the initial radical formed is a 3-amino-3-carboxypropyl rather than the 5′-dA radical, while the iron–sulfur cluster is bound by a three-cysteine motif using cysteine residues from three different domains (rather than from a single domain as in the canonical RSS). Although these superfamilies are not discussed further in this report, additional information, including sequence similarity networks (SSNs) and other data about these noncanonical superfamilies, is available from the Structure-Function Linkage Database (SFLD) (Akiva et al., 2014). Fig. 2 shows a comparison of a canonical RSS structure with those of these three noncanonical superfamilies. An additional superfamily has been described in the

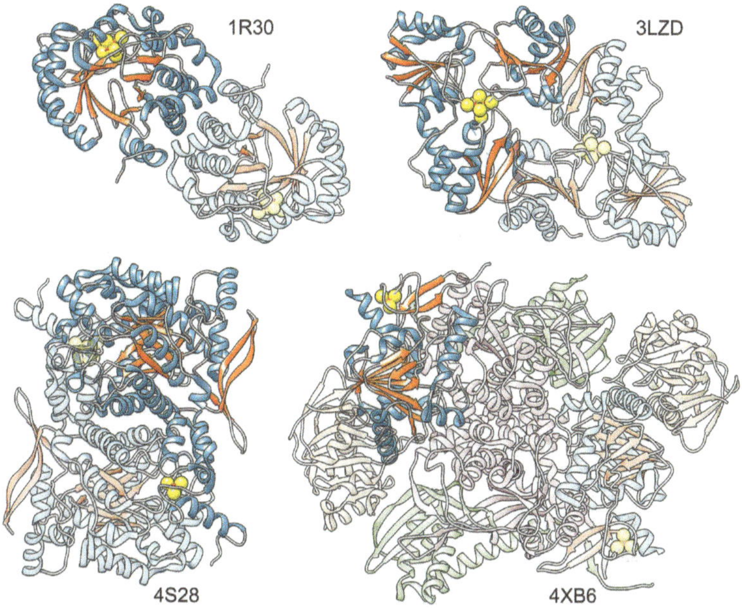

Fig. 2 Comparison of a canonical RSS structure with structures from unrelated superfamilies of other fold types whose members catalyze RSS-like chemistry. Chains containing the $[Fe_4S_4]$ cluster are colored by secondary structure, with helices in *blue* and strands in *orange*; one copy of the chain per structure is highlighted. The sulfur atoms from the $[Fe_4S_4]$ clusters and from their adjacent cysteine residues are shown as *spheres*. *Left to right top row*: canonical RSS, biotin synthase, PDB: 1R30; radical SAM 3-amino-3-carboxypropyl radical-forming superfamily, diphthamide synthetase, PDB: 3LZD; *bottom row*: radical SAM phosphomethylpyrimidine synthase superfamily, phosphomethylpyrimidine synthase, PDB: 4S28; and radical SAM phosphonate metabolism superfamily, PDB: 4XB6. The physiological unit of all of these structures is a homo-2-mer except for 4XB6, which is a hetero-8-mer.

literature that produces a 5′-dA radical but differs in other important details, such as in the source of the radical. For example, methylmalonyl-coenzyme A mutase (Mancia et al., 1996) is a member of a class of enzymes that uses coenzyme B12 (adenosylcobalamin) as a cofactor to generate the 5′-dA radical.

This chapter reports new findings regarding sequence–structure–function relationships in the canonical RSS that can be uniquely accessed from large-scale comparisons of their sequences. SSNs annotated with different types of functional and other information (Atkinson, Morris, Ferrin, & Babbitt, 2009) are used to summarize these relationships across the entire superfamily. Guided by the subgroupings emerging from these comparisons, we generated a classification of the RSS that includes known functions (knowns) along with the majority of members that are of unknown function (unknowns). For sequences sufficiently similar to characterized knowns, the large-scale context that resulted may provide clues useful for predicting some of their functional features. This global context can also be used to guide identification of informative targets for biochemical and structural characterization of unknowns.

The results of this work, along with additional data and information about the RSS, are available from the SFLD at http://sfld.rbvi.ucsf.edu/. SSNs for the subgroups and families curated in the SFLD, and other information, can be downloaded and interactively visualized and studied using the freely available Cytoscape software (Shannon et al., 2003).

2. RESULTS AND DISCUSSION

In this study, we first describe the RSS from a structural perspective, including the known variations across MDAs that typify the superfamily. Next, we provide a global view of sequence similarity relationships among the RSS using SSNs to illustrate the subgroups into which we partitioned these sequences to establish a comprehensive classification of the entire superfamily based on sequence similarity (in contrast to the majority of previously published RSS classifications that are based on various descriptions of reaction similarity). Mapping the SSNs with functional information describing the coverage of known functions and solved structures reveals how little we know about the breadth of chemical and structural variation across the RSS, while mapping the types of life in which members are found supports previous suggestions that the RSS is of ancient origin. Finally, we describe differences between similarity-based and reaction-based classifications of the RSS and discuss the implications of these differences for function prediction of unknowns.

2.1 Structural Overview

Until recently, it was thought that only two folds were represented in the canonical RSS superfamily, a full triose phosphate isomerase (TIM) barrel type ($(\beta/\alpha)_8$, CATH (Dawson, Sillitoe, Lees, Lam, & Orengo, 2017) topology 3.20.20.70) and a variant three-quarter barrel ($(\beta/\alpha)_6$, CATH domain 3.80.30.20). As the number of characterized structures has grown, greater structural diversity has become apparent, with topologies ranging from the full TIM barrel to several additional variants of pruned-down barrels. Some examples are given in Fig. 3.

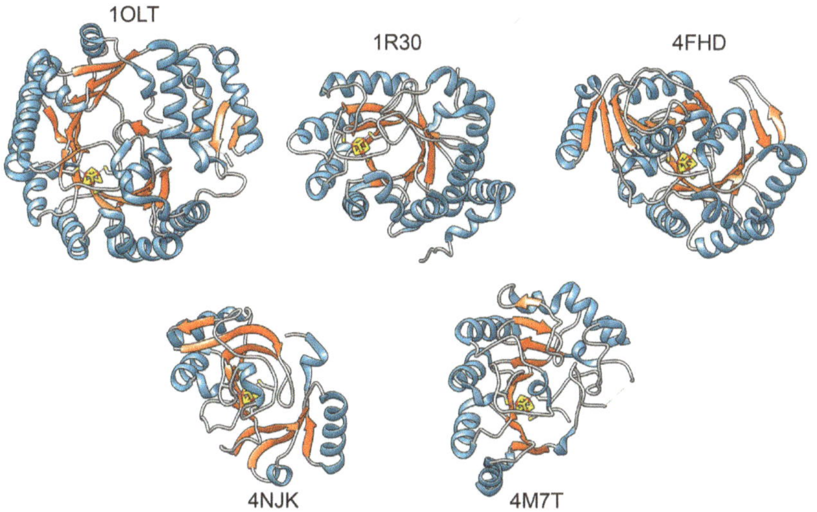

Fig. 3 Structural examples of some full-length RSS members of varied architectures. Structures are aligned to show the $[Fe_4S_4]$ clusters in a similar orientation. For structures with multiple chains, only chain A is shown. Secondary structure coloring is the same as in Fig. 2. *Top row*: 1OLT, coproporphyrinogen-III oxidase, $(\beta/\alpha)_6$; 1R30, biotin synthase, $(\beta/\alpha)_8$; 4FHD, spore photoproduct lyase, $(\beta/\alpha)_6$; *Bottom row*: 4NJK, 7-carboxy-7-deazaguanine synthase, (β_6/α_3); 4M7T, 2-deoxy-scillo-inosamine dehydrogenase (β_5/α_4).

Much previous work has provided in-depth information about some of the varied domain types that make up these proteins (for example, Dowling et al., 2012; Grell et al., 2015; Vey & Drennan, 2011). A superfamily-wide description of RSS component domains and their structural roles as they are defined and discussed in this chapter is provided in Box 1.

The shortest full-length RSS enzyme that is not a fragment is 46 residues in length, while the longest is over 2000 residues (UniProt, 2017). Much of this large variation in size can be ascribed to the large proportion of RSS enzymes represented by multidomain structures, many of them made up of accessory domains required for RSS function or that are involved in tailoring function in some full-length proteins. Both types of accessory domains are unrelated to the core superfamily domain and are homologous to each other only within specific subgroup(s) or families. For example, the B12-dependent enzymes require Vitamin B12, and so include a cobalamin-binding domain fused to the radical SAM domain. Our analysis, along with that of the InterPro resource (data not shown) suggests that most RSS proteins are on average composed of two or more domains.

BOX 1 Definitions of RSS structural domains used in this work

The canonical RSS described in this work is defined based on sequence and structural similarity rather than by chemical similarity. Each protein includes several structural elements or domains that make up its multiple domain architecture.

Superfamily domain: Structural domain that contains the core structural motif shown in Fig. 1A that functions in binding the iron–sulfur cluster involved in activation of *S*-adenosyl methionine. It is conserved throughout the canonical superfamily and represents the minimum structural motif required for superfamily membership (with at least one known exception; see Footnote[a]). It ranges from 46 to 250 residues in length. The great majority of RSS members have at least one of these motifs; some members have two or three of them.

Functional domain: The superfamily domain plus accessory domain(s) and other inserts or extensions required for tailoring RS functionality for specific roles. These accessory domains may be conserved in a subset of RSS enzymes (such as the SPASM/Twitch domain-containing enzymes) but are not conserved across the entire superfamily. Their lengths range from 46 to 1449 residues. For some RSS members, the superfamily and functional domains are the same.

Full-length RSS protein: This represents the complete amino acid sequence and may include domains with other functions than those performed by the RSS functional domain. These "extra" domains may be conserved within a subgroup or family but are not conserved across the entire RSS. They may be fused with the RSS functional domain in the full-length RSS polypeptide; alternatively, they can be found as separate enzymes in some organisms.

Full-length RSS proteins range from 46 to just over 2000 residues. They can be composed of only a superfamily or a functional domain, or as a longer fusion protein, as described earlier.

The variations in MDA structures represented in the RSS can be estimated from Pfam data, which describe proteins in terms of their constituent domains. Although the PFAM model (PF04055) itself represents only the RSS domain, it is accompanied by a prediction of the additional domain types found in the full-length sequences of the canonical RSS. Table 1 reports on the number of MDAs they predict. The highest confidence MDA predictions are those that are found in multiple sequences, rather than the large number of predicted MDAs that are represented by only one superfamily sequence.

Fig. 4 shows the primary MDAs predicted by Pfam as a network which is centered on the core RSS domain, along with the MDA variants to

Table 1 Number of MDAs Represented by 1 or More Proteins in InterPro

# Predicted MDAs	Minimum # of Proteins/MDA
20	1000
46	200
60	100
73	50
89	25
831	1

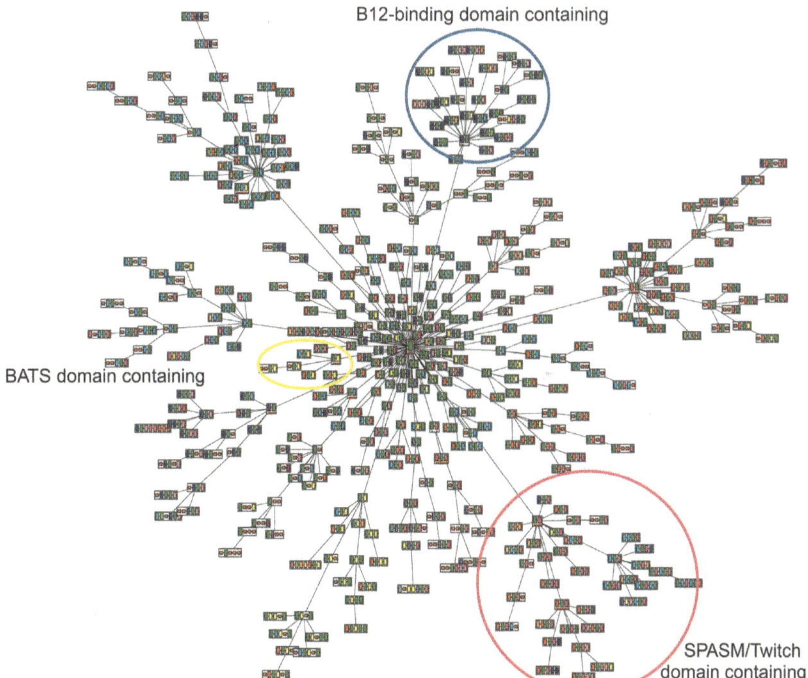

Fig. 4 Predicted domain architectures created using ArchSchema (Tamuri & Laskowski, 2010). Shown are 435 major architecture types of the more than 1500 representative domain architectures predicted for the 63,785 representative RSS protein sequences in Pfam version 27. These architectures represent 171 distinct domains. The *central green rectangle underlined in red* in the figure represents the core superfamily domain shared by all members of the canonical RSS, which is repeated in each MDA image shown. Edges connecting individual domains distinguish each complete MDA. The domain (*rectangle*) connecting each cluster to the larger MDA network is also *underlined in red*. The *circled clusters* represent the SPASM/Twitch-like domain (*magenta*), BATS-like domain (*yellow*) and B12-binding-like domain (*blue*) clusters.

which it links. As highlighted in Fig. 4, some of these MDAs have multiple domain patterns that cluster together based on the domains they have in common.

A wide literature has addressed the notion that functional diversity within enzyme superfamilies can be achieved through many routes (see, for example, Brown & Babbitt, 2014; Furnham et al., 2012). The analysis reported here identifies sequence and structural features that unite the RSS proteins on the scale of the superfamily and describes key types of variation known to confer structural and functional specificity. These findings are consistent with and extend similar conclusions made by others for smaller groups of RSS members. Viewed from this global perspective, the key structure–function relationship of the RSS reveals a conserved "plug and play" superfamily domain embedded in many different MDAs. Together with the conserved core motif delivered by the superfamily domain, additional MDA domains enable and distinguish the catalytic and specificity features of the varied reactions of the superfamily. Although this model is undoubtedly an oversimplification with respect to the refined mechanistic understanding of the interplay of structural and chemical features for several unique RSS reactions, it provides a useful foundation for creation of a global classification of the superfamily based on sequence similarity, further refined using available structural and functional information.

2.2 A New Classification of the RSS

As of December 2017, the RSS curated in the SFLD contains 113,776 unique sequences. To guide our mapping of structure–function relationships across the superfamily, we generated a representative SSN (see Section 3.2 for details), followed by a "divide and conquer" strategy to define subgroups of similarity-based clusters. One or more biochemically or structurally characterized enzymes can be associated with 20 of these subgroups; 22 additional subgroups containing no known reactions or structures were also identified and classified in the SFLD as "uncharacterized subgroups." The SSN shown in Fig. 5 provides a summary view of these Level 1 subgroups of our RSS classification. Together, they form the basis for the new classification of the RSS described in this work. Sequences that were too diverse to be confidently assigned to either named or uncharacterized subgroups were not analyzed further.

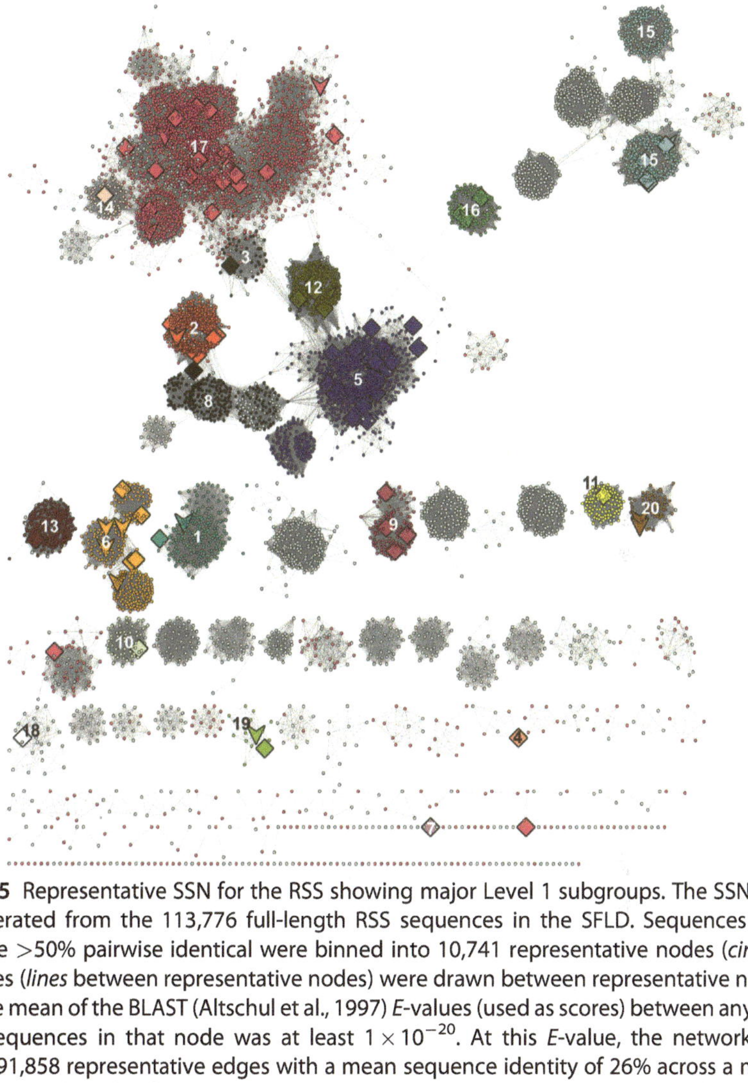

Fig. 5 Representative SSN for the RSS showing major Level 1 subgroups. The SSN was generated from the 113,776 full-length RSS sequences in the SFLD. Sequences that share >50% pairwise identical were binned into 10,741 representative nodes (*circles*). Edges (*lines* between representative nodes) were drawn between representative nodes if the mean of the BLAST (Altschul et al., 1997) *E*-values (used as scores) between any pair of sequences in that node was at least 1×10^{-20}. At this *E*-value, the network has 13,591,858 representative edges with a mean sequence identity of 26% across a mean alignment length of 300 residues The networks are laid out using the prefuse force directed layout in Cytoscape. Twenty subgroups are denoted by distinct colors and numbered according to Supplementary Table S1. Colored nodes are further specified by size and shape: Large nodes are colored if they include at least one experimentally characterized sequence. Diamond-shaped nodes specify that at least one of the sequences in that node has been experimentally characterized (but not structurally characterized). Nodes shaped like a *downward arrow* have at least one protein that has been structurally characterized. *Small circular colored* nodes have been assigned to a subgroup but are comprised entirely of sequences of unknown function. *Small gray circular* nodes have not been assigned to a subgroup and are comprised entirely of sequences of unknown function. The 22 largest of these entirely gray clusters are

(Continued)

The SFLD defines subgroups at different levels of detail (e.g., Level 1, Level 2, Level 3 subgroups) as sets of homologous sequences for which the similarities among all members in each subgroup are greater within that subgroup than they are to any sequence in another subgroup. Other information, such as structural variations, MDA organization, and detailed variations in subgroup-specific SAM-binding motifs add support for these subgroup boundaries. In particular, as the iron–sulfur cluster-binding motifs are highly conserved in the RSS, they are easy to identify so that small differences within each motif and its flanking sequences help to distinguish subgroups. As warranted by available information, more detailed examination of primary (Level 1) subgroups allows curation of additional subgroup levels within it, each representing subsets of sequences that share more detailed sequence, structural, or functional features than are shared across the parent subgroup (see Supplementary Table S1 in the online version at https://doi.org/10.1016/bs.mie.2018.06.004).

At the most granular level of the classification, the SFLD defines "reaction families" that represent sets of sequences within a superfamily or subgroup for which good evidence suggests that all family members catalyze the same reaction using a similar mechanism (Holliday, Brown, et al., 2017). Each reaction family is curated using one or more "founder" enzymes that have been biochemically (and often, structurally) characterized. Along with the founder enzyme(s), expert curators may assign to these families uncharacterized sequences that are sufficiently similar to the founder sequence and that conserve functionally distinguishing residues (see, for example, Brown, Gerlt, Seffernick, & Babbitt, 2006; Holliday, Brown, et al., 2017).

Although the use of computational annotation transfer expands the set of sequences that can be functionally assigned, our confidence in these assignments is still limited by the lack of direct experimental evidence of the annotated activity. To aid users in evaluating the confidence of these

Fig. 5—Cont'd curated in the SFLD as "Uncharacterized Radical SAM Subgroups." Some small colored clusters and singletons randomly laid out at the bottom of the image are not labeled with a number because they belong to a larger numbered cluster of the same color but fail to meet the *E*-value cutoff for drawing edges connecting them to that cluster. (The largest subgroup, subgroup 17, provides an example. In this visualization, both the large cluster at the *top left* and the smaller clusters and singletons that are colored *magenta* near the bottom of the image belong to subgroup 17, but these nodes are too diverse to be connected to the main subgroup because their similarities fall below the *E*-value threshold (1×10^{-20}) used for drawing edges to the main subgroup. Note that each representative node may contain many individual sequences (see Section 2.3).

assignments, these sequences are annotated in the SFLD with the IEA (Inferred from Electronic Annotation) evidence code to differentiate them from sequences with the IDA (Inferred from Direct Assay) evidence code. Even using this electronic annotation transfer protocol, more than 50,000 RSS sequences fail to meet our criteria for assigning them to reaction families.

The 20 colored and numbered subgroups shown in Fig. 5 are designated in this work as Level 1 subgroups. Within some of these, an additional 17 subgroups have been classified in the SFLD as Level 2 or Level 3 subgroups that are children of a Level 1 subgroup. Our classification also includes 101 reaction families for which 85 chemical reactions that have been described in the primary literature. This report is accompanied by an online version of Supplementary Table S1 at https://doi.org/10.1016/bs.mie.2018.06.004. It is divided into two parts. Supplementary Table S1a provides a list of the RSS reaction families curated in the SFLD along with the names of their known overall chemical transformations and other features that include accession numbers from several public sequence and structure databases, conserved residues and motifs, and the known types of life to which the majority of members of each subgroup belongs. Supplementary Table S1b shows images of the overall reactions. Both parts of Supplementary Table S1 are keyed to the subgroups shown in Fig 5.

To examine the contribution of the highly conserved RSS superfamily domain to the topology of the SSN generated from the full-length sequences shown in Fig. 5, we generated an SSN using only this typically shorter domain. The resulting network (not shown) is substantially similar to that of Fig. 5, indicating that the conserved RSS superfamily domain dominates the sequence signal captured by the all-by-all BLAST comparisons. This interpretation is consistent with other work suggesting that the superfamily domain provides the foundational structural machinery required for radical SAM chemistry, while decorations/domain additions to the superfamily and functional domains enable the wide variation in function for which the superfamily is known, for example, (Dowling et al., 2012; Vey & Drennan, 2011).

Fig. 6 shows secondary structure diagrams for the core RSS domains from several of the subgroups classified in the SFLD. The [Fe_4S_4]-AdoMet-binding motif is highlighted in each. This view indicates that not only are the overall topologies of these proteins varied, but the *position* of the motif in these sequences varies as well, illustrating yet another way in which nature has expanded the use of this plug and play motif to support

Table S1a Detailed Metadata for Radical SAM Superfamily Subgroups Classified in the SFLD as Keyed to Fig. 5

SFLD Subgroup #	*L1 Subgroup Color (RGB) in Fig. 5*	*L1 Subgroup*	*L2 Subgroup*	*L3 Subgroup*	*SFLD Family*	*Representative UniprotKB*
1	0,153,153	7-carboxy-7-deazaguanine synthase like	-	-	7-carboxy-7-deazaguanine synthase, Cx3CxxC type	O31677
				-	7-carboxy-7-deazaguanine synthase, Cx14CxxC type	A0A0H3KB22
				-	ExsD-like	O54060
2	255,0,0	anaerobic coproporphyrinogen-III oxidase like	-	-	cyclopropanase (jaw5-like)	A0A060PWX2
			-	-	heme degradation proteins (HutW/ChuW-like)	V0VQG0
			-	-	HemN-like, clustered with heat shock genes	P9WP73
			-	-	HemN-like, clustered with nucleoside-triphosphate RdgB	P52062
			-	-	Oxygen-independent coproporphyrinogen-III oxidase 1 (HemN-like)	P32131
			-	-	Oxygen-independent coproporphyrinogen-III oxidase 2 (HemZ-like)	Q796V8
			methyltransferase (Class C)	-	3-methyl-2-indolic acid methyltransferase (NocN/NosN like)	C6FX53
				-	pyrimidine ring methyltransferase (bleomycin-like antibiotic biosynthesis)	Q9FB10
				-	aromatic C-methyltransferase (yatakemycin antibiotic biosynthesis)	I3NN68

Representative PDB	Abbreviation	EC Number	Reaction Type	Primary PMID(s)	Protein Name	Known Conserved Residues	Domain Architecture	Type of Life Representation
-	QueE	4.3.99.3	complex rearrangement	19354300	7-carboxy-7-deazaguanine synthase, Cx3CxxC type	C34, C38, C41, D42, S43	IPR007197	Bacteria
4NJK	QueE_X14	4.3.99.3	complex rearrangement	11222759	7-carboxy-7-deazaguanine synthase, Cx14CxxC type	F27, C31, C46, C49, D50, T51, E116, H204	PF13394	Bacteria
-	ExsD	-	unknown	11222759	ExsD	C38, C42, C45	IPR007197	Bacteria
-	Jaw5	5.-.-.-	complex rearrangement	24756819	cyclopropanase, JAW5	C68, C76, C79	IPR007197	Bacteria
-	ChuW	-	complex rearrangement	27791000	heme degradation protein (HutW/ChuW-like)	Y61, C67, C71, C74	IPR007197	Bacteria
-	HemN_HSG	-	unknown	25646457	non-canonical HemN-like protein found with heat shock genes	Y9, H11, C15, C19, C22	IPR007197	Bacteria
-	HemN_RdgB	-	unknown	25646457	non-canonical HemN-like protein, found sith RdgB gene	Y10, H12, C16, C20, C23	IPR007197	Bacteria
1OLT	HemN	1.3.99.22	complex rearrangement	12114526	oxygen-independent coproporphyrinogen-III oxidase 1	Y56, H58, C62, C66, C69, C71	IPR007197 IPR010723	Bacteria, Fungi
-	HemZ	1.3.99.-	complex rearrangement	10498703	oxygen-independent coproporphyrinogen-III oxidase 2	Y174, C180, C184, C187	IPR007197	Bacteria
-	NosN	2.1.1.-	methylation	19678698	3-methyl-2-indolic acid methyltransferase	C15, C19, C22	IPR007197	Bacteria
-	blm	2.1.1.-	methylation	17216057	pyrimidine ring methyltransferase (bleomycin-like antibiotic biosynthesis)	C35, C39, C42	IPR007197	Bacteria
-	YtkT	2.1.1.-	methylation	22612591	aromatic C-methyltransferase (yatakemycin-like antibiotic biosynthesis)	C72, C76, C79	IPR007197	Bacteria

Continued

Table S1a Detailed Metadata for Radical SAM Superfamily Subgroups Classified in the SFLD as Keyed to Fig. 5—cont'd

SFLD Subgroup #	*L1 Subgroup Color (RGB) in Fig. 5*	*L1 Subgroup*	*L2 Subgroup*	*L3 Subgroup*	*SFLD Family*	*Representative UniprotKB*
3	51,51,51	antiviral proteins	-	-	Viperin	O70600
4	255,153,153	AviX12-like	-	-	avilamycin epimerase	Q93KV6
5	0,0,255	B12-binding domain containing	-	-	bacteriocin maturation	Q3ME29
			-	-	paromamine deoxygenase	Q2MFI7
			-	-	hopanetetrol cyclitol ether synthase	U1XSN3
			-	-	ladderane biosynthesis	Q1Q0N1
			-	-	anaerobic magnesium-protoporphyrin-IX monomethyl ester cyclase	P26168
			-	-	OxsB-like	O24770
			-	-	valine cyclopropanase	D2KTX6
			methyltransferase (Class B)	-	bacteriochlorophyll C12 methyltransferase	Q8KCU0
				-	bacteriochlorophyll C8 methyltransferase	Q8KBK9
				-	5″-pyrrole methyltransferase	Q8GHB6
				-	chondrochloren A methyltransferase (CndI-like)	B9ZUJ4
				-	phosphonoacetaldehyde methylase (fom3-like)	Q56184
				-	fortimicin-like methyltransferase	Q50258
				-	gentamicin-like N-methyltransferase (GntE-like)	Q6QVU0
				-	gentamicin C-6′ methyltransferase (GenK-like)	Q70KE5

Representative PDB	Abbreviation	EC Number	Reaction Type	Primary PMID(s)	Protein Name	Known Conserved Residues	Domain Architecture	Type of Life Representation
-	Viperin	-	unknown	29925952	3′-deoxy-3′, 4′-didehydro-CTP (ddhCTP) synthase	C82, C86, C89	IPR007197	Fungi, Vertebrates, Invertebrates
-	AviX12	5.1.3.-	complex rearrangement	16537546	avilamycin epimerase	C156, C160, C163	IPR007197	Bacteria
-	Ava_1133	-	unknown	21223593	bacteriocin maturation protein	C271, C279, C282	IPR007197	Bacteria, Archaea
-	AprD4	1.-.-.-	dehydrogenase	26879038	paromamine deoxygenase	C206, C210, C214	IPR007197	Bacteria
-	HpnJ	4..-.-.-	complex rearrangement	24888970	hopanetetrol cyclitol ether synthase	C206, C210, C213	IPR007197	Bacteria
-	Kuste2803	-	unknown	19220888	radical SAM protein involved in ladderine biosynthesis	C178, C182, C185	IPR006158 IPR007197	Bacteria
-	BchE		complex rearrangement	10841582	magnesium-protoporphyrin IX monomethyl ester anaerobic oxidative cyclase	C204, C208, C211	IPR006158 IPR007197	Bacteria
-	OxsB	-	complex rearrangement	22172915	radical SAM protein involved in contraction of an adenosine ribose ring	C313, C318, C321	IPR006158 IPR007197	Bacteria
-	Swb7	5.-.-.-	dehydrogenase	19514719	valine cyclopropanase	C183, C187, C190	IPR007197	Bacteria
-	BchR	2.1.1.-	methylation (B12)	19297239	bacteriochlorophyll C12 methyltransferase	C206, C210, C213	IPR007197	Bacteria
-	BchQ	2.1.1.-	methylation (B12)	19297239	bacteriochlorophyll C8 methyltransferase	C162, C166, C169	IPR006158 IPR007197	Bacteria
-	CloN6	2.1.1.-	methylation (B12)	12898629	clorobiocin pyrrole-2-carboxy methyltransferase	C219, C223, C226	IP***R006158 IPR007197	Bacteria, Archaea
-	CndI	2.1.1.-	methylation (B12)	19171307	chondrochloren A methyltransferase	C268, C272, C275	IPR006158 IPR007197	Bacteria
-	Fom3	2.1.1.-	methylation (B12)	17220970	2-hydroxyethyl phosphonate methyltransferase	C282, C286, C289	IPR006158 IPR007197	Bacteria
-	Fms7	2.1.1.-	methylation (B12)	7490235	fortimicin KL1 methyltransferase	C243, C247, C250	IPR006158 IPR007197	Bacteria
-	GntE	2.1.1.-	methylation (B12)	18533111	gentamicin C-methyltransferase	C316, C321, C324	IPR007197	Bacteria
-	GenK	2.1.1.-	methylation (B12)	23679096	gentamicin C-6′ methyltransferase	C250, C254, C257	IPR006158 IPR007197	Bacteria

Continued

Table S1a Detailed Metadata for Radical SAM Superfamily Subgroups Classified in the SFLD as Keyed to Fig. 5—cont'd

SFLD Subgroup #	*L1 Subgroup Color (RGB) in Fig. 5*	*L1 Subgroup*	*L2 Subgroup*	*L3 Subgroup*	*SFLD Family*	*Representative UniprotKB*
				-	hopanoid C2-methyltransferase (HpnP-like)	B3QHD1
				-	hopanoid C3-methyltransferase (HpnR-like)	Q60AV6
				-	P-methyltransferase (PhpK-like)	Q5IW50
				-	pactamycin C-methyltransferase (PctJ-like)	A8R0J3
				-	pactamycin C-methyltransferase (PctN-like)	A8R0J7
				-	pactamycin C-methyltransferase (PctO-like)	A8R0J8
				-	thioacetal methyltransferase (Swb9-like)	D2KTX8
				-	carbapenem intermediate methyltransferase (ThnK-like)	F8JND9
				-	carbapenem intermediate methyltransferase (ThnL-like)	F8JNE0
				-	carbapenem intermediate methyltransferase (ThnP-like)	F8JNE4
				-	tryptophan 2-C-methyltransferase	C0JRZ9
			B12-binding methylthio transferase like	-	None yet assigned	Q58275
6	**255,153,0**	BATS domain containing	biotin synthase like	-	biotin synthase	P12996
			HMD cofactor maturase (HmdB-like)	-	HMD cofactor maturase (HmdB-like)	Q58195
			HydE/PylB-like	-	[FeFe] hydrogenase maturase (HydE-like)	Q9X0Z6

Representative PDB	Abbreviation	EC Number	Reaction Type	Primary PMID(s)	Protein Name	Known Conserved Residues	Domain Architecture	Type of Life Representation
-	HpnP	2.1.1.-	methylation (B12)	20421508	hopanoid C2-methyltransferase	C189, C193, C196	IPR006158 IPR007197	Bacteria, Archaea, Fungi
-	HpnR	2.1.1.-	methylation (B12)	22826256	hopanoid C3-methyltransferase	C195, C199, C202	IPR006158 IPR007197	Bacteria
-	PhpK	2.1.1.-	methylation (B12)	21950770	P-methyltransferase	C254, C258, C261	IPR006158 IPR007197	Bacteria
-	PctJ	2.1.1.-	methylation (B12)	17827660	pactamycin C-methyltransferase, PctJ	C236, C240, C243	IPR006158 IPR007197	Bacteria
-	PctN	2.1.1.-	methylation (B12)	17827660	pactamycin C-methyltransferase, PctN	C224, C228, C231	IPR006158 IPR007197	Bacteria
-	PctO	2.1.1.-	methylation (B12)	17827660	pactamycin C-methyltransferase, PctO	C224, C228, C231	IPR006158 IPR007197	Bacteria
-	Swb9	2.1.1.-	methylation (B12)	19514719	thioacetal methyltransferase	C212, C216, C219	IPR006158 IPR007197	Bacteria
-	ThnK	2.1.1.-	methylation (B12)	26240322	carbapenem intermediate methyltransferase (ThnK)	C206, C210, C213	IPR006158 IPR007197	Bacteria
-	ThnL	2.1.1.-	methylation (B12)	21263049	carbapenem intermediate methyltransferase (ThnL)	C198, C202, C205	IPR006158 IPR007197	Bacteria
-	ThnP	2.1.1.-	methylation (B12)	21263049	carbapenem intermediate methyltransferase (ThnP)	C209, C213, C216	IPR006158 IPR007197	Bacteria
-	TsrM	2.1.1.106	methylation (B12)	19246004	tryptophan 2-C-methyltransferase	C253, C257, C260	IPR006158 IPR007197	Bacteria
-	-	2.-.-.-	-	-	None yet assigned, no families identified to date	C177, C181, C184	IPR006158 IPR007197	Archaea
1R30	BioB	2.8.1.6	sulfur insertion	8142361	biotin synthase	C53, C57, C60, C97, C128, C188. R280	IPR007197 IPR010722	All
-	HmdB	-	metallocofactor assembly	19897660	HMD cofactor maturase	C92, C98, C101	IPR007197 IPR010722	Methanogenic Archaea, Bacteria
3IIX	HydE	-	metallocofactor assembly	18400755	[FeFe] hydrogenase maturase	C63, C67, C70, R54, T134, R155	IPR007197 IPR010722	Bacteria, Invertebrates, Plants

Continued

Table S1a Detailed Metadata for Radical SAM Superfamily Subgroups Classified in the SFLD as Keyed to Fig. 5—cont'd

SFLD Subgroup #	L1 Subgroup Color (RGB) in Fig. 5	L1 Subgroup	L2 Subgroup	L3 Subgroup	SFLD Family	Representative UniprotKB
				-	3-methylornithine synthase (PylB-like)	Q46E78
			cleavage of the Ca-Cb bond in aromatic amino acids	-	2-iminoacetate synthase (ThiH-like)	P30140
				-	3-methyl-2-indolic acid synthase (NosL-like)	C6FX51
				-	[Fe] hydrogenase maturase (HydG-like)	Q6PSL4
7	204,153,255	DesII-like	-	-	dTDP-4-amino-4, 6-dideoxy-D-glucose ammonia-lyase (DesII-like)	Q9ZGH1
8	0,0,0	elongater protein-like	-	-	ELP3-like	Q02908
9	153,0,153	F420, menaquinone cofactor biosynthesis	7,8-didemethyl-8-hydroxy-5-deazariboflavin synthase	-	((2,3,4,5-tetrahydroxypentyl)amino) dihydropyrimidine-2, 4-dione synthase (CofH-like)	Q58826
				-	7,8-didemethyl-8-hydroxy-5-deazariboflavin synthase (CofG-like)	Q57888
			menaquinone synthesis involved	-	cyclic dehypoxanthine futalosine synthase (mqnC-like)	Q9XAP2
				-	aminofutalosine synthase (mqnE-like)	Q5SK48
10	204,255,204	FeMo cofactor biosynthesis protein	-	-	FeMo cofactor biosynthesis protein (nifB-like)	P11067
11	255,255,0	lipoyl synthase like	-	-	lipoyl synthase	P60716

Representative PDB	Abbreviation	EC Number	Reaction Type	Primary PMID(s)	Protein Name	Known Conserved Residues	Domain Architecture	Type of Life Representation
3T7V	PylB	5.4.99.58	B12-like rearrangement	21455182	methylornithine synthase	F64, C71, C75, C78, D112, S146, Y169, E171, R190, R235, M237, S277, D279	IPR007197	Bacteria, Archaea
-	ThiH	4.1.99.19	complex rearrangement	17969213	2-iminoacetate synthase	C85, C89, C92	IPR007197 IPR010722	Bacteria, Fungi, Archaea
4R34	NosL	4.-.-.-	complex rearrangement	25196319	3-methyl-2-indolic acid synthase	C95, C99, C102	IPR010722	Bacteria
4RTB	HydG	-	metallocofactor assembly	15082711	Fe-hydrogenase assembly protein	C179, C183, C186	IPR010722	Bacteria, Invertebrates, Plants
-	DesII	4.3.1.30	B12-like rearrangement	20121093	dTDP-4-amino-4, 6-dideoxy-D-glucose ammonia-lyase	C141, C145, C148	IPR013785	Bacteria
-	ELP3	-	unknown	15769872 15780940 15138274	Elongator complex protein 3	C108, C118, C121	IPR007197 IPR016181	All
-	CofH	-	complex rearrangement	25781338	((2,3,4, 5-tetrahydroxypentyl) amino) dihydropyrimidine-2, 4-dione synthase (FO synthase subunit 2)	C59, C63, C66	IPR007197	Archaea, Bacteria, Plants
-	CofG	-	complex rearrangement	25781338	7,8-didemethyl-8-hydroxy-5-deazariboflavin synthase (FO synthase subunit 1)	C49, C53, C56	IPR007197	Archaea, Bacteria, Plants
-	MqnC	1.21.98.1	complex rearrangement	23763543 18801996	cyclic dehypoxanthine futalosine synthase	C70, C74, C77	IPR007197	Bacteria, Archaea
-	MqnE	2.5.1.120	complex rearrangement	24083939	aminofutalosine synthase	C69, C73, C76	IPR007197	Bacteria, Archaea
-	NifB	-	metallocofactor assembly	23019652	FeMo cofactor biosynthesis protein (nifB)	C67, C71, C74	IPR007197 IPR003731	Bacteria, Archaea
4U0P	LipA	2.8.1.8	sulfur insertion	15157071	lipoyl synthase	C68, C73, C79, C94, C98, C101	IPR031691 IPR007197	All

Continued

Table S1a Detailed Metadata for Radical SAM Superfamily Subgroups Classified in the SFLD as Keyed to Fig. 5—cont'd

SFLD Subgroup #	*L1 Subgroup Color (RGB) in Fig. 5*	*L1 Subgroup*	*L2 Subgroup*	*L3 Subgroup*	*SFLD Family*	*Representative UniprotKB*
12	**102,102,0**	methylthiotransferase	-	-	(dimethylallyl)adenosine tRNA methylthiotransferase (MiaB-like)	Q9WZC1
			-	-	CDK5RAP1	Q96SZ6
			-	-	ribosomal protein S12 methylthiotransferase (RimO-like)	P0AEI4
			-	-	threonylcarbamoyladenosine tRNA methylthiotransferase	P54462
13	102,0,0	methyltransferase (Class A)	-	-	adenosine C2 methyltransferase (RlmN-like)	P36979
				-	adenosine C8 methyltransferase (Cfr-like)	Q9FBG4
14	**255,204,204**	methyltransferase (Class D)	-	-	7,8-dihydro-6-hydroxymethylpterin dimethyl transferase	Q58036
15	**0,255,255**	organic radical-activating enzymes	-	-	4-hydroxyphenylacetate decarboxylase activase	Q84F14
			activating enzymes, group 1	-	anaerobic ribonucleoside-triphosphate reductase activase	P0A9N8
			activating enzymes, group 2	-	benzylsuccinate synthase activase	O87941
				-	glycerol dehydratase activase	Q8GEZ7

Representative PDB	Abbreviation	EC Number	Reaction Type	Primary PMID(s)	Protein Name	Known Conserved Residues	Domain Architecture	Type of Life Representation
-	MiaB	2.8.4.3	sulfur insertion	12766153	tRNA-2-methylthio-N (6)-dimethylallyl adenosine synthase	C10, C46, C79, C150, C154, C157	IPR013848 IPR007197 IPR002792	Bacteria, Invertebrates, Plants, Vertebrates, Fungi
-	CDK5RAP1	2.8.4.-	sulfur insertion	22422838	CDK5 regulatory subunit-associated protein 1	C258, C262, C265	IPR013848 IPR007197 IPR002792	Bacteria, Invertebrates, Plants, Vertebrates, Fungi
2QGQ	RimO	2.8.4.4	sulfur insertion	19736993	[ribosomal protein S12] (aspartate89-C3)-methylthiotransferase	C17, C53, C82, C150, C154, C157	IPR013848 IPR007197 IPR002792	All
-	YqeV	2.8.4.5	sulfur insertion	20584901	tRNA (N6-L-threonylcarbamoyladenosine 37-C2)-methylthiotransferase	C11, C47, C78, C153, C157, C160	IPR013848 IPR007197 IPR002792	Bacteria, Invertebrates, Plants, Vertebrates
3RFA	RlmN	2.1.1.192	methylation	24806349	23S rRNA (adenine^{2503}-C^{2})-methyltransferase	E105, C118, C125, C129, C132, C355	IPR007197	Bacteria, Fungi, Invertebrates, Archaea, Plants
-	Cfr	2.1.1.224	methylation	19144912	23S rRNA (adenine^{2503}-C^{8})-methyltransferase	C105, C112, C116, C119, C338	IPR007197	Bacteria
-	MJ0619	2.1.1.-	methylation	25002541	7,8-dihydro-6-hydroxymethylpterin dimethylase	C73, C77, C80, C98, C102, C105	IPR007197	Bacteria, Archaea
-	HPD-AE	1.97.1.-	activase	16878993	[4-hydroxyphenylacetate decarboxylase]-activating enzyme	C34, C38, C41	IPR007197	Bacteria
-	nrdG	1.97.1.-	activase	11389585	[anaerobic ribonucleoside-triphosphate reductase]-activating enzyme	C26, C30, C33	IPR013785	Bacteria, Fungi, Invertebrates, Archaea
-	BssD	1.97.1.-	activase	9632263	[benzylsuccinate synthase]-activating enzyme	C29, C33, C36	IPR007197	Bacteria
-	dhaB2	1.97.1.-	activase	12704244	[Glycerol dehydratase]-activating enzyme	C32, C36, C39	IPR007197	Bacteria

Continued

Table S1a Detailed Metadata for Radical SAM Superfamily Subgroups Classified in the SFLD as Keyed to Fig. 5—cont'd

SFLD Subgroup #	*L1 Subgroup Color (RGB) in Fig. 5*	*L1 Subgroup*	*L2 Subgroup*	*L3 Subgroup*	*SFLD Family*	*Representative UniprotKB*
				-	pyruvate formate-lyase activase	P0A9N4
				-	YjjI activase	P39409
16	0,153,0	PLP-dependent	-	-	arginine 2,3-aminomutase	Q841K7
				-	glutamate 2,3-aminomutase	A4J6G2
				-	L-lysine 2,3-aminomutase (LAM-like)	Q9XBQ8
				-	L-lysine 2,3-aminomutase (yjeK like)	P39280
17	255,0,255	SPASM/twitch domain containing	main SPASM domain containing	heme carboxy lyase like	heme D1 biosynthesis (NirJ-like)	A0A1I5E523
					alternative heme biosynthesis C (AhbC-like)	B8J367

Representative PDB	Abbreviation	EC Number	Reaction Type	Primary PMID(s)	Protein Name	Known Conserved Residues	Domain Architecture	Type of Life Representation
3CB8	PFL-AE	1.97.1.4	activase	18852451	[formate-C-acetyltransferase 1]-activating enzyme	C29, C33, C23	IPR007197	Bacteria, Invertebrates, Plants, Fungi, Archaea
-	YjjI-AE	1.97.1.-	activase	22144874	putative glycyl-radical activating enzyme	C31, C35, C38	IPR007197	Bacteria
-	AAM	5.4.3.-	B12-like rearrangement	12964155	arginine 2,3-aminomutase	Y122, C134, C138, C141, R211, K348, R143, R125	IPR007197	Bacteria
-	GAM	5.4.3.9	B12-like rearrangement	17222594	glutamate 2, 3-aminomutase	Y164, C176, C180, C183, R249, K388, R185, R167, R187, K344, N381	IPR007197 IPR025895	Bacteria
2A5H	LAM	5.4.3.2	B12-like rearrangement	16166264	L-lysine 2,3-aminomutase	Y113, C125, C129, C132, R134, R198, K337, R116, D293, D330	IPR007197 IPR025895	Bacteria, Archaea
-	yjeK	5.4.3.2	B12-like rearrangement	17042480	L-lysine 2,3-aminomutase	Y108, C120, C124, C127, R111, R129, R192, K332	IPR007197	Bacteria, Invertebrates, Fungi
-	NirJ	-	unknown	20420837	radical SAM superfamily protein involved in heme D1 biosynthesis, NirJ	C38, C42, C45, C342, C345, C351, C373	IPR007197 IPR023885	Bacteria
-	AhbC	3.5.1.-	lyase	24713144	didecarboxysiroheme deacetylase	C50, C54, C57, C353, C356, C362, C384	IPR007197 IPR023885	Bacteria, Archaea

Continued

Table S1a Detailed Metadata for Radical SAM Superfamily Subgroups Classified in the SFLD as Keyed to Fig. 5—cont'd

SFLD Subgroup #	*L1 Subgroup Color (RGB) in Fig. 5*	*L1 Subgroup*	*L2 Subgroup*	*L3 Subgroup*	*SFLD Family*	*Representative UniprotKB*
					alternative heme biosynthesis D (AhbD-like)	Q46CH7
					mycofactocin peptide maturase (MftC-like)	P9WJ79
				cyclic pyranopterin phosphate synthase (MoaA-like)	cyclic pyranopterin phosphate synthase (MoaA-like)	P69848
				-	coenzyme PQQ synthesis protein E (PqqE-like)	P27507
				dehydrogenase like	anaerobic Cys-type sulfatase-maturating enzyme	Q0TTH1
					anaerobic Ser-type sulfatase-maturating enzyme	Q9X758

Representative PDB	Abbreviation	EC Number	Reaction Type	Primary PMID(s)	Protein Name	Known Conserved Residues	Domain Architecture	Type of Life Representation
-	AhbD	4.1.1.-	lyase	24669201	Fe-coproporphyrin III decarboxylase	C19, C23, C26, C312, C315, C321, C324	IPR007197 IPR023885	Bacteria, Archaea
-	MftC	4.1.1.-	lyase	21223593	C-terminal tyrosine in MftC decarboxylase	C30, C34, C37, C310, C313, C319, C323	IPR007197 IPR023885	Bacteria
1TV8	MoaA	4.1.99.18	complex rearrangement	15317939	C-terminal tyrosine decarboxylase	C24, C28, Y30, C31, R71, G75, S126, K163, M197, C261, C264, C278	IPR007197 IPR010505	All
-	PqqE	5.-.-.-	complex rearrangement	18371220	coenzyme PQQ synthesis protein E	C22, C26, C29, C306, C309, C319, C321	IPR007197 IPR023885	Bacteria
4K37	AnSME	1.1.99.-	dehydrogenase	18408004 17335281 16766528	anaerobic sulfatase-maturating enzyme	C15, C19, C22, C255, C261, C276, C317, C320, C326, C330	IPR007197 IPR023885	Bacteria
-	atsB	1.1.99.-	dehydrogenase	12419807	anaerobic sulfatase-maturating enzyme	C35, C39, C42, C270, C276, C291, C331, C334, C340, C344	IPR007197 IPR023885	Bacteria, Archaea

Continued

Table S1a Detailed Metadata for Radical SAM Superfamily Subgroups Classified in the SFLD as Keyed to Fig. 5—cont'd

SFLD Subgroup #	*L1 Subgroup Color (RGB) in Fig. 5*	*L1 Subgroup*	*L2 Subgroup*	*L3 Subgroup*	*SFLD Family*	*Representative UniprotKB*
				thioether bond formation requiring two auxiliary iron-sulfur clusters	quinohemoprotein amine dehydrogenase maturation protein (QhpD-like)	A1B2Q7
					tte1186 maturase	Q8RAM6
				thioether bond formation requiring one auxiliary iron-sulfur cluster	sporulation killing factor protein (SkfB-like)	O31423
					antilisterial bacteriocin subtilosin biosynthesis protein (AlbA-like)	P71011
					thuricin CD (TrnCD-like)	C2TQ82
					thurincin H (ThnB-like)	D0QZJ5
			BtrN-like SPASM domain containing	-	2-deoxy-scyllo-inosamine dehydrogenase (BtrN-like)	Q8G907
				-	tungsten cofactor oxidoreductase radical SAM maturase	Q51741
			-	-	adenosyl-hopene transferase	U1XZD1

Representative PDB	Abbreviation	EC Number	Reaction Type	Primary PMID(s)	Protein Name	Known Conserved Residues	Domain Architecture	Type of Life Representation
-	QhpD	5.4.99.-	sulfur insertion	25778402	quinohemoprotein amine dehydrogenase maturation protein, QphD	C112, C116, C119, C353, C371, C410, C413, C419, C423, C442	IPR007197 IPR023885	Bacteria
-	AslB3	5.4.99.-	sulfur insertion	27007615	tte1186 maturation protein	C104, C108, C111	IPR007197 IPR023885	Bacteria
-	SkfB	5.4.99.-	sulfur insertion	23282011	sporulation killing factor protein	C117, C121, C124, C380, C385, C387	IPR007197	Bacteria
-	AlbA	5.4.99.-	sulfur insertion	22366720	antilisterial bacteriocin subtilosin biosynthesis protein	C129, C133, C136, C408, C414, C417	IPR007197 IPR023885	Bacteria, Archaea
-	TrnCD	5.4.99.-	sulfur insertion	21760885	thuricin CD	C103, C107, C110	IPR007197 IPR023885	Bacteria
-	ThnB	5.4.99.-	sulfur insertion	21786372	thurincin H, ThnB	C131, C135, C138, C412, C415, C421, C244	IPR007197 IPR023885	Bacteria
4M7T	BtrN	1.1.99.38	dehydrogenase	24048029	2-deoxy-scyllo-inosamine dehydrogenase	C16, C20, C23, C152, C169, C187, C189, C232, C235	IPR013785 IPR023885	Bacteria
-	Cmo	-	unknown	7642512	tungsten cofactor oxidoreductase radical SAM maturase	C12, C16, C19	IPR007197 IPR023885	Bacteria, Archaea
-	HpnH	2.5.1.-	complex rearrangement	24888970	adenosyl-hopene transferase	C38, C42, C45, C255, C272, C303, C306	IPR007197 IPR022563	Bacteria

Continued

Table S1a Detailed Metadata for Radical SAM Superfamily Subgroups Classified in the SFLD as Keyed to Fig. 5—cont'd

SFLD Subgroup #	*L1 Subgroup Color (RGB) in Fig. 5*	*L1 Subgroup*	*L2 Subgroup*	*L3 Subgroup*	*SFLD Family*	*Representative UniprotKB*
			-	-	cytosylglucuronate decarboxylase	Q841K9
			-	-	KxxxW cyclic peptide maturase (StrB-like)	A0A0M4HYP2
			-	-	neomycin C-like epimerase	Q53U14
			-	-	UDP-N-acetyl-tunicamine-uracil synthase (TunB-like)	E5KJ95
			-	-	pcfB-like	Q6E3K8
18	**255,255,255**	spectinomycin biosynthesis (SpcY-like)	-	-	spectinomycin biosynthesis (SpcY-like)	Q9S1L5
19	**0,255,0**	spore photoproduct lyase like	-	-	spore photoproduct lyase 1	A4IQU1
				-	spore photoproduct lyase 2	Q97L63
20	**153,102,0**	tRNA wybutosine-synthesizing	-	-	tRNA wybutosine-synthesizing	Q57705

Representative PDB	Abbreviation	EC Number	Reaction Type	Primary PMID(s)	Protein Name	Known Conserved Residues	Domain Architecture	Type of Life Representation
-	BlsE	4.1.1.-	lyase	23874663	cytosylglucuronate decarboxylase	C31, C35, C38, C255, C274, C276	IPR007197 IPR023885	Bacteria
-	StrB	-	complex rearrangement	17921293	KxxxW cyclic peptide maturase, StrB	C117, C121, C124, C406, C109, C415, C419	IPR007197 IPR023885	Bacteria
-	NeoN	5.1.3.-	complex rearrangement	25230155	neomycin C epimerase	C26, C30, C33, C226, C247, C271, C274	IPR007197	Bacteria
-	TunB	4.1.99.-	complex rearrangement	22172915	UDP-N-acetyl-tunicamine-uracil synthase	C18, C22, C25	IPR007197	Bacteria
-	PcfB	-	lyase	15574930	radical SAM protein, pcfB	C21, C25, C28	IPR007197	Bacteria
-	SpcY	1.-.-.-	dehydrogenase	19058170	TDP-actinospectose synthase (SpcY)	C24, C28, C31	IPR007197	Bacteria
4FHD	SPL1	4.1.99.14	complex rearrangement	22761404	spore photoproduct lyase 1	C90, C94, C97, Y98, C140	-	Bacteria
-	SPL2	4.1.99.14	complex rearrangement	19178276	spore photoproduct lyase 2	C88, C92, C95, Y96, C94	IPR007197	Bacteria
2YX0	TYW1	4.1.3.44	complex rearrangement	17727881	tRNA 4-demethylwyosine synthase	K41, C26, C39, C52, C62, C66, C69, R221	IPR007197 IPR013917	Plants, Vertebrates, Fungi, Invertebrates, Archaea

Table S1b Reaction Families and Overall Reactions Associated With SFLD Subgroups of the Radical SAM Superfamily as Keyed to Fig. 5

SFLD Sub-group #	*L1 Subgroup Color (RGB) in Fig. 5*	*SFLD Family*	*Overall Chemical Transformation (reactions are not necessarily balanced)*
1	0,153,153	7-carboxy-7-deazaguanine synthase, Cx3CxxC type	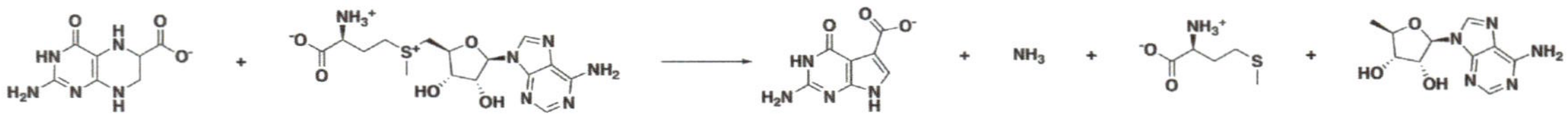
		7-carboxy-7-deazaguanine synthase, Cx14CxxC type	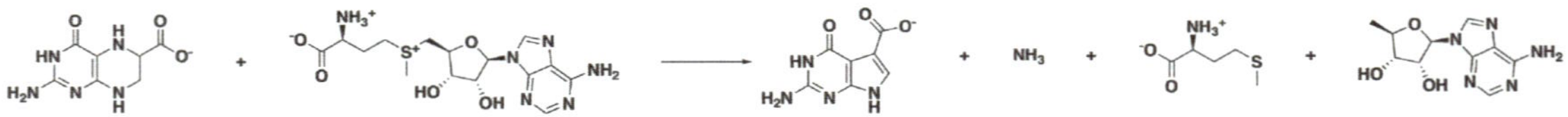
		ExsD-like	Unknown
2	255,0,0	cyclopropanase (jaw5-like)	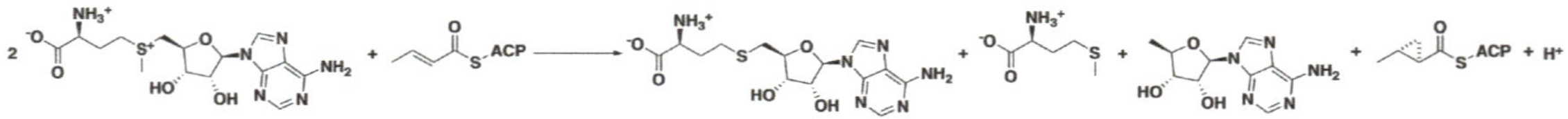 Catalyzes five successive cyclopropanations to form jawsamycin. The catalytic cycle involves the first cyclopropanation followed by chain elongation (performed by Jaw4 and Jaw6) and subsequent cyclopropanations until the final ACP-bound molecule is formed
		heme degradation proteins (HutW/ChuW-like)	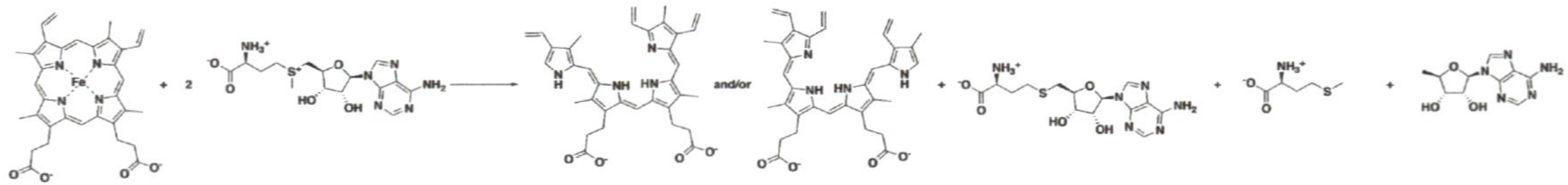
		HemN-like, clustered with heat shock genes	Unknown
		HemN-like, clustered with nucleoside-triphosphate RdgB	Unknown

		Oxygen-independent coproporphyrinogen-III oxidase 1 (HemN-like)	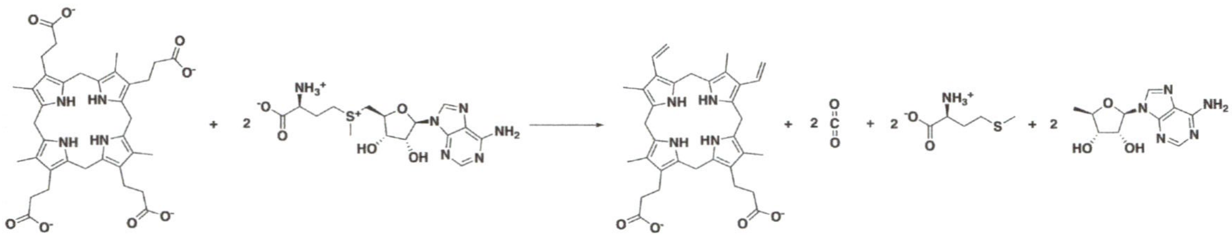
		Oxygen-independent coproporphyrinogen-III oxidase 2 (HemZ-like)	Involved in the biosynthesis of a porphyrin-containing compound*
		3-methyl-2-indolic acid methyltransferase (NocN/NosN like)	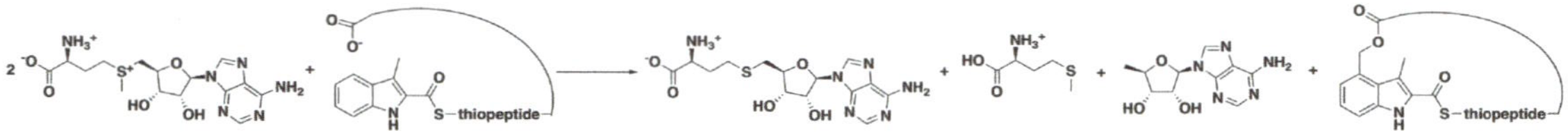
		pyrimidine ring methyltransferase (bleomycin-like antibiotic biosynthesis)	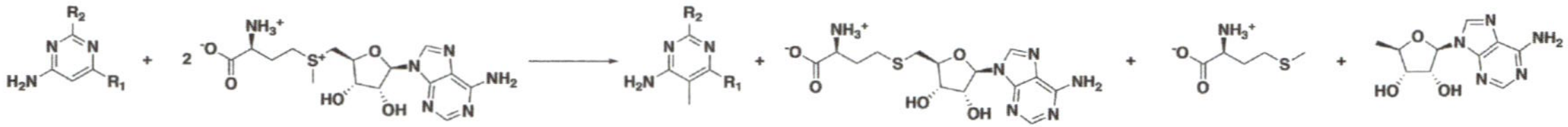
		aromatic C-methyltransferase (yatakemycin antibiotic biosynthesis)	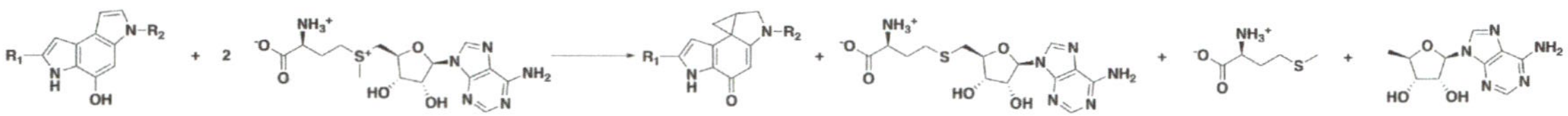
3	51,51,51	Viperin	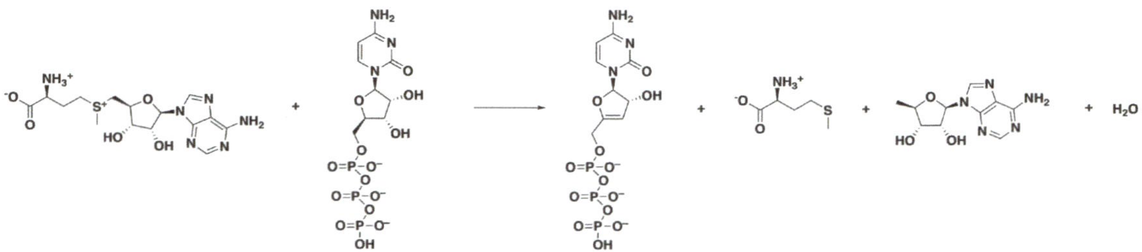

Continued

Table S1b Reaction Families and Overall Reactions Associated With SFLD Subgroups of the Radical SAM Superfamily as Keyed to Fig. 5—cont'd

SFLD Sub-group #	*L1 Subgroup Color (RGB) in Fig. 5*	*SFLD Family*	*Overall Chemical Transformation (reactions are not necessarily balanced)*
4	255,153,153	avilamycin epimerase	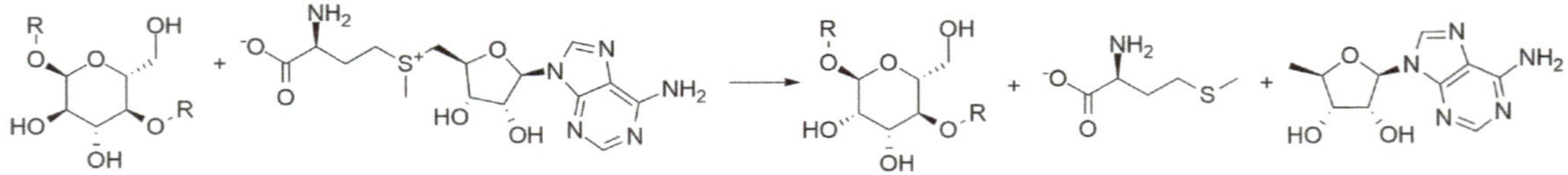
5	0,0,255	bacteriocin maturation	Unknown
		paromamine deoxygenase	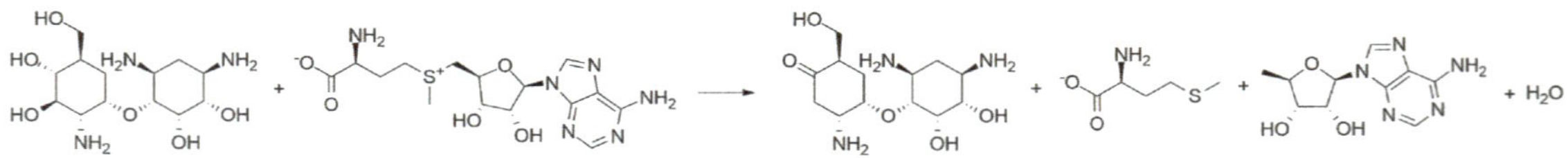
		hopanetetrol cyclitol ether synthase	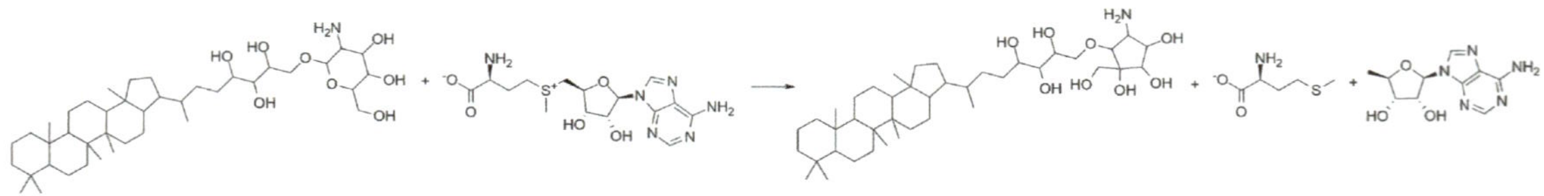
		ladderane biosynthesis	Thought to be involved in one of the steps in ladderine biosynthesis
		anaerobic magnesium-protoporphyrin-IX monomethyl ester cyclase	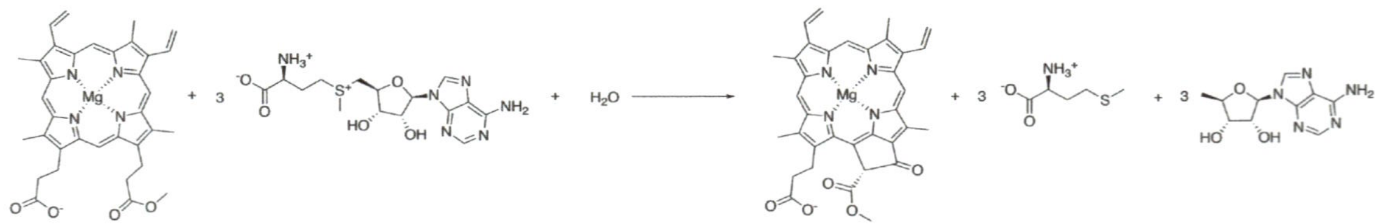

	OxsB-like	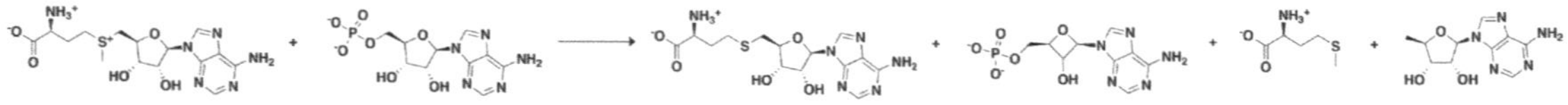
	valine cyclopropanase	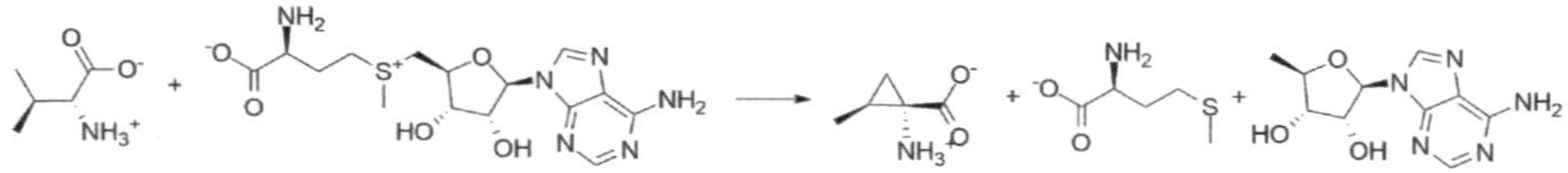
	bacteriochlorophyll C12 methyltransferase	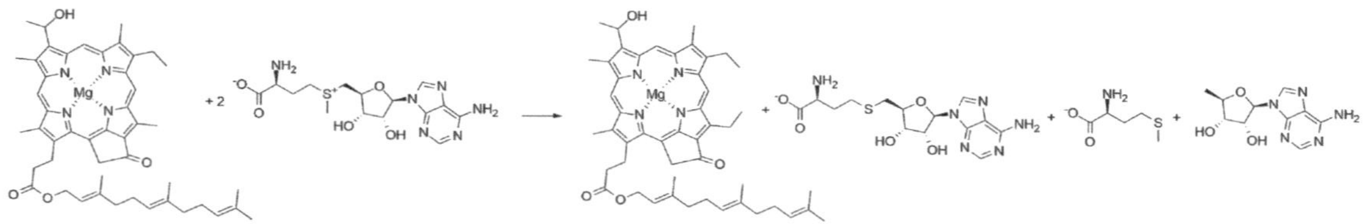
	bacteriochlorophyll C8 methyltransferase	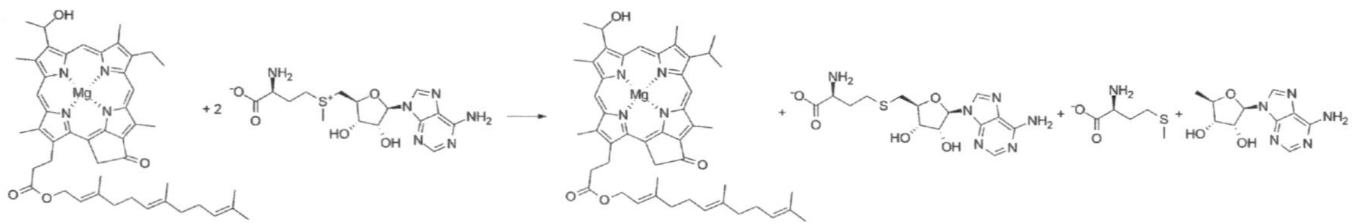

Continued

Table S1b Reaction Families and Overall Reactions Associated With SFLD Subgroups of the Radical SAM Superfamily as Keyed to Fig. 5—cont'd

SFLD Sub-group #	*L1 Subgroup Color (RGB) in Fig. 5*	*SFLD Family*	*Overall Chemical Transformation (reactions are not necessarily balanced)*
		5″-pyrrole methyltransferase	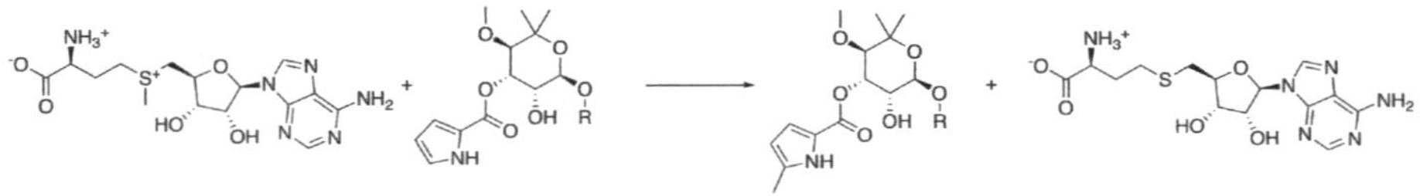
		chondrochloren A methyltransferase (CndI-like)	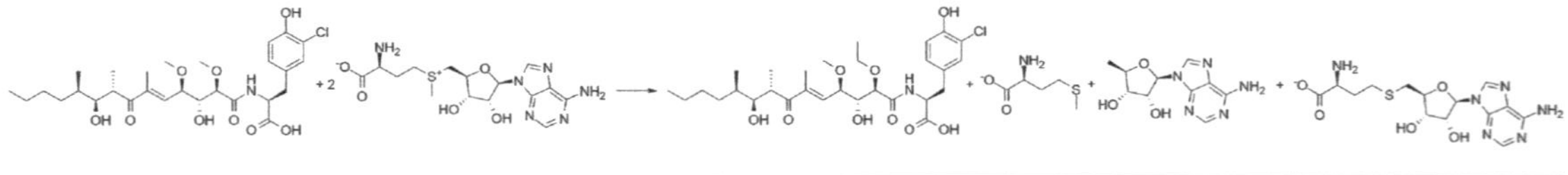
		phosphonoacetaldehyde methylase (fom3-like)	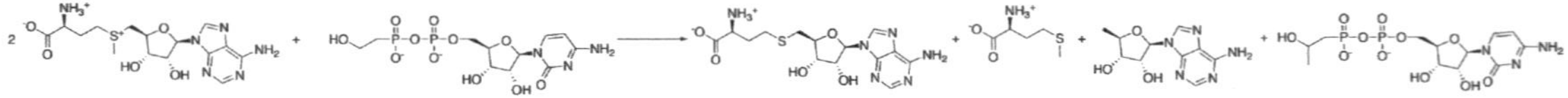
		fortimicin-like methyltransferase	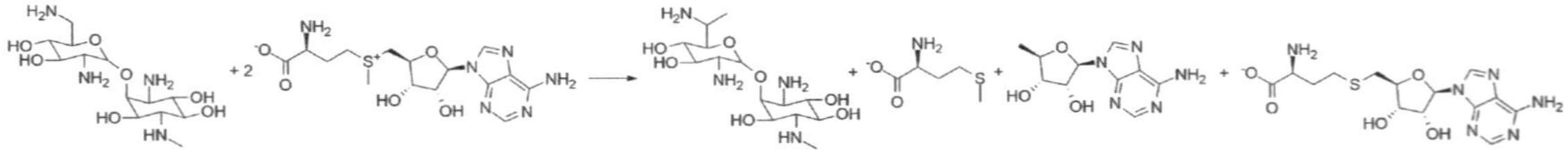
		gentamicin-like N-methyltransferase (GntE-like)	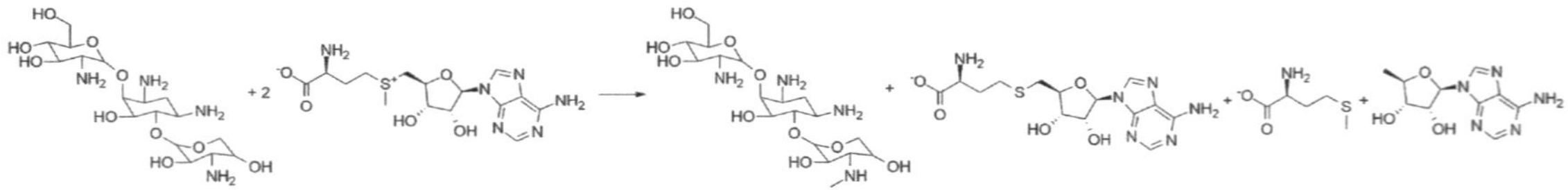

[redacted]	gentamicin C-6′ methyltransferase (GenK-like)	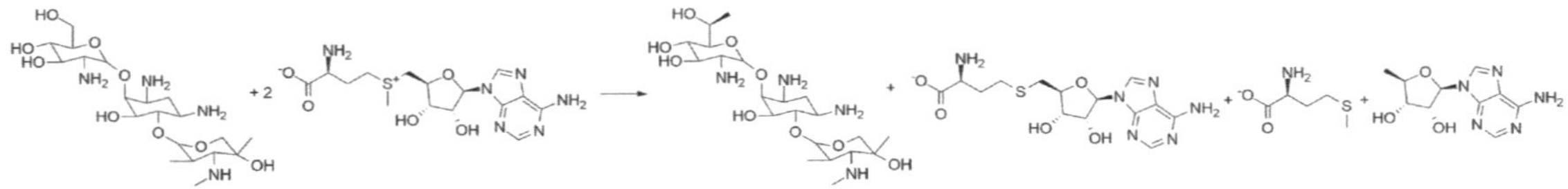
	hopanoid C2-methyltransferase (HpnP-like)	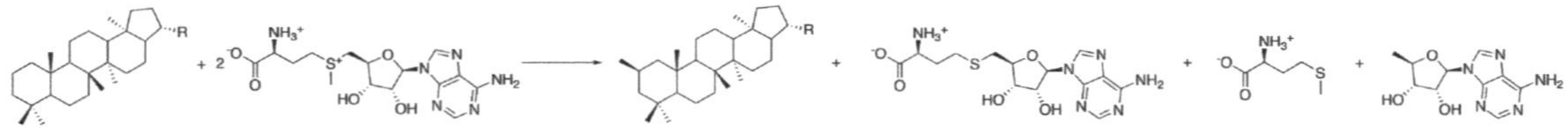
	hopanoid C3-methyltransferase (HpnR-like)	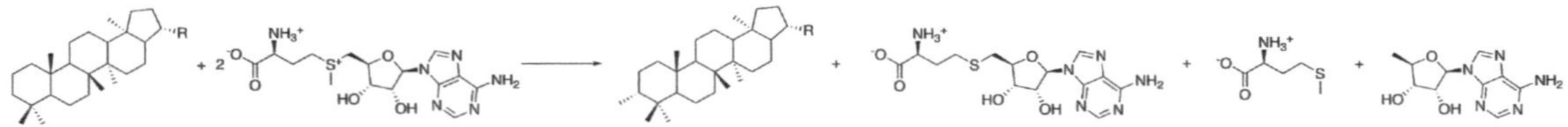
	P-methyltransferase (PhpK-like)	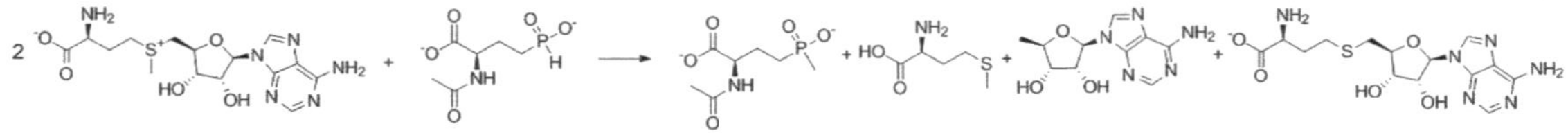 Methylates a phosphinate phosphorus atom during phosphinothricin biosynthesis
	pactamycin C-methyltransferase (PctJ-like)	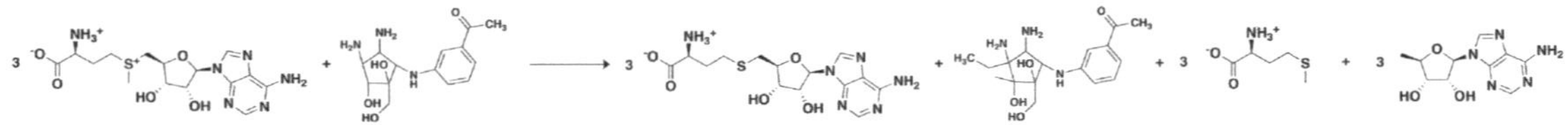
	pactamycin C-methyltransferase (PctN-like)	Each of these enzymes is involved in the methylation at one of the three positions of the cyclopentane ring during pactamycin biosynthesis*
	pactamycin C-methyltransferase (PctO-like)	

Continued

Table S1b Reaction Families and Overall Reactions Associated With SFLD Subgroups of the Radical SAM Superfamily as Keyed to Fig. 5—cont'd

SFLD Sub-group #	*L1 Subgroup Color (RGB) in Fig. 5*	*SFLD Family*	*Overall Chemical Transformation (reactions are not necessarily balanced)*
		thioacetal methyltransferase (Swb9-like)	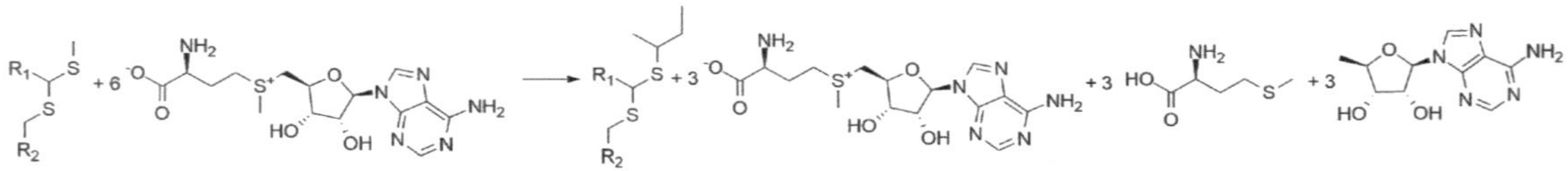 Iterative methylation of the thioacetal bridge to form (+) norcoronamic acid
		carbapenem intermediate methyltransferase (ThnK-like)	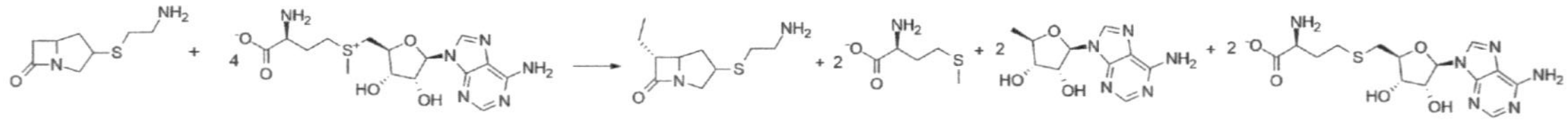 Shown to perform the sequential double methylation of the beta-lactam ring in carbapenem formation
		carbapenem intermediate methyltransferase (ThnL-like)	Involved in the methylation of one of the carbapenem intermediates*
		carbapenem intermediate methyltransferase (ThnP-like)	Involved in the methylation of one of the carbapenem intermediates*
		tryptophan 2-C-methyltransferase	 Methylates the C2 position in tryptophan in thiostrepton A biosynthesis
		None yet assigned	Methylthiotransferase

6	**255,153,0**	biotin synthase	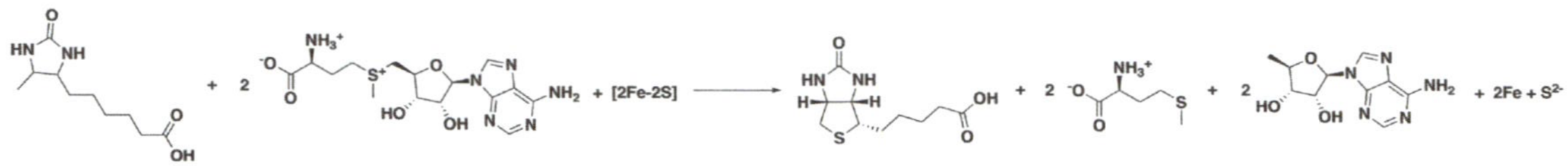
		HMD cofactor maturase (HmdB-like)	Involved in the maturation of the HMD cofactor*
		[FeFe] hydrogenase maturase (HydE-like)	Involved in the maturation of the [FeFe]-hydrogenase cofactor*
		3-methylornithine synthase (PylB-like)	
		2-iminoacetate synthase (ThiH-like)	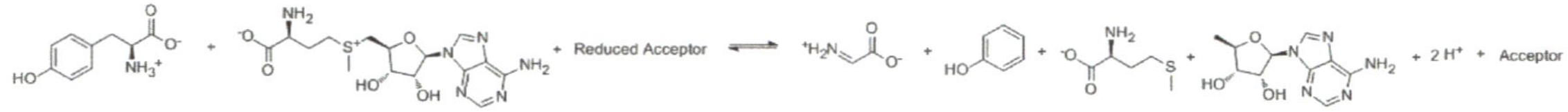
		3-methyl-2-indolic acid synthase (NosL-like)	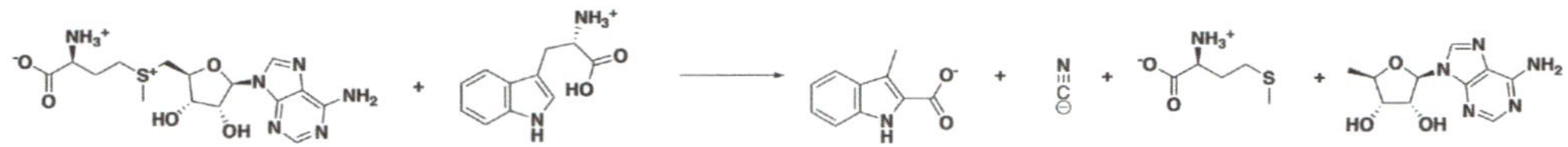
		[Fe] hydrogenase maturase (HydG-like)	Involved in the maturation of Fe-hydrogenase cofactor*
7	204,153,255	dTDP-4-amino-4,6-dideoxy-D-glucose ammonia-lyase (DesII-like)	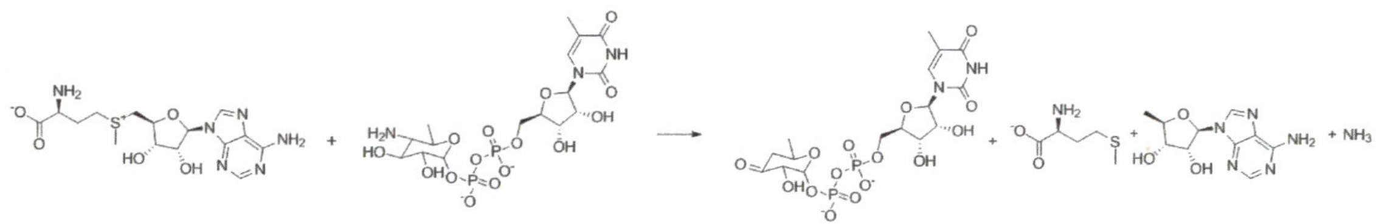

Continued

Table S1b Reaction Families and Overall Reactions Associated With SFLD Subgroups of the Radical SAM Superfamily as Keyed to Fig. 5—cont'd

SFLD Subgroup #	*L1 Subgroup Color (RGB) in Fig. 5*	*SFLD Family*	*Overall Chemical Transformation (reactions are not necessarily balanced)*
8	0,0,0	ELP3-like	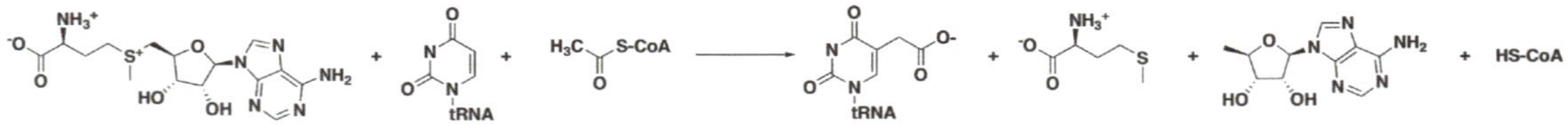
9	153,0,153	((2,3,4,5-tetrahydroxypentyl)amino) dihydropyrimidine-2,4-dione synthase (CofH-like)	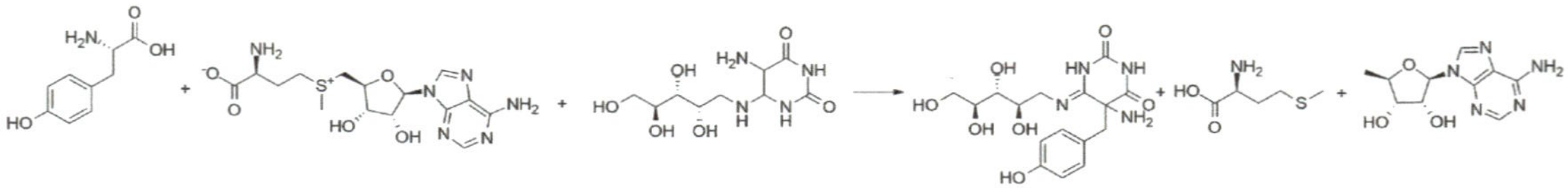 First half reaction involved in the synthesis of 7,8-didemethyl-8-hydroxy-5-deazariboflavin. The product from this half reaction is transferred to the CofG active site where it undergoes the second half reaction
		7,8-didemethyl-8-hydroxy-5-deazariboflavin synthase (CofG-like)	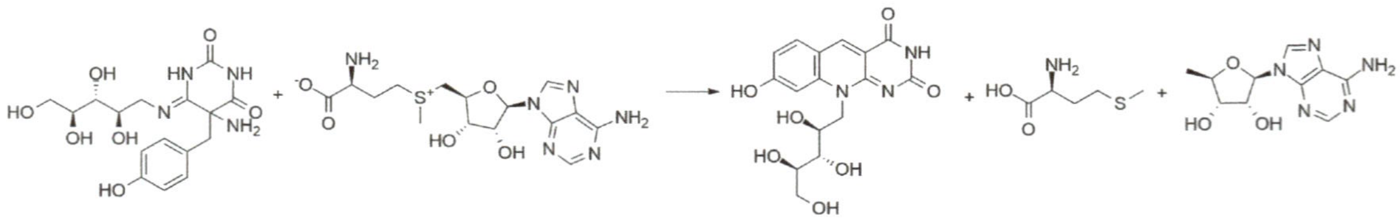 Second half reaction involved in the synthesis of 7,8-didemethyl-8-hydroxy-5-deazariboflavin
		cyclic dehypoxanthine futalosine synthase (mqnC-like)	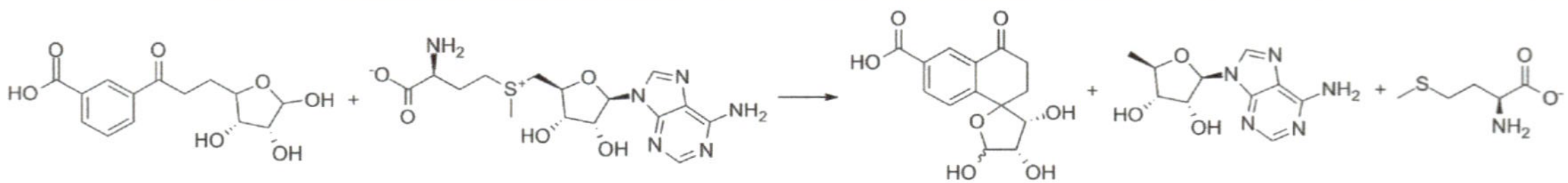

		aminofutalosine synthase (mqnE-like)	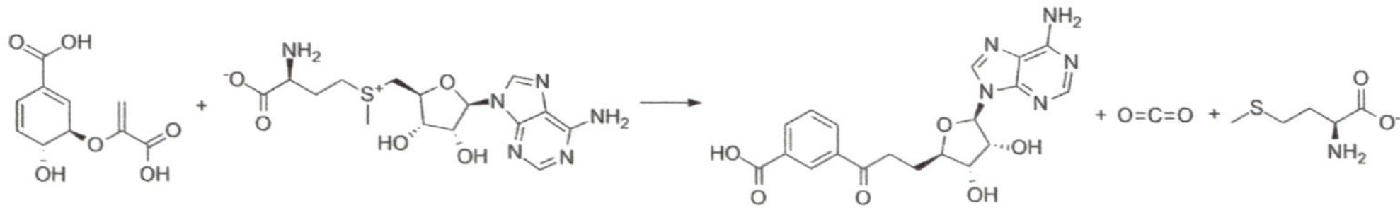
10	**204,255,204**	FeMo cofactor biosynthesis protein (nifB-like)	Involved in carbon insertion into the nitrogenase M-cluster*
11	**255,255,0**	lipoyl synthase	
12	**102,102,0**	(dimethylallyl)adenosine tRNA methylthiotransferase (MiaB-like)	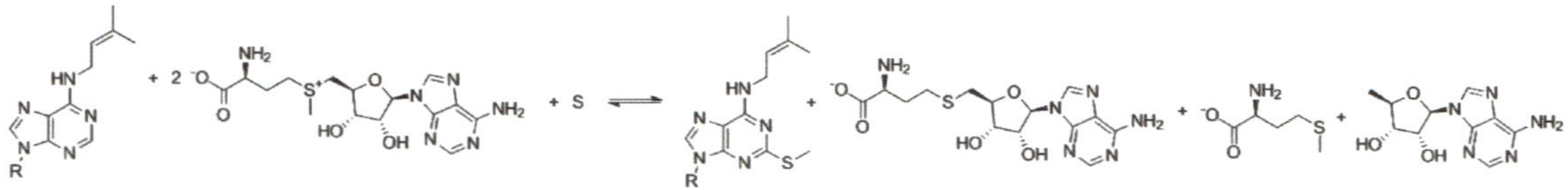
		CDK5RAP1	Inhibits CDK5 activation by CDK5R1. This protein is a methylthiotransferase that acts on nuclear and mitochondrial RNA
		ribosomal protein S12 methylthiotransferase (RimO-like)	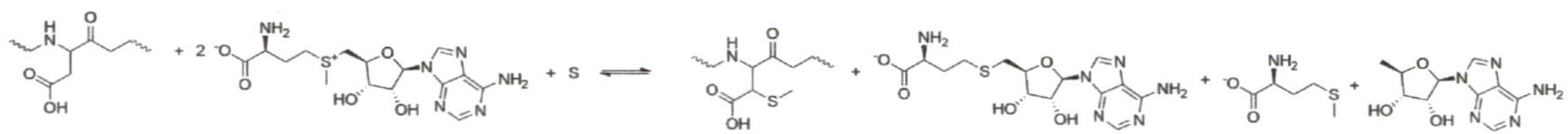
		threonylcarbamoyladenosine tRNA methylthiotransferase	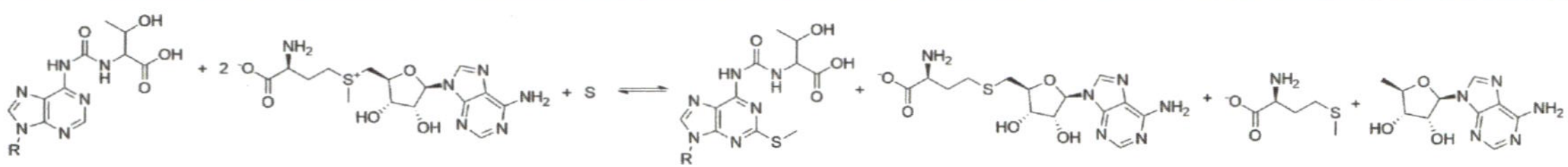

Continued

Table S1b Reaction Families and Overall Reactions Associated With SFLD Subgroups of the Radical SAM Superfamily as Keyed to Fig. 5—cont'd

SFLD Sub-group #	*L1 Subgroup Color (RGB) in Fig. 5*	*SFLD Family*	*Overall Chemical Transformation (reactions are not necessarily balanced)*
13	102,0,0	adenosine C2 methyltransferase (RlmN-like)	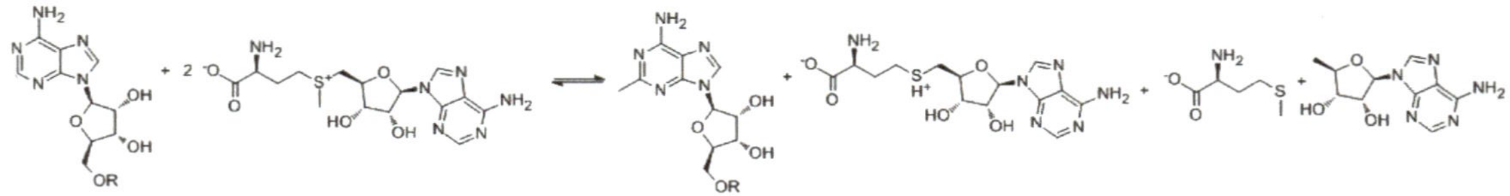
		adenosine C8 methyltransferase (Cfr-like)	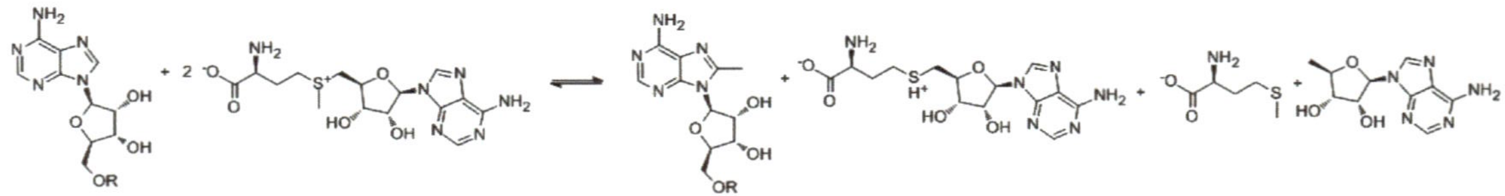
14	**255,204,204**	7,8-dihydro-6-hydroxymethylpterin dimethyltransferase	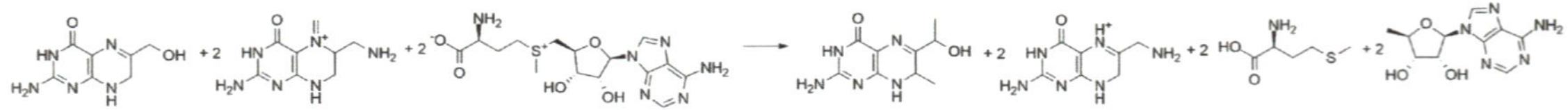 Involves methylenetetrahydromethanopterin as the methyl donor
15	0,255,255	4-hydroxyphenylacetate decarboxylase activase	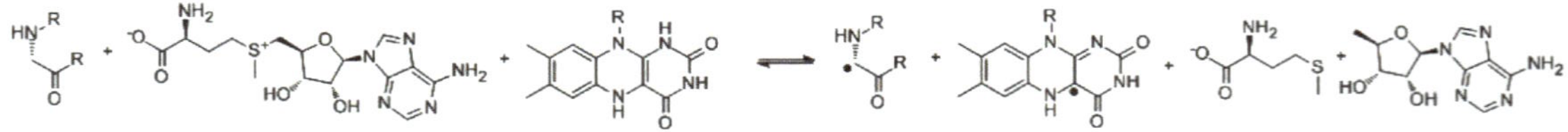
		anaerobic ribonucleoside-triphosphate reductase activase	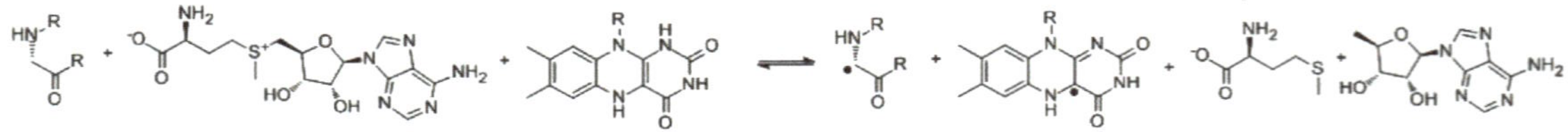
		benzylsuccinate synthase activase	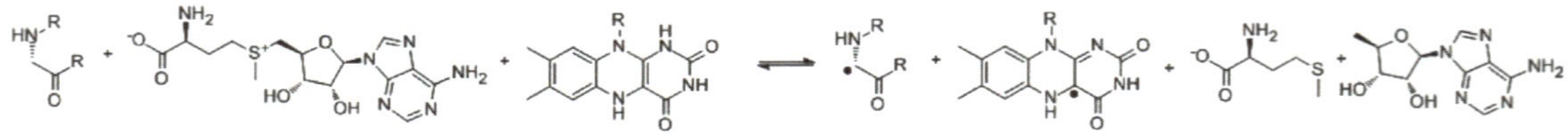

		glycerol dehydratase activase	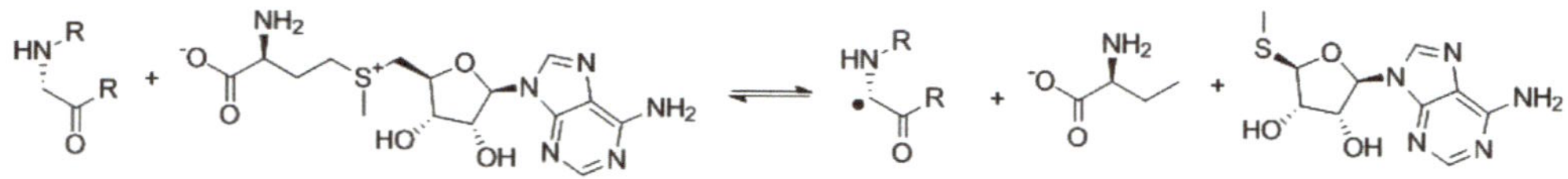
		pyruvate formate-lyase activase	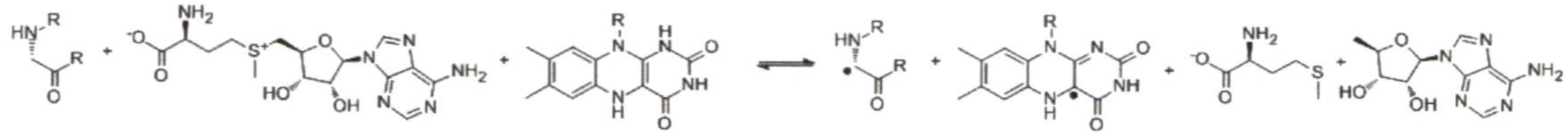
		YjjI activase	Activates an unidentified glycyl-radical enzyme*. Mechanism proposed to be similar to the other activating enzymes
16	0,153,0	arginine 2,3-aminomutase	
		glutamate 2,3-aminomutase	
		L-lysine 2,3-aminomutase (LAM-like)	
		L-lysine 2,3-aminomutase (yjeK like)	

Continued

Table S1b Reaction Families and Overall Reactions Associated With SFLD Subgroups of the Radical SAM Superfamily as Keyed to Fig. 5—cont'd

SFLD Sub-group #	*L1 Subgroup Color (RGB) in Fig. 5*	*SFLD Family*	*Overall Chemical Transformation (reactions are not necessarily balanced)*
17	255,0,255	heme D1 biosynthesis (NirJ-like)	Unknown. Involved in the biosynthesis of heme D1. Given similarity to other members of this subgroup, probably involved in decarboxylation
		alternative heme biosynthesis C (AhbC-like)	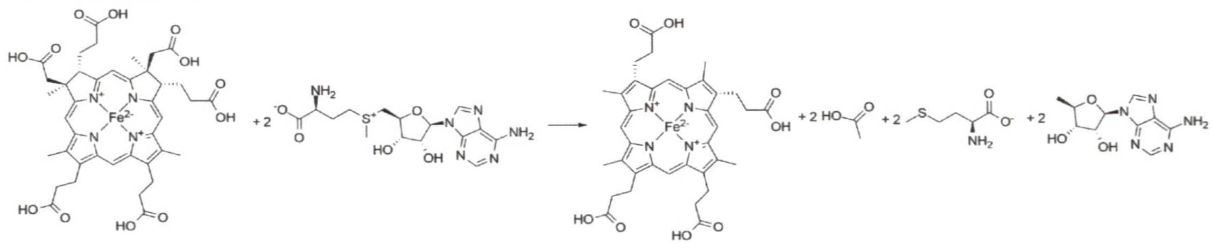
		alternative heme biosynthesis D (AhbD-like)	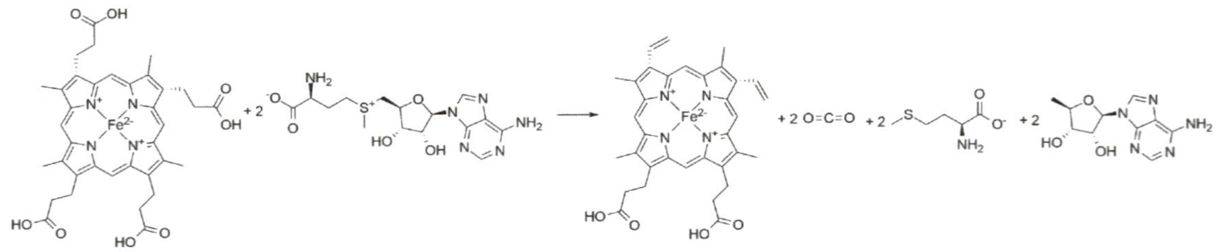
		mycofactocin peptide maturase (MftC-like)	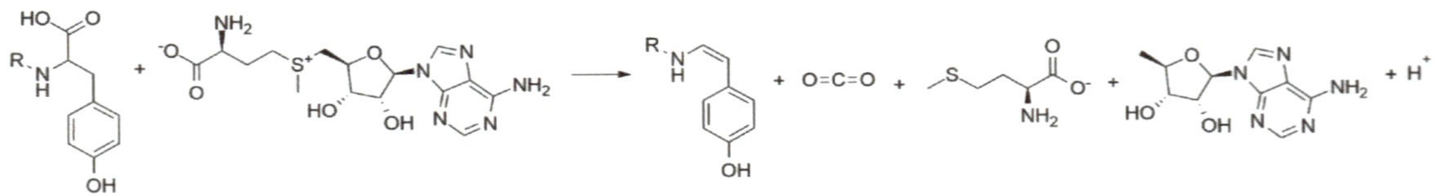

	cyclic pyranopterin phosphate synthase (MoaA-like)	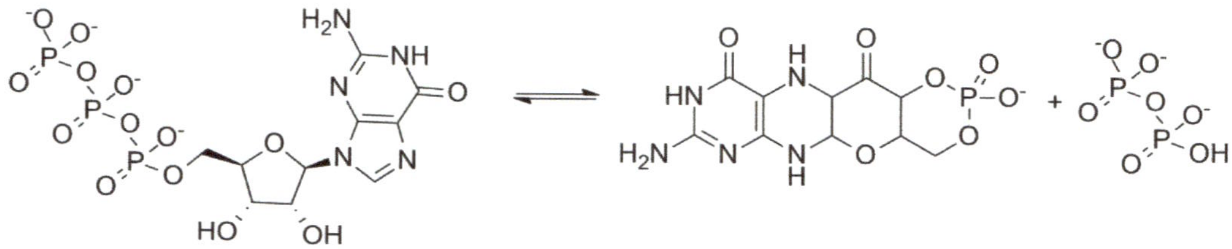
	coenzyme PQQ synthesis protein E (PqqE-like)	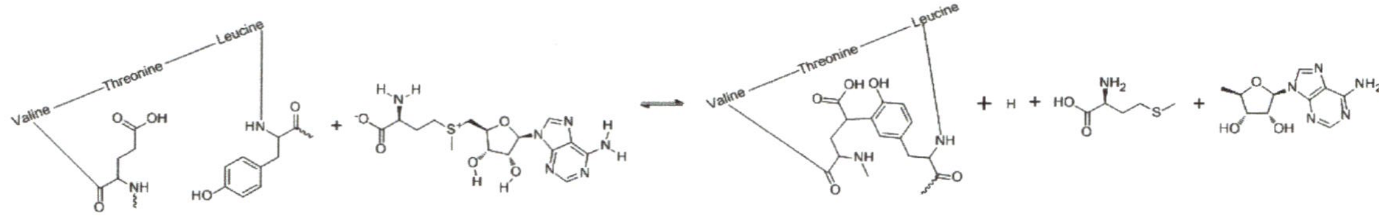 Formation of C-C linkage between Tyr and Glu is the formation of the PQQ cofactor
	anaerobic Cys-type sulfatase-maturating enzyme	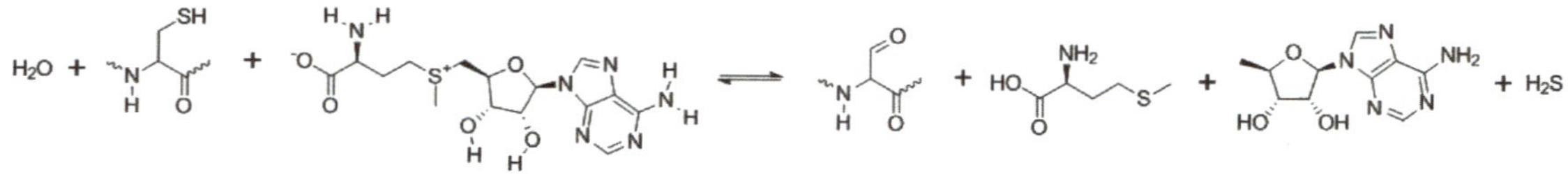
	anaerobic Ser-type sulfatase-maturating enzyme	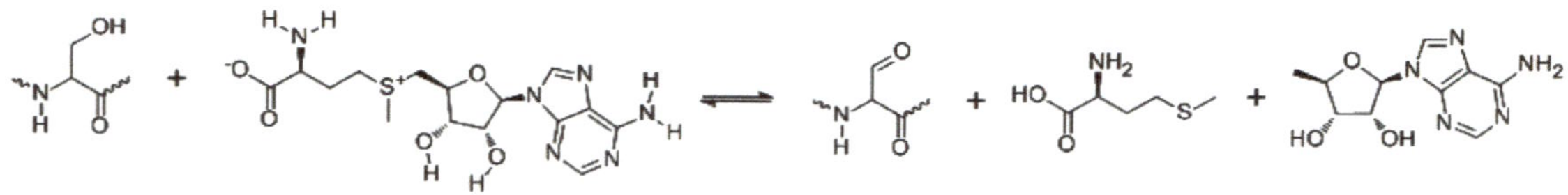

Continued

Table S1b Reaction Families and Overall Reactions Associated With SFLD Subgroups of the Radical SAM Superfamily as Keyed to Fig. 5—cont'd

SFLD Sub-group #	L1 Subgroup Color (RGB) in Fig. 5	SFLD Family	Overall Chemical Transformation (reactions are not necessarily balanced)
		quinohemoprotein amine dehydrogenase maturation protein (QhpD-like)	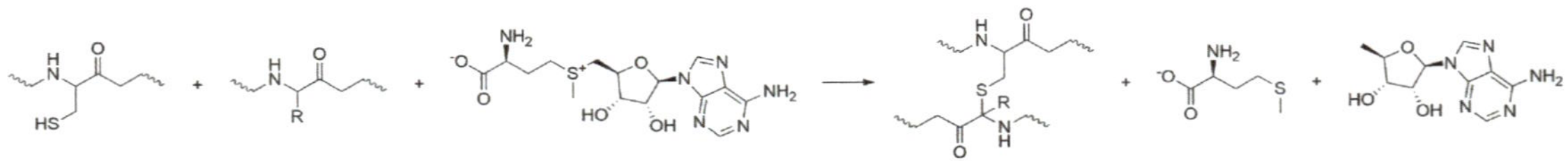 Involved in the formation of sequential thioether cross-links between the sulfur atom of a Cys residue and the methylene carbon atom of an Asp or Glu residue
		tte1186 maturase	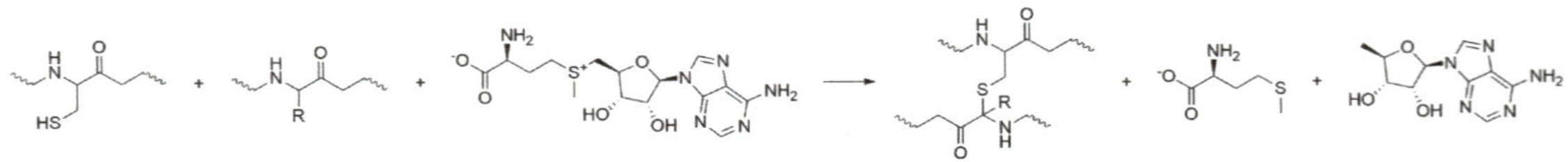 Thioether bond formation between a cysteine and the C-alpha of a threonine residue in tte1186a
		sporulation killing factor protein (SkfB-like)	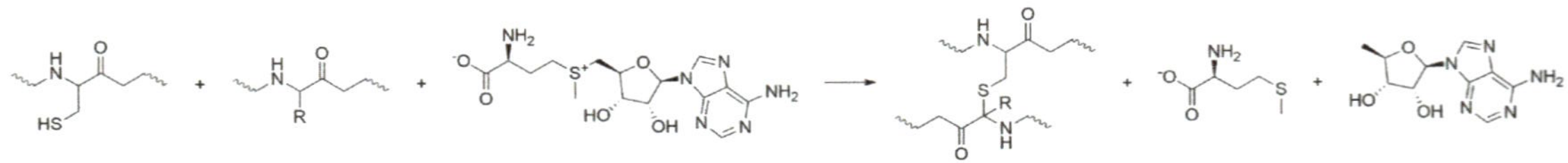 Thioether bond formation between cysteine and C-alpha of methionine
		antilisterial bacteriocin subtilosin biosynthesis protein (AlbA-like)	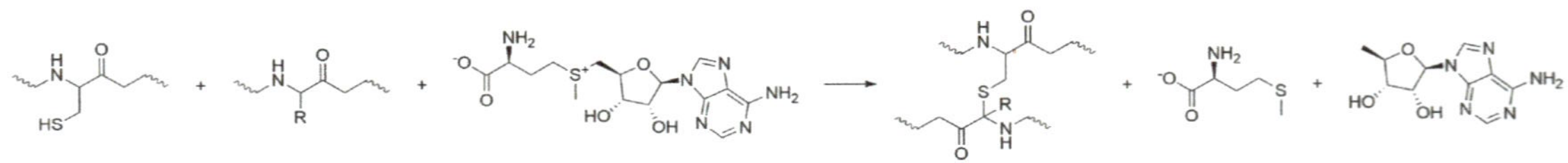 Thioether bond formation between cysteine and C-alpha a threonine or phenylalanine

thuricin CD (TrnCD-like)	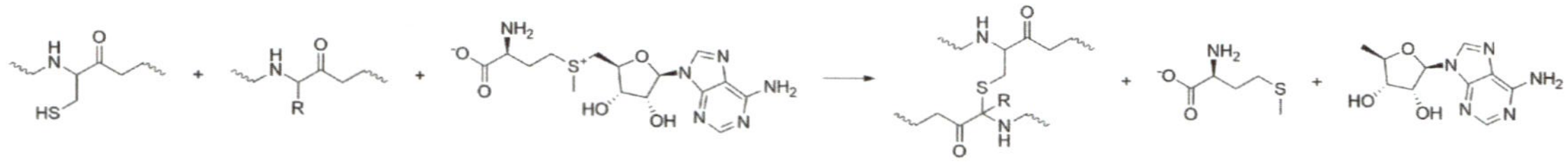 Thioether bond formation between cysteine and C-alpha and serine, threonine, alanine, and tyrosine residues in the Trna and Trnb polypeptides
thurincin H (ThnB-like)	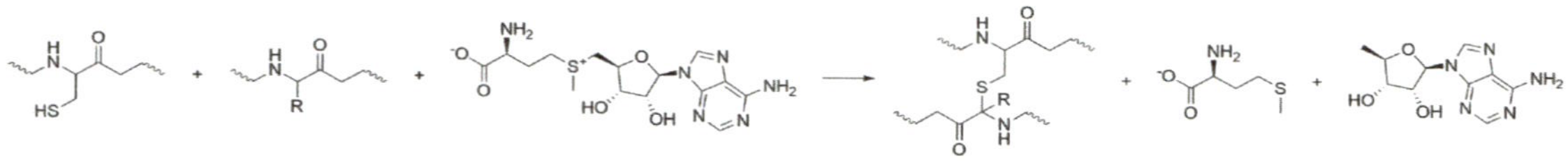 Catalyses the thioether bond formation between cysteine and C-alpha of a serine, threonine, and asparagine
2-deoxy-scyllo-inosamine dehydrogenase (BtrN-like)	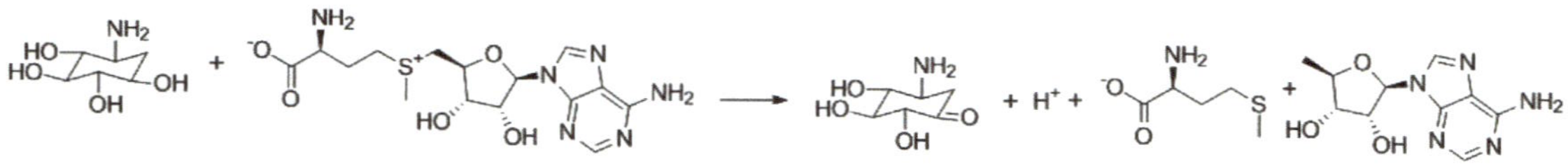
tungsten cofactor oxidoreducase radical SAM maturase	Unknown
adenosyl-hopene transferase	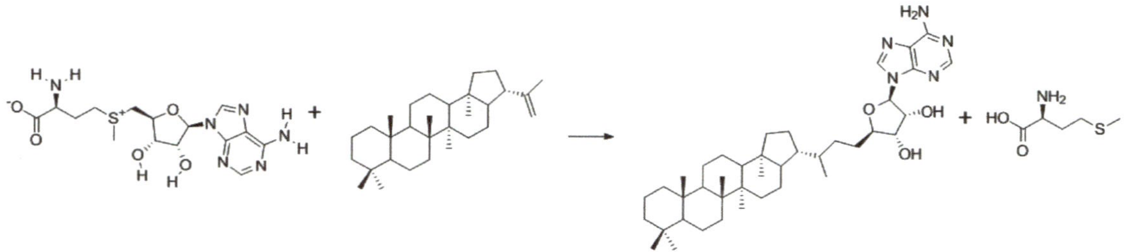

Continued

Table S1b Reaction Families and Overall Reactions Associated With SFLD Subgroups of the Radical SAM Superfamily as Keyed to Fig. 5—cont'd

SFLD Subgroup #	*L1 Subgroup Color (RGB) in Fig. 5*	*SFLD Family*	*Overall Chemical Transformation (reactions are not necessarily balanced)*
		cytosylglucuronate decarboxylase	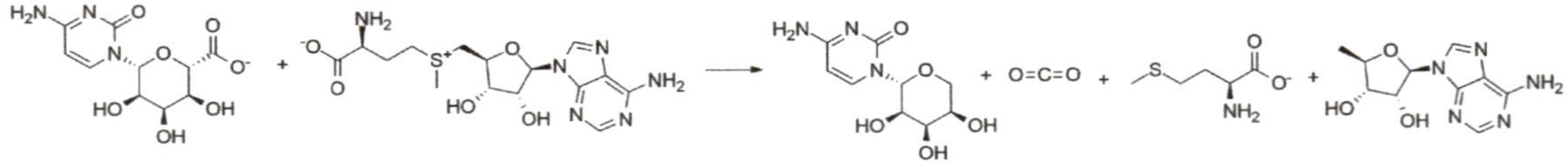
		KxxxW cyclic peptide maturase (StrB-like)	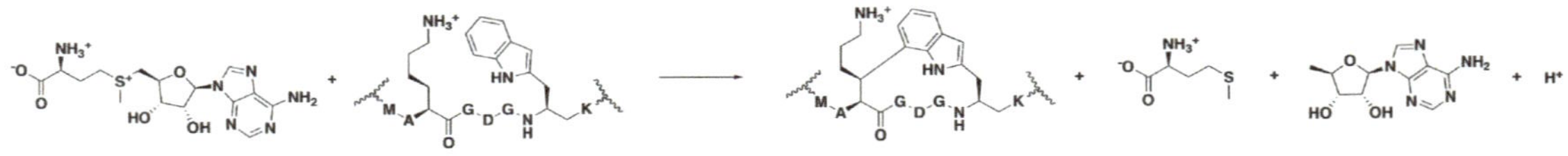
		neomycin C-like epimerase	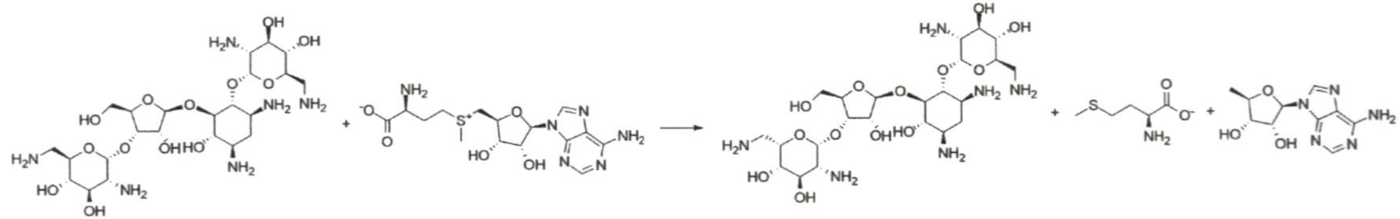
		UDP-N-acetyl-tunicamine-uracil synthase (TunB-like)	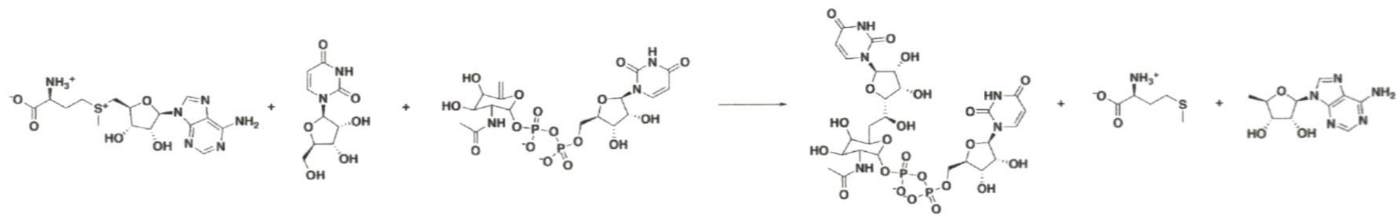
		pcfB-like	Radical cleavage of the N-terminus of propionicin F*

18	**255,255,255**	spectinomycin biosynthesis (SpcY-like)	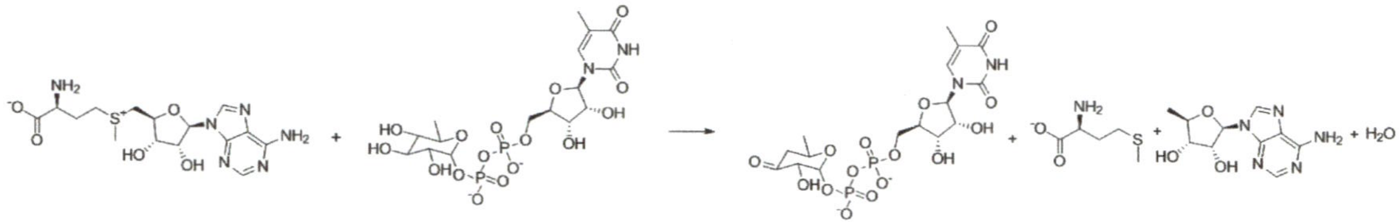
19	**0,255,0**	spore photoproduct lyase 1	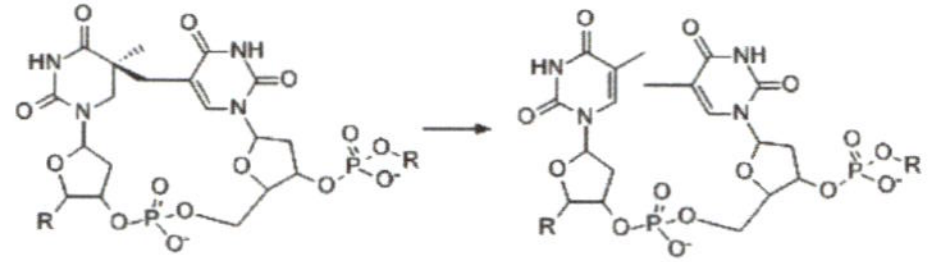
		spore photoproduct lyase 2	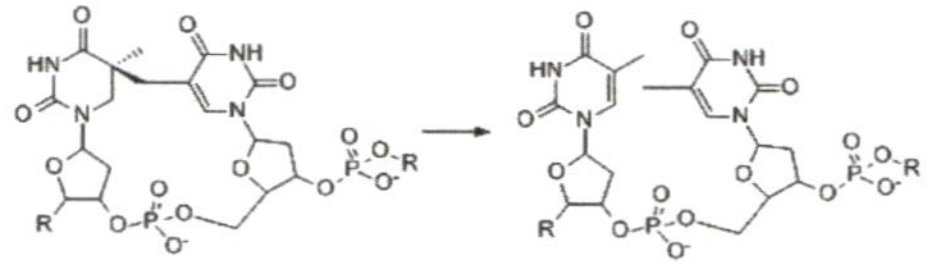
20	**153,102,0**	tRNA wybutosine-synthesizing	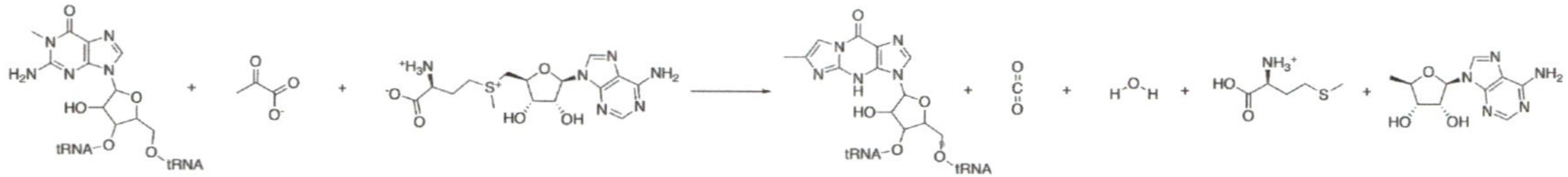

*Proposed reaction.

broadly different reactions. Although the topologies shown are associated with different subgroups, due to the still poor structural coverage of the superfamily, we cannot assert that each typifies the majority of the structures of its associated subgroup.

For each image shown in Fig. 6, the full names, abbreviations (in parentheses), PDB identifiers, and Level 1 subgroup numbers (in square brackets) are 7-carboxy-7-deazaguanine synthase (QueE), 4NJK, [1]; oxygen-independent coproporphyrinogen-III oxidase 1 (HemN) 1OLT, [2]; biotin synthase (BioB) 1R30, [6]; [Fe] hydrogenase maturase (HydG) 4WCX, [6]; lipoyl synthase (LipA) 4U0P, [11]; ribosomal protein S12 (aspartate89-C3)-methylthiotransferase (RimO) 2QGQ, [12]; 23S rRNA (adenine2503-C2)-methyltransferase (RlmN), 3RFA, [13]; pyruvate formate-lyase activating enzyme (PFL-AE) 3C8F, [15]; L-lysine 2,3-aminomutase (LAM) 2AH5 [16]; cyclic pyranopterin phosphate synthase (MoaA) 1TV8, [17, Level 3]; anaerobic Cys-type sulfatase-maturating enzyme (AnSME) 4K36, [17, Level 3];

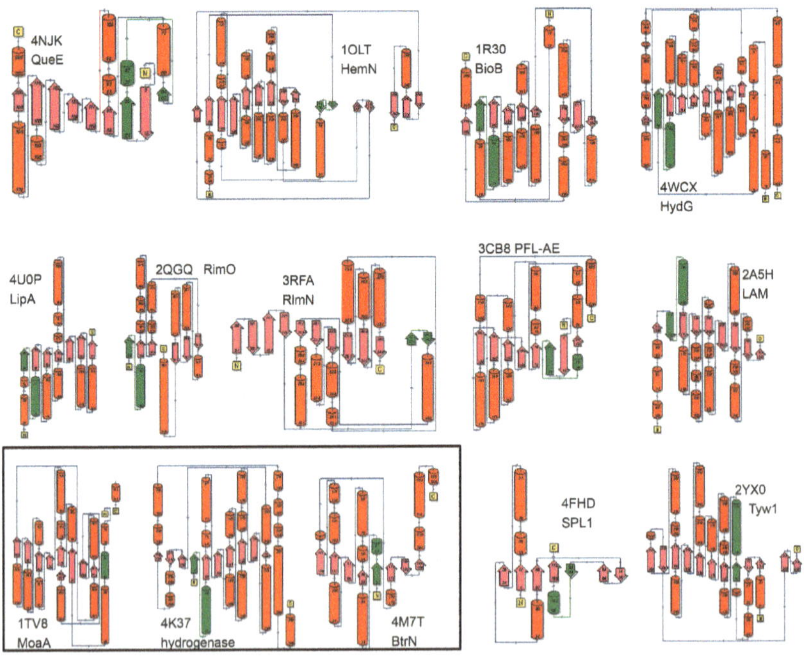

Fig. 6 Secondary structure topologies of representative radical SAM superfamily domains. Images for several RSS subgroups created using the PDBSum website (de Beer, Berka, Thornton, & Laskowski, 2014). *Red*: helices, *pink*: beta strands, *green*: $[Fe_4S_4]$-AdoMet-binding motif. The common abbreviations of the enzyme names and their PDB identifiers are shown on the figure.

2-deoxy-scyllo-inosamine dehydrogenase (BtrN) 4M7T, [17, Level 2]; spore photoproduct lyase 1 (SPL1) 4FHD, [19]; tRNA 4-demethylwyosine synthase (Tyw1) 2YX0 [20]. These and other data, expanded to include all RSS subgroups of our classification, are provided in Supplementary Table S1 in the online version at https://doi.org/10.1016/bs.mie.2018.06.004.

2.3 A Large-Scale View of Sequence–Structure–Function Relationships

The representative network shown in Fig. 5 provides an estimate of the breadth of experimental coverage for RSS subgroups. Although this visualization may appear to infer that several of these numbered Level 1 subgroups include many characterized functions and structures, this first-pass interpretation could lead to the incorrect inference that a significant proportion of RSS members are knowns. This results because the figure was intended to highlight how broadly and in which subgroups experimentally characterized enzymes sample the RSS sequence space, rather than to convey the proportion of that sequence space that has been experimentally characterized. Table 2 provides a direct count of the number of representative nodes and the total number of sequences in each of the 20 numbered Level 1 subgroups classified in the SFLD. There are 9137 representative nodes in the 20 numbered subgroups shown in Fig. 5 containing a total of 97,276 sequences. As detailed later, the overwhelming majority of these sequences have not been experimentally characterized in any way. (The number of sequences presented in Table 2 includes 97,276 sequences of the 113,776 sequences in the RSS, as the remainder are sufficiently diverse that they could not be confidently assigned to a named subgroup or even to one of the "uncharacterized" subgroups in the SFLD. None of these remainder sequences have characterized functions.)

Another indicator of the proportion of RSS sequences that are unknowns is the large amount of gray space in Fig. 5. There are 8254 small gray representative nodes (out of 10,741 total nodes in the figure) that are entirely made up of unknowns. Although these gray nodes are not included in Table 2, the number of sequences in each of these 22 "uncharacterized" subgroups in the SFLD range from 44 to 2635.

Especially daunting for achieving a realistic estimate of the functional and structural variation in the RSS, the proportion of characterized structures and molecular functions reported in the primary literature also remains remarkably small. Fewer than 4000 RSS members (less than 4% of the total number of sequences in the RSS) are described as experimentally

Table 2 Number of Representative Nodes and Sequences in Each Level 1 Subgroup Curated in the SFLD

Level 1 Subgroup #	Subgroup Name	# Rep Nodes	# Sequences
1	7-Carboxy-7-deazaguanine synthase like	220	4126
2	Anaerobic coproporphyrinogen-III oxidase like	1098	12,939
3	Antiviral proteins	57	312
4	AviX12-like	3	7
5	B12-binding domain containing	1615	6858
6	BATS domain containing	242	8239
7	DesII-like	1	12
8	Elongater protein like	267	2626
9	F420, menaquinone cofactor biosynthesis	154	3271
10	FeMo cofactor biosynthesis protein	68	813
11	Lipoyl synthase like	113	5416
12	Methylthiotransferase	828	14,884
13	Methyltransferase (Class A)	315	6632
14	Methyltransferase (Class D)	64	572
15	Organic radical-activating enzymes	587	7847
16	PLP-dependent	162	2679
17	SPASM/twitch domain containing	3251	18,807
18	Spectinomycin biosynthesis (SpcY like)	5	6
19	Spore photoproduct lyase like	14	437
20	tRNA wybutosine-synthesizing	73	793

characterized in the reviewed section of the UniProt Knowledge Base (UniProt, 2017), Swiss-Prot. (Swiss-Prot offers the largest high quality compilation of known protein functions currently available.) Thus, the experimentally determined proportion of RSS enzymes is likely to remain much smaller than that of the unknowns. Likewise, the structural coverage of the RSS is also extraordinarily low, with less than 100 structures representing even fewer unique proteins in the world-wide PDB (Berman, Henrick, & Nakamura, 2003).

Our knowledge of the extent of structural variation in the canonical RSS is also small due to the still poor structural sampling of its sequence space. Our attempts to use modeling of RSS to ascertain independently the range of fold variation that comprise the RSS functional domain failed to identify additional fold types (data not shown). As with other investigations of this superfamily, this analysis was limited by the poor availability of structurally characterized RSS enzymes for use as modeling templates.

2.4 Ancient Lineage of the RSS

It has been previously suggested that the RSS is of ancient origin, based in part on its ubiquity across the biosphere (see, for example, Holliday et al., 2007; Vey & Drennan, 2011). The global analysis of the RSS reported here reveals the broad extent of RSS members in all three Kingdoms of life, providing additional support for this notion. The SSN shown in Fig. 7 indicates that the vast majority of RSS members are bacterial, while archaeal sequences represent the next largest proportion. Although a small proportion compared to bacteria, archaeal sequences are found globally across the network, likely indicating evolutionary emergence of RSS enzymes and divergence into major subtypes prior to the archaeal–eubacterial split. Few members are found in eukaryotic organisms and these are largely grouped together in a small number of subgroups. Some of these eukaryotic enzymes have been associated with disease-causing mutations in humans, for example, the MoaA (Hanzelmann & Schindelin, 2004) and LipA (Baker et al., 2014) families of the cyclic pyranopterin phosphate Level 3 subgroup of the SPASM/Twitch Level 1 subgroup and the Level 1 LipA subgroup, respectively.

As another indication of the ancient origins of the superfamily, many of the reactions RSS enzymes are known to catalyze are fundamental to all types of life. For example, the RSS is involved in the biosynthesis of over half of the known organic cofactors. Table 3 shows some of the Level 1 subgroups and their constituent families involved in the biosynthesis of some of these cofactors, along with the types of life in which these enzymes are found.

The RSS also contains a large number of families responsible for the modification and repair of DNA and RNA, another fundamental and

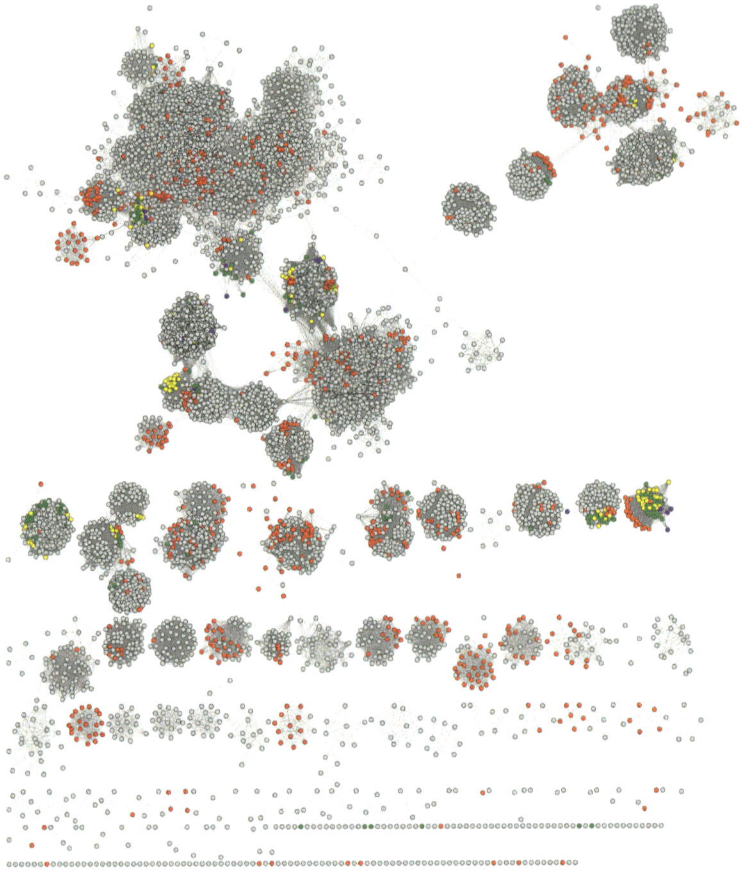

Fig. 7 RSS SSNs mapped with type of life. The same representative network shown in Fig. 5 except that node coloring is by type of life as defined in the SFLD. Representative nodes are colored by the dominant type of life in each. Bacteria: *gray*, Archaea: *red*, Invertebrates: *yellow*, Vertebrates: *blue*, Plants: *green*.

ancient requirement for life. For example, the methylthiotransferase subgroup 12, MTTases, catalyze a C–H to C–S bond conversion in the methylthiolation of tRNA. Four families can be assigned to this subgroup: MiaB-like, CDK5RAP1, RimO, and MtaB (Anantharaman, Koonin, & Aravind, 2001). MTTases appear to be made up of an N-terminal MTTase domain, a central radical-generating fold and a TRAM domain (an acronym representing two named domains, TRM2 and the MiaB), found at the C-terminal end. The TRAM domain can bind to an RNA substrate and appears to be important for substrate recognition (Lee et al., 2009; UniProt, 2017). In addition to the radical-generating $[Fe_4S_4]$ cluster domain

Table 3 RSS Enzymes Involved in Biosynthesis of Organic Cofactors

L1 Subgroup Name and (#)	Family	Cofactor	Type of Life[a]
Anaerobic coproporphyrinogen-III oxidase (2)	Heme degradation proteins (HutW/ChuW)	Heme (degradation)	B
	Oxygen-independent coproporphyrinogen-III oxidase 1 (HemN)	Heme (biosynthesis)	BPVI
	Oxygen-independent coproporphyrinogen-III oxidase 2 (HemZ)		B
B12-binding domain (5)	Anaerobic magnesium-protoporphyrin-IX monomethyl ester cyclase (BchE)	Bacteriochlorophyll/anaerobically synthesizsed chlorophyll	B
BATS domain containing (6)	2-Iminoacetate synthase (ThiH)	Thiamine	ABPI
	[Fe] Hydrogenase maturase (HydG)	[Fe] Hydrogenase metallocofactor assembly	BPI
	Biotin synthase (BioB)	Biotin	ABPVI
	HMD cofactor maturase (HmdB)	5,10-Methenyltetrahydromethanopterin hydrogenase cofactor	AB
	[FeFe] Hydrogenase maturase (HydE)	[FeFe] Hydrogenase metallocofactor assembly	BPI

Continued

Table 3 RSS Enzymes Involved in Biosynthesis of Organic Cofactors—cont'd

L1 Subgroup Name and (#)	Family	Cofactor	Type of Life
F420, menaquinone cofactor biosynthesis (9)	((2,3,4,5-Tetrahydroxypentyl)amino) dihydropyrimidine-2,4-dione synthase (CofH)	8-Hydroxy-5-deazaflavin (Coenzyme F420)	ABP
	7,8-Didemethyl-8-hydroxy-5-deazariboflavin synthase (CofG)		ABPV
	Aminofutalosine synthase (mqnE)	Menaquinone via futalosine (two steps along the same pathway)	AB
	Cyclic dehypoxanthine futalosine synthase (MqnC)		ABPV
FeMo cofactor biosynthesis protein (10)	FeMo cofactor biosynthesis protein (nifB)	FeMo cofactor	AB
Lipoyl synthase (11)	Lipoyl synthase (LipA)	Lipoyl	ABPVI
SPASM/twitch domain containing (17)	Coenzyme PQQ synthesis protein E (PqqE)[a]	Pyrroloquinoline quinone (PQQ)	B
	Cyclic pyranopterin phosphate synthase (MoaA)	Molybdenum cofactor	ABPVI
	Alternative heme biosynthesis C (AhbC)	Heme B (two steps along the same pathway)	AB
	Alternative heme biosynthesis D (AhbD)		
	C-terminal tyrosine decarboxylase (MftC)	Mycofactocin peptide maturase	B
	Heme D1 biosynthesis (NirJ)	Heme d1	B
	Tungsten cofactor oxidoreductase radical SAM maturase (Cmo)	Tungsten-containing cofactors	AB

[a]Type of life: *B*, bacteria; *P*, plants; *V*, vertebrates; *I*, invertebrates; *A*, archaea.

found in the middle of the protein, the N-terminal MTTase domain contains three cysteines that bind a second [Fe_4S_4] cluster, which could be involved in the thiolation reaction. Within the RSS, the TRAM domain is unique to the MTTase family and is not found in any other known superfamily members. This family is found in all types of life, reflecting the critical role of the reaction it catalyzes.

Other families in the RSS that have representation across all domains of life include the elongator protein-like family from the subgroup of the same name, thought to catalyze the tRNA wobble uridine modification at C5 of tRNA, and the Class A methyltransferase families of subgroup 13, adenosine C2 methyltransferase (RlmN-like) and adenosine C8 methyltransferase (Cfr-like). These subgroup 13 proteins utilize two SAM molecules, one as a methyl donor and one as the source of the 5′-dA radical. The viperin familiy, involved in antiviral activity, is also ubiquitous and has recently been determined to catalyze the reaction 3′-deoxy-3′,4′-didehydro-CTP (ddhCTP) synthase (Gizzi et al., 2018).

Together, these observations lend additional support to previous suggestions that RSS ancestors evolved early in the history of life due to their ability to provide a catalytically simple but powerful mechanism for activating atoms that are typically unreactive in biological organisms. This fundamental synthetic power, and the roles these enzymes play in the biosynthesis, degradation, modification, and repair of molecules essential to life can now be confirmed on the global scale of the superfamily. Further, breadth of their coverage in the biosphere adds further confirmation that the lineage of the RSS is indeed ancient.

2.5 Classification of the RSS Based on Sequence Similarity versus Chemical Similarity

Divergence of ancestral proteins that use the conserved mechanistic machinery of the RSS superfamily domain has led to the wide variety of overall chemical transformations that can be broadly categorized into several general types of chemistry, for example, the insertion of a sulfur atom, complex rearrangements, creation of a glycyl radical, or methylation (see Supplementary Table S1 in the online version at https://doi.org/10.1016/bs.mie.2018.06.004 for details). Fig. 8 illustrates how some of these major reaction types map to the SSN. As indicated by the figure, these general reaction types do not cluster discretely with specific subgroups. Even for the knowns of the B12-binding methylthiotransferase-like reaction type (yellow diamonds), which appear to map discretely to the main cluster of subgroup 5, several

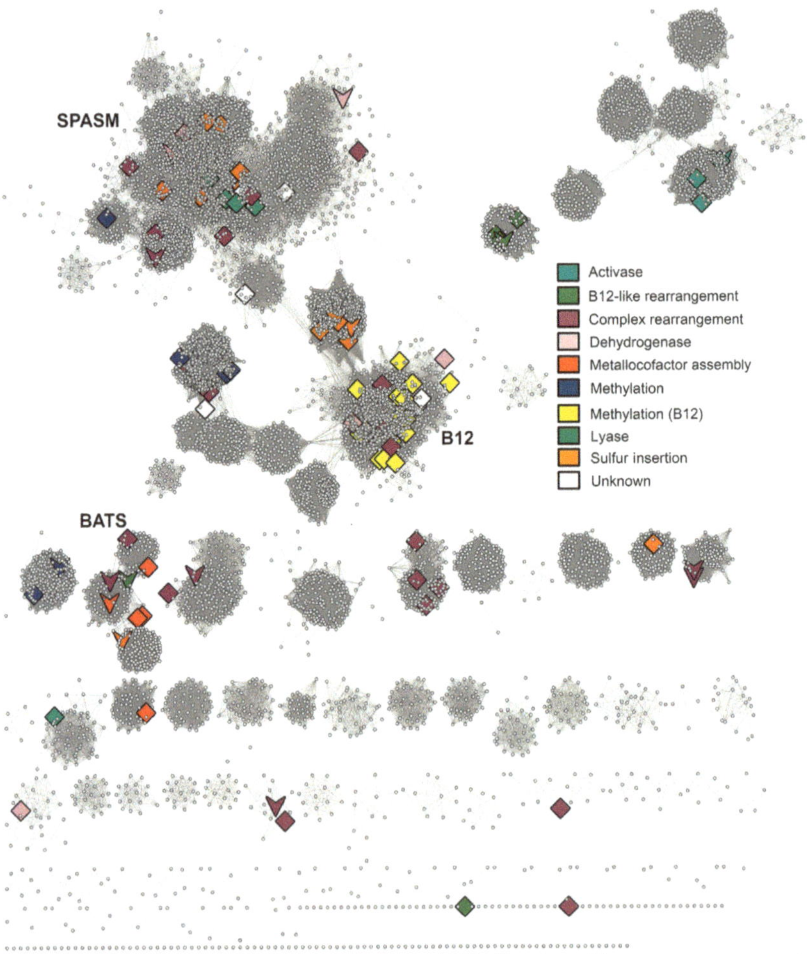

Fig. 8 SSN of RSS mapped with general types of RSS chemistry. The same representative network shown in Fig. 5 except that the highlighted nodes and coloring are by general reaction type as indicated in the key. *Diamonds*: large representative nodes include at least one functionally characterized member colored by dominant reaction type as shown in the accompanying key. *Downward arrows*: representative nodes include at least one structurally characterized member.

other reaction types map to this subgroup as well. Supplementary Table S1 in the online version at https://doi.org/10.1016/bs.mie.2018.06.004 describes 27 known chemical reactions in this subgroup to which only a total of 1307 sequences have been assigned to a family in the SFLD. Another 5551 sequences in subgroup 5 remain classified only at the subgroup level. These disconnects between sequence similarity-derived clustering and the types of reactions found in those clusters suggest that not even these very general reaction types can be assigned to unknowns based only on sequence similarity (although accurate annotation transfer of reaction type from characterized enzymes to closely related sequences in the same SFLD family can be done with some confidence, especially if key specificity-determining residues are conserved).

Other types of chemical classifications fail to track well with our sequence similarity-based classification as well (not shown). For example, the stoichiometry and utilization of SAM differ widely across the RSS. This view identifies three classes that differ with respect to the fate of SAM (Booker, 2009). Class I enzymes utilize SAM catalytically. Class II represents the glycyl radical-activating class, i.e., those reactions that abstract a hydrogen atom from a glycine residue in a protein substrate. Class III enzymes, by far the largest set of RSS enzymes currently known, utilize SAM stoichiometrically. However, our results indicate that SAM is utilized both stoichiometrically and catalytically within a subgroup. Even for two enzymes in the Level 2 subgroup of the Level 1 BATS domain-containing subgroup, PylB utilizes SAM as a true cofactor, whereas HydE utilizes it stoichiometrically.

Another way to map between reaction type and similarity groupings is provided by the Gene Ontology (Holliday, Davidson, Akiva, & Babbitt, 2017), which is used in many large resources to describe similarities in enzyme function. Here, enzyme reactions can be defined using the Enzyme Nomenclature Commission classification (Tipton, 1994), which defines enzyme reactions by EC number. Our automated comparison of known overall chemical reactions of the RSS using EC-BLAST (Rahman, Cuesta, Furnham, Holliday, & Thornton, 2014), confirms that similarity in EC classification also fails to track with sequence similarity. Issues with annotating unknowns with the EC number of the most similar characterized enzyme has been raised previously for other enzyme superfamilies (Babbitt, 2003).

The general disconnect between chemical classifications of RSS reactions and the similarity-based classification presented in this work suggests that functional and mechanistic prediction of newly discovered RSS sequences may not be asserted with confidence without biochemical characterization. One way to address this challenge is for experiment and bioinformatics to work together to achieve breakthroughs in resolving these limitations. For example, many recent studies have used multiple types of orthogonal information to obtain functional clues about unknowns (Radivojac et al., 2013). The context provided by SSNs along with information such as genome context, structural models, and in silico docking have been especially powerful in suggesting functional properties for many proteins (for example, Calhoun et al., 2018; Hermann et al., 2007; Kalyanaraman et al., 2008; Mashiyama et al., 2014; Zhao et al., 2013). In addition, the extensive variations in MDAs for the majority of the RSS proteins provides a somewhat unique advantage for functional inference for this and other superfamilies with complex MDAs, such as the haloacid dehalogenase superfamily (Burroughs, Allen, Dunaway-Mariano, & Aravind, 2006). As fusion proteins can be used to provide key evidence of functional association (Marcotte et al., 1999), a superfamily-wide strategy focusing on broad-scale mining of the MDA data such as shown in Fig. 4, especially if used conjunction with other orthogonal data (Gerlt, Babbitt, Jacobson, & Almo, 2012), may aid in forming useful functional hypotheses for unknowns that share conserved aspects of their MDAs.

2.6 Targeting Unknowns for Experimental Characterization

As the number of RSS sequences discovered in genome and metagenome projects continues to grow, the proportion of the RSS that can be experimentally characterized will continue to diminish. This scarcity of experimental data has significant consequences for predicting functions of unknowns. Functional annotation in public databases now depends on computational annotation transfer and is still heavily based on the assumption that sequence similarity between unknowns and the most similar protein(s) that have been experimentally characterized provides adequate support for annotation transfer. While this usually works well for closely related sequences, it can also lead to high levels of misannotation, particularly in functionally diverse enzyme superfamilies such as the RSS (Schnoes, Brown, Dodevski, & Babbitt, 2009). Likewise, the even greater scarcity of characterized structures limits our knowledge of unexplored regions of the RSS sequence space (Fig. 5). Our attempts to use modeling of RSS sequences to ascertain the range of fold variants that comprise the superfamily failed, as noted in Section 2.3, as the current scarcity of structures and their uneven breadth of coverage available for use as modeling templates diminishes our ability to even ask this question.

Conservation of functionally important residues, including active site features and associated mechanistic knowledge, has long provided important clues about specific functional features of unknowns in many superfamilies and families (for example, Brown et al., 2006; Holliday, Brown, et al., 2017).

Table 2 and Fig. 5 allow estimation of the current state of experimental coverage for the RSS, showing that some subgroups (such as the B12-binding domain subgroup 5 in Fig. 5) are broadly covered relative to other subgroups. Many other numbered subgroups have far fewer characterized enzymes; the 22 uncharacterized subgroups have none at all. As variations in active site conservation patterns, genome context, structural features, or biological phenotypes are available to inform experimental design, it may be possible to expand the coverage of these neglected subgroups. Especially relevant for the RSS, the rich variation among its MDAs could support a global analysis of accessory domains to gain new functional clues about RSS unknowns.

Comparison of conservation patterns has aided in the discrimination of RSS subgroups and families as well. For example, of the more than 5000 proteins that can be assigned to the BioB-like Level 2 subgroup (of the BATs domain subgroup 6), only about 3200 can be annotated with that reaction based on the presence of functionally important active site residues. The remainder are missing one or more of the residues considered critical for performing the canonical BioB reaction (Betz et al., 2015; Holliday, Davidson, et al., 2017), leading to the assertion that a large proportion of enzymes annotated with the BioB function in the Gene Ontology database (Ashburner et al., 2000) have been found to be misannotated (Holliday, Davidson, et al., 2017).

Broadening functional knowledge across the RSS requires as a necessary first step identifying unknowns that are likely to represent fruitful targets for experimental characterization. While superfamily members associated with important biological phenotypes or that are of particular interest for other reasons will remain key motivations for characterization, additional strategies to investigate unknowns throughout the broader sequence space are needed. The subgroup classification presented here identifies the least characterized subgroups of the RSS; examination of differences between their conservation patterns and those of better characterized subgroups will be useful in choosing targets likely to represent new reactions.

Many new strategies are now available for deeper evaluation of unknowns. As noted earlier, large-scale in silico modeling can lead to functional hypotheses that can be validated and studied in detail using targeted biochemical and structural characterization, while in silico docking of metabolites and genome context predictions can be evaluated using many types of experimental approaches. Our subgroup classification, along with the interactive versions of subgroup and family SSNs available for download from the SFLD, was created in part to provide a global context for target selection in the RSS and to facilitate application of the varied strategies now available to assign their functions.

3. METHODS

3.1 Collection of RSS Sequences

To initiate populating the RSS for the SFLD, we collected the full-length sequences in September 2012 associated with Pfam model PF04055 and InterPro family IPR007197, removed duplicate sequences and resolved other differences. This set was last updated using the SFLD automated update protocol on 7/9/14. Functional domains of this representative sequence set superfamily were last updated on 11/22/17.

3.2 Representative Networks

The full set of 113,776 sequences was clustered using CD-HIT (Li & Godzik, 2006) at 50% pairwise identity, resulting in 10,741 representative nodes. Edges were drawn between representative nodes if the BLAST *E*-value (used as a score) was $\leq 1 \times 10^{-20}$. The representative network shown in Figs. 5, 7, and 8 was calculated following the Pythoscape infrastructure (Barber & Babbitt, 2012) where similarity between each pair of representative nodes was calculated as the mean of the all-by-all comparison scores between the sequences in each representative node pair. The networks were visualized using the Prefuse force directed layout (mean edge *E*-value) in Cytoscape (Smoot, Ono, Ruscheinski, Wang, & Ideker, 2011). As the calculated network represents a multidimensional set of $n-1$ pairwise comparisons where n is the number of sequences, these data must be compressed for visualization in two dimensions. Other work has indicated that two-dimensional visualization provides an acceptable estimate of the similarity relationships in thresholded SSNs (see Atkinson et al., 2009, Supplementary Fig. 1 statistics).

3.3 Determining Subgroups and Families

RSS subgroups were defined semiautomatically using as a guide representative SSNs annotated with a list of RSS reactions currently known in the primary literature. These initial subgroup boundaries were refined based on manual inspection of the network across a range of *E*-value thresholds and further informed from MDA data and available structural information. The aim was to choose an *E*-value score threshold for drawing edges that resulted in sequence clusters that minimally split nodes with similar reactions while maximally splitting those of different reactions. The final mean *E*-value score shown in Fig. 5 and used to visualize the named subgroups curated in the SFLD is 1×10^{-20}. Nodes of the same color that ended up in different clusters in Fig. 5 reflect the extent to which this could not be achieved and likely resulted from several issues that include uneven rates of evolution of subgroups. For example, SFLD curation of the highly divergent Level 1 subgroup 17, which represents proteins with a SPASM/Twitch domain, required the creation of Level 2 and Level 3 subgroupings as the entire Level 1 subgroup could not be united at a statistically significant score using a single hidden Markov Model (HMM) (Eddy, 2011). This diversity is also reflected in Fig. 5: the *E*-value chosen to distinguish the majority of RSS subgroups was unable to cluster all of the SPASM/Twitch domain subgroup members together. In contrast, at this *E*-value threshold, all of the representative nodes of subgroup 1 cluster together due to the greater similarity among members of this subgroup relative to those of subgroup 17.

In theory, the assignment of families in the SFLD is relatively simple, requiring at minimum, at least one member that has been biochemically characterized, with at least some residues and features identified that are known to be functionally important in that reaction. However, transfer of function is problematic for representative nodes, each of which may contain substantially divergent sequences that may catalyze reactions that are different from that catalyzed by the characterized "founder" enzyme. To address this issue, we computed and examined for each subgroup (at the most detailed subgroup level available) SSNs in which each node represents a single sequence (data not shown). This allowed a detailed comparison of all sequences in a subgroup for conservation as well as differences among key residues known to be functionally important in the characterized member(s). Family assignments were then made guided by these data.

3.4 Annotation of the RSS in the SFLD

Annotation of subgroups and families is supported by several types of information, including multiple sequence alignments (MSAs) created using Clustal Omega (Sievers et al., 2011) as the sizes of their sequence sets allowed. HMMs used in curation of subgroups and families were created using HMMER 3. Each family is annotated with at least one reference from the primary literature and its overall chemical transformation (where known). Where possible, the conserved functional residues are reported (in the majority of cases, these are associated with the $[Fe_4S_4]$-binding motif).

Detailed annotation of the RSS is provided in Supplementary Table S1 in the online version at https://doi.org/10.1016/bs.mie.2018.06.004 of this work. Key data and information is also provided for download from the SFLD (see http://sfld.rbvi.ucsf.edu). These include MSAs and interactive SSNs at the subgroup and family levels that can be visualized and manipulated using Cytoscape. Sequence sets, chemical reactions, and other data are also available.

Supplementary data to this article can be found online at https://doi.org/10.1016/bs.mie.2018.06.004.

ACKNOWLEDGMENTS

Support for this work acknowledges NIH R01 GM60595 (P. Babbitt), NIH R01 GM-122595 (S. Booker), and NSF DBI-1356193 (P. Babbitt and G. Holliday). Some of the results described in this chapter were initially developed as part of a workshop on the RSS sponsored by the Enzyme Function Initiative with support from NIH U54GM093342 (J. Gerlt). The SFLD was developed as a joint project of the Babbitt lab with support by NIH R01GM60595 and NSF Grants DBI-0234768 and DBI-0640476 (P. Babbitt), NSF DBI-1356193 (P. Babbitt and G. Holliday), and The Resource for Biocomputing, Visualization, & Informatics (T. Ferrin) with support from NIGMS P41GM103311. Support for development of the SFLD was also provided NIH U54GM093342 and P01GM07790 (J. Gerlt). The authors thank David Mischel and Benjamin Polacco for technical support of the SFLD, including implementation of new code required of the back-end and graphical user interface, and Kathy Clement for help with generating graphics images of SSNs.

AUTHOR CONTRIBUTIONS

G.L.H performed the analysis, directed the research, and wrote the manuscript. E.C.M. generated the structure comparisons provided in Figs. 2 and 3 and assisted with analysis and proof reading of the manuscript. E.A. and S.D.B. assisted with analysis and proof reading of the manuscript. S.C., U.P., and A.S. performed the 3D-structure prediction. S.J.B. provided expertise in RSS enzymology and in assigning functions based on the literature. P.C.B. oversaw the project and wrote the manuscript.

COMPETING FINANCIAL INTERESTS STATEMENT

None.

REFERENCES

Akiva, E., Brown, S., Almonacid, D. E., Barber, A. E., 2nd, Custer, A. F., Hicks, M. A., et al. (2014). The structure-function linkage database. *Nucleic Acids Research*, *42*(Database issue), D521–D530. https://doi.org/10.1093/nar/gkt1130.

Altschul, S. F., Madden, T. L., Schaffer, A. A., Zhang, J., Zhang, Z., Miller, W., et al. (1997). Gapped BLAST and PSI-BLAST: A new generation of protein database search programs. *Nucleic Acids Research*, *25*(17), 3389–3402.

Anantharaman, V., Koonin, E. V., & Aravind, L. (2001). TRAM, a predicted RNA-binding domain, common to tRNA uracil methylation and adenine thiolation enzymes. *FEMS Microbiology Letters*, *197*(2), 215–221.

Ashburner, M., Ball, C. A., Blake, J. A., Botstein, D., Butler, H., Cherry, J. M., et al. (2000). Gene ontology: Tool for the unification of biology. The gene ontology consortium. *Nature Genetics*, *25*(1), 25–29. https://doi.org/10.1038/75556.

Atkinson, H. J., Morris, J. H., Ferrin, T. E., & Babbitt, P. C. (2009). Using sequence similarity networks for visualization of relationships across diverse protein superfamilies. *PLoS One*, *4*(2), e4345. https://doi.org/10.1371/journal.pone.0004345.

Babbitt, P. C. (2003). Definitions of enzyme function for the structural genomics era. *Current Opinion in Chemical Biology*, 7(2), 230–237.

Baker, P. R., 2nd, Friederich, M. W., Swanson, M. A., Shaikh, T., Bhattacharya, K., Scharer, G. H., et al. (2014). Variant non ketotic hyperglycinemia is caused by mutations in LIAS, BOLA3 and the novel gene GLRX5. *Brain*, *137*(Pt. 2), 366–379. https://doi.org/10.1093/brain/awt328.

Barber, A. E., 2nd, & Babbitt, P. C. (2012). Pythoscape: A framework for generation of large protein similarity networks. *Bioinformatics*, *28*(21), 2845–2846. https://doi.org/10.1093/bioinformatics/bts532.

Barr, I., Latham, J. A., Iavarone, A. T., Chantarojsiri, T., Hwang, J. D., & Klinman, J. P. (2016). Demonstration that the radical S-adenosylmethionine (SAM) enzyme PqqE catalyzes de novo carbon-carbon cross-linking within a peptide substrate PqqA in the presence of the peptide chaperone PqqD. *The Journal of Biological Chemistry*, *291*(17), 8877–8884. https://doi.org/10.1074/jbc.C115.699918.

Benjdia, A., Heil, K., Barends, T. R., Carell, T., & Schlichting, I. (2012). Structural insights into recognition and repair of UV-DNA damage by spore photoproduct lyase, a radical SAM enzyme. *Nucleic Acids Research*, *40*(18), 9308–9318. https://doi.org/10.1093/nar/gks603.

Berman, H., Henrick, K., & Nakamura, H. (2003). Announcing the worldwide Protein Data Bank. *Nature Structural Biology*, *10*(12), 980. https://doi.org/10.1038/nsb1203-980.

Betz, J. N., Boswell, N. W., Fugate, C. J., Holliday, G. L., Akiva, E., Scott, A. G., et al. (2015). [FeFe]-hydrogenase maturation: Insights into the role HydE plays in dithiomethylamine biosynthesis. *Biochemistry*, *54*(9), 1807–1818. https://doi.org/10.1021/bi501205e.

Blaszczyk, A. J., Silakov, A., Zhang, B., Maiocco, S. J., Lanz, N. D., Kelly, W. L., et al. (2016). Spectroscopic and electrochemical characterization of the iron-sulfur and cobalamin cofactors of TsrM, an unusual radical S-adenosylmethionine methylase. *Journal of the American Chemical Society*, *138*(10), 3416–3426. https://doi.org/10.1021/jacs.5b12592.

Booker, S. J. (2009). Anaerobic functionalization of unactivated C-H bonds. *Current Opinion in Chemical Biology*, *13*(1), 58–73. https://doi.org/10.1016/j.cbpa.2009.02.036.

Broderick, J. B., Duffus, B. R., Duschene, K. S., & Shepard, E. M. (2014). Radical S-adenosylmethionine enzymes. *Chemical Reviews*, *114*(8), 4229–4317. https://doi.org/10.1021/cr4004709.

Brown, S. D., & Babbitt, P. C. (2014). New insights about enzyme evolution from large scale studies of sequence and structure relationships. *The Journal of Biological Chemistry*, *289*(44), 30221–30228. https://doi.org/10.1074/jbc.R114.569350.

Brown, S. D., Gerlt, J. A., Seffernick, J. L., & Babbitt, P. C. (2006). A gold standard set of mechanistically diverse enzyme superfamilies. *Genome Biology*, 7(1), R8. https://doi.org/10.1186/gb-2006-7-1-r8.

Burroughs, A. M., Allen, K. N., Dunaway-Mariano, D., & Aravind, L. (2006). Evolutionary genomics of the HAD superfamily: Understanding the structural adaptations and catalytic diversity in a superfamily of phosphoesterases and allied enzymes. *Journal of Molecular Biology*, *361*(5), 1003–1034. S0022-2836(06)00777-7 [pii]. https://doi.org/10.1016/j.jmb.2006.06.049.

Calhoun, S., Korczynska, M., Wichelecki, D. J., San Francisco, B., Zhao, S., Rodionov, D. A., et al. (2018). Prediction of enzymatic pathways by integrative pathway mapping. *eLife*, 7. https://doi.org/10.7554/eLife.31097.

Cicchillo, R. M., Iwig, D. F., Jones, A. D., Nesbitt, N. M., Baleanu-Gogonea, C., Souder, M. G., et al. (2004). Lipoyl synthase requires two equivalents of S-adenosyl-L-methionine to synthesize one equivalent of lipoic acid. *Biochemistry*, *43*(21), 6378–6386. https://doi.org/10.1021/bi049528x.

Cicchillo, R. M., Lee, K. H., Baleanu-Gogonea, C., Nesbitt, N. M., Krebs, C., & Booker, S. J. (2004). Escherichia coli lipoyl synthase binds two distinct [4Fe-4S] clusters per polypeptide. *Biochemistry*, *43*(37), 11770–11781. https://doi.org/10.1021/bi0488505.

Coquille, S., Roux, C., Mehta, A., Begley, T. P., Fitzpatrick, T. B., & Thore, S. (2013). High-resolution crystal structure of the eukaryotic HMP-P synthase (THIC) from Arabidopsis thaliana. *Journal of Structural Biology*, *184*(3), 438–444. https://doi.org/10.1016/j.jsb.2013.10.005.

Dawson, N. L., Sillitoe, I., Lees, J. G., Lam, S. D., & Orengo, C. A. (2017). CATH-Gene3D: Generation of the resource and its use in obtaining structural and functional annotations for protein sequences. *Methods in Molecular Biology*, *1558*, 79–110. https://doi.org/10.1007/978-1-4939-6783-4_4.

de Beer, T. A., Berka, K., Thornton, J. M., & Laskowski, R. A. (2014). PDBsum additions. *Nucleic Acids Research*, *42*(Database issue), D292–D296. https://doi.org/10.1093/nar/gkt940.

Dinis, P., Wieckowski, B. M., & Roach, P. L. (2016). Metallocofactor assembly for [FeFe]-hydrogenases. *Current Opinion in Structural Biology*, *41*, 90–97. https://doi.org/10.1016/j.sbi.2016.06.004.

Dowling, D. P., Vey, J. L., Croft, A. K., & Drennan, C. L. (2012). Structural diversity in the AdoMet radical enzyme superfamily. *Biochimica et Biophysica Acta*, *1824*(11), 1178–1195. https://doi.org/10.1016/j.bbapap.2012.04.006.

Eddy, S. R. (2011). Accelerated profile HMM searches. *PLoS Computational Biology*, 7(10), e1002195. https://doi.org/10.1371/journal.pcbi.1002195.

Finn, R. D., Attwood, T. K., Babbitt, P. C., Bateman, A., Bork, P., Bridge, A. J., et al. (2017). InterPro in 2017-beyond protein family and domain annotations. *Nucleic Acids Research*, *45*(D1), D190–D199. https://doi.org/10.1093/nar/gkw1107.

Finn, R. D., Coggill, P., Eberhardt, R. Y., Eddy, S. R., Mistry, J., Mitchell, A. L., et al. (2016). The Pfam protein families database: Towards a more sustainable future. *Nucleic Acids Research*, *44*(D1), D279–D285. https://doi.org/10.1093/nar/gkv1344.

Furnham, N., Sillitoe, I., Holliday, G. L., Cuff, A. L., Laskowski, R. A., Orengo, C. A., et al. (2012). Exploring the evolution of novel enzyme functions within structurally defined protein superfamilies. *PLoS Computational Biology*, *8*(3), e1002403. https://doi.org/10.1371/journal.pcbi.1002403.

Gerlt, J. A., Babbitt, P. C., Jacobson, M. P., & Almo, S. C. (2012). Divergent evolution in enolase superfamily: Strategies for assigning functions. *The Journal of Biological Chemistry*, *287*(1), 29–34. https://doi.org/10.1074/jbc.R111.240945.

Gizzi, A. S., Grove, T. L., Arnold, J. J., Jose, J., Jangra, R. K., Garforth, S. J., et al. (2018). A naturally occurring antiviral ribonucleotide encoded by the human genome. *Nature*, *558*(7711), 610–614. https://doi.org/10.1038/s41586-018-0238-4.

Grell, T. A., Goldman, P. J., & Drennan, C. L. (2015). SPASM and twitch domains in S-adenosylmethionine (SAM) radical enzymes. *The Journal of Biological Chemistry*, *290*(7), 3964–3971. https://doi.org/10.1074/jbc.R114.581249.

Hanzelmann, P., & Schindelin, H. (2004). Crystal structure of the S-adenosylmethionine-dependent enzyme MoaA and its implications for molybdenum cofactor deficiency in humans. *Proceedings of the National Academy of Sciences of the United States of America*, *101*(35), 12870–12875. https://doi.org/10.1073/pnas.0404624101.

Hermann, J. C., Marti-Arbona, R., Fedorov, A. A., Fedorov, E., Almo, S. C., Shoichet, B. K., et al. (2007). Structure-based activity prediction for an enzyme of unknown function. *Nature, 448*(7155), 775–779. https://doi.org/10.1038/nature05981.

Hiratsuka, T., Furihata, K., Ishikawa, J., Yamashita, H., Itoh, N., Seto, H., et al. (2008). An alternative menaquinone biosynthetic pathway operating in microorganisms. *Science, 321*(5896), 1670–1673. https://doi.org/10.1126/science.1160446.

Holliday, G. L., Brown, S. D., Akiva, E., Mischel, D., Hicks, M. A., Morris, J. H., et al. (2017). Biocuration in the structure-function linkage database: The anatomy of a superfamily. *Database: The Journal of Biological Databases and Curation, 2017*. https://doi.org/10.1093/database/bax045.

Holliday, G. L., Davidson, R., Akiva, E., & Babbitt, P. C. (2017). Evaluating functional annotations of enzymes using the gene ontology. *Methods in Molecular Biology, 1446*, 111–132. https://doi.org/10.1007/978-1-4939-3743-1_9.

Holliday, G. L., Thornton, J. M., Marquet, A., Smith, A. G., Rebeille, F., Mendel, R., et al. (2007). Evolution of enzymes and pathways for the biosynthesis of cofactors. *Natural Product Reports, 24*(5), 972–987. https://doi.org/10.1039/b703107f.

Kalyanaraman, C., Imker, H. J., Fedorov, A. A., Fedorov, E. V., Glasner, M. E., Babbitt, P. C., et al. (2008). Discovery of a dipeptide epimerase enzymatic function guided by homology modeling and virtual screening. *Structure, 16*(11), 1668–1677. https://doi.org/10.1016/j.str.2008.08.015.

Kamat, S. S., Williams, H. J., & Raushel, F. M. (2011). Intermediates in the transformation of phosphonates to phosphate by bacteria. *Nature, 480*(7378), 570–573. https://doi.org/10.1038/nature10622.

Knappe, J., & Sawers, G. (1990). A radical-chemical route to acetyl-CoA: The anaerobically induced pyruvate formate-lyase system of Escherichia coli. *FEMS Microbiology Reviews, 6*(4), 383–398.

LaMattina, J. W., Wang, B., Badding, E. D., Gadsby, L. K., Grove, T. L., & Booker, S. J. (2017). NosN, a radical S-adenosylmethionine methylase, catalyzes both C1 transfer and formation of the ester linkage of the side-ring system during the biosynthesis of nosiheptide. *Journal of the American Chemical Society, 139*(48), 17438–17445. https://doi.org/10.1021/jacs.7b08492.

Lanz, N. D., & Booker, S. J. (2012). Identification and function of auxiliary iron-sulfur clusters in radical SAM enzymes. *Biochimica et Biophysica Acta, 1824*(11), 1196–1212. https://doi.org/10.1016/j.bbapap.2012.07.009.

Laskowski, R. A., & Swindells, M. B. (2011). LigPlot+: Multiple ligand-protein interaction diagrams for drug discovery. *Journal of Chemical Information and Modeling, 51*(10), 2778–2786. https://doi.org/10.1021/ci200227u.

Lee, K. H., Saleh, L., Anton, B. P., Madinger, C. L., Benner, J. S., Iwig, D. F., et al. (2009). Characterization of RimO, a new member of the methylthiotransferase subclass of the radical SAM superfamily. *Biochemistry, 48*(42), 10162–10174. https://doi.org/10.1021/bi900939w.

Li, W., & Godzik, A. (2006). Cd-hit: A fast program for clustering and comparing large sets of protein or nucleotide sequences. *Bioinformatics, 22*(13), 1658–1659. https://doi.org/10.1093/bioinformatics/btl158.

Lotierzo, M., Tse Sum Bui, B., Florentin, D., Escalettes, F., & Marquet, A. (2005). Biotin synthase mechanism: An overview. *Biochemical Society Transactions, 33*(Pt. 4), 820–823. https://doi.org/10.1042/BST0330820.

Mahanta, N., Hudson, G. A., & Mitchell, D. A. (2017a). Correction to radical S-adenosylmethionine enzymes involved in RiPP biosynthesis. *Biochemistry, 56*(45), 6072. https://doi.org/10.1021/acs.biochem.7b01056.

Mahanta, N., Hudson, G. A., & Mitchell, D. A. (2017b). Radical S-adenosylmethionine enzymes involved in RiPP biosynthesis. *Biochemistry*, *56*(40), 5229–5244. https://doi.org/10.1021/acs.biochem.7b00771.

Mancia, F., Keep, N. H., Nakagawa, A., Leadlay, P. F., McSweeney, S., Rasmussen, B., et al. (1996). How coenzyme B12 radicals are generated: The crystal structure of methylmalonyl-coenzyme A mutase at 2 A resolution. *Structure*, *4*(3), 339–350.

Marcotte, E. M., Pellegrini, M., Ng, H. L., Rice, D. W., Yeates, T. O., & Eisenberg, D. (1999). Detecting protein function and protein-protein interactions from genome sequences. *Science*, *285*(5428), 751–753.

Mashiyama, S. T., Malabanan, M. M., Akiva, E., Bhosle, R., Branch, M. C., Hillerich, B., et al. (2014). Large-scale determination of sequence, structure, and function relationships in cytosolic glutathione transferases across the biosphere. *PLoS Biology*, *12*(4), e1001843. https://doi.org/10.1371/journal.pbio.1001843.

Miller, J. R., Busby, R. W., Jordan, S. W., Cheek, J., Henshaw, T. F., Ashley, G. W., et al. (2000). Escherichia coli LipA is a lipoyl synthase: In vitro biosynthesis of lipoylated pyruvate dehydrogenase complex from octanoyl-acyl carrier protein. *Biochemistry*, *39*(49), 15166–15178.

Moss, M., & Frey, P. A. (1987). The role of S-adenosylmethionine in the lysine 2,3-aminomutase reaction. *The Journal of Biological Chemistry*, *262*(31), 14859–14862.

Nicolet, Y., Rubach, J. K., Posewitz, M. C., Amara, P., Mathevon, C., Atta, M., et al. (2008). X-ray structure of the [FeFe]-hydrogenase maturase HydE from Thermotoga maritima. *The Journal of Biological Chemistry*, *283*(27), 18861–18872. https://doi.org/10.1074/jbc.M801161200.

Padovani, D., Thomas, F., Trautwein, A. X., Mulliez, E., & Fontecave, M. (2001). Activation of class III ribonucleotide reductase from *E. coli*. The electron transfer from the iron-sulfur center to S-adenosylmethionine. *Biochemistry*, *40*(23), 6713–6719.

Pierre, S., Guillot, A., Benjdia, A., Sandstrom, C., Langella, P., & Berteau, O. (2012). Thiostrepton tryptophan methyltransferase expands the chemistry of radical SAM enzymes. *Nature Chemical Biology*, *8*(12), 957–959. https://doi.org/10.1038/nchembio.1091.

Pilet, E., Nicolet, Y., Mathevon, C., Douki, T., Fontecilla-Camps, J. C., & Fontecave, M. (2009). The role of the maturase HydG in [FeFe]-hydrogenase active site synthesis and assembly. *FEBS Letters*, *583*(3), 506–511. https://doi.org/10.1016/j.febslet.2009.01.004.

Puehringer, S., Metlitzky, M., & Schwarzenbacher, R. (2008). The pyrroloquinoline quinone biosynthesis pathway revisited: A structural approach. *BMC Biochemistry*, *9*, 8. https://doi.org/10.1186/1471-2091-9-8.

Radivojac, P., Clark, W. T., Oron, T. R., Schnoes, A. M., Wittkop, T., Sokolov, A., et al. (2013). A large-scale evaluation of computational protein function prediction. *Nature Methods*, *10*(3), 221–227. https://doi.org/10.1038/nmeth.2340.

Rahman, S. A., Cuesta, S. M., Furnham, N., Holliday, G. L., & Thornton, J. M. (2014). EC-BLAST: A tool to automatically search and compare enzyme reactions. *Nature Methods*, *11*(2), 171–174. https://doi.org/10.1038/nmeth.2803.

Reyda, M. R., Dippold, R., Dotson, M. E., & Jarrett, J. T. (2008). Loss of iron-sulfur clusters from biotin synthase as a result of catalysis promotes unfolding and degradation. *Archives of Biochemistry and Biophysics*, *471*(1), 32–41. https://doi.org/10.1016/j.abb.2007.12.001.

Schnoes, A. M., Brown, S. D., Dodevski, I., & Babbitt, P. C. (2009). Annotation error in public databases: Misannotation of molecular function in enzyme superfamilies. *PLoS Computational Biology*, *5*(12), e1000605. https://doi.org/10.1371/journal.pcbi.1000605.

Shannon, P., Markiel, A., Ozier, O., Baliga, N. S., Wang, J. T., Ramage, D., et al. (2003). Cytoscape: A software environment for integrated models of biomolecular interaction networks. *Genome Research*, *13*(11), 2498–2504.

Sievers, F., Wilm, A., Dineen, D., Gibson, T. J., Karplus, K., Li, W., et al. (2011). Fast, scalable generation of high-quality protein multiple sequence alignments using Clustal Omega. *Molecular Systems Biology*, 7, 539. https://doi.org/10.1038/msb.2011.75.

Smoot, M. E., Ono, K., Ruscheinski, J., Wang, P. L., & Ideker, T. (2011). Cytoscape 2.8: New features for data integration and network visualization. *Bioinformatics*, *27*(3), 431–432. https://doi.org/10.1093/bioinformatics/btq675.

Sofia, H. J., Chen, G., Hetzler, B. G., Reyes-Spindola, J. F., & Miller, N. E. (2001). Radical SAM, a novel protein superfamily linking unresolved steps in familiar biosynthetic pathways with radical mechanisms: Functional characterization using new analysis and information visualization methods. *Nucleic Acids Research*, *29*(5), 1097–1106.

Tamuri, A. U., & Laskowski, R. A. (2010). ArchSchema: A tool for interactive graphing of related Pfam domain architectures. *Bioinformatics*, *26*(9), 1260–1261. https://doi.org/10.1093/bioinformatics/btq119.

Tao, M., Wang, L., Wendt-Pienkowski, E., George, N. P., Galm, U., Zhang, G., et al. (2007). The tallysomycin biosynthetic gene cluster from Streptoalloteichus hindustanus E465-94 ATCC 31158 unveiling new insights into the biosynthesis of the bleomycin family of antitumor antibiotics. *Molecular BioSystems*, *3*(1), 60–74. https://doi.org/10.1039/b615284h.

Tipton, K. F. (1994). Nomenclature Committee of the International Union of Biochemistry and Molecular Biology (NC-IUBMB). Enzyme nomenclature. Recommendations 1992. Supplement: Corrections and additions. *European Journal of Biochemistry*, *223*(1), 1–5.

UniProt Consortium. (2017). UniProt: The universal protein knowledgebase. *Nucleic Acids Research*, *45*(D1), D158–D169. https://doi.org/10.1093/nar/gkw1099.

Vey, J. L., & Drennan, C. L. (2011). Structural insights into radical generation by the radical SAM superfamily. *Chemical Reviews*, *111*(4), 2487–2506. https://doi.org/10.1021/cr9002616.

Vey, J. L., Yang, J., Li, M., Broderick, W. E., Broderick, J. B., & Drennan, C. L. (2008). Structural basis for glycyl radical formation by pyruvate formate-lyase activating enzyme. *Proceedings of the National Academy of Sciences of the United States of America*, *105*(42), 16137–16141. https://doi.org/10.1073/pnas.0806640105.

Wang, J., Woldring, R. P., Roman-Melendez, G. D., McClain, A. M., Alzua, B. R., & Marsh, E. N. (2014). Recent advances in radical SAM enzymology: New structures and mechanisms. *ACS Chemical Biology*, *9*(9), 1929–1938. https://doi.org/10.1021/cb5004674.

Yang, L., & Li, L. (2015). Spore photoproduct lyase: The known, the controversial, and the unknown. *The Journal of Biological Chemistry*, *290*(7), 4003–4009. https://doi.org/10.1074/jbc.R114.573675.

Young, A. P., & Bandarian, V. (2011). Pyruvate is the source of the two carbons that are required for formation of the imidazoline ring of 4-demethylwyosine. *Biochemistry*, *50*(49), 10573–10575. https://doi.org/10.1021/bi2015053.

Yu, Y., Duan, L., Zhang, Q., Liao, R., Ding, Y., Pan, H., et al. (2009). Nosiheptide biosynthesis featuring a unique indole side ring formation on the characteristic thiopeptide framework. *ACS Chemical Biology*, *4*(10), 855–864. https://doi.org/10.1021/cb900133x.

Zhang, Y., Zhu, X., Torelli, A. T., Lee, M., Dzikovski, B., Koralewski, R. M., et al. (2010). Diphthamide biosynthesis requires an organic radical generated by an iron-sulphur enzyme. *Nature*, *465*(7300), 891–896. https://doi.org/10.1038/nature09138.

Zhao, S., Kumar, R., Sakai, A., Vetting, M. W., Wood, B. M., Brown, S., et al. (2013). Discovery of new enzymes and metabolic pathways by using structure and genome context. *Nature*, *502*(7473), 698–702. https://doi.org/10.1038/nature12576.

CHAPTER TWO

Purification and Characterization of the Choline Trimethylamine-Lyase (CutC)-Activating Protein CutD

Smaranda Bodea*, Emily P. Balskus[†,1]
*Merck & Co., Inc., Boston, MA, United States
[†]Department of Chemistry and Chemical Biology, Harvard University, Cambridge, MA, United States
[1]Corresponding author: e-mail address: balskus@chemistry.harvard.edu

Contents

Abstract

Anaerobic choline deamination catalyzed by the glycyl radical enzyme choline trimethylamine-lyase (CutC) has emerged as a major route for trimethylamine (TMA) production within anaerobic environments, including the human gut. The association of this microbial metabolite and its downstream products with diseases such as atherosclerosis and chronic kidney disease has driven the need for a better molecular understanding of TMA-generating enzymes. Our previous work has shown that generating

Methods in Enzymology, Volume 606
ISSN 0076-6879
https://doi.org/10.1016/bs.mie.2018.04.012

the critical, glycine-centered radical species on CutC requires posttranslational modification by an *S*-adenosyl-L-methionine (SAM)-dependent radical-activating protein (CutD) harboring an oxygen-sensitive [4Fe–4S] cofactor. In this chapter, we describe our strategy to heterologously express and purify *Desulfovibrio alaskensis* G20 CutD in *Escherichia coli* and reconstitute its iron–sulfur center under anaerobic conditions. In addition, we present the methods we have developed to characterize the activity of CutD and utilize this enzyme in conjunction with purified CutC to gain an unprecedented insight into the anaerobic C–N cleavage of choline.

ABBREVIATIONS

5′-dA	5′-deoxyadenosine
5′-dA•	5′-deoxyadenosyl radical
AE	activating enzyme
BDE	bond dissociation energy
BDPA	α,γ-bisdiphenylene-β-phenylallyl
Cam	chloramphenicol
DTT	DL-dithiothreitol
EPR	electron paramagnetic resonance spectroscopy
GRE	glycyl radical enzyme
HPLC	high-performance liquid chromatography
IPTG	isopropyl β-D-1-thiogalactopyranoside
Kan	kanamycin
LC-MS	liquid chromatography-mass spectrometry
MRM	multiple reaction monitoring
NaDT	sodium dithionite
OPA	*o*-phthaldialdehyde
SAM	*S*-adenosyl-L-methionine
SDS-PAGE	sodium dodecyl sulfate polyacrylamide gel electrophoresis
TMA	trimethylamine
YADH	yeast alcohol dehydrogenase

1. INTRODUCTION

The glycyl radical enzyme (GRE) family has emerged as an important class of proteins for anaerobic microorganisms from the environment as well as from the human body. GREs play central roles in anoxic microbial metabolic pathways, such as nucleotide reduction (class III ribonucleotide reductase, RNR), carbon-source utilization (pyruvate formate-lyase, PFL; benzylsuccinate synthase, BSS; glycerol dehydratase, GDH; propanediol dehydratase, PD; choline trimethylamine-lyase, CutC), and amino acid metabolism (*trans*-4-hydroxy-L-proline dehydratase, *t*4LHypD; 4-hydroxyphenylacetate decarboxylase, HPAD) (Levin et al., 2017; Selmer, Pierik, & Heider, 2005). The impact of GRE-catalyzed reactions on human health has only recently begun to be appreciated. Several metagenomic and proteomic studies have found GREs to be among the most abundant protein families in the human gut microbiota (Kolmeder et al., 2012; Kurokawa et al., 2007). The products of GRE-catalyzed metabolism within the human gut have been associated with multiple disorders, ranging from heart disease to autism. For instance, anaerobic choline utilization by gut bacteria is catalyzed by CutC, resulting in the formation of trimethylamine (TMA) and acetaldehyde (Craciun & Balskus, 2012; Martínez-del Campo et al., 2015; Romano et al., 2017). An exclusively microbial product in the human body, TMA is the precursor of trimethylamine *N*-oxide, a metabolite strongly correlated with susceptibility to both chronic kidney disease and cardiovascular disease (Tang et al., 2015; Wang et al., 2011).

GREs perform chemically difficult reactions using a glycine-centered radical species generated by a separate activating enzyme (AE or activase), a [4Fe–4S] cluster-containing member of the radical *S*-adenosyl-L-methionine (SAM) protein family (Selmer et al., 2005). This activation process is thought to be specific for each functional class of GREs, since GREs and AEs are typically encoded adjacent to one another in microbial genomes (Shisler & Broderick, 2014). As members of the radical SAM enzyme superfamily, all AEs utilize a reduced $[4Fe–4S]^+$ cluster to convert SAM into a 5′-deoxyadenosyl radical (5′-dA$^{\bullet}$). Structural and spectroscopic studies of PFL-AE have recently revealed that within the enzyme active site, SAM is bound to the catalytic iron–sulfur center, and the carboxylate oxygen of SAM also interacts with an enzyme-bound monovalent cation (likely K^+), which can further control the reactivity of the enzyme (Shisler et al., 2017). Freeze-quenching spectroscopy studies of PFL-AE enabled the observation of a short-lived covalent bond between the 5′ carbon of the deoxyadenosyl group of SAM and the single iron atom of the iron–sulfur cluster, as a catalytically competent intermediate toward the formation of 5′-dA$^{\bullet}$ (Horitani et al., 2016). This 5′-dA$^{\bullet}$ intermediate is extremely reactive (C–H bond dissociation energy (BDE) ~99 kcal/mol) and thus capable of abstracting the C–H bond of the GRE active site glycine (C–H BDE ~79.3 kcal/mol) (Fig. 1), generating the active form of the GRE (Himo & Eriksson, 1998).

Due to the extreme oxygen sensitivity of the [4Fe–4S] cluster and difficulties in reconstituting these cofactors in vitro, it has been challenging to

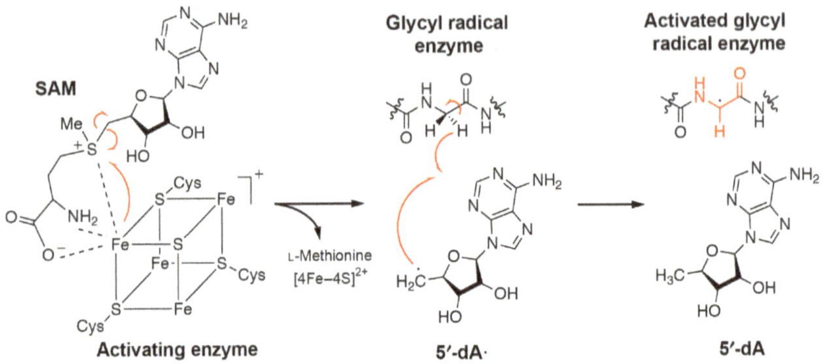

Fig. 1 General scheme for activation of a glycyl radical enzyme by a GRE-activating enzyme.

express GRE–AEs in heterologous hosts and to purify and reconstitute active forms of these enzymes. With the proper anaerobic equipment and techniques, some of these proteins have been amenable to in vitro studies: the activating proteins for class III RNR (Wei et al., 2014), PFL (Kulzer, Pils, Kappl, Huttermann, & Knappe, 1998), GDH (O'Brien et al., 2004), and *t*4LHypD (Levin et al., 2017) could be expressed and purified under aerobic conditions, with only their [4Fe–4S] cluster(s) needing to be reconstituted in an anaerobic environment. PFL-AE was also crystallized under anaerobic conditions, affording a unique insight into this class of enzymes (Vey et al., 2008). The AEs for HPAD (Selvaraj, Pierik, Bill, & Martins, 2013) and PD (Levin et al., 2017) have been overexpressed and purified only under anaerobic conditions, whereas benzylsuccinate synthase AE could not be overexpressed and purified despite multiple attempts (Funk, Marsh, & Drennan, 2015).

Maintaining oxygen-free conditions throughout overexpression and purification could be highly beneficial for HPAD-AE, PD-AE, and CutD, as a multiple sequence alignment suggests that they might possess additional [4Fe–4S] clusters (Craciun, Marks, & Balskus, 2014). In addition to the conserved [4Fe–4S]-binding CX_3CX_2C motif, each of these AEs contains additional "Clostridial"-type $CX_{2-5}CX_{2-4}CX_3C$ motifs that might bind auxiliary clusters. These motifs are typically present in ferredoxins, ubiquitous proteins in bacteria, and archaea that usually mediate electron transfer within proteins (Duee et al., 1994). In the case of AEs, it is likely that the additional [4Fe–4S] clusters could shuttle electrons from an electron donor (such as a reduced flavodoxin protein) to the SAM-bound [4Fe–4S] cluster, but generally, the roles of auxiliary [4Fe–4S] clusters in radical SAM enzymes are still poorly understood and challenging to study (Lanz & Booker, 2015).

In this chapter, we present strategies that we have developed for overproducing and purifying CutD from *Desulfovibrio alaskensis* G20 under anaerobic conditions, as a model for the characterization of other CutC AEs. We also describe how we used purified CutD in assays with purified CutC in order to detect glycyl radical generation by electron paramagnetic resonance spectroscopy (EPR) and product formation by high-performance liquid chromatography (HPLC) and liquid chromatography-mass spectrometry (LC-MS).

2. OVERPRODUCTION AND PURIFICATION OF CutD

We drew inspiration for the expression and purification of GRE activase CutD from *D. alaskensis* G20 from protocols for other radical SAM enzymes that might harbor additional [4Fe–4S] clusters (Selvaraj et al., 2013; Wecksler et al., 2009). As a result, the expression, purification, and reconstitution of CutD with a C-terminal His_6-tag were performed under strictly anaerobic conditions with coexpression of the *E. coli isc* operon *IscSUA-HscBA-Fd* (Tokumoto & Takahashi, 2001), which plays a main role in the assembly of [4Fe–4S] cofactors of cellular iron–sulfur proteins. Interestingly, the coexpression of the *isc* operon improved CutD yields by ~twofold, while the coexpression of the *suf* operon *sufABCDSE* (typically used to repair oxidatively damaged [4Fe–4S] cluster proteins) (Hanzelmann et al., 2004) interfered with the expression of CutD, providing greatly reduced yields of the protein. Maltose-binding protein-tagged and Strep-tagged CutD also purified in ~10-fold lower yields, so the C-His_6-tag construct was chosen for characterization.

2.1 Gene Cloning and Overexpression Strategy

The *cutD* gene (Dde_3281) from *D. alaskensis* G20 was cloned into a pET-29b(+) vector (Novagen) after PCR amplification as previously described (Craciun & Balskus, 2012). The pET system was chosen because it is a powerful platform for overexpressing proteins, allowing the user to induce protein overproduction at precise time points by the addition of isopropyl β-D-1-thiogalactopyranoside (IPTG). The vector containing the *isc* operon was constructed by Hanzelmann et al. by cloning the *E. coli iscSUA* genes and *E. coli hscBA-fdx* genes into two different sites of the pACYCDuet-1 vector (Hanzelmann et al., 2004). The Duet vectors can be cotransfected with pET vectors and are thus ideal for protein coexpression when the assembly of protein complexes is needed, as in the case of iron–sulfur cluster biogenesis. The two vectors confer resistance to different antibiotics (kanamycin (Kan) and chloramphenicol (Cam)), so a double antibiotic selection was performed to select for successfully transfected clones of *E. coli* BL21(DE3), a strain routinely used for protein overexpression.

2.2 Protocol for Overexpression of the *CutD* Gene

1. Transform chemically competent *E. coli* BL21(DE3) cells with pACYCDuet-1 vector harboring *E. coli iscSUA-hscBA-fdx* genes and pET-29b-CutD vector using standard procedures and at least 1 μg of each plasmid. Select transformants on Luria-Bertani (LB) plates containing 50 μg/mL Kan and 50 μg/mL Cam.
2. Use a single colony to inoculate 50 mL of LB medium with 50 μg/mL Kan and 50 μg/mL Cam, then incubate the culture overnight at 37°C with 150 rpm shaking.
3. Dilute the starter culture 1:100 into a screw-capped 2.8-L flask (Corning®) containing 2 L of LB medium supplemented with 50 μg/mL Kan, 50 μg/mL Cam, 2 m*M* iron (III) ammonium citrate, and 0.5% (w/v) glucose. Incubate the culture at 37°C with shaking at 175 rpm until reaching an OD_{600} = 0.35–0.4, at which point place the flask at room temperature and sparge the culture with argon gas for ~30 min.
4. Upon addition of L-cysteine (2 m*M* final concentration) and sodium fumarate (20 m*M* final concentration), sparge the culture with argon gas until it reaches an OD_{600} = 0.5–0.6, then induce protein expression by adding 500 μ*M* IPTG (Teknova), cap the flask tightly, and wrap the outside cap with electrical tape. Bring the flask into an Mbraun glove box and incubate the culture at 15°C overnight (~16–18 h) without shaking.
5. Harvest cells by centrifugation in 250-mL Corning® polypropylene centrifuge tubes (3000 × *g* for 20 min) and utilize the resulting cell paste immediately in the purification process.

2.3 Protocol for Purifying CutD Under Anaerobic Conditions

Equipment

- All purification steps were performed in an anaerobic chamber at 11°C (MBraun glove box), with the exception of cell lysis and centrifugation.

- Sonication for disrupting *E. coli* cells overexpressing CutD was performed with a Branson Sonifier S-450D equipped with a 1/2 in. horn inside a glove box (Coy Laboratory Products) under an atmosphere of 96% nitrogen and 4% hydrogen, at 4°C.
- Standard sodium dodecylsulfate (SDS) denaturing electrophoresis equipment from Bio-Rad and 4%–15% Tris–HCl polyacrylamide gels were used for protein electrophoresis.
- To maintain anaerobic conditions of protein mixtures outside of a glove box, Corning® 50-mL polypropylene tubes with plug seal caps with vinyl tape-sealed edges were used.

Buffers

- Lysis buffer (500 mL, made anoxic by sparging with argon gas for at least 30 min): 50 m*M* HEPES pH 8, 200 m*M* NaCl, 10 m*M* $MgCl_2$.
- Elution buffer (200 mL, made anoxic by sparging with argon gas for at least 30 min): 50 m*M* HEPES pH 8, 200 m*M* NaCl, 10 m*M* $MgCl_2$, 200 m*M* imidazole.
- Dialysis buffer (3 L, made anoxic by sparging with argon gas for at least 1 h): 50 m*M* HEPES pH 8, 50 m*M* NaCl, and 10% (v/v) glycerol.

1. Resuspend cells harvested from 2 L of culture in 40 mL of anoxic lysis buffer supplemented with half of a SIGMAFAST™ protease inhibitor cocktail tablet (Sigma-Aldrich), 8 mg of chicken egg lysozyme, and 5 m*M* DL-dithiothreitol (DTT).
2. Lyse the cells using a sonifier inside a 4°C oxygen-free chamber at 25% amplitude 10 s on, 30 s off, for 5 min.
3. Clarify the lysate by centrifugation (10,000 × *g* for 30 min) and incubate the supernatant with 2 mL of Ni-NTA resin (previously sparged with nitrogen) and 5 m*M* imidazole for 2 h at 4 °C under anaerobic conditions.
4. Centrifuge the lysate–resin mixture (3000 × *g* for 5 min), bring the tube inside the glove box, and discard the unbound fraction.
5. Resuspend the Ni-NTA resin inside the glove box in 10 mL of lysis buffer supplemented with 5 m*M* imidazole and load mixture into a glass column.
6. Elute the resin-bound protein using a stepwise imidazole gradient in lysis buffer (10, 25, 50, 100, 200 m*M* imidazole), collecting 2 mL fractions.
7. Use SDS-PAGE analysis (4%–15% Tris–HCl gel) to ascertain the presence and purity of protein in each fraction (Fig. 2). After combining the fractions containing CutD, proceed immediately to the reconstitution step.

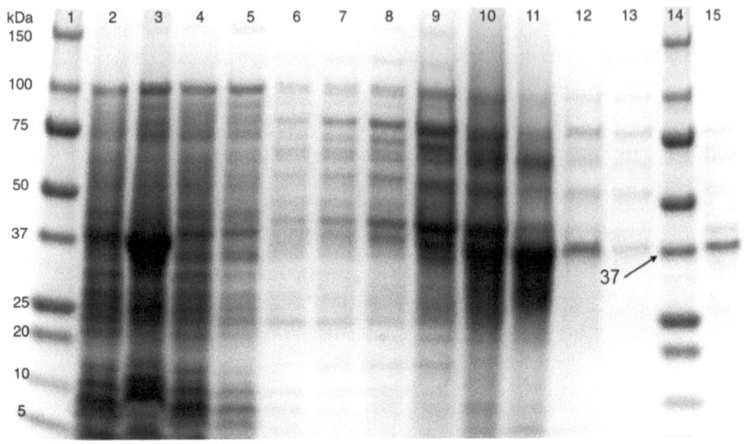

Fig. 2 SDS-PAGE of CutD from *D. alaskensis* G20 purification. Lanes are as follows: (1) Precision Plus Protein™ All Blue Standards (Bio-Rad), (2) soluble lysate, (3) pellet, (4) flow-through, (5) wash, (6–13) eluate with 10, 10, 25, 50, 100, 200, 200, 200 m*M* imidazole, respectively, (14) protein standards, (15) combined eluate from lanes 9–13.

2.4 Procedure for Chemical Reconstitution of Isolated CutD

1. Working under anaerobic conditions, transfer the fractions containing CutD into a 50-mL conical tube equipped with a small stir bar. Prepare each reconstitution reagent (DTT, Na_2S, $Fe(NH_4)_2(SO_4)_2$) by introducing the solid powder or crystals inside the anaerobic chamber and dissolving them using anoxic water.
2. While stirring slowly, add the following reagents, incubating for 5 min before each addition: 2 m*M* DTT (from 200 m*M* stock), 0.2 m*M* Na_2S (from 300 m*M* stock), and 0.25 m*M* $Fe(NH_4)_2(SO_4)_2$ (from 10 m*M* stock, in aliquots of ~50 μL). Incubate the reconstitution mixture at 11°C overnight (13–16 h) inside the glove box.
3. Filter reconstituted CutD through a 25-mm, 0.22-μm pore-size Acrodisc syringe filter with HT Tuffryn Membrane (Pall Life Sciences) to remove particulates. Transfer the clear solution to a 12- to 30-mL Slide-A-Lyzer™ Dialysis Cassette (Thermo Fischer Scientific) with a 3.5-K molecular weight cutoff filter.
4. Dialyze the protein solution for at least 2 h against three 1 L batches of 50 m*M* HEPES pH 8, 50 m*M* NaCl, and 10% (v/v) glycerol, under anaerobic conditions, at 11°C.
5. Concentrate the protein solution using Spin-X® UF 6 mL centrifugal concentrators with a 5000-Da molecular weight cutoff membrane (Corning®) placed inside 50-mL Corning® centrifuge tubes with plug seal caps.

6. Aliquot the concentrated protein into 0.5-mL cryogenic vials and place in 18 × 150 mm Hungate tubes (Chemglass) sealed with butyl stoppers and aluminum seals. Freeze the tubes by immersing into liquid nitrogen and store at −80°C. This procedure routinely afforded 9 mg CutD (~90% pure) per L of bacterial culture.

3. CHARACTERIZATION OF THE IRON–SULFUR CLUSTERS IN CutD

3.1 UV–vis Spectrum of Reconstituted CutD

The UV–vis spectrum of reconstituted CutD displayed features typical of $[4Fe–4S]^{2+}$ clusters: a broad peak at 380–420 nm which decreased in intensity upon reduction to $[4Fe–4S]^{+}$ using excess sodium dithionite (NaDT) (Fig. 3) (Wecksler et al., 2009). The CutD extinction coefficient at 410 nm prior to reduction with 10 equivalents of NaDT was $\sim 12.6\,\mathrm{m}M^{-1}\,\mathrm{cm}^{-1}$, suggesting that not all iron–sulfur centers had been properly reconstituted ($[4Fe–4S]^{2+}$ $\varepsilon_{410} \sim 15\,\mathrm{m}M^{-1}\,\mathrm{cm}^{-1}$) (Wecksler et al., 2009). Analyses of CutD-bound iron and sulfide using previously published methods (Craciun et al., 2014) indicated the possibility of as many as two [4Fe–4S] clusters (8.4 mol of Fe and 7.6 mol of S per mole of CutD), although we cannot rule out adventitious binding of iron.

1. After the reconstitution of purified CutD, dilute the protein solution inside the anaerobic chamber to 10 μ*M* with anoxic buffer (50 m*M* HEPES pH 8, 50 m*M* NaCl). Treat one-half of the protein solution with a 10-fold excess of NaDT for 15 min.
2. Measure the absorbance of both protein solutions over a 250- to 800-nm range using a quartz 96-well plate and a PowerWave HT Microplate Spectrophotometer (BioTek).
3. Analyze the data using Gen5 Data Analysis Software, subtracting the absorbance for buffer from the absorbance readings for CutD solutions. Calculate the molar absorptivity coefficients by dividing the absorbance to the protein concentration and path length (1 cm).

3.2 Protocol for EPR Analysis of Iron–Sulfur Clusters

The EPR spectrum of NaDT-reduced CutD was consistent with the UV–vis spectral data. Despite the apparent splitting noticeable in the (A) and (B) spectra from Fig. 4 (which could be due to a superimposition of signals),

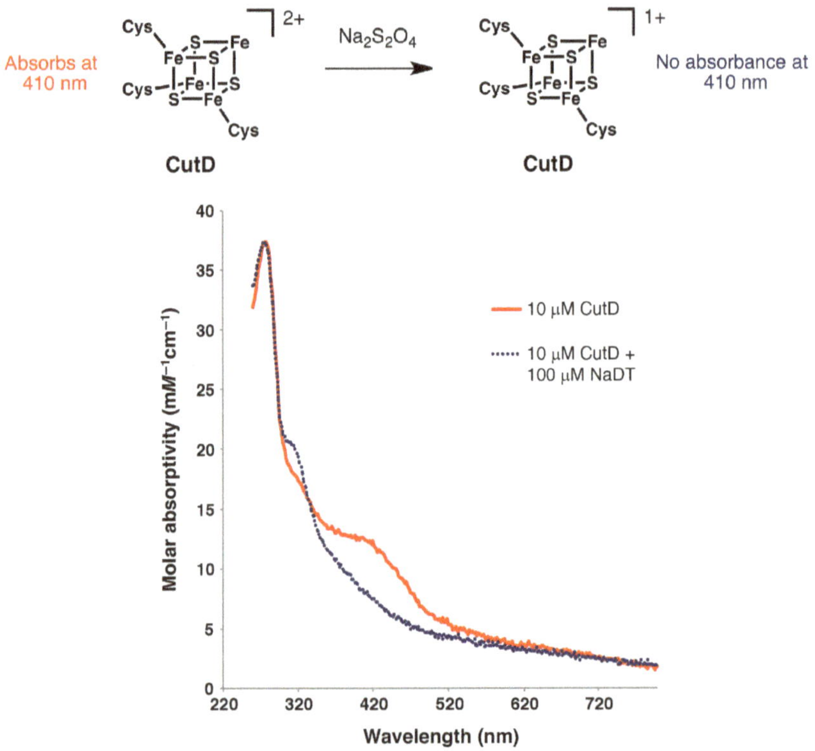

Fig. 3 UV–vis spectrum of purified CutD before (*red*) and after (*blue*) reduction with sodium dithionite (NaDT). *Reprinted with permission from Craciun, S., Marks, J. A., & Balskus, E. P. (2014). Characterization of choline trimethylamine-lyase expands the chemistry of glycyl radical enzymes.* ACS Chemical Biology, *9(7), 1408–1413. Copyright 2014 American Chemical Society.*

the signal with $g_{||}$ of 2.045 and $g_{\perp}$ of 1.94 can be unambiguously attributed to a $[4Fe–4S]^+$ cluster because it broadens and decreases in intensity at higher (40 K) temperatures, as shown in spectrum (C). The *g* values and temperature dependence of this signal further confirm the presence of $[4Fe–4S]^+$ centers in reduced CutD, while its intensity (0.8 spins per CutD monomer, measured using a Cu-EDTA standard) is indicative of incomplete reconstitution of the clusters. Interestingly, the shape and amplitude of the EPR signal do not seem to change upon addition of excess SAM, an observation that has been made for other glycyl radical-activating proteins (Kulzer et al., 1998; Ollagnier et al., 1997; Yu, Blaser, Andrei, Pierik, & Selmer, 2006).

1. Prepare the standard used for quantification of the signal arising from the iron–sulfur clusters of CutD by combining 800 μ*M* $CuSO_4$ with 8 m*M* EDTA. Place 200 μL of standard solution in a cap-sealed EPR tube with

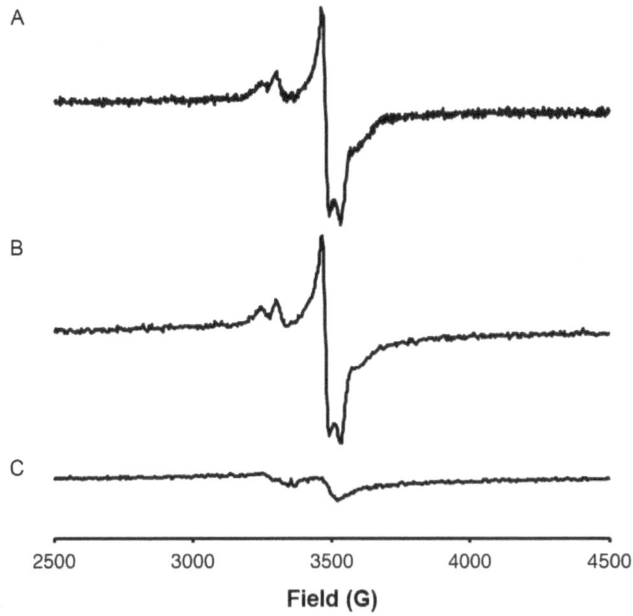

Fig. 4 EPR characterization of the iron–sulfur cluster in CutD. (A) Spectrum at 9 K of NaDT-reduced CutD. (B) Spectrum at 9 K of NaDT-reduced CutD incubated for <5 min with excess SAM. (C) Spectrum at 40 K of NaDT-reduced CutD. *Reprinted with permission from Craciun, S., Marks, J. A., & Balskus, E. P. (2014). Characterization of choline trimethylamine-lyase expands the chemistry of glycyl radical enzymes.* ACS Chemical Biology, *9(7), 1408–1413. Copyright 2014 American Chemical Society.*

4 mm outer diameter and 8″ length (Wilmad Lab-Glass, 734-LPV-7). Freeze the solution by slowly immersing the tube in liquid N_2.

2. Prepare the reduced CutD samples by incubating 50 m*M* HEPES pH 8, 50 m*M* NaCl, 40 μ*M* CutD, and 4 m*M* NaDT at room temperature (22°C) in a total volume of 200 μL under anaerobic conditions for 20 min prior to transferring to an EPR tube, sealing it, and freezing the assay in liquid N_2.
3. In order to determine the effect of SAM on the EPR signal, add 4 m*M* SAM to one of the reduced CutD samples and freeze the sample within less than 5 min.
4. Record perpendicular mode X-band EPR spectra for the Cu(II)-EDTA standard and all CutD samples on an EPR instrument (such as Bruker ElexSysE500) fitted with a cryostat for measurements at 9–40 K. Example of parameter settings: temperature, 9–40 K, center field, 3500 G; range, 2000 G; microwave power, 2 mW; microwave frequency, 9.45 MHz; modulation amplitude, 0.4 mT; modulation

frequency, 100 kHz; time constant, 20.48 ms; conversion time, 20.48 ms; scan time, 20.97 s; receiver gain, 60 dB (for protein samples) or 30 dB (for standard); number of scans, 4. Calibrate the magnetic field with a standard sample of α,γ-bisdiphenylene-β-phenylallyl (BDPA), $g = 2.0026$ (Bruker). Perform data acquisition with Xepr software (Bruker).

5. Calculate the spin concentration for CutD by comparing the area of the spectra acquired with that of the Cu(II)-EDTA standard, without considering the difference in receiver gain (which is already accounted for by the EPR spectrometer).

3.3 Generation of SAM Cleavage Products Monitored by HPLC

Having confirmed that CutD can harbor a $[4Fe–4S]^{+}$ cluster upon reduction with NaDT, we proceeded in examining the activity of reduced CutD toward SAM. We chose to perform these experiments in the absence of CutC to allow for the accumulation of reaction products, 5′-deoxyadenosine and L-methionine, which were detected by HPLC using purchased standards. We observed generation of these products under the chosen conditions, which is in agreement with the results of similar assays using previously characterized GRE-activating enzymes (Shisler & Broderick, 2014) (Fig. 5).

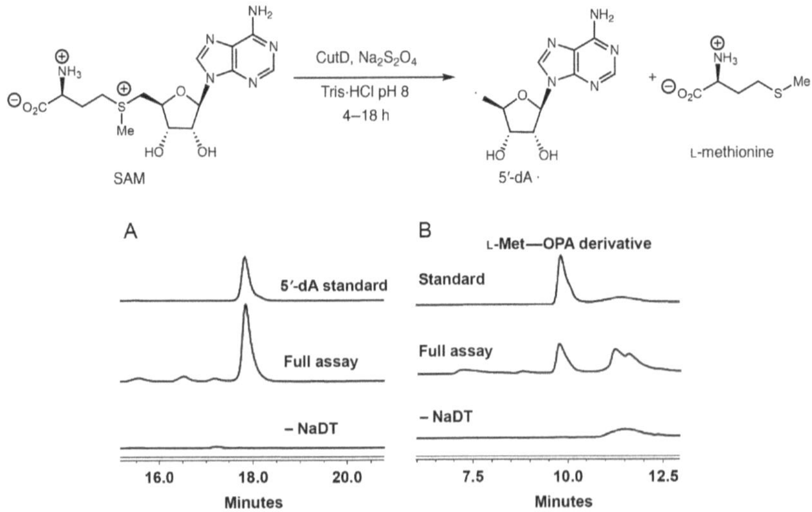

Fig. 5 HPLC assays for characterization of SAM cleavage products generated in the presence of CutD. (A) Detection of 5′-deoxyadenosine (5′-dA) arising from quenching of 5′-dA• by the solvent. (B) Detection of L-methionine as an OPA derivative. *Reprinted with permission from Craciun, S., Marks, J. A., & Balskus, E. P. (2014). Characterization of choline trimethylamine-lyase expands the chemistry of glycyl radical enzymes.* ACS Chemical Biology, *9(7), 1408–1413. Copyright 2014 American Chemical Society.*

Protocol for HPLC detection of 5′-deoxyadenosine (dA)

1. Assay mixtures should contain 25 m*M* Tris–HCl buffer pH 8, 50 m*M* NaCl, 16 m*M* NaDT, 40 μ*M* CutD, and 1 m*M* SAM in a total volume of 100 μL. Working under anaerobic conditions, combine all components except for SAM in order to allow CutD to reduce for 15 min. Upon addition of SAM, incubate the mixtures at room temperature in the glove box for 4 h.
2. Upon removing the assay mixtures from the anaerobic chamber, quench the reactions with 2.5 μL formic acid and centrifuge (10,000 × *g* for 10 min) to remove precipitated protein.
3. Analyze by HPLC an aliquot (20 μL) of the supernatant and a 50-μ*M* solution of commercial 5′-deoxyadenosine dissolved in buffer. For example, use a Dionex Ultimate 3000 instrument (Thermo Scientific), equipped with a Phenomenex Luna analytical C18 reverse phase column (5 μm, 4.6 × 250 mm, 100 Å), preceded by a HAIPEEK guard cartridge holder (Higgins Analytical) containing a CLIPEUS C18 (5 μm, 2 cm × 3.2 mm) cartridge (Higgins Analytical).
4. Utilize a solvent system consisting of 40 m*M* ammonium acetate pH 6 (A) and acetonitrile (B), and apply the following elution conditions: 0% B for 2 min, 0%–25% B over 25 min, 25%–0% B over 2 min, 0% B for 5 min, setting the flow rate to 1 mL/min. Detect products by UV at 216 and 260 nm, comparing the traces with those for the 5′-deoxyadenosine standard.

Protocol for HPLC detection of L-methionine

1. Assay mixtures should contain 25 m*M* Tris–HCl buffer pH 8, 50 m*M* NaCl, 12 m*M* NaDT, 30 μ*M* CutD, and 1 m*M* SAM in a total volume of 100 μL. Working under anaerobic conditions, combine all components except for SAM in order to allow CutD to reduce for 15 min. Upon addition of SAM incubate the reactions at room temperature in the glove box for 18 h (overnight).
2. Upon removing the assay mixtures from the anaerobic chamber, quench with 200 μL cold methanol, cool on ice for 10 min, and centrifuge (10,000 × *g* for 10 min) to remove precipitated protein.
3. Remove most of the organic solvent in vacuo, concentrating the mixtures to ~100 μL.
4. Mix a 20-μL aliquot of each assay or of a 500-μ*M* solution of commercial L-methionine standard with 140 μL *o*-phthaldialdehyde (OPA) reagent (Sigma-Aldrich; containing 2 μL of 2-mercaptoethanol per mL of incomplete reagent), incubating for 5–6 min.
5. Analyze an aliquot (40 μL) of each derivatization mixture by HPLC using a Dionex Acclaim C18 reverse phase column (3 μm,

2.1 × 150 mm, 120 Å), preceded by the same guard cartridge as for the 5′-dA assay.

6. Utilize a solvent system consisting of water (A) and acetonitrile (B) and apply the following elution conditions with a flow rate of 0.25 mL/min: 0% B for 1 min, 0%–70% B over 15 min, 70%–0% B over 5 min, 0% B for 2 min. Detect the OPA-derivatized L-methionine by UV at 338 nm.

4. ACTIVITY OF CutD TOWARD CutC

4.1 Protocol for Quantification of Extent of CutC Activation by EPR

We performed additional EPR experiments to investigate the reactivity of reconstituted CutD toward purified CutC from *D. alaskensis* G20. Truncating the first 52 amino acids (−52 aa) was found to improve stability and expression in *E. coli*, and all experiments described were performed using this construct (expression and purification detailed in Craciun et al., 2014). Spectra of anaerobic reaction mixtures containing CutD, CutC, SAM, and NaDT taken at 77 K revealed a signal centered at $g = 2.0037$ and characterized by a twofold splitting of ~1.46 mT (Fig. 6). The line shape of this signal and its g value closely resemble those of the glycine-centered

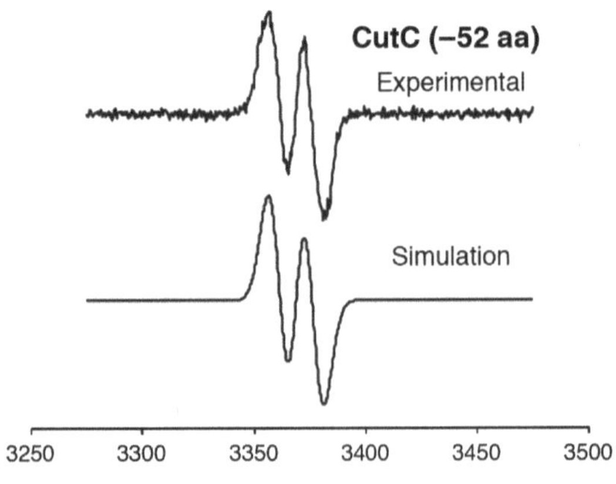

Fig. 6 EPR experimental spectrum of activated CutC (−52 aa) and simulation assuming an isotropic signal with $g = 2.0035$, linewidth 1.01 mT, and hyperfine coupling A = 1.44 mT. *Reprinted with permission from Craciun, S., Marks, J. A., & Balskus, E. P. (2014). Characterization of choline trimethylamine-lyase expands the chemistry of glycyl radical enzymes.* ACS Chemical Biology, *9(7), 1408–1413. Copyright 2014 American Chemical Society.*

radicals from activated PFL (Wagner, Frey, Neugebauer, Schafer, & Knappe, 1992) and class III RNR (Sun et al., 1996). The level of activation we observed after optimizing the assay conditions was 0.18–0.21 radicals per polypeptide, whereas the maximum yield of activated GRE (such as PFL) reported was 0.5 radicals per polypeptide (Shisler & Broderick, 2014). Table 1 describes the parameters surveyed for glycyl radical formation on CutC and their respective outcomes. The activation step was optimized with CutC (−52 aa) using a coupled, spectrophotometric assay for acetaldehyde, as described for steady-state kinetics measurement of choline cleavage by CutC (Section 4.2).

Table 1 Parameters Surveyed for the Optimization of Glycyl Radical Formation on CutC

Parameter Surveyed for Activation Assay	Outcome
Incubation time (30, 60, or 90 min)	Selected 60 min as incubation time due to reproducible results; only a modest increase was observed at 90 min, as shown in Fig. 7
Buffer (25 m*M* Tris pH 8 vs 50 m*M* potassium phosphate pH 8)	Selected 50 m*M* potassium phosphate pH 8 due to twofold improved stability of the glycyl radical
Ratio of CutC to CutD (from 1:0.4 to 1:4)	Found that highest amounts of CutD afford best results, probably due to incomplete reconstitution of [4Fe–4S] clusters
Using size exclusion chromatography-purified CutC	Did not make a difference
Using size exclusion chromatography-purified CutD	Did not make a difference, and avoiding this step might prevent potential oxygen-mediated damage to the [4Fe–4S] clusters
Reductant: sodium dithionite; *E. coli* flavodoxin (Hall, Jordan-Starck, Loo, Ludwig, & Matthews, 2000) reduced with NaDT; acriflavin/bicine; methyl viologen; titanium (III) citrate	Moderate amounts of NaDT were found to afford best results
Concentration of NaDT (2–20 equivalents with respect to CutC)	10–20 equivalents afforded best results
Concentration of SAM (10–80 equivalents with respect to CutC)	20 equivalents afforded best results

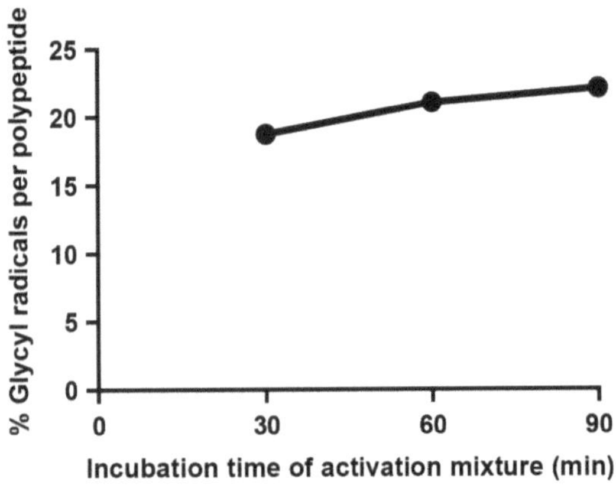

Fig. 7 Stability of the glycyl radical of CutC (−52 aa), assessed by varying the incubation time of the activation mixture and measuring glycyl radical content by electron paramagnetic resonance spectroscopy. *Reprinted with permission from Bodea, S., Funk, M. A., Balskus, E. P., & Drennan C. L. (2016). Molecular basis of C-N bond cleavage by the glycyl radical enzyme choline trimethylamine-lyase.* Cell Chemical Biology, *23(10), 1206–1216.*

1. Sparge a CutC (−52 aa) stock solution purified under aerobic conditions (Craciun et al., 2014) with argon gas for 15 min, while keeping the solution on ice. After introducing it into the glove box, remove a small aliquot and measure the concentration using a NanoDrop 2000 UV–vis Spectrophotometer (Thermo Scientific) and the ExPASy protparam tool-calculated extinction coefficient for CutC ($\varepsilon = 128{,}870\ M^{-1}\ \mathrm{cm}^{-1}$).
2. Prepare activation assay mixtures in a glove box in 1.5-mL polypropylene Eppendorf tubes, containing 50 m*M* potassium phosphate pH 8, 50 m*M* potassium chloride, 150 μ*M* NaDT, 200 μ*M* SAM, 10 μ*M* CutC (−52 aa) dimer, and 30 μ*M* CutD in a total volume of 200 μL. Incubate a mixture of buffer, CutD and NaDT, for 20 min prior to addition of SAM and CutC, then carry out glycyl radical formation for 1 h at room temperature.
3. Transfer the activation reaction mixtures to a cap-sealed EPR tube, remove from the glove box, and freeze the tubes in liquid N_2.
4. Prepare a $K_2(SO_3)_2NO$ standard by dissolving solid $K_2(SO_3)_2NO$ under anaerobic conditions in anoxic 0.5 *M* $KHCO_3$ and diluting to a final concentration of 0.3–0.5 m*M*. This standard is not stable for a long time (Murib & Ritter, 1952), so it must be prepared before each set of EPR measurements. To account for any decomposition during

dissolution, measure the concentration of a small aliquot of the standard at 248 nm ($\varepsilon = 1690\ M^{-1}\ cm^{-1}$), using a NanoDrop 2000 UV–vis Spectrophotometer.

5. Record X-band EPR spectra at 77 K by fitting the instrument with a quartz dewar (Wilmad Lab-Glass) filled with liquid N_2. Example of parameter settings: temperature, 77 K; center field, 3370 G; range, 200 G; microwave power, 20 μW; microwave frequency, 9.45 MHz; modulation amplitude, 0.4 mT; modulation frequency, 100 kHz; time constant, 20.48 ms; conversion time, 20.48 ms; scan time, 20.97 s; receiver gain, 60 dB (for protein samples) or 30 dB (for standard); number of scans, 16. Calibrate the magnetic field with a standard sample of BDPA, $g = 2.0026$ (Bruker). Perform data acquisition with Xepr software (Bruker).
6. Perform spin concentration measurements for CutC activation assays by numerically calculating the double integral of the simulated spectra and comparing the area with that of a $K_2(SO_3)_2NO$ standard, without considering the difference in receiver gain (which is typically already accounted for by the EPR spectrometer).

4.2 Protocol for Kinetic Analysis of Choline Cleavage by CutC Upon Activation by CutD

In order to assess the catalytic parameters of CutC-catalyzed choline cleavage and determine the effect of different components of the activation assay, we performed a continuous spectrophotometric-coupled assay in which the formation of acetaldehyde from choline by activated CutC was coupled to the reduction of acetaldehyde by NADH-dependent yeast alcohol dehydrogenase (YADH), as previously described for GDH (O'Brien et al., 2004). The kinetic parameters observed for the truncated CutC (−52 aa) variant from *D. alaskensis* G20 (Fig. 8, $K_M = 0.13 \pm 0.01$ mM, $k_{cat} = 157 \pm 2\,s^{-1}$, glycyl radical normalized $k_{cat} = 377 \pm 5\,s^{-1}$) were in accordance to those reported for other GREs (Craciun et al., 2014), and consistent with the role of CutC in bacterial primary metabolism (Craciun & Balskus, 2012).

1. Carry out the activation of CutC under anaerobic conditions as described in Section 4.1 in a total volume of 50 μL.
2. After the glycyl radical formation, dilute the assay mixture 100-fold with buffer (50 mM potassium phosphate pH 8, 50 mM potassium chloride) and add into wells of a 96-well plate containing buffer, 200 μM NADH, 0.4 μM YADH, and choline (0–4 mM), such that the final concentration of CutC dimer is 2.5 nM and the total volume of the assay mixture is 200 μL.

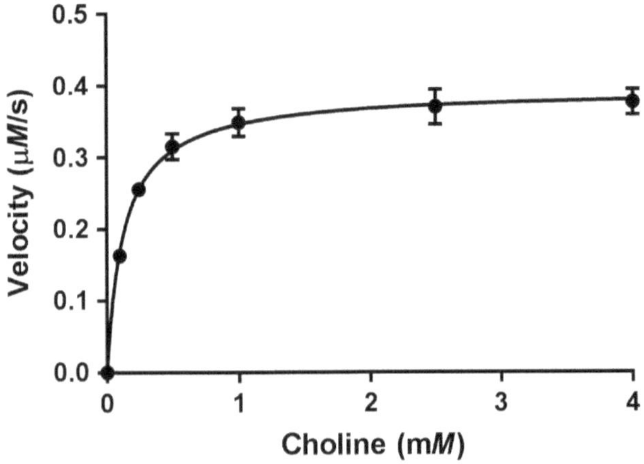

Fig. 8 Michaelis–Menten kinetics of CutC (−52 aa) from *D. alaskensis* G20. *Reprinted with permission from Craciun, S., Marks, J. A., & Balskus, E. P. (2014). Characterization of choline trimethylamine-lyase expands the chemistry of glycyl radical enzymes.* ACS Chemical Biology, *9(7), 1408–1413. Copyright 2014 American Chemical Society.*

3. Monitor the NADH absorbance at 340 nm for 7 min with reads 20 s apart using a PowerWave HT Microplate Spectrophotometer (BioTek) placed inside an Mbraun glove box at room temperature. Perform path length correction such that absorbance values are reported for a path length of 1 cm.
4. Analyze the data using Gen5 Data Analysis Software and calculate kinetic parameters using GraphPad Prism6.

4.3 Protocol for LC-MS/MS-Based Detection of TMA From CutC-Catalyzed Choline Cleavage

Another method to examine the reactivity of activated CutC is by using an end point assay for detecting the TMA arising from CutC-catalyzed choline cleavage. We developed an LC-MS/MS detection method for the detection and quantification of TMA, which can be applied to both enzymatic assays (Fig. 9), as well as whole bacterial cultures.

1. As described in Section 4.1, prepare activation assay mixtures containing 25 m*M* Tris buffer pH 8, 50 m*M* sodium chloride, 200 μ*M* NaDT, 200 μ*M* SAM, 10 μ*M* CutC dimer, and 40 μ*M* CutD in a total volume of 50 μL.

2. After activating CutC, add choline to a final concentration of 20 m*M* and incubate the assay mixtures in an Mbraun glove box at room temperature for a time interval >5 min.
3. Dilute an aliquot of 4 μL of each reaction mixture into buffer R (20% 5 m*M* ammonium formate pH 4 and 80% acetonitrile). Centrifuge this mixture at 10,000 × *g* for 10 min to remove particulates, then dilute 20-fold with buffer R, and transfer to an LC-MS vial.
4. For quantification of TMA, prepare a standard curve by spiking concentrations of 0–20 m*M* TMA in an activation assay mixture that was incubated under aerobic conditions such that the glycyl radical of CutC was completely inactivated (Craciun et al., 2014). Dilute the standards with buffer R in the same manner as the samples.
5. Analyze an aliquot of each sample (3 μL) by LC-MS/MS, using a precursor-product ion pair transition of *m/z* 60.1 → *m/z* 45. A collision energy of 21 V was optimal when using an Agilent 6410 Triple Quadrupole LC/MS instrument with electrospray ionization source, operated in multiple reaction monitoring mode and positive mode. Perform the LC analysis, for example, using a Kinetex (Phenomenex) HILIC column (2.6 μm, 30 mm × 2.1 mm, 100 Å), preceded by a C4

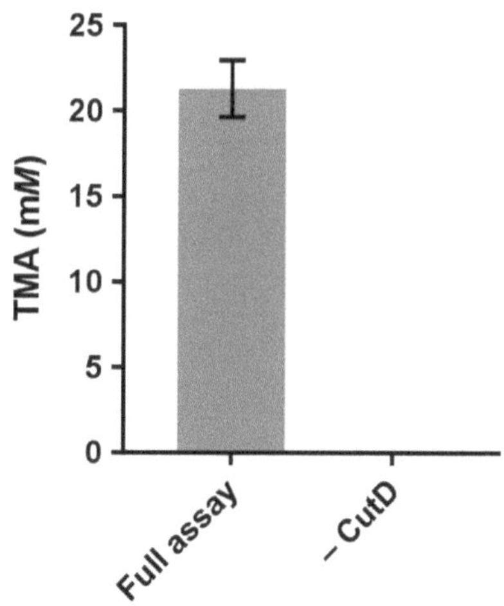

Fig. 9 LC-MS/MS detection of TMA production in assays conducted with 20 m*M* choline in Tris buffer with CutC (−52 aa). Each assay was performed in triplicate, and the bar represents the mean ± standard deviation of three replicates. N/D = not detected. *Reprinted with permission from Bodea, S., Funk, M. A., Balskus, E. P., & Drennan C. L. (2016). Molecular basis of C-N bond cleavage by the glycyl radical enzyme choline trimethylamine-lyase.* Cell Chemical Biology, *23(10), 1206–1216.*

precolumn (3.5 μm, 2.0 mm × 20 mm) and the following LC conditions: 80% B for 1.6 min, a gradient decreasing to 40% B over 2.4 min, 40% B for 2 min, a gradient increasing to 80% B over 2 min, and 80% B for 3 min (solvent A = 5 m*M* ammonium formate in water with 0.1% formic acid, solvent B = 100% acetonitrile, flow rate 0.3 mL/min).

5. CONCLUSIONS

The protocols we have developed for purifying CutC-activating protein CutD have allowed us to successfully reconstitute in vitro the activity of CutC, a disease-associated GRE. The insights we have gained into the activity of these two proteins can further enable research teams to pursue the discovery of small molecule inhibitors of anaerobic microbial choline metabolism. Such molecules would be valuable tools for understanding the role of this metabolic pathway in the human gut and its connection to disease.

ACKNOWLEDGMENTS

We thank the Klinman group (UC Berkeley) for the plasmids encoding the *isc* and *suf* systems, as well as A. Chan and K. Bu for assistance with cloning and protein purification, and M.J.T. Wilding, M.B. Brophy, and Y. Wei for advice concerning EPR.

REFERENCES

Craciun, S., & Balskus, E. P. (2012). Microbial conversion of choline to trimethylamine requires a glycyl radical enzyme. *Proceedings of the National Academy of Sciences of the United States of America*, *109*(52), 21307–21312.

Craciun, S., Marks, J. A., & Balskus, E. P. (2014). Characterization of choline trimethylamine-lyase expands the chemistry of glycyl radical enzymes. *ACS Chemical Biology*, *9*(7), 1408–1413.

Duee, E. D., Fanchon, E., Vicat, J., Sieker, L. C., Meyer, J., & Moulis, J. M. (1994). Refined crystal-structure of the 2[4Fe-4S] ferredoxin from *Clostridium acidurici* at 1.84-Å resolution. *Journal of Molecular Biology*, *243*(4), 683–695.

Funk, M. A., Marsh, E. N., & Drennan, C. L. (2015). Substrate-bound structures of benzylsuccinate synthase reveal how toluene is activated in anaerobic hydrocarbon degradation. *Journal of Biological Chemistry*, *290*(37), 22398–22408.

Hall, D. A., Jordan-Starck, T. C., Loo, R. O., Ludwig, M. L., & Matthews, R. G. (2000). Interaction of flavodoxin with cobalamin-dependent methionine synthase. *Biochemistry*, *39*(35), 10711–10719.

Hanzelmann, P., Hernández, H. L., Menzel, C., García-Serres, R., Huynh, B. H., Johnson, M. K., et al. (2004). Characterization of MOCS1A, an oxygen-sensitive iron-sulfur protein involved in human molybdenum cofactor biosynthesis. *Journal of Biological Chemistry*, *279*(33), 34721–34732.

Himo, F., & Eriksson, L. A. (1998). Catalytic mechanism of pyruvate formate-lyase (PFL). A theoretical study. *Journal of the American Chemical Society*, *120*(44), 11449–11455.

Horitani, M., Shisler, K., Broderick, W. E., Hutcheson, R. U., Duschene, K. S., Marts, A. R., et al. (2016). Radical SAM catalysis via an organometallic intermediate with an Fe-[5'-C]-deoxyadenosyl bond. *Science*, *352*(6287), 822–825.

Kolmeder, C. A., de Been, M., Nikkilä, J., Ritamo, I., Mättö, J., Valmu, L., et al. (2012). Comparative metaproteomics and diversity analysis of human intestinal microbiota testifies for its temporal stability and expression of core functions. *PLoS One*, 7(1), e29913.

Kulzer, R., Pils, T., Kappl, R., Huttermann, J., & Knappe, J. (1998). Reconstitution and characterization of the polynuclear iron-sulfur cluster in pyruvate formate-lyase-activating enzyme—Molecular properties of the holoenzyme form. *Journal of Biological Chemistry*, *273*(9), 4897–4903.

Kurokawa, K., Itoh, T., Kuwahara, T., Oshima, K., Toh, H., Toyoda, A., et al. (2007). Comparative metagenomics revealed commonly enriched gene sets in human gut microbiomes. *DNA Research*, *14*(4), 169–181.

Lanz, N. D., & Booker, S. J. (2015). Auxiliary iron-sulfur cofactors in radical SAM enzymes. *Biochimica et Biophysica Acta*, *1853*(6), 1316–1334.

Levin, B. J., Huang, Y. Y., Peck, S. C., Wei, Y., Martínez-del Campo, A., Marks, J. A., et al. (2017). A prominent glycyl radical enzyme in human gut microbiomes metabolizes *trans*-4-hydroxy-L-proline. *Science*, *355*(6325), pii: eaai8386.

Martínez-del Campo, A., Bodea, S., Hamer, H. A., Marks, J. A., Haiser, H. J., Turnbaugh, P. J., et al. (2015). Characterization and detection of a widely distributed gene cluster that predicts anaerobic choline utilization by human gut bacteria. *MBio*, *6*(2), pii: e00042-15.

Murib, J. H., & Ritter, D. M. (1952). Decomposition of nitrosyl disulfonate ion. I. Products and mechanism of color fading in acid solution. *Journal of the American Chemical Society*, *74*(13), 3394–3398.

O'Brien, J. R., Raynaud, C., Croux, C., Girbal, L., Soucaille, P., & Lanzilotta, W. N. (2004). Insight into the mechanism of the B_{12}-independent glycerol dehydratase from *Clostridium butyricum*: Preliminary biochemical and structural characterization. *Biochemistry*, *43*(16), 4635–4645.

Ollagnier, S., Mulliez, E., Schmidt, P. P., Eliasson, R., Gaillard, J., Deronzier, C., et al. (1997). Activation of the anaerobic ribonucleotide reductase from *Escherichia coli*—The essential role of the iron-sulfur center for *S*-adenosylmethionine reduction. *Journal of Biological Chemistry*, *272*(39), 24216–24223.

Romano, K. A., Martínez-del Campo, A., Kasahara, K., Chittim, C. L., Vivas, E. I., Amador-Noguez, D., et al. (2017). Metabolic, epigenetic, and transgenerational effects of gut bacterial choline consumption. *Cell Host & Microbe*, *22*(3), 279–290.

Selmer, T., Pierik, A. J., & Heider, J. (2005). New glycyl radical enzymes catalysing key metabolic steps in anaerobic bacteria. *Biological Chemistry*, *386*(10), 981–988.

Selvaraj, B., Pierik, A. J., Bill, E., & Martins, B. M. (2013). 4-Hydroxyphenylacetate decarboxylase activating enzyme catalyses a classical *S*-adenosylmethionine reductive cleavage reaction. *Journal of Biological Inorganic Chemistry*, *18*(6), 633–643.

Shisler, K. A., & Broderick, J. B. (2014). Glycyl radical activating enzymes: Structure, mechanism, and substrate interactions. *Archives of Biochemistry and Biophysics*, *546*, 64–71.

Shisler, K. A., Hutcheson, R. U., Horitani, M., Duschene, K. S., Crain, A. V., Byer, A. S., et al. (2017). Monovalent cation activation of the radical SAM enzyme pyruvate formate-lyase activating enzyme. *Journal of the American Chemical Society*, *139*(34), 11803–11813.

Sun, X. Y., Ollagnier, S., Schmidt, P. P., Atta, M., Mulliez, E., Lepape, L., et al. (1996). The free radical of the anaerobic ribonucleotide reductase from *Escherichia coli* is at glycine 681. *Journal of Biological Chemistry*, *271*(12), 6827–6831.

Tang, W. H. W., Wang, Z., Kennedy, D. J., Wu, Y., Buffa, J. A., Agatisa-Boyle, B., et al. (2015). Gut microbiota-dependent trimethylamine *N*-oxide (TMAO) pathway contributes to both development of renal insufficiency and mortality risk in chronic kidney disease. *Circulation Research*, *116*(3), 448–455.

Tokumoto, U., & Takahashi, Y. (2001). Genetic analysis of the *isc* operon in *Escherichia coli* involved in the biogenesis of cellular iron-sulfur protein. *Journal of Biochemistry*, *130*(1), 63–71.

Vey, J. L., Yang, J., Li, M., Broderick, W. E., Broderick, J. B., & Drennan, C. L. (2008). Structural basis for glycyl radical formation by pyruvate formate-lyase activating enzyme. *Proceedings of the National Academy of Sciences of the United States of America*, *105*(42), 16137–16141.

Wagner, A. F. V., Frey, M., Neugebauer, F. A., Schafer, W., & Knappe, J. (1992). The free-radical in pyruvate formate-lyase is located on glycine-734. *Proceedings of the National Academy of Sciences of the United States of America*, *89*(3), 996–1000.

Wang, Z., Klipfell, E., Bennett, B. J., Koeth, R., Levison, B. S., Dugar, B., et al. (2011). Gut flora metabolism of phosphatidylcholine promotes cardiovascular disease. *Nature*, *472*(7341), 57–63.

Wecksler, S. R., Stoll, S., Tran, H., Magnusson, O. T., Wu, S. P., King, D., et al. (2009). Pyrroloquinoline quinone biogenesis: Demonstration that PqqE from *Klebsiella pneumoniae* is a radical *S*-adenosyl-L-methionine enzyme. *Biochemistry*, *48*(42), 10151–10161.

Wei, Y., Funk, M. A., Rosado, L. A., Baek, J., Drennan, C. L., & Stubbe, J. (2014). The class III ribonucleotide reductase from Neisseria bacilliformis can utilize thioredoxin as a reductant. *Proceedings of the National Academy of Sciences of the United States of America*, *111*(36), E3756–E3765.

Yu, L., Blaser, M., Andrei, P. I., Pierik, A. J., & Selmer, T. (2006). 4-Hydroxyphenylacetate decarboxylases: Properties of a novel subclass of glycyl radical enzyme systems. *Biochemistry*, *45*(31), 9584–9592.

CHAPTER THREE

QueE: A Radical SAM Enzyme Involved in the Biosynthesis of 7-Deazapurine Containing Natural Products

Julia K. Lewis*, Nathan A. Bruender†, Vahe Bandarian*,[1]
*Department of Chemistry, University of Utah, Salt Lake City, UT, United States
†Department of Chemistry and Biochemistry, St. Cloud State University, St. Cloud, MN, United States
[1]Corresponding author: e-mail address: vahe@chem.utah.edu

Contents

Abstract

7-Carboxy-7-deazaguanine (CDG) is a common intermediate in the biosynthesis of 7-deazapurine-containing natural products. The biosynthesis of CDG from GTP requires three enzymes: GTP cyclohydrolase I, 6-carboxy-5,6,7,8-tetrahydropterin (CPH_4) synthase,

Methods in Enzymology, Volume 606
ISSN 0076-6879
https://doi.org/10.1016/bs.mie.2018.05.001

and CDG synthase (QueE). QueE is a member of the radical *S*-adenosyl-L-methionine (SAM) superfamily and catalyzes the SAM-dependent radical-mediated ring contraction of CPH_4 to generate CDG. This chapter focuses on methods to reconstitute the activity of QueE in vitro.

ABBREVIATIONS

Amp ampicillin
$AmSO_4$ ammonium sulfate
CDG 7-carboxy-7-deazaguanine
CPH_4 6-carboxy-5,6,7,8-tetrahydropterin
dAdo· 5′-deoxyadenosyl radical
DTT dithiothreitol
IPTG isopropyl β-D-1-thiogalactopyranoside
Kan kanamycin
KPi potassium phosphate
LB lysogeny broth
PIPES piperazine-*N*,*N*′-bis(2-ethanesulfonic acid)
PMSF phenylmethane sulfonyl fluoride
SAM *S*-adenosyl-L-methionine

1. INTRODUCTION

Pyrrolopyrimidine moieties are components of a structurally diverse group of natural products that share a 7-deazapurine core. These compounds are found in all kingdoms of life and their function(s) range from secondary metabolites produced by *Actinomyces*, to the modified bases, queuosine and archaeosine, in tRNA (see McCarty & Bandarian, 2012 for review). The first example of a deazapurine, toyocamycin, was isolated on the basis of its biological activity in 1956 (Nishimura, Katagiri, Sato, Mayama, & Shimaoka, 1956). To date, >30 compounds with deazapurine cores have been observed in Nature (McCarty, Krebs, & Bandarian, 2012).

Prior to elucidation of the biosynthetic pathway to deazapurines, radiotracer isotope experiments had demonstrated that deazapurines were derived from purines in a pathway that retains some of the carbons of the sugar, and C-2 from the base, but removes C-8 (Buff & Dairman, 1975; Burg & Brown, 1968; Elstner & Suhadolnik, 1971, 1975; Guroff & Strenkoski, 1966; Kuchino, Kasai, Nihei, & Nishimura, 1976; Reynolds & Brown, 1962, 1964; Smulson & Suhadolnik, 1967; Suhadolnik & Uematsu, 1970; Uematsu & Suhadolnik, 1970). Despite the structural diversity, it now appears that the deazapurine moiety is biosynthesized in three steps from guanosine triphosphate (GTP) (McCarty et al., 2012; McCarty, Somogyi, & Bandarian, 2009; McCarty, Somogyi, Lin, Jacobsen, & Bandarian, 2009; Miles, Roberts, McCarty, & Bandarian, 2014; Reader, Metzgar, Schimmel, & de Crécy-Lagard, 2004), and modified as appropriate by additional species-specific tailoring enzymes (Bai, Fox, Lacy, Van Lanen, & Iwata-Reuyl, 2000; Dowling et al., 2016; Frey, McCloskey, Kersten, & Kersten, 1988; Lee, Van Lanen, & Iwata-Reuyl, 2007; McCarty et al., 2012; Miles, McCarty, Molnar, & Bandarian, 2011; Miles, Myers, Kincannon, Britt, & Bandarian, 2015; Nelp, Astashkin, Breci, McCarty, & Bandarian, 2014; Nelp & Bandarian, 2015; Nelp, Song, Wysocki, & Bandarian, 2016; Okada, Harada, & Nishimura, 1976; Okada et al., 1979; Phillips et al., 2008; Reuter, Slany, Ullrich, & Kersten, 1991; Shindo-Okada, Okada, Ohgi, Goto, & Nishimura, 1980; Song, Nelp, Bandarian, & Wysocki, 2015; Van Lanen et al., 2005; Watanabe et al., 1997) (Fig. 1). In the first step of the reaction GTP cyclohydrolase I (GCH I) converts GTP to 7,8-dihydroneopterin triphosphate (H_2NTP), which is subsequently modified to 6-carboxy-5,6,7,8-tetrahydropterin (CPH_4) by CPH_4 synthase (QueD), prior to undergoing a ring contraction to form 7-carboxy-7-deazaguanine (CDG) catalyzed by CDG synthase (McCarty et al., 2012; McCarty, Somogyi, & Bandarian, 2009; McCarty, Somogyi, Lin, et al., 2009; Phillips et al., 2008). CDG likely serves as the starting point for all known deazapurines.

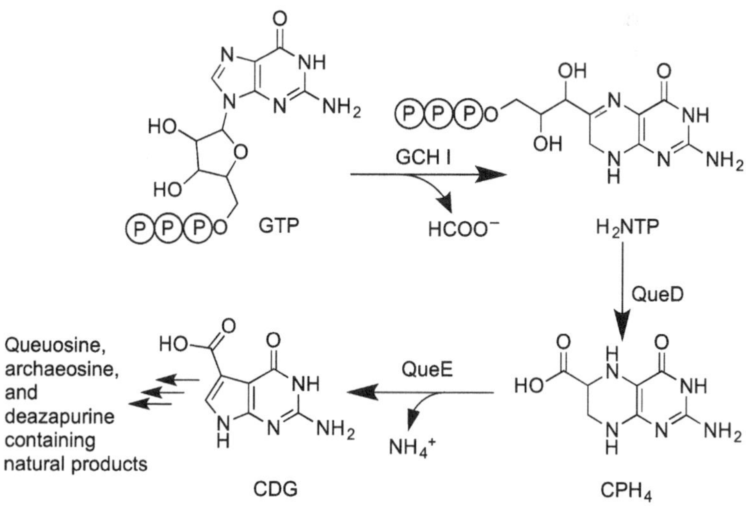

Fig. 1 Biosynthesis of the 7-deazapurine core. The 7-deazapurine core is obtained in three steps from GTP in Nature.

QueE catalyzes the radical-mediated ring contraction of CPH_4 to form CDG (McCarty, Somogyi, & Bandarian, 2009). It is a member of the radical *S*-adenosyl-L-methionine (SAM) enzymes, which were initially identified on the basis of a conserved CxxxCxxC motif (Frey, Hegeman, & Ruzicka, 2008; Sofia, 2001). The Cys sidechains in the conserved motif coordinate three iron atoms of a site-differentiated 4Fe—4S cluster. The fourth iron to binds SAM through its amino and carboxylate moieties (Berkovitch, Nicolet, Wan, Jarret, & Drennan, 2003; Walsby, Ortillo, Broderick, Broderick, & Hoffman, 2002; Walsby et al., 2005). In the +1 oxidation state, the iron–sulfur cluster reductively cleaves SAM to generate 5′-deoxyadenosyl radical (5′-dAdo·), which initiates chemistry by H-atom abstraction in most radical SAM enzymes (Moss & Frey, 2001). Recent bioinformatic studies suggest that radical SAM enzymes constitute a wide superfamily of >100,000 members (Brown & Babbitt, 2015) (Fig. 2).

In QueE, the reversible cleavage of SAM generates 5′-dAdo· which initiates that ring contraction in CPH_4 by abstracting a H-atom from C-6 of the substrate (Fig. 3) (McCarty et al., 2012). The substrate radical is thought to rearrange to a product radical that is quenched by H-atom abstraction from dAdo. Stereospecific abstraction of the C-7 *proR* hydrogen and elimination of the amino group complete the catalytic cycle and reforming SAM. In addition to detailed biochemical studies, QueE has been quite amenable to structural studies. Multiple high-resolution crystal structures of the protein provide detailed snapshots of the active site of the protein in the ± various ligands, including the cofactor, substrate, and substrate analogs (Bruender et al., 2017; Bruender, Young, & Bandarian, 2015; McCarty et al., 2012; McCarty, Somogyi, Lin, et al., 2009; Wilcoxen, Bruender, Bandarian, & Britt, 2018) (Fig. 4).

Studies with QueE require access to a number of reagents. As with other radical SAM enzymes, QueE activity requires the presence of a means of reducing the 4Fe—4S cluster to the catalytically active +1 state (Figs. 2 and 5). Dithionite is often used in the lab, but it has been demonstrated that higher activities are achieved with biological reducing systems, such as flavodoxin (Fld)/flavodoxin reductase (FPR)/NADPH (Bruender et al., 2015). However, it is also possible to bypass the FPR/NADPH and use dithionite-reduced Fld to achieve efficient reduction of the enzyme. In addition to a source of reductant, the substrate, CPH_4, is not commercially available and must be obtained by enzymatic synthesis from GTP using the biosynthetic enzymes. This chapter will focus on the methods to obtain each of these components to reconstitute in vitro activity of recombinant

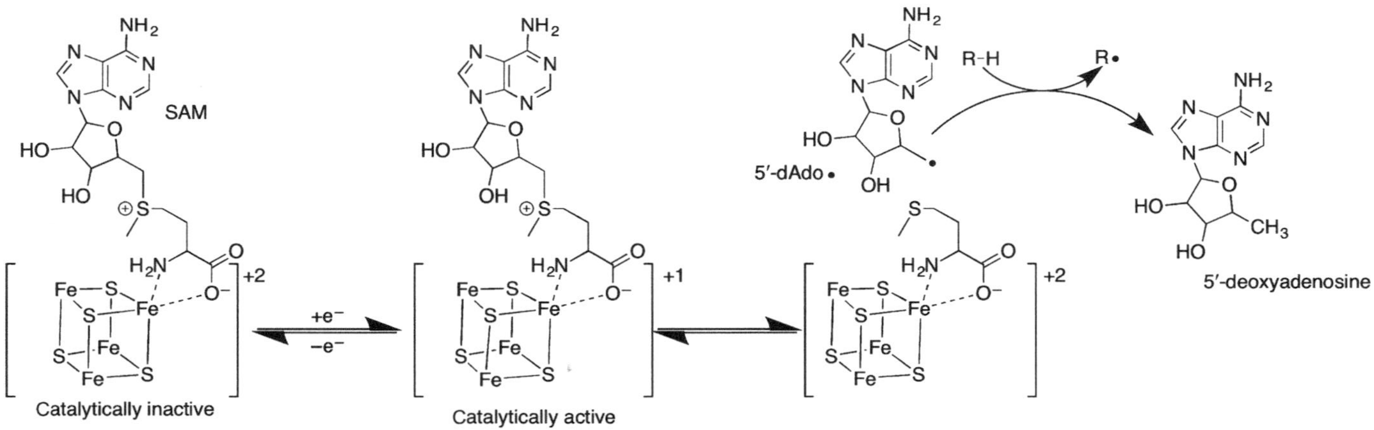

Fig. 2 Reductive cleavage of SAM. The [4Fe-4S] cluster in RS enzymes reductively cleaves SAM to 5′-dAdo·, which initiates the catalytic cycle.

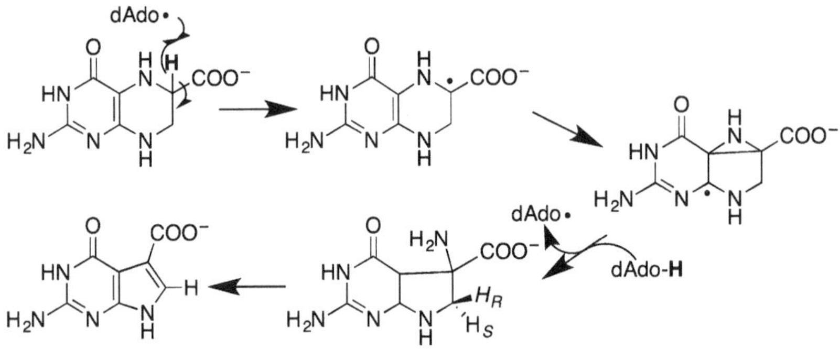

Fig. 3 Radical-mediated ring contraction mechanism of QueE. The dAdo· abstracts a hydrogen atom from the C-6 position of CPH_4 which is thought to undergo rearrangement to a *gem*-amino carboxylate product. The product maintains the C-7 *proS* proton of the substrate.

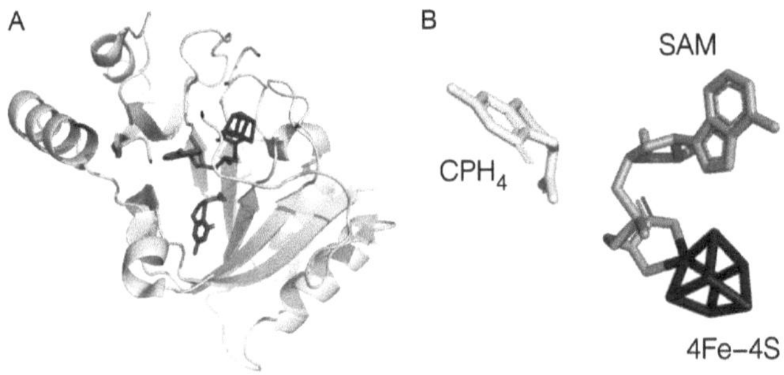

Fig. 4 Crystal structure of QueE. The overall structure of QueE with SAM, CPH_4, and [4Fe-4S] cluster (A), and close-up of the active site (B). SAM is bound to the cluster through its α-amino and α-carboxylate moieties.

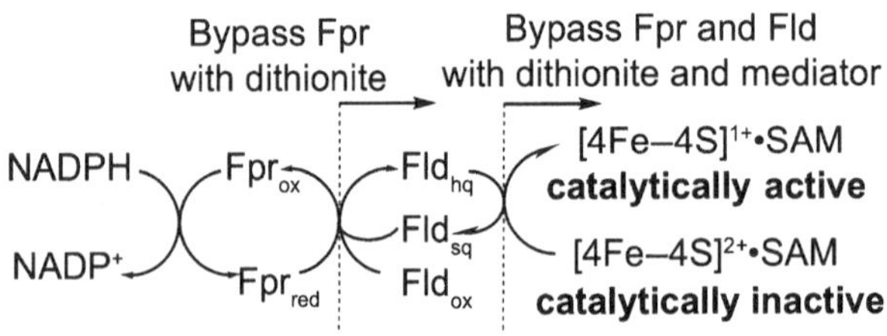

Fig. 5 Biological reduction pathways. The biological reduction pathway from NADPH or the chemical reduction via dithionite. *Reprinted from Bruender, N. A., Young, A. P., & Bandarian, V. (2015). Chemical and biological reduction of the radical SAM enzyme CPH4 synthase.* Biochemistry, *54(18), 2903–2910* with permission from American Chemical Society.

QueE. We note that while the foregoing discussion will focus on the *Bacillus subtilis* enzyme, the methods have also been applied successfully to the *B. multivorans* protein as well (Dowling et al., 2013) and are therefore general to QueE proteins.

2. EXPRESSION AND PURIFICATION OF QueE

The 4Fe—4S cluster that is required in all RS enzymes is often oxygen sensitive, requiring the enzymes to be purified and studied under strictly anaerobic conditions. Since QueE is highly overexpressed in *E. coli* hosts, a single step purification under anoxic conditions, using the His_6 affinity tag step suffices to purify sufficient protein for biochemical, structural, and spectroscopic studies.

2.1 Expression of QueE

The gene of *queE* was amplified from genomic DNA of *B. subtilis* (ATCC 23857D) and ligated into pET28a using the *Nde*I and *BamH*III restriction sites to make the plasmid pRM78 (McCarty, Somogyi, Lin, et al., 2009). The pET28a plasmid encodes kanamycin (Kan) resistance and expression is induced with isopropyl β-D-1-thiogalactopyranoside (IPTG). We find that it is necessary to express *queE* with pDB1282, which contains the genes *iscSUA-hscB-hscA-fdx-orf3* from *Azotobacter vinelandii* that are involved in assembly of the iron–sulfur clusters in nitrogenase, to enhance cluster content (Frazzon, Fick, & Dean, 2002). The pDB1282 plasmid contains ampicillin (Amp) resistance and the expression of the *isc* operon is induced with arabinose.

Protocol for Expression of *QueE* Gene

1. Cotransform BL21(DE3) *E. coli* cells with plasmids pRM78 and pDB1282. Plate cells on lysogeny broth (LB) agar plates containing 100 μg/mL Amp and 34 μg/mL Kan.
2. Prepare the LB media in 2.8-L Fernbach flasks by dissolving 10 g of tryptone, 5 g of yeast extract, and 5 g of NaCl in 1 L of deionized water. Autoclave 30 min and cool to room temperature before use.
3. Select a single colony from the transformation plate and inoculate 100 mL LB media containing both 100 μg/mL Amp and 34 μg/mL Kan (see step 1). Grow the starter culture at 37°C overnight with shaking at 200 rpm.
4. Use the overnight culture to inoculate the expression cultures. Add 10 mL of overnight culture to the 1 L of fresh LB in 2.8-L Fernbach flasks

supplemented with 100 μg/mL Amp and 34 μg/mL Kan. Allow the cells to grow at 37°C with shaking at 175 rpm to an optical density at 600 nm (OD_{600}) of ~0.3 and induce expression of the *isc* genes from the pDB1282 plasmid by addition of 0.5 g/L of solid arabinose. Also add 50 μ*M* (final concentration) $FeCl_3$ to ensure that enough iron is present for reconstitution of the cluster. Incubate the cultures with continued shaking at 37°C to OD_{600} of ~0.5 prior to inducing expression of QueE with IPTG (100 μ*M* final concentration). Shake the cultures at 120 rpm for 6 h after inducing expression of QueE.

5. Harvest the cells at 5000 × *g*, flash freeze in liquid nitrogen, and store at −80°C until use.

2.2 Purification of QueE

Purification of QueE should be performed in an anaerobic chamber under a nitrogen atmosphere. We use a Coy chamber, which typically has an atmosphere composed of 97% nitrogen and 3% hydrogen. Anaerobicity is achieved by allowing H_2 to react with O_2 on the surface of the palladium catalyst within the chamber to generate water. Excess moisture in the chamber is removed by both desiccant packs and a Coy dehumidifier. The anaerobicity of the chamber is monitored by a Coy cam-12 detector. The chamber is in a laboratory which is maintained at 22°C. Materials to be introduced into the chamber pass through an antechamber and made anaerobic by repeated cycling between vacuum and either N_2 or a 95%N_2/5%H_2 gas mix. We have modified the default cycle program for the antechamber so that it is vacuumed to 19.9 in Hg and refilled with N_2 four times prior to final round of evacuation and filling with mixed gas (95%N_2/5%H_2). To facilitate purifications steps, we have placed an ÄKTAprime plus inside the chamber. All tubes, tips, and other plasticware are placed in the glove box at least 24 h prior to use. Water and concentrated buffer stocks are prepared aerobically and stirred in the glove box for 12 h to exchange the gas. Cell disruption is carried out on ice and the centrifugation steps are at 4°C. Cells are disrupted by a Branson sonifier as indicated below. We maintain the horn of the sonifier in the glove box and attach it to a controller outside the glove box.

Protocol for Purification of QueE

1. Place 5–10 g of frozen cell pellet in a stainless steel beaker with a stir bar and 1 m*M* phenylmethane sulfonyl fluoride (PMSF) in the middle of a crystallization dish filled with ice and cycle into the glove box.

2. Add deoxygenated lysis solution (4 mL/g of wet cell paste) containing 0.05 *M* KPi (pH 7.4), 0.5 *M* KCl, 0.05 *M* imidazole, and 10 m*M* dithiothreitol (DTT).
3. While the suspension is stirred, lyse cells a total of 10 min, with 15 s bursts at 50% power that are followed by 45 s rests to allow the solution to chill in between.
4. Transfer the lysate into an Oakridge centrifuge tube with O-rings, tightly seal the tube inside the glove box, and remove from the chamber.
5. Centrifuge at 18,500 × *g* for 35 min at 4°C to pellet cell debris.
6. Charge a 5 mL HisTrapHP column (GE Healthcare) with $NiSO_4$ according to manufacturer's directions and equilibrate with lysis solution (step 2).
7. Transfer the centrifuge tube into the anaerobic chamber and carefully decant the supernatant.
8. Using a 50 mL SuperLoop, load the protein using an ÄKTAprime FPLC onto the column. Wash with lysis solution (step 2) until the UV trace returns to baseline (~50 mL). Elute the protein with a 50 mL gradient from 0.05 to 0.5 *M* imidazole in lysis buffer.
9. Pool the dark brown fractions containing QueE and exchange into a solution containing 0.05 *M* PIPES–NaOH (pH 7.4) and 10 m*M* DTT using a BioRad 10DG desalting column.
10. Concentrate to a volume of ~5–10 mL using an Amicon concentrator under N_2 with an YM-10 membrane (Millipore). N_2 can be brought into the glove box through a tube that is fed through a side nipple on the Coy chamber. Freeze the resulting protein in small aliquots in liquid nitrogen and store at −80°C. A typical SDS-PAGE of purified QueE is shown in Fig. 6.
11. Quantify the QueE stock by a Bradford assay (BioRad) with bovine serum albumin (BSA) as standard. We have empirically determined a Bradford concentration correction factor of 0.67 for QueE, which is based on the absolute amino acid content (McCarty et al., 2012).

2.3 Chemical Reconstitution of QueE

While as purified QueE has some intact 4Fe—4S cluster, it is necessary to carry out a chemical reconstitution of QueE under anaerobic conditions to obtain cofactor-replete enzyme. We have empirically determined that efficient reconstitution can be achieved when QueE is incubated with 8 *M* excess of iron ($FeCl_3$) and sulfide (Na_2S) as described below.

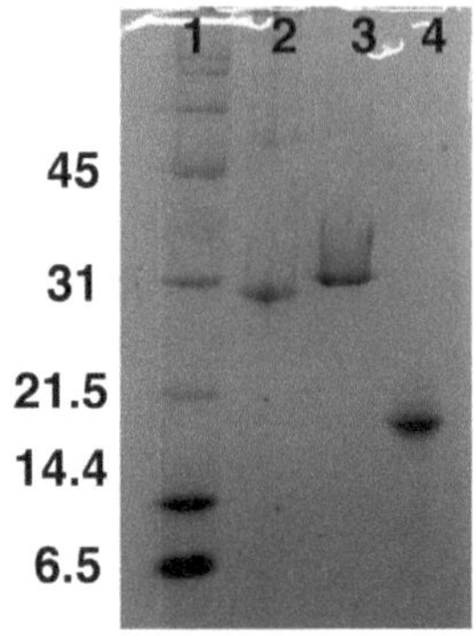

Fig. 6 SDS-PAGE of purified proteins. The gel shows (1) ladder, (2) QueE, (3) FPR, and (4) YkuN.

1. Weigh out ~20 mg of Na_2S and $FeCl_3$ into separate glass scintillation vials. Cap, and cycle into the anaerobic chamber.
2. Dissolve the Na_2S in water and the $FeCl_3$ in saturated sodium bicarbonate to a final concentration of 100 m*M*.
3. Add 8 *M* equivalents of $FeCl_3$ and Na_2S to the protein based on its corrected concentration determined by the Bradford assay (Section 2.2, step 11). Dropwise add the $FeCl_3$ and allow it to stir for about 3 min before adding the Na_2S.
4. Allow the reconstitution to proceed at room temperature for 4 h after Na_2S has been added.
5. At the end of the 4 h, the solution often has some precipitate. Remove the precipitate by centrifugation prior to desalting with a BioRad 10DG desalting column into solution of 0.05 *M* PIPES–NaOH (pH 7.4) containing 10 m*M* DTT.
6. Concentrate the desalted QueE using an Amicon concentrator under N_2 with an YM-10 membrane (Millipore) to a minimal volume (~1 mL).
7. Freeze the protein in liquid nitrogen as small aliquots and store at −80°C. Typical yield from this procedure is ~150 mg of QueE from 5 g of wet cell paste.
8. As needed, QueE can be further purified by gel filtration on a HiPrep 16/60 Sephacryl S-200 high-resolution column (GE Healthcare). Equilibrate the column with solution containing 0.05 *M* PIPES–NaOH (pH 7.4), 0.15 *M* NaCl, and 10 m*M* DTT. We usually calibrate the column with chymotrypsinogen A (25 kDa), ovalbumin (43 kDa), BSA (67 kDa), and aldolase (158 kDa).

9. Inject 1 mL of concentrated QueE onto the equilibrated column and collect fractions while washing with the loading buffer.
10. Analyze the fractions with an SDS-PAGE gel and compare to standards to determine the oligomeric state. We only collect fractions that correspond to dimeric QueE.
11. The fractions containing QueE are pooled and concentrated in the same manner as in step 6. The concentrated QueE is then aliquoted, flash frozen in liquid nitrogen, and stored at −80°C until use.

3. EXPRESSION AND PURIFICATION OF YkuN

The reduction of the Fe–S cluster in QueE to initiate the catalytic cycle can be accomplished by a number of methods (see Fig. 5). The simplest route is to use a strong reductant, such as sodium dithionite. However, we have shown that the biological reducing system, such as flavodoxin/flavodoxin reductase is a better choice for QueE, as significantly improves the activity (Bruender et al., 2015). We generally use the *B. subtilis* flavodoxin homolog, YkuN, in our assays to pass electrons to QueE either from dithionite, or flavodoxin reductase and NADPH.

3.1 Expression of *B. subtilis* Flavodoxin YkuN

The gene *ykuN* was amplified by PCR from *B. subtilis* subsp. *subtilis* str. 168. The gene was A-tailed and ligated into the pGEM Easy-T vector (Promega) for blue and white screening. The gene was then moved to the pET29a expression vector to generate pNB466 (Bruender et al., 2015).

Protocol for Expression of *B. subtilis* Flavodoxin YkuN

1. BL-21 (DE3) *E. coli* cells were transformed with pNB466 and plated onto LB agar plates containing 34 μg/mL Kan. Incubate at 37°C overnight.
2. Pick a colony from the LB agar plate to inoculate a 100 mL overnight starter culture containing 34 μg/mL Kan. Incubate at 37°C overnight with shaking at 200 rpm.
3. Add 10 mL of the overnight culture to 1 L of fresh LB in 2.8-L Fernbach flasks containing 1 L LB each and supplemented with 34 μg/mL Kan. Incubate at 37°C with shaking at 175 rpm until they reach OD_{600} ~ 0.5.
4. Induce expression of *ykuN* with 0.1 m*M* IPTG (final concentration) and add riboflavin to a final concentration of 0.1 m*M* at the time of induction to ensure that cofactor-replete protein is obtained. Allow cells to express the protein at 37°C overnight while shaking at 175 rpm.

5. Collect the cells after ~18 h by centrifugation at 5000 × *g* for 15 min at 4°C.
6. Flash freeze wet cell pastes in liquid nitrogen and store at −80°C until use.

3.2 Purification of YkuN

The expression vector for YkuN produces native protein, which is readily purified by successive anion exchange and hydrophobic interactions chromatographic steps as described below.

Protocol for the Purification of YkuN

1. In a stainless steel beaker with a stir bar, add 40 g of frozen cell paste and solution (4 mL/g of wet cell paste) containing 0.05 *M* Tris–HCl (pH 7.2), 1 m*M* EDTA, 2 m*M* DTT, 1 m*M* PMSF. Place the beaker in the middle of a crystallization dish filled with ice.
2. While the suspension is stirred, lyse cells with a Branson digital sonifier for total of 10 min, with 15 s bursts at 50% power that are followed by 45 s rests to allow the solution to chill in between.
3. Centrifuge at 18,000 × *g* for 35 min to remove cell debris.
4. Equilibrate a DEAE-Sepharose column (GE Healthcare) (11 × 3 cm) with 3 column volumes of the lysis solution (0.05 *M* Tris–HCl (pH 7.2), 1 m*M* EDTA, and 2 m*M* DTT) on an ÄKTAprime FPLC.
5. Load the cleared lysate via a 50 mL SuperLoop onto the column. Wash the column with the loading buffer until the UV absorbance is at the baseline. Elute YkuN with a 0.85 L linear gradient 0–1 *M* KCl in lysis solution.
6. Pool fractions containing YkuN on the basis of SDS-PAGE analysis of the yellow fractions (Flavodoxin (YkuN) binds FMN which is a yellow chromophore).
7. Slowly add an equal volume of 50 m*M* Tris–HCl (pH 7.0) containing 3 *M* ammonium sulfate ($AmSO_4$) and 2 m*M* DTT to the pooled YkuN fractions over 30 min.
8. Equilibrate a butyl-Sepharose column (GE Healthcare) (11 × 3 cm) with solution containing 0.05 *M* Tris–HCl (pH 7.0), 1.5 *M* $AmSO_4$, and 2 m*M* DTT on an ÄKTAprime FPLC.
9. Using a 50 mL SuperLoop load YkuN onto the column and wash with the loading buffer until the UV absorbance returns to baseline. Elute the protein with a 0.7 L linear gradient from the 1.5–0 *M* $AmSO_4$ in loading solution.

10. Identify YkuN-containing fractions by SDS-PAGE gel of yellow fractions. Fractions containing YkuN were pooled and dialyzed against buffer containing 20 m*M* HEPES-NaOH (pH 7.5), 0.1 *M* KCl, and 2 m*M* DTT at 4°C overnight.
11. Concentrate the protein to ~1 mL using Amicon concentrator under N_2 with an YM-10 membrane (Millipore). Freeze YkuN in small aliquots in liquid nitrogen and store at −80°C until use. See Fig. 6 for representative SDS-PAGE of purified YkuN.

4. EXPRESSION AND PURIFICATION OF FPR

FPR is used in tandem with YkuN to biologically reduce the [4Fe—4S] cluster of QueE by shuttling electrons that are ultimately derived from NADPH to QueE. This section describes expression and purification of FPR.

4.1 Expression of the *fpr* Gene

The gene for *E. coli fpr* was synthesized in codon-optimized form (Genscript), and cloned into *Hin*dIII and *Nde*I restriction sites of pET28a for expression of His_6-tagged protein. The sequence for the synthetic gene and the corresponding protein is shown in Fig. 7.

Protocol for *fpr* Gene Expression

1. Transform HMS174(DE3) *E. coli* cells with the *fpr* expression plasmid and plate onto an LB agar plate containing 34 μg/mL Kan. Put the LB agar plate at 37°C overnight.

*ATG*GGCAGCAGCCATCATCATCATCATCACAGCAGCGGCCTGGTGCCGCGCGGCAGC**CATATG**
GCTGATTGGGTAACAGGCAAAGTCACTAAAGTGCAGAACTGGACCGACGCCCTGTTTAGTCTCA
CCGTTCACGCCCCCGTGCTTCCGTTTACCGCCGGGCAATTTACCAAACTTGGCCTTGAAATCGAC
GGCGAACGCGTCCAGCGCGCCTACTCCTATGTAAACTCGCCCGATAATCCCGATCTGGAGTTTTA
CCTGGTCACCGTCCCCGATGGCAAATTAAGCCCACGACTGGCGGCACTGAAACCAGGCGATGAA
GTGCAGGTGGTTAGCGAAGCGGCAGGATTCTTTGTGCTCGATGAAGTGCCGCACTGCGAAACGC
TATGGATGCTGGCAACCGGTACAGCGATTGGCCCTTATTTATCGATTCTGCAACTAGGTAAAGATT
TAGATCGCTTCAAAAATCTGGTCCTGGTGCACGCCGCACGTTATGCCGCCGACTTAAGCTATTTGC
CACTGATGCAGGAACTGGAAAAACGCTACGAAGGAAAACTGCGCATTCAGACGGTGGTCAGTCG
GGAAACGGCAGCGGGGTCGCTCACCGGACGGATACCGGCATTAATTGAAAGTGGGGAACTGGA
AAGCACGATTGGCCTGCCGATGAATAAAGAAACCAGCCATGTGATGCTGTGCGGCAATCCACAG
ATGGTGCGCGATACACAACAGTTGCTGAAAGAGACCCGGCAGATGACGAAACATTTACGTCGCC
GACCGGGCCACATGACAGCGGAGCATTACTGGTAAT**AAGCTT**

Fig. 7 Sequence of codon-optimized FPR. The cloning sites are shown in **bold** and the translation start site is in *italic*.

2. Select a colony from plate to inoculate a 0.1 L starter culture containing 34 μg/mL Kan and incubate at 37°C overnight with shaking at 200 rpm.
3. Inoculate six 2.8-L Fernbach flasks containing 1 L LB each with 34 μg/mL Kan with 10 mL of overnight culture. Shake the cultures at 175 rpm at 37°C.
4. At an OD_{600} ~0.5 add 0.1 m*M* (final concentration) IPTG to induce expression. As with YkuN, add 0.1 m*M* riboflavin to ensure that cofactor-replete protein is obtained.
5. Harvest the cells after ~6 h by centrifugation at 5000 × *g* for 15 min at 4°C.
6. Flash freeze the cell paste in liquid nitrogen and store at −80°C until use.

4.2 Purification of FPR

FPR is produced as a His_6-tagged protein, which simplifies purification using Ni-NTA affinity chromatography. FPR is a flavoprotein and fractions containing the enzyme are readily identified by the bright yellow color due to the presence of the flavin chromophore and by SDS-PAGE.

Protocol for FPR Purification

1. Place ~5 g of cell paste and a stir bar in a stainless steel beaker. Add lysis buffer containing 0.02 *M* KPi (pH 7.2) and 0.5 *M* NaCl at a ratio of ~4 mL buffer per g of cell paste. Add solid PMSF to obtain 1 m*M* final concentration. Put the beaker in the center of a recrystallization dish filled with ice.
2. While stirring, sonicate the cells with 48 bursts of 15 s each at 50% power for 12 min with 45 s rest to minimize heating of the sample.
3. Transfer the lysate to Oakridge tubes and centrifuge at 18,000 × *g* for 35 min.
4. Charge a 1 mL Histrap column (GE Healthcare) with $NiSO_4$ according to manufacturer's protocol and equilibrate a 1 mL Histrap (GE Healthcare) with lysis buffer on an ÄKTAprime FPLC.
5. Load the clarified lysate onto the column using a 50 mL SuperLoop.
6. Wash the column with ~5 mL of lysis buffer until the UV absorbance is near baseline and stable.
7. Elute FPR with a 5 mL linear gradient in lysis buffer from 0.05 to 0.5 *M* imidazole. Pool the yellow fractions containing FPR.
8. Dialyze the pooled fraction against 4 L of 0.02 *M* HEPES–NaOH (pH 7.4) at 4°C for ~16 h. Change the dialysis buffer twice.
9. Concentrate the protein with Amicon concentrator with an YM-10 membrane (Millipore) under N_2. Use liquid nitrogen to flash freeze the protein aliquots and store at −80°C until use. See Fig. 6 for representative SDS-PAGE of purified FPR.

5. ENZYMATIC SYNTHESIS OF CPH_4 WITH GCH I AND QueD FROM GTP

The substrate for QueE, CPH_4, is not commercially available but can be synthesized enzymatically using GCH I and QueD (see Fig. 1). In this section, we describe protocols for preparation of GCH I and QueD, and the enzymatic preparation of CPH_4 from GTP. These methods can be adapted for the synthesis of deuterated isotopomers of CPH_4 with label at C7*proR* or C7*proS* (McCarty et al., 2012) (Fig. 8). The [7*R*-D] isotope of CPH_4 is generated by completing the first step of the reaction of GTP with GCH I in D_2O, followed by QueD in H_2O. The [7*S*-D] isotope of CPH_4 is made using 97% [U-D] GTP (Cambridge Isotope Laboratories, Inc.) as the starting material with the GCH I and QueD reactions run in H_2O. The C6-labeled isotope is obtained by reacting sepiapterin (Sigma), with QueD in D_2O. Fig. 8 summarizes the strategies for enzymatic synthesis of isotopologs of CPH_4.

5.1 Expression and Purification of GCH I

The gene for GCH I (*folE*) was amplified from the *E. coli* W3110 genomic DNA and cloned into the *Hind*III and *Nde*I site of pET28a to generate pJB163. GCH I is purified via anion exchange chromatography and hydrophobic interaction chromatography.

Protocol for Expression of GCH I

1. Transform BL21 (DE3) *E. coli* cells with pJB163 and plate onto an LB agar plate containing 34 μg/mL Kan. Incubate the plate at 37°C overnight.
2. Select a colony from the LB agar plate and use the colony to inoculate a 0.1 L starter culture containing 34 μg/mL Kan. Allow the culture to grow at 37°C overnight.
3. Use 10 mL of starter culture to inoculate six 2.8-L Fernbach flasks, each having 1 L of LB with 34 μg/mL Kan and 100 μ*M* $ZnSO_4$. Grow at 37°C and 200 rpm.
4. At an OD_{600} ~ 0.7 induce the cells with 100 μ*M* IPTG.
5. Harvest the cells after 4 h of growth by centrifugation at 5000 × *g* at 4°C. Flash freeze the cell paste in liquid nitrogen and store at −80°C until use.

Protocol for Purification of GCH I

1. GCH I is purified at 4°C under aerobic conditions using an ÄKTA FPLC.
2. Weigh out ~10 g of cell paste and place in a stainless steel beaker with a stir bar. Dissolve in lysis buffer (4 mL/g) containing 0.02 *M* Tris–HCl (pH 8.0), 10 m*M* DTT, and supplemented with 1 m*M* PMSF. Place beaker in the middle of a recrystallization dish packed with ice.

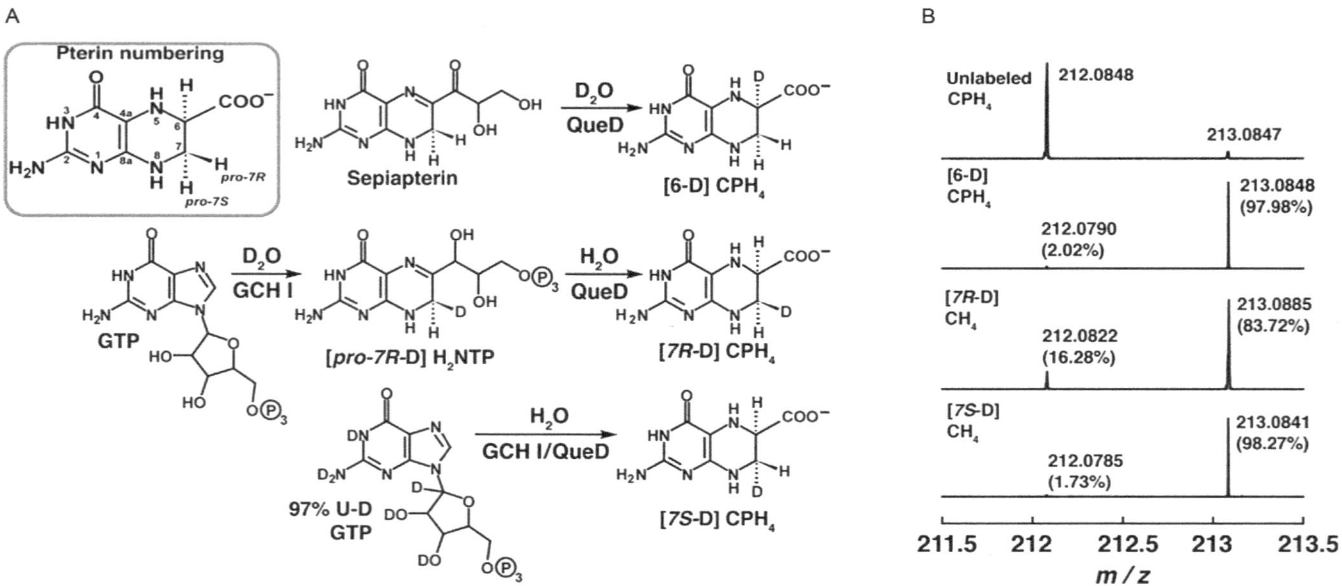

Fig. 8 Enzymatic synthesis of isotopologs of CPH_4. (A) Isotopologs of CPH_4 can be prepared enzymatically by carrying out the GCH I and QueD reactions in H_2O or D_2O. (B) Mass spectra of the synthetic isotopologs. *Reprinted from McCarty, R. M., Krebs, C., & Bandarian, V. (2012). Spectroscopic, steady-state kinetic, and mechanistic characterization of the radical SAM enzyme QueE, which catalyzes a complex cyclization reaction in the biosynthesis of 7-deazapurines.* Biochemistry, *52(1), 188–198 with permission from American Chemical Society.*

3. While stirring the cells, sonicate using a Branson digital sonifier with 40 bursts at 50% power for 15 s, followed by 45 s of rest to avoid heating the sample.
4. Pour the cell lysate into Oakridge tubes and centrifuge for 1 h at 18,000 × *g* and 4°C.
5. Equilibrate a Q-Sepharose column (GE Healthcare) (2.6 × 14.5 cm) with ~3 column volumes of the lysis buffer.
6. Use a SuperLoop to inject the clarified lysate onto the column.
7. Wash the column with ~2 column volumes of loading buffer until the UV trace returns to near baseline.
8. Elute the protein with 0.85 L linear gradient from 0 to 0.5 *M* KCl in loading buffer.
9. Identify the fractions containing GCH I by SDS-PAGE.
10. Pool the fractions containing GCH I and concentrate to ~50 mL using an Amicon concentrator with an YM-10 membrane (Millipore) under N_2.
11. Equilibrate a butyl-Sepharose column (GE Healthcare) (11 × 3 cm) with loading buffer containing 0.02 *M* Tris–HCl (pH 8.0), 1 *M* $AmSO_4$, and 10 m*M* DTT.
12. To the pooled fractions of GCH I from step 9 add an equal volume of buffer containing 20 m*M* Tris–HCl (pH 8.0), 2 *M* $AmSO_4$, and 10 m*M* DTT. This solution should be added slowly over a period of 30 min, while the pooled fractions are stirring on a stir plate.
13. Use a SuperLoop to inject the fractions of GCH I onto the column. After loading, wash the column with the loading buffer until the UV trace has returned to baseline.
14. Elute the protein over a 0.7 L linear from 1 to 0 *M* $AmSO_4$ in loading solution.
15. Identify and pool fractions containing GCH I by SDS-PAGE.
16. Dialyze the GCH I fractions against 4 L of buffer containing 20 m*M* Tris–HCl (pH 8.0) and 10 m*M* DTT at 4°C for ~16 h. Dialyze against fresh buffer a second time.
17. Concentrate the GCH I to a minimal volume using an Amicon concentrator under N_2 gas with a YM-10 membrane (Millipore).
18. Aliquot the protein and flash freeze with liquid nitrogen. Store at −80°C until use.

5.2 Expression of QueD

Native QueD from *E. coli* is expressed from a pET29-based plasmid with *ygcM* (b2765) cloned between *Hind*III and *Nde*I sites (McCarty, Somogyi, & Bandarian, 2009). The gene, *ygcM*, was PCR amplified from *E. coli* W3110 chromosomal DNA.

Protocol for Expression of QueD

1. Transform BL21(DE3) with vector containing *E. coli QueD* gene and incubate on an LB plate containing 34 μg/mL Kan at 37°C overnight.
2. Inoculate 6 L of LB media containing 34 μg/mL Kan with 10 mL from a starter culture prepared from a single colony picked from the transformation plate.
3. Grow at 37°C and 200 rpm to OD_{600} ~0.5. Induce the cells with 0.1 m*M* IPTG. Since QueD requires Zn^{2+} for activity, supplement the cultures with 100 μL of 1 *M* $ZnSO_4$ (dissolved in water) to give a final concentration of 100 μ*M*.
4. Continue to allow the cells to grow at 37°C for 4 h after induction.
5. Collect cells by centrifugation at 5000 × *g* at 4°C. Use liquid nitrogen to flash freeze the cell paste and store at −80°C until use.

5.3 Purification of QueD

QueD is purified by a combination of anion exchange and hydrophobic interaction chromatographic steps as follows. All purification steps are carried out 4°C under aerobic conditions.

1. In a stainless steel beaker containing a stir bar suspend ~10 g of cell paste in 40 mL of loading buffer containing 0.02 *M* Tris–HCl (pH 8) and 10 m*M* DTT, and supplemented with 1 m*M* PMSF. Place the beaker in a recrystallization dish packed with ice.
2. Lyse the cells by sonication while stirring using a Branson digital sonifier at with 48 bursts at 50% amplitude 45 s rests to allow the mixture to remain cold.
3. Clear the lysate by centrifugation at 18,000 × *g* for 35 min at 4°C.
4. Equilibrate a Q-Sepharose column (GE Healthcare) (2.6 × 14.5 cm) with loading buffer. Load the cleared lysate onto the column, wash with ~3 column volumes of the loading buffer and elute with a 0.8 L linear gradient from 0 to 0.5 *M* NaCl in the loading buffer.
5. Identify fractions containing QueD by SDS-PAGE analysis. Pool the fractions and slowly add solid $AmSO_4$ to 1 *M* over 20 min while stirring.

6. Equilibrate a butyl-sepharose CL-4B column (GE Healthcare) (2.6×14.5 cm) with a butyl loading buffer containing 0.02 *M* Tris·HCl (pH 8), 1 *M* $AmSO_4$, and 10 m*M* DTT.
7. Load the QueD onto the column, wash with ~2 column volumes of the loading buffer and elute with a 0.7 L linear gradient from 1 to 0 *M* $AmSO_4$ in loading buffer.
8. Identify fractions containing QueD by SDS-PAGE analysis, pool, and concentrate in an Amicon pressure concentration device fitted with an YM-10 membrane (Millipore).
9. Dialyze the protein against 4 L of 0.02 *M* HEPES–NaOH (pH 7.5) at 4°C for ~16 h with one buffer change.
10. The protein can be concentrated as necessary by Amicon filtration as before.
11. Freeze the protein in small aliquots in liquid nitrogen and store at −80°C.

5.4 Enzymatic Synthesis of CPH_4

CPH_4 is synthesized by the combined action of GCH I and QueD from GTP. The resulting CPH_4 is readily purified by anion exchange chromatography on DEAE-Sepharose. CPH_4 is light and oxygen sensitive. Therefore, the synthesis and purification of CPH_4 are carried out inside a Coy anaerobic chamber (5%H_2/95%N_2) and in the dark.

1. Set up a synthesis mixture that contains 0.05 *M* Tris–HCl (pH 8.0), 10 m*M* DTT, 0.5 m*M* GTP, 20 μ*M* GCH I, and 20 μ*M* QueD in a total volume of 20 mL. Stir the reaction 12 h under anaerobic conditions. The flask is kept dark by covering in foil.
2. Equilibrate a DEAE-Sepharose (11×3 cm) column (GE Healthcare) with 10 m*M* ammonium bicarbonate. The column and the fractions are also kept dark by covering in foil.
3. Load the reaction mixture onto the column and elute the CPH_4 isocratically with 10 m*M* ammonium bicarbonate.
4. CPH_4 typically elutes ~0.45 L and can be detected by UV absorbance at 280 nm of the eluate.
5. Fractions containing CPH_4 are pooled and lyophilized. The lyophilization vessel is also kept dark by covering with foil.
6. The lyophilized CPH_4 is resuspended in a minimal volume of 0.05 *M* Tris–HCl (pH 8.0) and flash frozen in small aliquots in liquid nitrogen and stored at −80°C until use.
7. CPH_4 is quantified using the extinction coefficient of tetrahydrobiopterin at 297 nm at pH 8.0 ($\varepsilon_{297} = 8.71 \times 10^3\ M^{-1}\ cm^{-1}$).

6. TYPICAL ASSAY CONDITIONS FOR QueE

QueE is assayed under anaerobic conditions using YkuN/Fpr/NADPH to reduce the protein. The conversion of CPH_4 to CDG is followed by LC–MS analysis of the reaction. CDG is quantified by comparison of the peak in the UV chromatograms to standards.

6.1 Protocol for Assaying QueE In Vitro

QueE is assayed in the anaerobic chamber using solutions that are prepared inside the chamber with anaerobic water. Assays are typically carried out in 0.05 *M* PIPES–KOH (pH 7.4) containing 10 m*M* DTT, 2 m*M* $MgSO_4$, and 2 m*M* SAM. The reduction of QueE is achieved by including 2 m*M* NADPH, 4 μ*M* FPR, and 10 μ*M* YkuN. Alternatively, dithionite (10 m*M*) and YkuN can also be used to bypass FPR/NADPH. The reactions at various time points can be quenched with addition of 30% (w/v) trichloroacetic acid (TCA), to a final concentration of 3% (w/v) TCA. The reaction mixtures are then analyzed by LC–MS.

7. LC–MS ANALYSIS OF TURNOVER BY CDG SYNTHASE

CDG is readily detected and quantified by HPLC methods. We use the UV trace to quantify the product and the MS detector to ensure the identity of the peak. Reverse phase chromatography at 0.2 mL/min on a C18 column is carried out on a Vanquish UPLC that is interfaced with an LTQ Orbitrap XL instrument (ThermoFisher) operated in positive ion mode, scanning *m/z* 75–1000 at a resolution of 100,000.

Protocol for LC–MS Detection of CDG

1. Centrifuge the quenched reaction mixtures to remove any precipitated protein.
2. Equilibrate a ThermoFisher Hypersil GOLD C18 column (2.1 × 150 mm, particle size of 1.9 μm) in Optima water (ThermoFisher) containing 0.1% Optima TFA (ThermoFisher).
3. Inject an aliquot (60 μL) of the reaction mixture and elute with a gradient to 100% Optima acetonitrile (ThermoFisher) containing 0.1% TFA with the following program: 0% acetonitrile from 0 to 3 min; 0%–10% acetonitrile from 3 to 38 min; 100% acetonitrile from 38.1 to 40.1 min; and 0% acetonitrile from 40.1 to 43.1 min to reequilibrate the column.

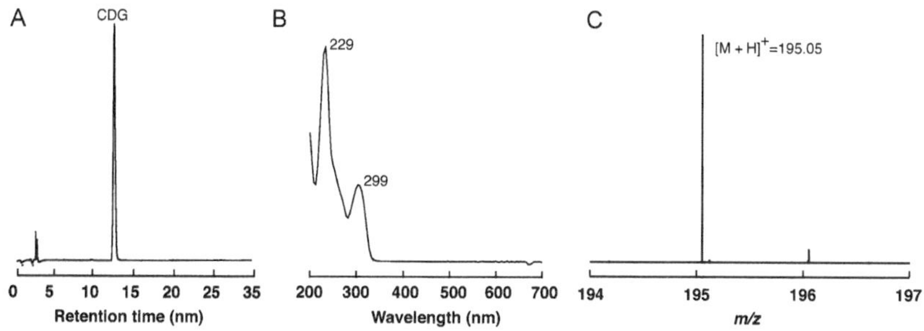

Fig. 9 LC–MS analysis of turnover by QueE. Production of QueE is monitored at 254 nm using a diode array detector (A). CDG has characteristic spectral features at 229 and 299 nm (B). The in-line mass spectrometer allows confirmation of the mass of the CDG (C).

4. A representative chromatogram is shown in Fig. 9. Under these conditions, CDG elutes at 13 min. The UV–visible spectrum of the product has distinct features at 230 and 300 nm. Identity of the peak as CDG is confirmed by the extracted ion chromatogram at m/z 195.05.
5. Quantify the CDG by comparison to known standards using the area under the UV–visible trace at 254 nm corresponding to CDG.

8. CONCLUSIONS

The wealth of structural and biochemical data available on QueE makes this an ideal system for exploring structure/function relationships in a radical SAM enzyme. In addition to cofactor-replete enzyme, studies of QueE require access to a source of reductant and substrate. In this chapter, we have described methods to all other components that are necessary for in vitro studies of the enzyme.

ACKNOWLEDGMENTS

This work was supported by the National Institutes of Health (GM072623 and GM120638).

REFERENCES

Bai, Y., Fox, D. T., Lacy, J. A., Van Lanen, S. G., & Iwata-Reuyl, D. (2000). Hypermodification of tRNA in thermophilic archaea. *Journal of Biological Chemistry*, *275*(37), 28731–28738.

Berkovitch, F., Nicolet, Y., Wan, J. T., Jarret, J. T., & Drennan, C. L. (2003). Crystal structure of biotin synthase, an S-adenosylmethionine-dependent radical enzyme. *Science*, *303*, 76–79.

Brown, S., & Babbitt, P. (2015). Using the structure-function linkage database to characterize functional domains in enzymes. *Current Protocols in Bioinformatics*, *48*, 2.10.1–2.10.16.

Bruender, N. A., Grell, T. A. J., Dowling, D. P., McCarty, R. M., Drennan, C. L., & Bandarian, V. (2017). 7-Carboxy-7-deazaguanine synthase: A radical S-adenosyl-L-methionine enzyme with polar tendencies. *Journal of the American Chemical Society*, *139*(5), 1912–1920.

Bruender, N. A., Young, A. P., & Bandarian, V. (2015). Chemical and biological reduction of the radical SAM enzyme CPH_4 synthase. *Biochemistry*, *54*(18), 2903–2910.

Buff, K., & Dairman, W. (1975). Biosynthesis of biopterin by two clones of mouse neuroblastoma. *Molecular Pharmacology*, *11*, 87–93.

Burg, A. W., & Brown, G. M. (1968). Biosynthesis of folic acid 8. Purification and properties of the enzyme that catalyzes the production of formate from carbon atom 8 of guanosine triphosphate. *Journal of Biological Chemistry*, *243*(9), 2349–2358.

Dowling, D. P., Bruender, N. A., Young, A. P., McCarty, R. M., Bandarian, V., & Drennan, C. L. (2013). Radical SAM enzyme QueE defines a new minimal core fold and metal-dependent mechanism. *Nature Chemical Biology*, *10*(2), 106–112.

Dowling, D. P., Miles, Z. D., Köhrer, C., Maiocco, S. J., Elliott, S. J., Bandarian, V., et al. (2016). Molecular basis of cobalamin-dependent RNA modification. *Nucleic Acids Research*, *44*, 9965–9976.

Elstner, E. F., & Suhadolnik, R. J. (1971). The biosynthesis of the nucleoside antibiotics 9. Purification and properties of guanosine triphosphate 8-formylhydroloase that catalyzes productino of formic acid from the ureido carbon of guanosine triphophate. *Journal of Biological Chemistry*, *246*(22), 6973–6981.

Elstner, E. F., & Suhadolnik, R. J. (1975). Guanosine triphosphate-8-formylhydrolase. *Methods in Enzymology*, *43*, 515–520.

Frazzon, J., Fick, J. R., & Dean, D. R. (2002). Biosynthesis of iron-sulphur clusters is a complex and highly conserved process. *Biochemical Society Transactions*, *30*, 680–685.

Frey, P. A., Hegeman, A. D., & Ruzicka, F. J. (2008). The radical SAM superfamily. *Critical Reviews in Biochemistry and Molecular Biology*, *43*(1), 63–88.

Frey, B., McCloskey, J., Kersten, W., & Kersten, H. (1988). New function of vitamin B_{12}: Cobamide-dependent reduction of epoxyqueuosine to queuosine in tRNAs of *Esherichia coli* and *Salmonella typhimurium*. *Journal of Bacteriology*, *170*(5), 2078–2082.

Guroff, G., & Strenkoski, C. A. (1966). Biosynthesis of pteridines and of phenylalanine hydroxylase cofactor in cell-free extracts of *Pseudomonas* species (ATCC 11299a). *Journal of Biological Chemistry*, *241*(20), 2220–2227.

Kuchino, Y., Kasai, H., Nihei, K., & Nishimura, S. (1976). Biosynthesis of the modified nucleoside Q in transfer RNA. *Nucleic Acids Research*, *3*(2), 393–398.

Lee, B. W. K., Van Lanen, S. G., & Iwata-Reuyl, D. (2007). Mechanistic studies of *Bacillus subtilis* QueF, the nitrile oxidoreductase involved in queuosine biosynthesis. *Biochemistry*, *46*(44), 12844–12854.

McCarty, R. M., & Bandarian, V. (2012). Biosynthesis of pyrrolopyrimidines. *Bioorganic Chemistry*, *43*, 15–25.

McCarty, R. M., Krebs, C., & Bandarian, V. (2012). Spectroscopic, steady-state kinetic, and mechanistic characterization of the radical SAM enzyme QueE, which catalyzes a complex cyclization reaction in the biosynthesis of 7-deazapurines. *Biochemistry*, *52*(1), 188–198.

McCarty, R. M., Somogyi, A., & Bandarian, V. (2009). *Escherichia coli* QueD is a 6-carboxy-5,6,7,8-tetrahydropterin synthase. *Biochemistry*, *48*(11), 2301–2303.

McCarty, R. M., Somogyi, A., Lin, G., Jacobsen, N. E., & Bandarian, V. (2009). The deazapurine biosynthetic pathway revealed: In vitro enzymatic synthesis of PreQ(0) from guanosine 5′-triphosphate in four steps. *Biochemistry*, *48*(18), 3847–3852.

Miles, Z. D., McCarty, R. M., Molnar, G., & Bandarian, V. (2011). Discovery of epoxyqueuosine (oQ) reductase reveals parallels between halorespiration and tRNA modification. *PNAS*, *108*(18), 7368–7372.

Miles, Z. D., Myers, W. K., Kincannon, W. M., Britt, R. D., & Bandarian, V. (2015). Biochemical and spectroscopic studies of epoxyqueuosine reductase: A novel iron–sulfur cluster- and cobalamin-containing protein involved in the biosynthesis of queuosine. *Biochemistry*, *54*(31), 4927–4935.

Miles, Z. D., Roberts, S. A., McCarty, R. M., & Bandarian, V. (2014). Biochemical and structural studies of 6-carboxy-5,6,7,8- tetrahydropterin synthase reveal the molecular basis of catalytic promiscuity within the tunnel-fold superfamily. *Journal of Biological Chemistry*, *289*(34), 23641–23652.

Moss, M., & Frey, P. A. (2001). The role of S-adenosylmethionine in the lysine 2,3-aminomutase reaction. *Journal of Biological Chemistry*, *262*(31), 14859–14862.

Nelp, M. T., Astashkin, A. V., Breci, L. A., McCarty, R. M., & Bandarian, V. (2014). The alpha subunit of nitrile hydratase is sufficient for catalytic activity and post-translational modification. *Biochemistry*, *53*(24), 3990–3994.

Nelp, M. T., & Bandarian, V. (2015). A single enzyme transforms a carboxylic acid into a nitrile through an amide intermediate. *Angewandte Chemie International Edition*, *54*(36), 10627–10629.

Nelp, M. T., Song, Y., Wysocki, V. H., & Bandarian, V. (2016). A protein-derived oxygen is the source of the amide oxygen of nitrile hydratases. *The Journal of Biological Chemistry*, *291*(15), 7822–7829.

Nishimura, H., Katagiri, K., Sato, K., Mayama, M., & Shimaoka, N. (1956). Toyocamycin, a new anti-candida antibiotics. *Journal of Antibiotics*, *9*, 60–62.

Okada, N., Harada, F., & Nishimura, S. (1976). Specific replacement of Q base in the anticodon of tRNA by guanine catalyzed by a cell-free extract of rabbit reticulocytes. *Nucleic Acids Research*, *3*(10), 2593–2603.

Okada, N., Noguchi, S., Kasai, H., Shindo-Okada, N., Ohgi, T., Goto, T., et al. (1979). Novel mechanism of post-translation modification of tRNA. *Journal of Biological Chemistry*, *254*(8), 3067–3073.

Phillips, G., Yacoubi, E. B., Lyons, B., Alvarez, S., Iwata-Reuyl, D., & de Crecy-Lagard, V. (2008). Biosynthesis of 7-deazaguanosine-modified tRNA nucleosides: a new role for GTP cyclohydrolase I. *Journal of Bacteriology*, *190*(24), 7876–7884.

Reader, J. S., Metzgar, D., Schimmel, P., & de Crécy-Lagard, V. (2004). Identification of four genes necessary for biosynthesis of the modified nucleoside queuosine. *Journal of Biological Chemistry*, *279*(8), 6280–6285.

Reuter, K., Slany, R., Ullrich, F., & Kersten, H. (1991). Structure and organization of *Escherichia coli* genes involved in biosynthesis of the deazaguanine derivative queuine. *Journal of Bacteriology*, *173*(7), 2256–2264.

Reynolds, J. J., & Brown, G. M. (1962). Enzymatic formation of the pteridine moiety of folic acid from guanine compounds. *Journal of Biological Chemistry*, *237*, 2713–2715.

Reynolds, J. J., & Brown, G. M. (1964). The biosynthesis of folic acid. IV. Enzymatic synthesis of dihydrofolic acid from guanine and ribose compounds. *Journal of Biological Chemistry*, *239*(1), 317–325.

Shindo-Okada, N., Okada, N., Ohgi, T., Goto, T., & Nishimura, S. (1980). Transfer ribonucleic acid guanine transglycosylase isolated from rat liver. *Biochemistry*, *19*(2), 395–400.

Smulson, M. E., & Suhadolnik, R. J. (1967). The biosynthesis of the 7-deazaadenine ribonucleoside, tubercidin, bv *Streptomyces tubercidicus*. *Journal of Biological Chemistry*, *242*(12), 2872–2876.

Sofia, H. J. (2001). Radical SAM, a novel protein superfamily linking unresolved steps in familiar biosynthetic pathways with radical mechanisms: Functional characterization using new analysis and information visualization methods. *Nucleic Acids Research*, *29*(5), 1097–1106.

Song, Y., Nelp, M. T., Bandarian, V., & Wysocki, V. H. (2015). Refining the structural model of a heterohexameric protein complex: Surface induced dissociation and ion mobility provide key connectivity and topology information. *ACS Central Science*, *1*(9), 477–487.

Suhadolnik, R. J., & Uematsu, T. (1970). Biosynthesis of the pyrrolopyrimidine nucleoside antibiotic, toyocamycin. *Journal of Biological Chemistry*, *245*(17), 4365–4371.

Uematsu, T., & Suhadolnik, R. J. (1970). Nucleoside antibiotics. VI. Biosynthesis of the pyrrolopyrimidine nucleoside antibiotic toyocamycin by *Streptomyces rimosus*. *Biochemistry*, *9*(5), 1260–1266.

Van Lanen, S. G., Reader, J. S., Swairjo, M. A., de Crécy-Lagard, V., Lee, B., & Iwata-Reuyl, D. (2005). From cyclohydrolase to oxidoreductase: Discovery of nitrile reductase activity in a common fold. *PNAS*, *102*(12), 4264–4269.

Walsby, C. J., Ortillo, D., Broderick, W. E., Broderick, J. B., & Hoffman, B. M. (2002). An anchoring role for FeS clusters: Chelation of the amino acid moiety of S-adenosylmethionine to the unique iron site of the [4Fe – 4S] cluster of pyruvate formate-lyase activating enzyme. *Journal of the American Chemical Society*, *124*(38), 11270–11271.

Walsby, C. J., Ortillo, D., Yang, J., Nnyepi, M. R., Broderick, W. E., Hoffman, B. M., et al. (2005). Spectroscopic approaches to elucidating novel iron – sulfur chemistry in the "Radical-SAM" protein superfamily. *Inorganic Chemistry*, *44*(4), 727–741.

Watanabe, M., Matsuo, M., Tanaka, S., Akimoto, H., Asahi, S., Nishimura, S., et al. (1997). Biosynthesis of archaeosine, a novel derivative of 7-deazaguanosine specific to archaeal tRNA, proceeds via a pathway involving base replacement on the tRNA polynucleotide chain. *Journal of Biological Chemistry*, *272*(32), 20146–20151.

Wilcoxen, J., Bruender, N. A., Bandarian, V., & Britt, R. D. (2018). A radical intermediate in *Bacillus subtilis* QueE during turnover with the substrate analogue 6-carboxypterin. *Journal of the American Chemical Society*, *140*(5), 1753–1759.

CHAPTER FOUR

TYW1: A Radical SAM Enzyme Involved in the Biosynthesis of Wybutosine Bases

Anthony P. Young, Vahe Bandarian[1]
Department of Chemistry, University of Utah, Salt Lake City, UT, United States
[1]Corresponding author: e-mail address: vahe@chem.utah.edu

Contents

Methods in Enzymology, Volume 606
ISSN 0076-6879
https://doi.org/10.1016/bs.mie.2018.04.024

Abstract

Transfer RNA is extensively modified by the actions of a variety of enzymes. The radical *S*-adenosyl-L-methionine enzyme TYW1 modifies tRNA^{Phe} forming the characteristic tricyclic ring via the condensation of carbons 2 and 3 of pyruvate. This chapter details methods that are required for studies of TYW1.

ABBREVIATIONS

Amp ampicillin
$AmSO_4$ ammonium sulfate
ATP adenosine 5′-triphosphate
dAdo$^{\bullet}$ 5′-deoxyadenosyl radical
DEPC diethyl pyrocarbonate
DTT dithiothreitol
EDTA ethylenediaminetetraacetic acid
imG-14 4-demethylwyosine
IPTG isopropyl β-D-1-thiogalactopyranoside
Kan kanamycin
Kpi potassium phosphate
LB lysogeny broth
m^1G *N*-methylguanosine
PIPES piperazine-*N*,*N*′-bis(2-ethanesulfonic acid)
PMSF phenylmethane sulfonyl fluoride
RS radical SAM
SAM *S*-adenosyl-L-methionine
tRNA transfer RNA

1. INTRODUCTION

Transfer RNA (tRNA) in all kingdoms of life is extensively modified by introduction of either simple changes that are installed by a single enzyme, or hypermodifications that require the coordinated actions of multiple enzymes (Cantara et al., 2011; Machnicka et al., 2013). Many modifications are ubiquitous and are conserved in both structure and position across nearly all organisms. Position 37 of $tRNA^{Phe}$ in archaea and eukaryotes bears a fluorescent and acid-labile tricyclic base, wyosine (Blobstein, Grunberger, Weinstein, & Nakanishi, 1973). Wyosine derivatives were initially discovered in yeast but have been found subsequently in other eukaryotes as well as in archaea (Kasai et al., 1976; Kasai, Yamaizumi, Kuchino, & Nishimura, 1979; McCloskey et al., 1987; RajBhandary & Chang, 1968; RajBhandary et al., 1967; RajBhandary, Stuart, Hoskinson, & Khorana, 1968). The biosynthetic pathway to the tricyclic wyosine derivatives was established in *Saccharomyces cerevisiae* using deletion strains (Fig. 1) (Noma, Kirino, Ikeuchi, & Suzuki, 2006; Waas, de Crécy-Lagard, & Schimmel, 2005). Subsequent studies have identified additional enzymes that are responsible for the biosynthesis of structurally related derivatives (de Crécy-Lagard et al., 2010; Kato et al., 2011; Noma et al., 2010, 2006).

Wyosine bases are all derived from modification of a genetically encoded guanosine nucleotide (Li, Nakanishi, Grunberger, & Weinstein, 1973) (Fig. 1), with the tricyclic base, 4-demethylwyosine (imG-14), serving as the precursor to all wyosine bases. ImG-14 is obtained in two steps that involve methylation of the G by TRM5 to form *N*-methylguanosine (m^1G) (Björk et al., 2001; Droogmans & Grosjean, 1987), followed by formation of the imidazoline ring containing imG-14 base by the action of a radical *S*-adenosyl-L-methionine (SAM) enzyme, TYW1 (Noma et al., 2006; Waas et al., 2005). ImG-14 is further tailored by the actions of TYW2–5 and TRM5b (de Crécy-Lagard et al., 2010; Noma et al., 2010, 2006; Waas et al., 2005). While SAM plays a role in each of these steps, the reaction catalyzed by TYW1 is unique in that SAM is used to activate the substrate rather than as a group donor.

TYW1 is a member of the radical SAM (RS) superfamily, which was identified on the basis of a conserved CxxxCxxC motif (Sofia, Chen, Hetzler, Reyes-Spindola, & Miller, 2001). The thiolates in this conserved sequence bind three atoms of a site-differentiated 4Fe–4S cluster, leaving the fourth iron with an open coordination site to bind SAM via its α-carboxyl and α-amino moieties (Berkovitch, Nicolet, Wan, Jarrett, & Drennan, 2004; Walsby et al., 2002). In the catalytically active +1 oxidation state, the iron–sulfur cluster catalyzes reductive cleavage of the bond between the 5′ carbon of ribose and the sulfonium moiety in SAM forming methionine and 5′-deoxyadenosyl radical (dAdo$^{\bullet}$). The dAdo$^{\bullet}$ initiates radical-mediated reactions by H-atom abstraction. While this paradigm is common to most RS enzymes, variations have been noted. For example, some members of the RS superfamily have been found to bind the cluster through a noncanonical sequence (Dowling et al., 2013; Kim et al., 2013; Zhang et al., 2010), others reductively cleave

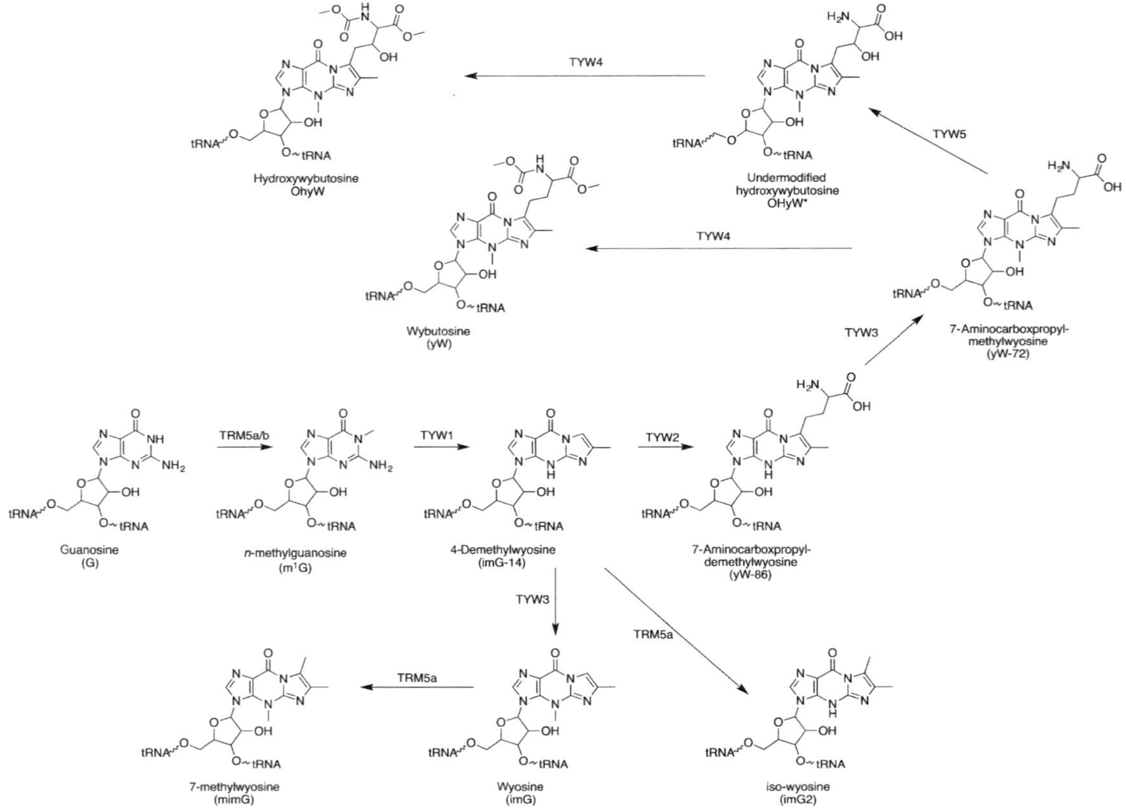

Fig. 1 Biosynthesis of all known wyosine bases. The biosynthetic pathway for wyosine derivatives starting from guanosine.

SAM to generate aminocarboxypropyl radical (Demick & Lanzilotta, 2011), and another group promotes radical additions with dAdo$^\bullet$ rather than H-atom abstraction (Mahanta, Fedoseyenko, Dairi, & Begley, 2013). We note that while TYW2 catalyzes transfer of aminocarboxylpropyl group of SAM to 7-aminocarboxypropyl demethylwyosine, this enzyme is not a member of the RS superfamily and the transfer occurs by polar mechanisms (Noma et al., 2006; Umitsu et al., 2009).

While gene deletion experiments had established a role for TYW1 in the conversion of m^1G to imG-14, the mechanism of this transformation had been enigmatic (Noma et al., 2006; Waas et al., 2005). The formation of the imidazoline ring from m^1G requires two new carbon atoms, the source of which was not known. The breakthrough that has allowed mechanistic studies of TYW1 was the identification of pyruvate as the substrate (Young & Bandarian, 2011). Stable isotope labeling experiments have since tracked the fate of carbon and hydrogen atoms and show that C2 and C3 are incorporated into the imidazoline base (Fig. 2) (Young & Bandarian, 2011, 2015).

Two high-resolution structures of TYW1 are available in the literature (Goto-Ito et al., 2007; Suzuki et al., 2007); however, neither structure reveals electron density for the anticipated RS cluster. Sequence alignments suggest that in addition to the CxxxCxxC motif, TYW1 has additional conserved Cys residues (Fig. 3), which appear to be clustered in the active site (Goto-Ito et al., 2007; Suzuki et al., 2007). Spectroscopic studies have revealed two magnetically distinct 4Fe–4S sites in TYW1, one of which is presumably the RS cluster and the other the second cluster, a so-called auxiliary cluster, bound to the additionally conserved Cys residues (Kathirvelu et al., 2017; Perche-Letuvée et al., 2012). There is a growing number of RS enzymes that catalyze complex transformations with the aid of both RS and auxiliary clusters. However, in most cases, the role of the auxiliary cluster remains unknown.

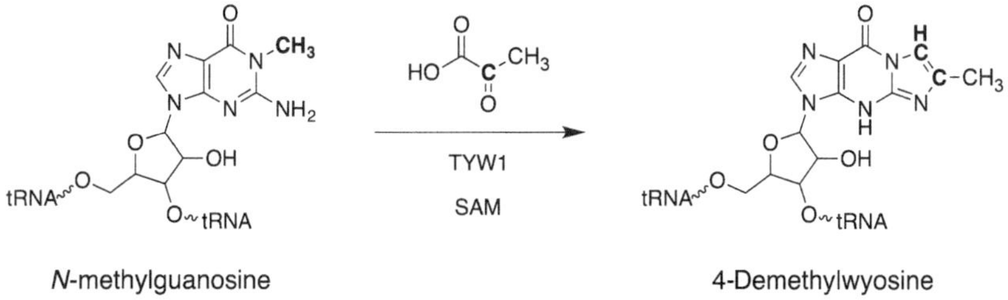

Fig. 2 Source of atoms in imG-14. TYW1 incorporates C2 and C3 from pyruvate into m^1G to form imG-14.

```
M. jannaschii    3 ------------PEEIYKILRKQRYQIDG-HTAVKLCGWVRKKMLEDKNCYKSKFYGIET
P. horikoshii   19 ----------NMPKEVAELFRKQHYEIVGRHSGVKLCHWLKKSLTEGRFCYKQKFYGIHS
M. barkeri      21 NRKEDTASLPFDIPDFKTLLKKQGYALAGRHSAVKTCLWLRHAMNGQGFCYKSKFYGVQS
H. sapiens     345 GRNEDGERRAMITPALREALTKQGYQLIGSHSGVKLCRWTKSMLRGRGGCYKHTFYGIES
S. cerevisiae  409 GADEIKQMVAKDSP-TYKNLTKQGYKVIGSHSGVKICRWTKNELRGKGSCYKKSLFNIAS
consensus      481                      ** * . * *..** * * .  .     ***   . . .

M. jannaschii   50 HRCIQCTPSVIWCQQNCIFCWRVLPRDIGIDISQIKEPKWEEPEVVYEKILAMHKRIIMG
P. horikoshii   69 HRCLQMTPVLAWCTHNCIFCWRPMENFLGTELP----QPWDDPAFIVEESIKAQRKLLIG
M. barkeri      81 HRCLQMTPTLM-CNQRCLFCWRPTEVPVPAP------EEWDSPEKIVEESIACQRKLITG
H. sapiens     405 HRCMETTPSLA-CANKCVFCWRHHTNPVGTEWR----WKMDQPEMILKEAIENHQNMIKQ
S. cerevisiae  468 SRCMELTPSLA-CSSKCVFCWRHGTNPVSKNWR----WEVDEPEYILENALKGHYSMIKQ
consensus      541  **.  ** .  *   *.****     .            . *  .    .      ..

M. jannaschii  110 YAGVLDRVGEKKFKEALEPKHVAISLSGEPTLYPYLDELIKIFHKNGFTTFVVSNGILTD
P. horikoshii  125 YKGNPK-VDKKKFEEAWNPTHAAISLSGEPMLYPYMGDLVEEFHKRGFTTFIVTNGTIPE
M. barkeri     134 FGGSPN-ALKERWLEGNEPNNVAISLSGEPTFYPYLPELIKEYEKRGFTTFLVTNGTVPE
H. sapiens     460 FKGVPG-VKAERFEEGMTVKHCALSLVGEPIMYPEINRFLKLLHQCKISSFLVTNAQFPA
S. cerevisiae  523 MRGVPG-VIAERFAKAFEVRHCALSLVGEPILYPHINKFIQLLHQKGITSFLVCNAQHPE
consensus      601   *        ..  .      *.** ***  ** .   .        ..*.* *.

M. jannaschii  170 VIEKIE-----PTQLYISLDAYDLDSYRRICGG-KKEYWESILNTLDILKEK---KRTCI
P. horikoshii  184 RLEEMIKEDKLPTQLYVSITAPDIETYNSVNIPMIPDGWERILRFLELMRDL--PTRTVV
M. barkeri     193 MFAKVS-----PSQLYMSLDAPDLETYLKVCQPRSPALWEKINESLSLMKQK--KSRTVI
H. sapiens     519 EIRNLE----PVTQLYVSVDASTKDSLKKIDRPLFKDFWQRFLDSLKALAVK--QQRTVY
S. cerevisiae  582 ALRNIV----KVTQLYVSIDAPTKTELKKVDRPLYKDFWERMVECLEILKTVQNHQRTVF
consensus      661     .       .***.*. *        .        *      *  .       **

M. jannaschii  221 RTTLIRGYND-DILKFVELYERADVHFIELKSYMHVGYSQ---KRLKKEDMLQHDEILKL
P. horikoshii  242 RLTLVKGENMHSPEKYAKLILKARPMFVEAKAYMFVGYSR---NRLTINNMPSHQDIREF
M. barkeri     246 RTTLVKGENLFKPEAYARLMEKASPDFVEIKAYMHLGFSR---LRLDRSAMPTHEEVLEF
H. sapiens     573 RLTLVKAWNVDELQAYAQLVSLGNPDFIEVKGVTYCGESSA--SSLTMAHVPWHEEVVQF
S. cerevisiae  638 RLTLVKGFNMGDISAYADLVQRGLPSFIEVKGATFSGSSDGNGNPLTMQNIPFYEECVNF
consensus      721 * **... *      .  *   .   *.* *     * *      *    .    .

M. jannaschii  277 AKMLDEN-----SSYKLIDDSEDSRVALLQNENRKI--NPK---L---------------
P. horikoshii  299 AEALVKH----LPGYHIEDEYEPSRVVLIMRDDVDPQGTGVEGRFIKH------------
M. barkeri     303 SQELAKH-----LGYEVADESEISRVVLLSKDGRKSPVNGKKISC---------------
H. sapiens     631 VHELVDL----IPEYEIACEHEHSNCLLIAHRKFKIVVNGGHGSIITA------------
S. cerevisiae  698 VKAFTTELQRRGLHYDLAAEHAHSNCLLIADTKFKI--NGEWHTHIDFDKFFVLLNSGKD
consensus   781               * .  .   *   *.

H. sapiens     675 ---------------------------SRSSSRNMKIVVDQKRSAQRIIWPELLTGHYLV
S. cerevisiae  756 FTYMDYLEKTPEWALFGNGGFAPGNTRVYRKDKKKQNKENQETTTRETPLPPIPA-----
consensus      841

H. sapiens     708 PVKEALIPRTQDIRERTNQRLFLDVEII
S. cerevisiae      ----------------------------
consensus      901
```

Fig. 3 Sequence alignment of TYW1. The alignment of TYW1 sequences from *Methanocaldococcus jannaschii* (*M. jannaschii*), *Pyrococcus horikoshii* (*P. horikoshii*), *Methanosarcina barkeri* (*M. barkeri*), *Homo sapiens* (*H. sapiens*), and *Saccharomyces cerevisiae* (*S. cerevisiae*) is shown. The cysteine residues that coordinate the iron–sulfur clusters are highlighted in *black boxes*. The N-terminal sequence of both *H. sapiens* and *S. cerevisiae* is truncated, and only the regions of the protein that are similar to the archaeal protein are shown.

A current mechanistic paradigm for TYW1 that is consistent with biochemical and spectroscopic studies is shown in Fig. 4. Pyruvate is proposed to be bound to the active site by forming a Schiff base to a conserved and catalytically essential Lys residue (Young & Bandarian, 2011, 2015). Reductively cleaved SAM provides dAdo$^{\bullet}$ to initiate catalysis by abstracting a

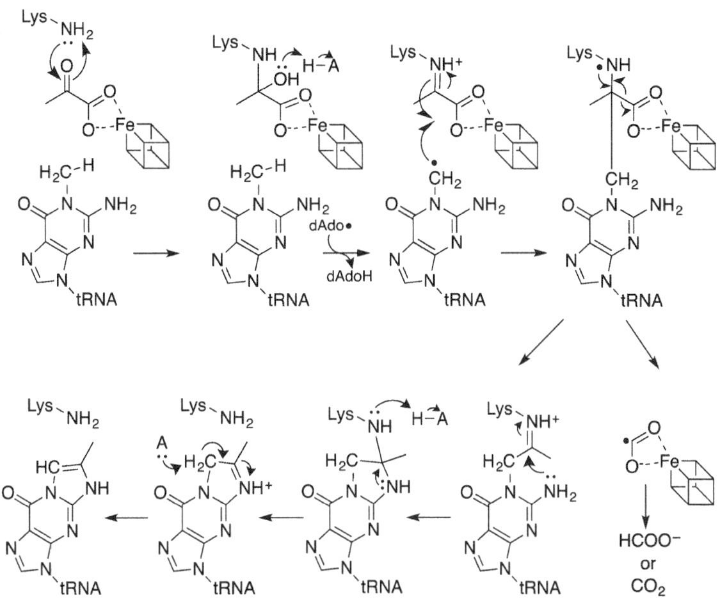

Fig. 4 Mechanistic paradigm for TYW1. Reductive cleavage of SAM forms 5′-dAdo•, which is used to activate the substrate by H-atom transfer. Pyruvate is proposed to bind to TYW1 as a Schiff base to a conserved Lys residue. The auxiliary cluster is shown as a *cube*.

hydrogen atom from m^1G. The radical adds to the immobilized pyruvate, which subsequently undergoes decarboxylation. Transimination to form the imidazoline ring completes the catalytic cycle. The auxiliary cluster may active the pyruvate for decarboxylation by serving as either an electron sink or donor to the departing "COO^-."

Studies of TYW1, in addition to cofactor-replete TYW1, also require production of m^1G containing tRNA substrate. This chapter will outline our methods for preparation and characterization of TYW1.

2. EXPRESSION AND PURIFICATION OF TYW1

The extreme oxygen sensitivity of RS enzymes often necessitates use of purification tags that allow for rapid enrichment of the recombinant enzyme from crude mixtures. All in vitro studies with TYW1 have utilized homologs from thermophilic organisms (Goto-Ito et al., 2007; Kathirvelu et al., 2017; Perche-Letuvée et al., 2012; Suzuki et al., 2007; Young & Bandarian, 2011, 2015), which permit the use of a heat step that denatures many of the proteins in the expression host. After this step, hydrophobic interactions and affinity chromatography using the genetically encoded purification tag allow pure TYW1 that is free nucleic acid contamination to be obtained. Methods for the expression and purification of TYW1 from *Methanocaldococcus jannaschii* are presented in this section.

2.1 Expression of TYW1 Gene

To maximize the expression of TYW1, the gene for TYW1 from *M. jannaschii* was obtained in codon-optimized form (GenScript) (see Fig. 5 for codon-optimized sequence) and cloned between the *Nde*I and *Hind*III restriction sites of a modified pET28 vector, containing a *Tobacco Etch Virus nuclear-inclusion-a endopeptidase* (TEV) protease cleavage site (Thoden & Holden, 2005), creating pAY613. This vector contains a T7 promoter that is inducible with IPTG (isopropyl β-D-1-thiogalactopyranoside) and a kanamycin (Kan) resistance marker. We often cotransform the expression host with the plasmid pDB1282 to aid in assembly of the Fe–S clusters. This plasmid contains *iscSUA-hscB-hscA-fdx-orf3* from *Azotobacter vinelandii* (Frazzon, Fick, & Dean, 2002), which have been shown to be involved in biosynthesis of iron–sulfur clusters. The genes are maintained on pARA13 plasmid, which has an ampicillin (Amp) resistance marker and is inducible by arabinose. Cotransformation with pDB1282 leads to a higher level of iron–sulfur cluster biogenesis proteins, which supplement the endogenous *Escherichia coli* iron–sulfur cluster biogenesis machinery.

Our general protocol entails overexpression of TYW1 in *E. coli* strain BL21(DE3) grown in lysogeny broth (LB) containing Kan and Amp to select for bacteria containing the plasmids for the expression of TYW1 and the *isc* operon. Protein expression is induced by the addition of IPTG (pAY613) and arabinose (pDB1282), and protein expression is carried out overnight at 37°C with decreased shaking rate.

2.1.1 Protocol for Expression of TYW1 *Gene*

1. Optimal expression is achieved by always starting from a freshly transformed *E. coli* cells. The plates are discarded after 1 week.
2. Make LB (Lennox) agar plates by combining 5 g of tryptone, 2.5 g yeast extract, 2.5 g NaCl, and 7.5 g agar with 500 mL of water. Autoclave for 30 min and allow to cool to approximately 50°C. Add 500 μL of 34 mg/mL Kan (in water) and 500 μL of 100 mg/mL Amp (in water) to the LB agar. Pour the plates and allow to solidify. This same recipe is used for preparation of LB liquid media, except that agar is excluded.

DNA sequence

ATGGGCAGCAGCCATCATCATCATCATCACAGCAGC**GAGAACCTGTACTTCCAGGGC***CATATGA*
TCCCGGAAGAAATCTATAAAATTCTGCGTAAACAGCGCTACCAAATTGATGGCCATACCGCAGT
GAAACTGTGCGGTTGGGTTCGTAAGAAAATGCTGGAAGACAAAAACTGTTACAAATCAAAATTC
TACGGCATCGAAACCCACCGCTGCATTCAGTGTACGCCGTCGGTTATCTGGTGCCAGCAAAATT
GCATTTTCTGTTGGCGTGTCCTGCCGCGATATTGGTATCGACATTTCACAAATCAAAGAACC
GAAATGGGAAGAACCGGAAGTGGTTTACGAAAAAATCCTGGCAATGCATAAACGTATCATCATG
GGCTACGCTGGTGTTCTGGATCGCGTCGGCGAGAAAAAATTTAAAGAAGCGCTGGAACCGAAAC
ATGTGGCCATCAGTCTGTCCGGTGAACCGACCCTGTATCCGTACCTGGATGAACTGATCAAAAT
TTTCCACAAAAACGGCTTTACCACGTTCGTCGTGAGCAATGGTATTCTGACCGATGTTATCGAA
AAAATTGAACCGACGCAGCTGTATATCAGTCTGGACGCGTATGATCTGGACTCCTACCGTCGCA
TTTGCGGCGGTAAAAAAGAATACTGGGAATCTATCCTGAACACCCTGGATATTCTGAAAGAGAA
AAAACGTACGTGTATCCGCACCACGCTGATTCGTGGCTATAATGATGACATCCTGAAATTTGTC
GAACTGTACGAACGCGCTGATGTGCATTTCATTGAACTGAAAAGCTATATGCACGTGGGTTACT
CTCAGAAACGTCTGAAAAAAGAAGATATGCTGCAACACGACGAAATCCTGAAACTGGCGAAAAT
GCTGGATGAAAACAGCTCTTATAAACTGATTGATGACTCAGAAGACTCGCGTGTTGCCCTGCTG
CAGAACGAAAATCGCAAAATTAATCCGAAACTGTAA*AAGCTT*

Expressed protein sequence

MGSSHHHHHHSS**ENLYFQG**HMIPEEIYKILRKQRYQIDGHTAVKLCGWVRKKMLEDKNCYKSKF
YGIETHRCIQCTPSVIWCQQNCIFCWRVLPRDIGIDISQIKEPKWEEPEVVYEKILAMHKRIIM
GYAGVLDRVGEKKFKEALEPKHVAISLSGEPTLYPYLDELIKIFHKNGFTTFVVSNGILTDVIE
KIEPTQLYISLDAYDLDSYRRICGGKKEYWESILNTLDILKEKKRTCIRTTLIRGYNDDILKFV
ELYERADVHFIELKSYMHVGYSQKRLKKEDMLQHDEILKLAKMLDENSSYKLIDDSEDSRVALL
QNENRKINPKL*

Cleaved protein sequence

GHMIPEEIYKILRKQRYQIDGHTAVKLCGWVRKKMLEDKNCYKSKFYGIETHRCIQCTPSVIWC
QQNCIFCWRVLPRDIGIDISQIKEPKWEEPEVVYEKILAMHKRIIMGYAGVLDRVGEKKFKEAL
EPKHVAISLSGEPTLYPYLDELIKIFHKNGFTTFVVSNGILTDVIEKIEPTQLYISLDAYDLDS
YRRICGGKKEYWESILNTLDILKEKKRTCIRTTLIRGYNDDILKFVELYERADVHFIELKSYMH
VGYSQKRLKKEDMLQHDEILKLAKMLDENSSYKLIDDSEDSRVALLQNENRKINPKL*

Fig. 5 Codon-optimized sequence of TYW1. The *top* shows the codon-optimized TYW1 sequence, expressed by pAY613 with the TEV protease site shown *bolded* and the restriction sites used to insert the gene in *italics*. The *middle* shows the translated protein sequence expressed by pAY613 with the TEV protease site shown *bolded*. The *bottom* shows the final sequence of TYW1 purified in this study with the extra amino acids not encoded by the original TYW1 gene are shown in *bold italics*.

3. Cotransform *E. coli* cell line BL21(DE3) T1 phage resistant (New England Biolabs C2527) with pAY613 and pDB1282. Plate transformation mixture on an LB (Lennox) agar plate containing 34 μg/mL Kan and 100 μg/mL Amp and incubate at 37°C overnight.
4. Make 1 L of LB as described in step 2. Transfer 125 mL to an autoclaved 250-mL Erlenmeyer flask aseptically.
5. Use a single colony from the transformation plate to inoculate a 125-mL culture of LB containing 34 μg/mL Kan and 100 μg/mL Amp in a 250-mL Erlenmeyer flask. Grow cells overnight at 37°C with shaker set to 225 rpm.

6. Prepare 6 L of LB (as described in step 2 above) and distribute equally into six 2.8-L Fernbach flasks.
7. Inoculate 1 L of LB (in 2.8-L Fernbach flasks) containing 34 μg/mL Kan and 100 μg/mL Amp with 10 mL of the overnight culture.
8. Shake the flasks at 175 rpm at 37°C to an optical density at 600 nm (OD_{600}) of ~0.7.
9. Add 0.5 mL of 0.1 *M* iron(III)chloride (in water) to give a concentration of 50 μ*M* along with 0.5 g solid arabinose (0.05% w/v) and induce expression by addition of 100 μL of 1 *M* IPTG (in water) to give a final concentration of 100 μ*M*.
10. Immediately upon induction, the shaking rate is reduced to 125 rpm and the cultures are incubated at 37°C for 18 h.
11. Harvest cells by centrifugation at 5000 × *g* and flash-freeze cell pellet in liquid nitrogen.
12. Typical yield from this procedure is 3 g wet cell paste from each Liter of media. Store cells at −80°C.

2.2 Purification of TYW1

Purification of TYW1 should be performed in an anaerobic chamber under a nitrogen atmosphere. We use a Coy chamber, which typically has an atmosphere composed of 97% nitrogen and 3% hydrogen. Anaerobicity is achieved by allowing H_2 to react with O_2 on the surface of the palladium catalyst within the chamber. Excess moisture in the chamber is removed by both a desiccant pack within the chamber and a Coy dehumidifier. The anaerobicity of the chamber is monitored by a Coy cam-12 detector. The chamber is in a laboratory which is maintained at 22°C. Samples are introduced into the glove box through an antechamber, which is automatically cycled through vacuum and gas filling cycles. We have modified the default cycle of the antechamber so that it is vacuumed to 19.9 inHg and refilled with N_2 four times prior to final vacuum and fill cycle where a mixed gas ($95\%N_2/5\%H_2$) is used. To facilitate purification steps, we have placed an AKTAprime plus inside the chamber. All tube, tips, and other plasticware are placed in the glove box at least 24 h prior to use. Water and concentrated buffer stocks are stirred in the glove box for 12 h. Cell disruption is carried out on ice, and the centrifugation steps are at 4°C. Cells are disrupted by a Branson sonifier as indicated below. We maintain the horn of the sonifier in the glove box and attach it to a controller maintained outside the glove box.

The purification of TYW1 from *M. jannaschii* is greatly facilitated by a heat denaturation step, which removes the majority of the contaminating native *E. coli* proteins improving protein quality. Next, hydrophobic chromatography is utilized to remove *E. coli* nucleic acids that copurify with TYW1 and potentially interfere with protein characterization. The final purification step is affinity chromatography utilizing the N-terminal His_6. This step serves to purify the protein by removing a small number of contaminants, but it also concentrates the protein from 0.2 to 0.05 L. At this point the purification of the His_6-TYW1 is complete, and protein may be frozen after desalting to remove imidazole and salt.

2.2.1 Protocol for Purification of His_6-TYW1 Protein

1. In an anaerobic chamber prepare the following buffers using water, solid salts, and concentrated buffer stocks that have been stirred in the chamber for 24 h: *lysis buffer*—20 m*M* tris(hydroxymethyl)aminomethane hydrochloride (pH 8) (Tris–HCl); *butyl column buffer*—20 m*M* Tris–HCl (pH 8) containing 1 *M* ammonium sulfate ($AmSO_4$); *HisTrap loading buffer*—50 m*M* potassium phosphate (pH 7.4) (Kpi) containing 0.5 *M* KCl, and 50 m*M* imidazole; *HisTrap elution buffer*—50 m*M* Kpi (pH 7.4) containing 0.5 *M* KCl, and 0.5 *M* imidazole; *desalting buffer*—50 m*M* piperazine-*N*,*N'*-bis(2-ethanesulfonic acid)–sodium hydroxide (pH 7.4) (PIPES–NaOH) containing 2 m*M* dithiothreitol (DTT).
2. Place 18 g of frozen cell pellet, a stir bar, and solid phenylmethane sulfonyl fluoride (PMSF) (to a final concentration of 1 m*M*) in a stainless steel beaker and transfer into the anaerobic chamber along with a crystallization dish packed with ice.
3. Add 0.15 L of anaerobic lysis buffer to the beaker and place beaker in the crystallization dish with the ice surrounding the beaker on all sides. Place crystallization dish on a stir plate and stir the suspension.
4. Disrupt cells using a Branson sonifier. Cell disruption is achieved with 48 bursts of 15 s each, with 60 s intervals to allow sample time remain cold. The amplitude of the bursts is kept at 50%.
5. Transfer lysate to a centrifuge tube that has a lid with an O-ring, cap the lid tightly, and centrifuge at 18,000 × *g* for 30 min at 4°C.
6. Transfer the centrifuge tube back into the anaerobic chamber and decant the supernatant into a 0.25-L bottle and cap the bottle tightly.
7. Remove the sealed bottle from the glove box and place in a water bath set to 80°C for 30 min.

8. Transfer the bottle into an ice bath for 15 min.
9. Reintroduce the bottle back into the anaerobic chamber and transfer the lysate to centrifuge tubes. Cap the lid tightly and centrifuge at 18,000 × *g* for 30 min at 4°C.
10. Reintroduce the centrifuge bottle back into the anaerobic chamber and transfer the clarified lysate to a measuring cylinder and record the volume.
11. Transfer the clarified lysate to a clean beaker containing a stir bar.
12. Slowly add solid $AmSO_4$ to the stirring lysate to a final concentration of 1 *M* and stir to completely dissolve.
13. All column chromatography steps are carried out using a AKTAprime plus instrument with a UV detector set to 280 nm.
14. Manually pack a butyl-Sepharose FF column (11 × 3 cm) (GE Healthcare) and equilibrate with butyl column buffer.
15. Load the lysate onto the equilibrated column and wash with the loading buffer until the absorbance has returned to baseline. This typically takes 2–3 column volumes.
16. Elute adsorbed protein with a step gradient to 100% lysis buffer and collect brown fractions.
17. Using the AKTAprime plus equilibrate a 5 mL HisTrap chelating HP column (GE Healthcare) charged with nickel chloride according to manufacturer's directions with HisTrap loading buffer.
18. Load the pooled fractions from the butyl column directly onto the equilibrated HisTrap chelating column. When all lysate is loaded wash the column with 2–3 column volumes of HisTrap loading buffer until the absorbance has returned to baseline.
19. Elute adsorbed protein with a *step* gradient to 100% HisTrap elution buffer, collect, and pool the protein that elutes.
20. Equilibrate a Bio-Gel P-6 DG gravity column (8 × 3 cm) with 100 mL of desalting buffer.
21. Load up to 35 mL of protein onto the equilibrated column at one time. Start collecting protein at 14 mL and stop collecting at 50 mL. If more than 35 mL of protein needs to be desalted, reequilibrate the column with 100 mL of desalting buffer and repeat.
22. At this point the His_6-TYW1 can be flash-frozen and stored at −80°C, or be treated with TEV protease to remove the purification tag.

2.3 Cleavage of the His_6 Tag With TEV Protease

Expression of TYW1 from the pAY613 plasmid places a TEV protease site between the His_6 affinity tag and the protein. Following cleavage, native

TYW1 can be obtained by chromatography of the sample on a HisTrap column to remove the tag, as well as uncleaved protein, and TEV protease. Expression and purification of the TEV protease are described in Section 3.

The buffer to perform the cleavage should contain DTT, which was introduced in the previous desalting step. The TEV protease cleavage is carried out overnight in the anaerobic chamber with stirring. Following chromatography on the HisTrap column, the protein is desalted to remove imidazole and place the protein in buffer that is compatible with either freezing or proceeding to the chemical reconstitution of the iron–sulfur clusters.

2.3.1 Protocol for TEV Protease Cleavage of the His_6 Tag

1. In an anaerobic chamber prepare the following buffers using water, salts, and concentrated buffer stocks that have been stirred in the chamber for 24 h: *HisTrap loading buffer*—50 m*M* Kpi (pH 7.4) containing 0.5 *M* KCl and 50 m*M* imidazole; *desalting buffer*—50 m*M* PIPES–NaOH (pH 7.4) containing 2 m*M* DTT.
2. Combine His_6-TYW1 purified as described in Section 2.2 in a beaker with a stir bar and ~1 mL of TEV protease (~120 μ*M*) and stir overnight at room temperature.
3. Equilibrate a 5-mL HisTrap chelating HP column (GE Healthcare) charged with nickel chloride according to manufacturer's directions with HisTrap loading buffer using the AKTAprime plus.
4. Load the TYW1 TEV mixture onto the equilibrated column and collect the protein that flows through the column.
5. When all the protein has been loaded, wash the column with HisTrap loading buffer until the absorbance returns to baseline, collecting all protein that elutes during the wash.
6. Pool all protein that did not bind to the column.
7. Equilibrate a Bio-Gel P-6 DG gravity column (8 × 3 cm) with 100 mL of desalting buffer.
8. Load up to 35 mL of protein onto the equilibrated column at one time. Start collecting protein at 14 mL and stop collecting at 50 mL. If more than 35 mL of protein needs to be desalted, reequilibrate the column with 100 mL of desalting buffer and repeat.
9. Protein eluting from the desalting column can be flash-frozen at −80°C or undergo chemical reconstitution.

2.4 Chemical Reconstitution of the Iron–Sulfur Clusters

TYW1 as isolated is inactive and requires chemical reconstitution to obtain cofactor-replete protein. We have determined empirically using the increase of absorbance at 420 nm that 4 h is the optimal reconstitution time for TYW1. At longer incubation times, we find significantly more aggregated protein upon gel filtration, suggesting that the protein may undergo irreversible loss of activity when reconstitution time is extended.

Chemical reconstitution of iron–sulfur clusters is performed in a reducing anaerobic environment in the presence of excess iron and sulfide. We have found that a 10-fold molar excess of $FeCl_3$ and Na_2S over the concentration of TYW1 is ideal. To determine the protein concentration a Bradford assay is performed. A correction factor, for the Bradford assay, is determined by amino acid analysis, we use UC Davis Genome Center Molecular Structure Facility. Following anaerobic reconstitution, any precipitate is removed by centrifugation, and the protein is concentrated and subjected to size exclusion chromatography, which separates the aggregated protein from monomeric TYW1.

2.4.1 Protocol for Chemical Reconstitution of Iron–sulfur clusters

1. In an anaerobic chamber prepare the following buffer using water, salts, and concentrated buffer stocks that have been stirred in the chamber for 24 h: *gel filtration buffer*—50 m*M* PIPES–NaOH (pH 7.4) containing 150 m*M* KCl, and 2 m*M* DTT.
2. To determine the quantity of Fe and S to use in the reconstitution, an accurate protein concentration is essential. We routinely use a Bradford assay (Boston BioProducts) with BSA as standard to quantify TYW1. An empirically determined correction factor of 0.32 (Young & Bandarian, 2015), which is based on the absolute amino acid content of the protein preparation, is applied to the Bradford assays.
3. Weigh out solid iron(III)chloride and sodium sulfide into separate glass vials and cycle into the anaerobic chamber.
4. Add enough anaerobic water to obtain solutions of iron(III)chloride and sodium sulfide at final concentration of 1 *M*.
5. Transfer the protein solution to a beaker with a stir bar and add 10-fold molar excess of iron(III)chloride, followed by sodium sulfide.
6. Stir the reconstitution mixture at room temperature for 4 h.
7. Aliquot the reconstitution mixture into multiple Eppendorf tubes and remove precipitated protein by centrifuging for 10 min at 16,000 × *g* at room temperature.
8. Combine the supernatants from the centrifugation.
9. Equilibrate a Bio-Gel P-6 DG gravity column (8 × 3 cm) with 100 mL of gel filtration buffer.

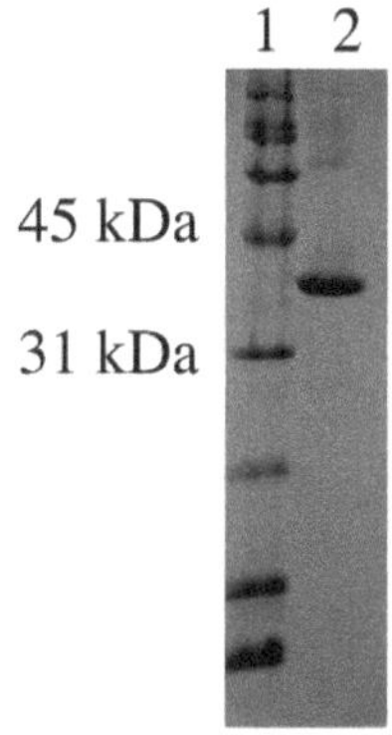

Fig. 6 SDS-PAGE gel showing purified TYW1. A gel showing TYW1 with BioRad broad range standard (cat # 161-0317) in lane 1 and TYW1 in lane 2.

10. Load up to 35 mL of protein onto the equilibrated column at one time. Start collecting protein at 14 mL and stop collecting at 50 mL. If more than 35 mL of protein needs to be desalted, reequilibrate the column with 100 mL of gel filtration buffer, and repeat.
11. Concentrate the protein using an Amicon concentrator with a YM-10 membrane (Millipore) under nitrogen gas to a volume of ~1 mL. This is readily accomplished by bringing in nitrogen gas through one of the nipple ports in the glove box. We find that this allows for rapid protein concentration without exposing the protein to oxygen.
12. Equilibrate a Sephacryl 16/60 S-200 column (GE Healthcare) in gel filtration buffer.
13. Desalt TYW1 in 1 mL aliquots over the equilibrated column.
14. Pool and concentrate the brown fractions that elute at 65 mL, corresponding to monomeric protein, using an Amicon concentrator with a YM-10 membrane under nitrogen gas.
15. Aliquot, flash-freeze, and store the concentrated protein at −80°C. Typical yield of homogeneous TYW1 is 1 mg/g wet cell paste. Fig. 6 shows a sample of TYW1 on a SDS-PAGE gel purified in this manner

3. EXPRESSION AND PURIFICATION OF TEV PROTEASE

The TEV protease recognizes the sequence ENLYFQ/G (Carrington & Dougherty, 1988; Kapust, Tözsér, Copeland, & Waugh, 2002). TYW1 is cloned into pET28t, which is a modified pET28a vector containing the TEV protease cut site in place of thrombin cut site (Thoden & Holden, 2005), such that cleavage would lead to amino acids GH on the N-terminus of TYW1. The expression vector for TEV protease (prTEV) and the cell line used for expression of TEV protease (SG13000009) were a gift from Professor Hazel Holden at the University of Wisconsin–Madison.

3.1 Expression of TEV Protease

Expression of TEV protease is carried out in an *E. coli* cell line SG13000009 containing pLacIRARE. The pLacIRARE plasmid already in the strain has chloramphenicol (Cam) resistance, while the prTEV plasmid has Amp resistance. Expression of TEV protease is carried out in LB with cells initially grown at 37°C and then the temperature is decreased to 16°C when protein expression is induced by addition of IPTG.

3.1.1 Protocol for Expression of TEV Protease

1. Make LB (Lennox) agar plates as described in Section 2.1.
2. Transform *E. coli* cell line SG1300009 pLacIRARE with prTEV and plated transformants on an LB agar plate containing 34 μg/mL Cam (dissolved in ethanol) and 100 μg/mL Amp. Incubate plates at 37°C overnight.
3. Make 1 L of LB as described in Section 2.1. Use a single colony to inoculate a 125 mL of LB in a 250-mL Erlenmeyer flask containing 34 μg/mL Cam and 100 μg/mL Amp. Flasks are grown at 37°C shaking at 225 rpm overnight.
4. Make 6 L of LB and distribute among six 2.8-L Fernbach flasks. Autoclave for 30 min and allow to cool.
5. Add 34 μg/mL Cam and 100 μg/mL Amp to each flask and inoculate with 10 mL of the overnight culture.
6. Allow the cultures to shake at 175 rpm and at 37°C to OD_{600} ~0.6.
7. Add 1 mL of a 1 *M* stock solution of IPTG (1 m*M* final concentration) to induce expression of the protein. At this point, reduce the temperature to 16°C and allow the cultures to grow for a further 18 h.
8. Harvest cells by centrifugation at 5000 × *g* and flash-freeze cell pellet in liquid nitrogen. Typical yield from this protocol is ~5 g/L media.

3.2 Purification of TEV Protease

The TEV protease used to cleave His_6-TYW1 is readily purified using a His_6 affinity tag as described below. The presence of the His_6 affinity tag allows TEV protease to be easily purified from *E. coli* and it also provides a method for purification of the protease when it is used to cleave the tag from a protein of interest.

The purification procedure involves lysing the cells and then passing the clarified lysate over a HisTrap column charged with nickel. Bound protein is eluted via a gradient and the fractions containing concentrated protein are dialyzed against HisTrap loading buffer. Finally, glycerol is added, and the protein is stored at −20°C.

3.2.1 Protocol for Purification of TEV Protease

1. Prepare the following buffers using water at 4°C: *lysis buffer*—50 m*M* Kpi (pH 7.4) containing 0.5 *M* KCl, 50 m*M* imidazole, and 20% glycerol (v/v); *elution buffer*—50 m*M* Kpi (7.4) containing 0.5 *M* KCl, 0.5 *M* imidazole, and 20% glycerol (v/v).
2. Place 30 g of frozen cells, a stir bar, and solid PMSF in a stainless steel beaker and add 0.15 L of lysis buffer.
3. Place beaker in a crystallization dish with ice surrounding the beaker on all sides. Place crystallization dish on a stir plate and adjust speed so that the cells are stirring.
4. Disrupt cells by 48 bursts of sonication for 15 s at 50% power, followed by 60 s breaks to avoid heating the samples.
5. Transfer lysate to a centrifuge tube that has a lid with an O-ring, cap the lid tightly, and centrifuge at 18,000 × *g* for 30 min at 4°C.
6. Using an AKTA FPLC at 4°C, equilibrate a 5-mL HisTrap chelating HP column (GE Healthcare) charged with nickel chloride, according to manufacturer's directions, with lysis buffer.
7. Transfer clarified lysate from centrifuge tube to a clean beaker.
8. Load clarified lysate onto the equilibrated column. When all lysate is loaded wash the column with lysis buffer (10-column volumes) until the absorbance has returned to baseline.
9. Elute adsorbed protein with a gradient to 100% elution buffer over 50 mL.
10. Identify TEV protease containing fractions by SDS-PAGE and pool the more concentrated fractions.
11. Dialyze pooled TEV protease against 4 L of lysis buffer.
12. Change dialysis buffer twice after at least 4 h.
13. Add glycerol to the pooled TEV protease to a final concentration of 50% (v/v), aliquot, and flash-freeze in liquid nitrogen. Store aliquots at −20°C.

4. EXPRESSION AND PURIFICATION OF TRM5

The first step in the biosynthesis of wybutosine is methylation of the G at position 37 of $tRNA^{Phe}$ (Björk et al., 2001; Droogmans &

Grosjean, 1987; Noma et al., 2010). The synthesis of a tRNA substrate for TYW1 using in vitro transcription requires a methyl transferase capable of methylating the tRNA. In our studies, we have routinely employed TRM5 from *M. jannaschii*. This section will describe expression and purification of *Mj*TRM5.

4.1 Expression of *Mj*TRM5 Gene

The gene for *Mj*TRM5, which was codon optimized for expression in *E. coli*, was synthesized (GenScript) and cloned between the *Nde*I and *Hind*III sites of pET29a vector to create pAY423 (see Fig. 7 for optimized gene sequence). This vector contains a T7 promoter that is inducible with IPTG and a Kan resistance marker. The protein is expressed in LB media at 37°C.

4.1.1 Protocol for Expression of TRM5 gene

1. Make LB (Lennox) agar plates containing 34 μg/mL Kan as described in Section 2.1.
2. Transform *E. coli* cell line BL21(DE3) with pAY423, which contains codon-optimized TRM5 gene in pET29a vector. Plate transformants on LB agar plates containing 34 μg/mL Kan and incubate at 37°C overnight.

DNA sequence

*CATATG*CCGCTGTGCCTGAAAATTAACAAAAAACATGGCGAACAGACCCGTCGCATTCTGATCG
AAAACAATCTGCTGAACAAAGATTACAAAATCACGAGCGAGGGTAACTATCTGTACCTGCCGAT
TAAAGATGTTGACGAAGATATCCTGAAATCTATCCTGAACATCGAATTCGAACTGGTGGACAAA
GAACTGGAAGAAAAGAAAATTATCAAAAAACCGAGTTTCCGTGAAATCATCTCCAAAAAATACC
GCAAAGAAATCGATGAAGGCCTGATCAGCCTGTCTTACGACGTGGTTGGTGATCTGGTGATTCT
GCAAATCAGTGACGAAGTTGATGAAAAAATCCGTAAAGAAATCGGCGAACTGGCATATAAACTG
ATTCCGTGCAAAGGCGTTTTTCGTCGCAAATCCGAAGTCAAGGGTGAATTCCGTGTGCGCGAAC
TGGAACATCTGGCTGGCGAAAACCGTACCCTGACGATTCACAAAGAAAATGGTTATCGCCTGTG
GGTTGATATCGCGAAAGTCTACTTTTCACCGCGTCTGGGCGGTGAACGTGCACGTATTATGAAA
AAAGTTTCGCTGAACGACGTCGTGGTTGATATGTTTGCGGGCGTGGGTCCGTTCAGCATTGCGT
GTAAAAACGCGAAGAAAATTTACGCAATTGATATCAATCCGCATGCTATTGAACTGCTGAAGAA
AAACATCAAACTGAACAAACTGGAACACAAAATCATCCCGATTCTGTCTGACGTCCGTGAAGTC
GATGTGAAAGGCAACCGCGTGATCATGAATCTGCCGAAATTTGCGCATAAATTCATTGACAAAG
CCCTGGATATCGTGGAAGAAGGCGGTGTTATTCACTATTACACCATCGGTAAAGACTTCGATAA
AGCCATCAAACTGTTCGAGAAAAAATGCGACTGTGAAGTTCTGGAAAAACGTATCGTCAAAAGC
TATGCACCGCGCGAATACATTCTGGCTCTGGATTTCAAAATCAATAAAAAATAA *AAGCTT*

Expressed protein sequence

MPLCLKINKKHGEQTRRILIENNLLNKDYKITSEGNYLYLPIKDVDEDILKSILNIEFELVDKE
LEEKKIIKKPSFREIISKKYRKEIDEGLISLSYDVVGDLVILQISDEVDEKIRKEIGELAYKLI
PCKGVFRRKSEVKGEFRVRELEHLAGENRTLTIHKENGYRLWVDIAKVYFSPRLGGERARIMKK
VSLNDVVVDMFAGVGPFSIACKNAKKIYAIDINPHAIELLKKNIKLNKLEHKIIPILSDVREVD
VKGNRVIMNLPKFAHKFIDKALDIVEEGGVIHYYTIGKDFDKAIKLFEKKCDCEVLEKRIVKSY
APREYILALDFKINKK*

Fig. 7 Codon-optimized gene for *Mj*TRM5 and the corresponding protein sequence. Cloning sites are shown in *italic*.

3. Make 1 L of LB as described in Section 2.1.
4. Use a single colony from the plate to inoculate a 125 mL of LB containing 34 μg/mL Kan in a 250-mL Erlenmeyer flask. Cells were grown overnight at 37°C with shaking speed set to 200 rpm.
5. Make 6 L of LB as described in Section 2.1.
6. Inoculate 1 L of LB containing 34 μg/mL Kan in a 2.8-L Fernbach flask with 10 mL of the overnight culture.
7. Grow the cells at 37°C with shaking set to 200 rpm.
8. At OD_{600} ~0.6 add 100 μL of IPTG from a 1 *M* stock to make a final concentration of 100 μ*M*.
9. Allow cells to grow for ~18 h.
10. Harvest cells by centrifugation at 5000 × *g* and flash-freeze cell paste in liquid nitrogen. Store at −80°C.

4.2 Purification of TRM5

The purification of TRM5 involves a heat step to remove the majority of *E. coli* proteins. A hydrophobic column is then used to remove heat stable *E. coli* proteins and nucleic acids that copurify with TYW1. The final step in the purification is dialysis to remove the $AmSO_4$ used in the hydrophobic chromatography step, resulting in pure protein capable of methylating the $tRNA^{Phe}$ transcript.

4.2.1 Protocol for Purification of TRM5

1. Prepare the following buffers using water at 4°C: *Lysis buffer*—20 m*M* Tris–HCl (pH 8), 2 m*M* DTT; *Butyl column buffer*—20 m*M* Tris–HCl (pH 8), 1.5 *M* $AmSO_4$, 2 m*M* DTT; *Butyl column elution buffer*—20 m*M* Tris–HCl (pH 8), 2 m*M* DTT; *Dialysis buffer*—20 m*M* Tris–HCl (pH 8), 50 m*M* KCl, 2 m*M* DTT.
2. Place 12 g of frozen cells, a stir bar, solid PMSF, and add 0.15 L of lysis buffer into a stainless steel beaker.
3. Place beaker into a crystallization dish surrounded on all sides by ice.
4. While stirring the frozen cells in buffer, sonify with 48 bursts of 15 s with sonifier at 50% amplitude, with 60 s rest to allow the solution to remain cold.
5. Transfer lysate to a centrifuge tube that has a lid with an O-ring, cap the lid tightly, and centrifuge at 18,000 × *g* for 30 min at 4°C.
6. Transfer the clarified lysate from the centrifuge tube to a glass bottle and seal the lid tightly.
7. Place the bottle in an 80°C water bath and incubate for 30 min.
8. Transfer the bottle to an ice water bath and incubate for 15 min.

9. Transfer the lysate to a centrifuge tube and centrifuge for 30 min at 18,000 × *g* at 4°C.
10. Transfer the clarified lysate to a measuring cylinder and determine the volume.
11. Add the lysate to a beaker containing a stir bar and stir over a stir plate while slowly adding solid $AmSO_4$ to a final concentration of 1.5 *M* over the course of 20 min.
12. Manually pack a butyl-Sepharose FF column (GE Healthcare) (11 × 3 cm) and equilibrate with butyl column buffer using an AKTA FPLC at 4°C.
13. Load the lysate onto the equilibrated column and wash with the loading buffer until the absorbance has returned to baseline. This typically takes 2–3 column volumes.
14. Elute adsorbed protein with a gradient to 100% butyl column elution buffer over 0.4 L.
15. Identify fractions containing pure TRM5 by SDS-PAGE.
16. Pool fractions with highest purity protein and transfer to dialysis tubing with a 10-kDa molecular weight cutoff.
17. Dialyze against 4 L of dialysis buffer for at least 4 h with three changes.
18. Concentrate the protein to a minimal volume using an Amicon stirred cell with a YM10 membrane (Millipore).
19. Aliquot and flash-freeze with liquid nitrogen. Store frozen aliquots at −80°C.

5. SYNTHESIS OF SUBSTRATE FOR TYW1 VIA IN VITRO TRANSCRIPTION

One method utilized to prepare substrate from TYW1 is in vitro transcription using a DNA oligonucleotide as the template with T7 RNA polymerase. A DNA oligonucleotide complementary to *M. jannaschii* MJ_t16 (phenylalanine tRNA gene) with the T7 promoter at the 3′ end is combined with an oligonucleotide consisting of the T7 promoter sequence. The two primers are annealed and then used as a template for recombinant T7 polymerase. This reaction is carried out at 37°C in the presence of inorganic pyrophosphatase, which hydrolyzes pyrophosphate into phosphate in order to stop it from binding $MgCl_2$ and forming a precipitate. The final step is incubation with DNase I to remove the template oligonucleotides. This section describes in vitro transcription reaction. Expression and purification of the T7 RNA polymerase used in these studies are described in Section 6.

5.1 Protocol for Synthesis of $tRNA^{Phe}$

1. Prepare diethyl pyrocarbonate (DEPC)-treated water to be used in the synthesis of tRNA and the purification of *S. cerevisiae* tRNA in the next section as follows: add 1 mL of DEPC to 1 L of water. Incubate for 2 h at 37°C and then autoclave for 15 min.
2. Obtain the following oligonucleotides nucleotides: 5′-TGGTGCCGCGGGGGTGATTTGAACACCCGACAACTGGA-TCTTCAGTCCAGCGTTCTCCCAGGCTGAACTACCGCGG-CTATAGTGAGTCGTATTAGGATCC-3′ (MJ_t16) and 5′-GGATCCTAATACGACTCACTATAG-3′ (T7 promoter) and resuspend them in DEPC-treated water.
3. Prepare the following buffer using water that has been DEPC treated: 5 × *transcription buffer*—200 m*M* Tris–HCl (pH 7.9), 30 m*M* magnesium chloride ($MgCl_2$), 50 m*M* DTT, 50 m*M* NaCl, 10 m*M* spermidine.
4. Anneal 10 μ*M* each of T7 promoter primer with (MJ_t16) primer using a thermocycler with the following conditions in a volume of 200 μL: 95°C for 20 s with a −0.5°C change in temperature for 34 cycles.
5. Set up a reaction containing the following in a 500 μL volume: 1 × transcription buffer, 0.75 μ*M* annealed primers, 10 m*M* DTT, 10 m*M* $MgCl_2$, 4 m*M* each of ribonucleotide triphosphates, 0.5 units inorganic pyrophosphatase (Sigma), and 10% (v/v) recombinant T7 RNA polymerase.
6. Incubate at 37°C for 3 h.
7. Add 20 units of RNase free DNase I.
8. Incubate at 37°C for 30 min.
9. Split reaction into two tubes with 260 μL in each tube.
10. Freeze or purify further as described below in Section 9.3.

6. EXPRESSION AND PURIFICATION OF T7 RNA POLYMERASE

A glycerol stock containing a Rosetta (DE3) *E. coli* strain transformed with a pET42a plasmid containing T7 RNA polymerase (constructed by Bevilacqua Lab at Penn State) was a generous gift from Prof. Squire Booker at The Penn State University. T7 RNA polymerase recognizes the T7 promoter and has a low error rate. The T7 RNA polymerase expressed from this plasmid contains a His-tag which aids in purification.

6.1 Expression of T7 RNA Polymerase

The T7 RNA polymerase gene is cloned downstream of an IPTG inducible promoter on a pET42a plasmid. The pET42a plasmid contains a gene for Amp resistance, while the Rosetta (DE3) strain has Cam resistance. The protein is expressed in LB at 37°C for 3 h.

6.1.1 Protocol for the Expression of T7 RNA Polymerase

1. Make LB (Lennox) agar plates containing 34 μg/mL Cam and 100 μg/mL Amp as described in Section 2. Streak out a loop of T7 RNA polymerase in Rosetta DE3 glycerol stock and incubate overnight at 37°C.
2. Make 1 L of LB as described in Section 2.
3. Use a single colony from the plate to inoculate a 125 mL of LB, containing 34 μg/mL Cam and 100 μg/mL Amp, in a 250-mL Erlenmeyer flask. Cells were grown overnight at 37°C with shaking speed set to 200 rpm.
4. Make 10 L of LB as described in Section 2 and distribute among 10 2.8-L Fernbach flask. Autoclave for 30 min and allow to cool.
5. Add 34 μg/mL Cam and 100 μg/mL Amp to each flask and inoculate with 10 mL of the overnight culture.
6. Grow the cells at 37°C with shaking set to 200 rpm.
7. At OD_{600} ~0.6 add 0.5 mL of 1 *M* IPTG to a final concentration of 500 μ*M*.
8. Allow cells to grow for ~3 h.
9. Harvest cells by centrifugation at 5000 × *g* and flash-freeze cell paste in liquid nitrogen. Store at −80°C.
10. Typical yield is 3.4 g per L of culture.

6.2 Purification of T7 Polymerase

The T7 RNA polymerase expressed from this plasmid has a His_6 affinity tag which greatly aids the purification. Protein is purified in a very similar manner to that used to purify TEV protease in Section 3.2 with a HisTrap chelating HP column charged with nickel utilized.

6.2.1 Protocol for Purification of T7 Polymerase

1. Prepare the following buffers using water at 4°C: *Lysis buffer*—50 m*M* Tris–HCl (pH 8) buffer containing 100 m*M* NaCl, 5 m*M* DTT, 10 m*M* imidazole, and 5% (v/v) glycerol; *Elution buffer*—50 m*M* Tris–HCl (pH 8) buffer containing 100 m*M* NaCl, 5 m*M* DTT, 400 m*M* imidazole, and 20% (v/v) glycerol; *Desalting buffer*—50 m*M* Tris–HCl (pH 8)

buffer containing 100 m*M* NaCl, 1 m*M* ethylenediaminetetraacetic acid (EDTA), 10 m*M* DTT, and 50% (v/v) glycerol.

2. Place 34 g of frozen cells, a stir bar, solid PMSF, and add 0.2 L of lysis buffer into a stainless steel beaker.
3. Place beaker into a crystallization dish surrounded on all sides by ice.
4. While stirring the frozen cells in buffer, sonify with 48 bursts of 15 s with sonifier at 50% amplitude, with 60 s rest to allow the solution to remain cold.
5. Transfer lysate to a centrifuge tube that has a lid with an O-ring, cap the lid tightly, and centrifuge at 18,500 × *g* for 30 min at 4°C.
6. Using an AKTA FPLC, equilibrate a 5-mL HisTrap chelating HP column charged with nickel chloride, according to manufacturer's directions, with lysis buffer.
7. Transfer clarified lysate from centrifuge tube to a clean beaker.
8. Load clarified lysate onto the equilibrated column. When all lysate is loaded wash the column with lysis buffer (20-column volumes) until the absorbance has returned to baseline.
9. Elute adsorbed protein with a gradient to 100% elution buffer over 100 mL.
10. Identify T7 polymerase containing fractions by SDS-PAGE and pool the fractions.
11. Concentrate to 3 mL using an Amicon stirred cell with an YM10 membrane (Millipore).
12. Equilibrate a BioRad Econo-Pac 10 DG desalting column by filling the column to the top with desalting buffer and allowing it to flow-through.
13. Add the 3 mL of concentrated protein to the desalting column and allow it to enter the column.
14. Elute the protein by adding 4 mL of desalting buffer and collecting the flow-through.
15. Aliquot the protein and store at −80°C.

7. ISOLATION OF SUBSTRATE FOR TYW1 FROM *S. CEREVISIAE* DELETION STRAIN

As an alternative to the synthetic substrate, TYW1 can be assayed with tRNA enriched for $tRNA^{Phe}$ obtained from a deletion strain of *S. cerevisiae*, where TYW1 has been deleted. The sequence of the genes encoding $tRNA^{Phe}$ from *S. cerevisiae* and *M. jannaschii* is shown in Fig. 8. The

```
M. jannaschii    1 GCCGCGGTAGTTCAGCCTGGGAGAACGCTGGACTGAAGATCCAGTTGTCGGGTGTTCAAA
S. cerevisiae    1 GCGGACTTAGCTCAGT-TGGGAGAGCGCCAGACTGAAGATCTGGAGGTCCTGTGTTCGAT
consensus        1 ** *   *** ****  *******.*** .*********** .*  ***  ******.*

M. jannaschii   61 TCACCCCCGCGGCACCA
S. cerevisiae   60 CCACAGAGTTCGCACCA
consensus       61  ***       ******
```

!

Fig. 8 Alignment of *M. jannaschii* and *S. cerevisiae* tRNA^{Phe}. The sequence of the *S. cerevisiae* tRNA^{Phe} is similar to that of *M. jannaschii*, allowing it to be substituted as substrate TYW1 from *M. jannaschii*. The site of modification is shown in *bold*.

tRNA^{Phe} isolated from the TYW1 (*YPL207W*) deletion strain will be fully modified with the exception of G37, which will contain an m^{1}G residue instead of the expected wybutosine.

7.1 Growth of *S. cerevisiae* Deletion Strain

S. cerevisiae is grown in YPD broth at 30°C in order to generate cell mass. Cells are grown to a high optical density, harvested by centrifugation, and flash-frozen with liquid nitrogen.

7.1.1 Protocol for Growth of S. cerevisiae *YPL207W Deletion Strain*

1. Make 1 L of YPD broth by combining 20 g of peptone, 10 g yeast extract, and 20 g glucose with 1 L of water in a 1-L bottle. Autoclave for 30 min and allow to cool. Transfer 0.1 L to an autoclaved 250-mL Erlenmeyer flask aseptically.
2. Inoculate a 0.1 L YPD broth culture inoculated with a loop of *S. cerevisiae* YPL207W deletion strain (YSC1-21-547084 Thermo Scientific).
3. Grow at 30°C with shaking at 200 rpm for approximately 48 h.
4. Make 6 L of YPD broth and distribute among six 2.8-L Fernbach flasks. Autoclave for 30 min and allow to cool. Inoculate each with 10 mL of the culture from step 3.
5. Grow at 30°C until the OD_{600} reaches ~10. This typically takes 72 h.
6. Harvest by centrifugation at 5000 × *g* and flash-freeze cell pellet with liquid nitrogen. This typically yields 20 g per L of culture.

7.2 Purification of Soluble RNA From *S. cerevisiae*

Soluble RNA is extracted from *S. cerevisiae* via a phenol extraction. Phenol permeabilizes the cell membrane and wall releasing the cytoplasmic contents. The aqueous layer is extracted, followed by an ethanol precipitation to collect RNA. The pellet is washed with 1 *M* NaCl to remove high-molecular weight RNA (von Ehrenstein, 1967), followed by an ethanol precipitation of the supernatant.

7.2.1 Protocol for Purifying Soluble RNA From S. cerevisiae *YPL207W* Deletion Strain

1. Prepare DEPC water as described in Section 5.1.
2. Prepare the following buffers using DEPC-treated water: *resuspension buffer*—10 m*M* Tris–HCl (pH 8) containing 10 m*M* $MgCl_2$ and 150 m*M* NaCl; *sodium acetate solution*—3 *M* sodium acetate (pH 5.2); *sodium chloride solution*—1 *M* NaCl.
3. Add resuspension buffer to cells (1 mL/1 g of wet cell paste) in a beaker containing a stir bar and stir for 2 h in order to resuspend the cells.
4. Add liquefied phenol (pH 4.3) (Fisher) (1 mL/1 g of wet cell paste) and stir the mixture 1 h.
5. Transfer to a centrifuge bottle and centrifuge at 1500 × *g* at room temperature for 10 min.
6. Transfer supernatant aqueous layer to a clean beaker and add liquefied phenol (pH 4.3) (1 mL/1 g of wet cell paste) and mix by inversion.
7. Transfer to a centrifuge bottle and centrifuge at 15,000 × *g* at 4°C for 30 min.
8. Extract the upper layer and determine the volume.
9. Add 0.1 volume of sodium acetate solution and 3 volumes of ethanol.
10. Incubate at −20°C for at least 2 h.
11. Transfer to a centrifuge bottle and centrifuge at 15,000 × *g* for 30 min at 4°C.
12. Discard the supernatant and resuspend the pellet in sodium chloride solution.
13. Centrifuge at 7000 × *g* for 30 min at 4°C.
14. Discard the supernatant.
15. Resuspend the pellet in DEPC-treated water, aliquot, and freeze.
16. Determine the concentration using A_{260} ($\varepsilon = 0.025$ (μg/mL)$^{-1}$ cm^{-1})

8. SYNTHESIS OF *S*-ADENOSYL-L-METHIONINE

While SAM is commercially available, it is a mixture of isomers where only one is biologically active. In order to prepare enantiomerically pure SAM, a cell-free extract from cells overexpressing SAM synthetase (Markham, Hafner, Tabor, & Tabor, 1980) DM22-(pK8) (generous gift from Prof. Doug Markham) is utilized. SAM synthetase catalyzes the reaction between Met and ATP to generate SAM. SAM is subsequently purified on a carboxymethyl cellulose column and frozen in small aliquots for subsequent assays (Iwig & Booker, 2004).

8.1 Growth of *E. coli* Strain Overexpressing SAM Synthetase

The *E. coli* strain DM22-(pK8) is used in the cell-free assays to synthesize SAM. The strain is grown for 12 h at 37°C and has oxytetracycline resistance.

8.1.1 Protocol for Growth of E. coli *Strain Overexpressing SAM Synthetase*

1. Make 1 L of LB as described in Section 2.
2. Inoculate a 0.125 L LB culture containing 35 μg/mL oxytetracycline (in ethanol) with a loop of *E. coli* strain DM22-(pK8).
3. Grow culture overnight at 37°C with shaking at 225 rpm.
4. Make 6 L of LB as described in Section 2 and distribute equally between six 2.8-L Fernbach flasks. Autoclave for 30 min and allow to cool.
5. Add 35 μg/mL oxytetracycline to each flask and inoculate with 10 mL of the overnight culture.
6. Grow at 37°C with shaking at 225 rpm for 12 h.

Harvest by centrifugation 5000 × *g* and flash-freeze cell paste in liquid nitrogen. Store at −80°C. The typical yield is 4 g of cell paste from each Liter of growth.

8.2 Preparation of Cell Lysates for Synthesis

The method used to produce SAM enzymatic involves lysate from the *E. coli* strain DM22-(pK8). The cells are lysed using a sonifier and then dialyzed against buffer in order to remove small molecules such as ATP and Met that may interfere with synthesis of isotopically enriched SAM.

8.2.1 Protocol for Cell Lysis

1. Prepare the following buffer using water at 4°C: *cell lysis buffer*—20 m*M* Tris–HCl (pH 8), 1 m*M* EDTA, and 50 μg/mL lysozyme.
2. Place 17 g of frozen cells and a stir bar in a stainless steel beaker and add 0.04 L of cell lysis buffer.
3. Stir at room temperature for 30 min.
4. Add solid PMSF to give a final concentration of 1 m*M* (7 mg).
5. Place beaker in a crystallization dish with ice surrounding the beaker on all sides. Place crystallization dish on a stir plate and adjust speed so that the cells are stirring.

6. Disrupt cells by 40 bursts of sonication for 30 s at 50% power, followed by 60 s breaks to avoid heating the samples.
7. Transfer lysate to a centrifuge tube that has a lid with an O-ring, cap the lid tightly, and centrifuge at 18,000 × *g* for 30 min at 4°C.
8. Transfer the clarified lysate to a dialysis bag and dialyze against 4 L of cell lysis buffer at 4°C for a minimum of 4 h with two changes.
9. Aliquot the lysate in 15 mL aliquots, flash-freeze with liquid nitrogen, and store at −80°C.

8.3 Synthesis of SAM

Enantiomerically pure SAM is synthesized from ATP and Met using a cell-free system. DM22-(pK8) cell lysate is incubated at room temperature for 5 h in the presence of ATP and methionine allowing for the incorporation of specific isotopes creating enriched SAM. Acetonitrile is included in the reaction mixture to prevent product inhibition.

Following the synthesis reaction SAM is purified on a cation exchange column, to which the positively charged SAM binds. SAM is eluted with a gradient to 0.2 *M* HCl. Fractions containing SAM are identified with HPLC analysis, pooled, and lyophilized.

8.3.1 Protocol for Synthesizing SAM

1. Prepare the following buffers using cold water: *loading buffer*—1 m*M* sodium acetate (pH 5); *elution buffer*—200 m*M* HCl.
2. Set up the synthesis reaction in a reaction containing 0.1 *M* Tris–HCl (pH 8), 50 m*M* KCl, 1 m*M* EDTA, 20% acetonitrile (v/v), 26 m*M* magnesium chloride, 13 m*M* ATP, 10 m*M* Met, and 12 mL of cell lysate, in a total volume of 0.1 L.
3. Incubate for 5 h with stirring at room temperature.
4. To quench the reaction add 12.1 *M* HCl dropwise to the reaction till the pH decreases to 5.
5. Place the mixture on ice for 15 min.
6. Transfer the reaction to a centrifuge tube and centrifuge at 26,500 × *g* for 30 min at 4°C.
7. Transfer the clarified reaction to a 1-L beaker and add 0.9 L of cold loading buffer.
8. Manually pack a column (5 × 18 cm) containing CM-52 column resin (Whatman).
9. Equilibrate a column (5 × 18 cm) containing CM-52 resin (Whatman) column with loading buffer using an AKTA FPLC.

10. Load the solution from step 7 onto the equilibrated column. Wash the column with the loading buffer until the absorbance returns to baseline using an AKTA FPLC.
11. Elute SAM material with a gradient to 100% elution buffer over 0.8 L.
12. We typically determine fractions containing SAM by HPLC analysis as follows. A 50 μL aliquot from each fraction is injected onto an Agilent 1100 HPLC. The fractions are analyzed using an Eclipse XDB-C18 reverse phase column (250 × 2.6 mm) equilibrated in 0.1% TFA:water (v/v) with a gradient to 10% acetonitrile:0.1% TFA (v/v) over 20 min with a flow rate of 0.75 mL/min with monitoring at 260 nm.
13. Pool fractions containing pure SAM and lyophilize.
14. Resuspend the lyophilized product in a minimal volume of water and determine concentration by UV–vis spectroscopy, using the extinction coefficient of adenosine at 260 nm (15.4 mM^{-1} cm^{-1}).

9. ASSAY CONDITIONS FOR TYW1

The conversion of m^1G to imG-14 catalyzed by TYW1 is best monitored by LC-MS. We typically include methyl viologen in the assays since early experiments on the system indicated that it appears to be required in vitro to mediate reduction of the protein with dithionite (Young & Bandarian, 2011).

The large size of the substrate makes direct detection of the modification difficult. However, the RNA can be extracted, digested to the nucleoside level by treating with P1 nuclease, phosphodiesterase, and alkaline phosphatase (Miles, McCarty, Molnar, & Bandarian, 2011; Pomerantz & McCloskey, 1990). The nucleosides are separated on a C-18 HPLC column and detected by in-line mass spectrometer. The modified base is often present in concentrations that are far below the canonical base. However, it is readily observed by extracted ion analysis of the mass spectral data.

There are two different methods that can be used to provide substrate for TYW1. One of the methods uses tRNA produced via in vitro transcription, which is then modified to the m^1G substrate by TRM5. The other method uses tRNA purified from a *S. cerevisiae* TYW1 deletion strain, which contains all the normal modifications found in $tRNA^{Phe}$ in *S. cerevisiae*, except for wybutosine. The protocol for both and subsequent analysis via LC-MS is described below.

9.1 Assay Using In Vitro Transcript as Substrate

The tRNA substrate prepared via in vitro transcription contains only the four standard RNA bases. In order to be a substrate for TYW1, G37 needs to be methylated using a SAM-dependent methyltransferase. TRM5 is employed to do this and can be used to make the substrate in situ in the presence of TYW1 and all the other components required for the assay.

9.1.1 Protocol for TYW1 Assay Using In Vitro Transcript Substrate

1. All steps should be carried out in an anaerobic chamber. We carry out the assays in the same chambers used to purify the protein described earlier.
2. All reagents should be dissolved in oxygen-free water when possible. Typical assay mixtures should contain 0.1 *M* Tris–HCl (pH 8), 0.1 *M* KCl, 4 m*M* DTT, 2 m*M* $MgCl_2$, 2 m*M* SAM, 20 μ*M* tRNA, 20 μ*M* TRM5, 1.5 m*M* methyl viologen, 10 m*M* sodium dithionite, 2 m*M* pyruvate, and 20 μ*M* TYW1.
3. Incubate the reaction mixtures at 50°C for desired period of time.
4. When the incubation time is over, the assays can be either frozen or the RNA can be extracted using the procedure described in Section 9.3.

9.2 Assay Using *S. cerevisiae* tRNA as Substrate

The other source of substrate is from a *S. cerevisiae* knockout strain. In this assay all the soluble RNA extracted from a *S. cerevisiae* YPL207w (TYW1) deletion strain is incubated in the presence of TYW1, reductant, etc. Following incubation the RNA is extracted and digested to the nucleoside level. The high level of RNA in the assay causes there to be a high background of nucleotides when they are analyzed via LC-MS; however, imG-14 is easily detected by looking at the extracted ion chromatograms at *m*/*z* 322.

9.2.1 Protocol for TYW1 Assay Using S. cerevisiae tRNA

1. All steps should be carried out in an anaerobic chamber.
2. All reagents should be dissolved in oxygen-free water when possible. Typical assay mixtures should contain 0.1 *M* Tris–HCl (pH 8), 4 m*M* DTT, 1.5 m*M* methyl viologen, 10 m*M* sodium dithionite, 2 m*M* SAM, 2 m*M* pyruvate, 2 μg yeast tRNA, and 20 μ*M* TYW1.
3. Incubate the reaction mixtures at 50°C for desired period of time.
4. When the incubation time is over, the assays can be either frozen or the RNA can be extracted using the procedure described in Section 8.3.

9.3 Extraction of Nucleic Acids From Reaction Mixtures

In order to analyze the product made in a TYW1 reaction the RNA is extracted from the reaction mixture using Qiazol, HiBind RNA minicolumns, and other Qiagen buffers. In this procedure the RNA is extracted using Qiazol and chloroform, followed by an ethanol precipitation. The precipitating RNA is loaded onto a HiBind RNA minicolumn and subsequently washed. Finally, purified RNA is eluted using DEPC-treated water, which can then be analyzed or used as required.

9.3.1 Protocol for RNA Extraction

1. Add 250 μL of Qiazol (Qiagen) to reaction and vortex.
2. Add 50 μL of chloroform to reaction and vortex.
3. Centrifuge for 5 min at 21,000 × *g* at room temperature.
4. Extract upper clear layer and combine with 400 μL of cold (−20°C) ethanol in a new tube.
5. Transfer contents of tube (including precipitate if present) to a HiBind RNA minicolumn (Omega bio-tek) inside a collection tube.
6. Centrifuge at 21,000 × *g* for 1 min at room temperature.
7. Discard flow-through in collection tube and add 700 μL RWT buffer (Qiagen) to the minicolumn.
8. Centrifuge at 21,000 × *g* for 1 min at room temperature.
9. Discard flow-through in collection tube and add 500 μL RPE buffer (Qiagen) to the minicolumn.
10. Centrifuge at 21,000 × *g* for 1 min at room temperature.
11. Discard flow-through in collection tube and add 500 μL RPE buffer (Qiagen) to the minicolumn.
12. Centrifuge at 21,000 × *g* for 1 min at room temperature.
13. Discard flow-through in collection tube and centrifuge empty minicolumn for 2 min at 21,000 × *g* at room temperature.
14. Discard collection tube and transfer minicolumn to a clean 1.5-mL tube. Add 50 μL of DEPC-treated water to the minicolumn and incubate at room temperature for 5 min.
15. Centrifuge at 21,000 × *g* for 1 min at room temperature.
16. Discard minicolumn and store the RNA sample at −20°C until needed.

9.4 Digestion of RNA to the Nucleoside Level

In order to detect imG-14, it is necessary to digest the RNA to the nucleoside level, creating a sample that can be analyzed on a C18 column. P1 nuclease and phosphodiesterase I are used to cleave the RNA backbone to form mononucleotides. Alkaline phosphatase is then used to dephosphorylate the mononucleotides, forming nucleosides that are detected via reverse phase chromatography coupled with mass spectrometry. This protocol is adapted from that of Crain (1990).

9.4.1 Protocol for Digestion to the Nucleoside level

1. Prepare the following buffers: *P1 buffer*—0.1 *M* ammonium acetate (pH 5.3); *PD buffer*—1 *M* ammonium bicarbonate; *AP buffer*—10 m*M* Tris–HCl (pH 7.5) containing 10 m*M* $MgCl_2$.
2. Add 5 μL of P1 buffer to 50 μL of sample.
3. Add 1 μL of P1 nuclease (from *Penicillium citrinum*) (Fisher) (1.4 U/μL) to the reaction.
4. Incubate for 6 h at 45°C.
5. Add 5 μL of PD buffer to the sample.
6. Add 3 μL of PD (from *Crotalus adamanteus*) (Sigma) (2 U/μL) to the reaction.
7. Incubate for 2 h at 37°C.
8. Add 100 μL of AP buffer to the sample.
9. Add 3 μL of fast AP (Thermo) (150 U/μL) to the sample.
10. Incubate for 2 h at 37°C.
11. Centrifuge the reaction for 20 min at 8000 × *g* with an YM-10 centrifugal concentrator to remove enzymes.

9.5 LC-MS Analysis of ImG-14

Nucleosides are analyzed using reverse phase chromatography (c18) on a Thermo Vanquish UHPLC interfaced with a Thermo LTQ OrbiTrap XL. The absorbance is monitored at 260 nm, with the mass range m/z 250–350 scanned. The mobile phase consists of a volatile ammonium acetate buffer, along with acetonitrile.

9.5.1 Protocol for Analyzing Samples via LC-MS

1. Prepare the following buffers using Optima grade (Fisher) water and acetonitrile: *buffer A*—50 m*M* ammonium acetate (pH 5); *buffer B*—40% acetonitrile (v/v), 60% water (v/v).
2. We typically separate analytes at 0.2 mL/min on a Thermo Vanquish UHPLC using a Thermo Hypersil gold C18 column (150 × 2.1 mm) 1.9 μ particle size. The UHPLC is interfaced with a Thermo LTQ OrbiTrap XL.
3. Preequilibrate the C18 column in buffer A. Set up the following separation program: 0–3 min, 0% buffer B; 3–3.25 min, 0%–0.2% buffer B;

3.25–3.5 min, 0.2%–0.8% buffer B; 3.5–3.75 min, 0.8%–3.2% buffer B; 3.75–4 min, 3.2%–5% buffer B; 4–7 min, 5%–25% buffer B; 7–10 min, 25%–50% buffer B; 10–12 min, 50%–75% buffer B; 12–12.1 min, 75%–100% buffer B; 12.1–15 min, 100% buffer B; 15–15.1 min, 100%–0% buffer B; 15.1–18 min, 0% buffer B.

4. The LTQ OrbiTrap XL is operated in positive ion mode with the FT analyzer set to 100,000 resolution. Typical settings are: capillary temperature 275°C, sheath gas flow 35, auxiliary gas flow 12, and a source voltage of 3 kV.
5. A 50 μL aliquot of the sample is typically injected.
6. Under these conditions, TYW1 elutes at 12 min and is detected in extracted ion chromatogram at *m/z* 322. A representative chromatogram using *S. cerevisiae* RNA as substrate is shown in Fig. 9.

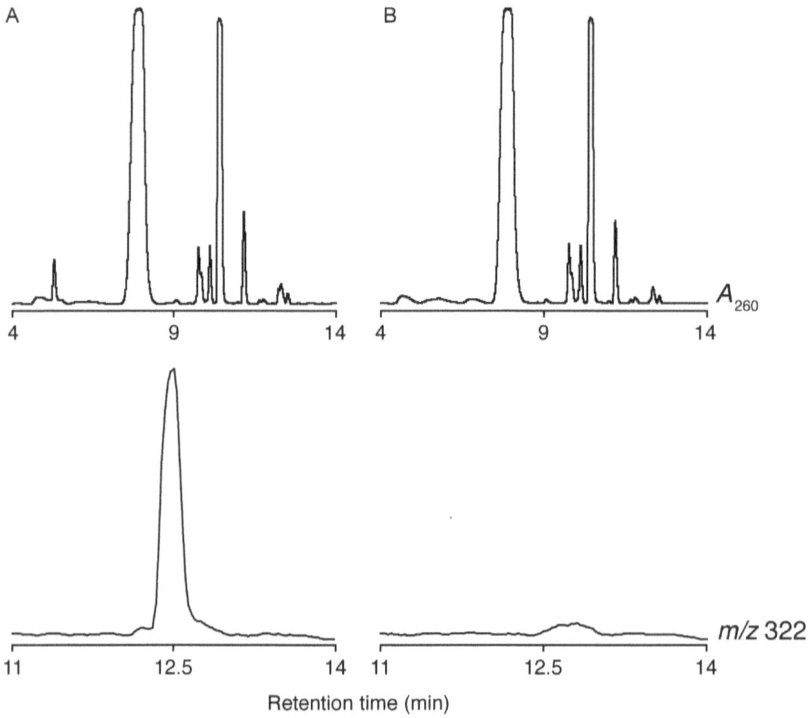

Fig. 9 The UV–vis and extracted ion chromatogram from a typical TYW1 assay. (A) The UV–vis absorbance trace at 260 nm (*upper trace*) and the extracted ion chromatogram at *m/z* 322 (*lower trace*) obtained when all components of the reaction are present. (B) UV–vis absorbance trace at 260 nm (*upper trace*) and the extracted ion chromatogram at *m/z* 322 (*lower trace*) when TYW1 is not present in the reaction. ImG-14 elutes at 12.5 min and is only observed when all reaction components are present.

10. CONCLUSIONS

The discovery of pyruvate as the source of the two carbon atoms that are required for the biosynthesis of the imidazoline ring of imG-14 has made it possible for mechanistic studies of this TYW1. The outline of the reaction has now been established by stable isotope studies with pyruvate (Young & Bandarian, 2011, 2015). The methods that are described in this chapter should allow detailed studies aimed at understanding the mechanism of this complex radical-mediated transformation.

ACKNOWLEDGMENTS

This work was supported by the National Institutes of Health (GM072623 and GM120638).

REFERENCES

Berkovitch, F., Nicolet, Y., Wan, J. T., Jarrett, J. T., & Drennan, C. L. (2004). Crystal structure of biotin synthase, an *S*-adenosylmethionine-dependent radical enzyme. *Science*, *303*, 76–79.

Björk, G. R., Jacobsson, K., Nilsson, K., Johansson, M. J. O., Byström, A. S., & Persson, O. P. (2001). A primordial tRNA modification required for the evolution of life? *The EMBO Journal*, *20*, 231–239.

Blobstein, S. H., Grunberger, D., Weinstein, I. B., & Nakanishi, K. (1973). Isolation and structure determination of the fluorescent base from bovine liver phenylalanine transfer ribonucleic acid. *Biochemistry*, *12*, 188–193.

Cantara, W. A., Crain, P. F., Rozenski, J., McCloskey, J. A., Harris, K. A., Zhang, X., et al. (2011). The RNA modification database, RNAMDB: 2011 Update. *Nucleic Acids Research*, *39*, D195–D201.

Carrington, J. C., & Dougherty, W. G. (1988). A viral cleavage site cassette: Identification of amino acid sequences required for tobacco etch virus polyprotein processing. *Proceedings of the National Academy of Sciences of the United States of America*, *85*, 3391–3395.

Crain, P. F. (1990). Preparation and enzymatic hydrolysis of DNA and RNA for mass spectrometry. *Methods in Enzymology*, *193*, 782–790.

de Crécy-Lagard, V., Brochier-Armanet, C., Urbonavičius, J., Fernandez, B., Phillips, G., Lyons, B., et al. (2010). Biosynthesis of wyosine derivatives in tRNA: An ancient and highly diverse pathway in archaea. *Molecular Biology and Evolution*, *27*, 2062–2077.

Demick, J. M., & Lanzilotta, W. N. (2011). Radical SAM activation of the B_{12}-independent glycerol dehydratase results in formation of 5′-deoxy-5′-(methylthio)adenosine and not 5′-deoxyadenosine. *Biochemistry*, *50*, 440–442.

Dowling, D. P., Bruender, N. A., Young, A. P., McCarty, R. M., Bandarian, V., & Drennan, C. L. (2013). Radical SAM enzyme QueE defines a new minimal core fold and metal-dependent mechanism. *Nature Chemical Biology*, *10*, 106–112.

Droogmans, L., & Grosjean, H. (1987). Enzymatic conversion of guanosine 3′ adjacent to the anticodon of yeast $tRNA^{Phe}$ to N^1-methylguanosine and the wye nucleoside: Dependence on the anticodon sequence. *The EMBO Journal*, *6*, 477–483.

Frazzon, J., Fick, J. R., & Dean, D. R. (2002). Biosynthesis of iron-sulphur clusters is a complex and highly conserved process. *Biochemical Society Transactions*, *30*, 680–685.

Goto-Ito, S., Ishii, R., Ito, T., Shibata, R., Fusatomi, E., Sekine, S. I., et al. (2007). Structure of an archaeal TYW1, the enzyme catalyzing the second step of wye-base biosynthesis. *Acta Crystallographica. Section D, Biological Crystallography, 63*, 1059–1068.

Iwig, D. F., & Booker, S. J. (2004). Insight into the polar reactivity of the onium chalcogen analogues of S-adenosyl-l-methionine. *Biochemistry, 43*, 13496–13509.

Kapust, R. B., Tözsér, J., Copeland, T. D., & Waugh, D. S. (2002). The P1' specificity of tobacco etch virus protease. *Biochemical and Biophysical Research Communications, 294*, 949–955.

Kasai, H., Goto, M., Ikeda, K., Zama, M., Mizuno, Y., Takemura, S., et al. (1976). Structure of wye (Yt base) and wyosine (Yt) from *Torulopsis utilis* phenylalanine transfer ribonucleic acid. *Biochemistry, 15*, 898–904.

Kasai, H., Yamaizumi, Z., Kuchino, Y., & Nishimura, S. (1979). Isolation of hydroxyl-Y base from rat liver $tRNA^{Phe}$. *Nucleic Acids Research, 6*, 993–999.

Kathirvelu, V., Perche-Letuvée, P., Latour, J.-M., Atta, M., Forouhar, F., Gambarelli, S., et al. (2017). Spectroscopic evidence for cofactor-substrate interaction in the radical-SAM enzyme TYW1. *Dalton Transactions, 11*, 134–139.

Kato, M., Araiso, Y., Noma, A., Nagao, A., Suzuki, T., Ishitani, R., et al. (2011). Crystal structure of a novel JmjC-domain-containing protein, TYW5 involved in tRNA modification. *Nucleic Acids Research, 39*, 1576–1585.

Kim, H. J., McCarty, R. M., Ogasawara, Y., Liu, Y.-N., Mansoorabadi, S. O., LeVieux, J., et al. (2013). GenK-catalyzed C-6′ methylation in the biosynthesis of gentamicin: Isolation and characterization of a cobalamin-dependent radical SAM enzyme. *Journal of the American Chemical Society, 135*, 8093–8096.

Li, H. J., Nakanishi, K., Grunberger, D., & Weinstein, I. B. (1973). Biosynthetic studies of the Y base in yeast phenylalanine tRNA. Incorporation of guanine. *Biochemical and Biophysical Research Communications, 55*, 818–823.

Machnicka, M. A., Milanowska, K., Osman Oglou, O., Purta, E., Kurkowska, M., Olchowik, A., et al. (2013). MODOMICS: A database of RNA modification pathways—2013 Update. *Nucleic Acids Research, 41*, D262–D267.

Mahanta, N., Fedoseyenko, D., Dairi, T., & Begley, T. P. (2013). Menaquinone biosynthesis: Formation of aminofutalosine requires a unique radical SAM enzyme. *Journal of the American Chemical Society, 135*, 15318–15321.

Markham, G. D., Hafner, E. W., Tabor, C. W., & Tabor, H. (1980). *S*-adenosylmethionine synthetase from *Escherichia coli*. *The Journal of Biological Chemistry, 256*, 9082–9092.

McCloskey, J. A., Crain, P. F., Edmonds, C. G., Gupta, R., Hashizume, T., Phillipson, D. W., et al. (1987). Structure determination of a new fluorescent tricyclic nucleoside from archaebacterial tRNA. *Nucleic Acids Research, 15*, 683–693.

Miles, Z. D., McCarty, R. M., Molnar, G., & Bandarian, V. (2011). Discovery of epoxyqueuosine (oQ) reductase reveals parallels between halorespiration and tRNA modification. *Proceedings of the National Academy of Sciences of the United States of America, 108*, 7368–7372.

Noma, A., Ishitani, R., Kato, M., Nagao, A., Nureki, O., & Suzuki, T. (2010). Expanding role of the jumonji C domain as an RNA hydroxylase. *The Journal of Biological Chemistry, 285*, 34503–34507.

Noma, A., Kirino, Y., Ikeuchi, Y., & Suzuki, T. (2006). Biosynthesis of wybutosine, a hyper-modified nucleoside in eukaryotic phenylalanine tRNA. *The EMBO Journal, 25*, 2142–2154.

Perche-Letuvée, P., Kathirvelu, V., Berggren, G., Clemancey, M., Latour, J.-M., Maurel, V., et al. (2012). 4-Demethylwyosine synthase from *Pyrococcus abyssi* is a radical *S*-adenosyl-L-methionine enzyme with an additional $[4Fe\text{-}4S]^{+2}$ cluster that interacts with the pyruvate co-substrate. *The Journal of Biological Chemistry, 287*, 41174–41185.

Pomerantz, S. C., & McCloskey, J. A. (1990). Analysis of RNA hydrolyzates by liquid chromatography-mass spectrometry. *Methods in Enzymology, 193*, 796–824.

RajBhandary, U. L., & Chang, S. H. (1968). Studies on polynucleotides LXXXII. Yeast phenylalanine transfer ribonucleic acid: Partial digestion with ribonuclease T_1 and derivation of the total primary structure. *The Journal of Biological Chemistry, 243*, 598–608.

RajBhandary, U. L., Chang, S. H., Stuart, A., Faulkner, R. D., Hoskinson, R. M., & Khorana, H. G. (1967). Studies on polynucleotides, LXVIII. The primary structure of yeast phenylalanine transfer RNA. *Proceedings of the National Academy of Sciences of the United States of America, 57*, 751–758.

RajBhandary, U. L., Stuart, A., Hoskinson, R. M., & Khorana, H. G. (1968). Studies on polynucleotides, LXXVIII. Yeast phenylalanine transfer ribonucleic acid: Terminal sequences. *The Journal of Biological Chemistry, 214*, 565–574.

Sofia, H. J., Chen, G., Hetzler, B. G., Reyes-Spindola, J. F., & Miller, N. E. (2001). Radical SAM, a novel protein superfamily linking unresolved steps in familiar biosynthetic pathways with radical mechanisms: Functional characterization using new analysis and information visualization methods. *Nucleic Acids Research, 29*, 1097–1106.

Suzuki, Y., Noma, A., Suzuki, T., Senda, M., Senda, T., Ishitani, R., et al. (2007). Crystal structure of the radical SAM enzyme catalyzing tricyclic modified base formation in tRNA. *Journal of Molecular Biology, 372*, 1204–1214.

Thoden, J. B., & Holden, H. M. (2005). The molecular architecture of human *N*-acetylgalactosamine kinase. *The Journal of Biological Chemistry, 280*, 32784–32791.

Umitsu, M., Nishimasu, H., Noma, A., Suzuki, T., Ishitani, R., & Nureki, O. (2009). Structural basis of AdoMet-dependent aminocarboxypropyl transfer reaction catalyzed by tRNA-wybutosine synthesizing enzyme, TYW2. *Proceedings of the National Academy of Sciences of the United States of America, 106*, 15616–15621.

von Ehrenstein, G. (1967). Isolation of sRNA from intact *Escherichia coli* cells. *Methods in Enzymology, 12*, 588–596.

Waas, W. F., de Crécy-Lagard, V., & Schimmel, P. (2005). Discovery of a gene family critical to wyosine base formation in a subset of phenylalanine-specific transfer RNAs. *The Journal of Biological Chemistry, 280*, 37616–37622.

Walsby, C. J., Hong, W., Broderick, W. E., Cheek, J., Ortillo, D., Broderick, J. B., et al. (2002). Electron-nuclear double resonance spectroscopic evidence that *S*-adenosylmethionine binds in contact with the catalytically active $[4Fe-4S]^+$ cluster of pyruvate formate-lyase activating enzyme. *Journal of the American Chemical Society, 124*, 3143–3151.

Young, A. P., & Bandarian, V. (2011). Pyruvate is the source of the two carbons that are required for formation of the imidazoline ring of 4-demethylwyosine. *Biochemistry, 50*, 10573–10575.

Young, A. P., & Bandarian, V. (2015). Mechanistic studies of the radical *S*-Adenosyl-L-methionine enzyme 4-demethylwyosine synthase reveal the site of hydrogen atom abstraction. *Biochemistry, 54*, 3569–3572.

Zhang, Y., Zhu, X., Torelli, A. T., Lee, M., Dzikovski, B., Koralewski, R. M., et al. (2010). Diphthamide biosynthesis requires an organic radical generated by an iron-sulphur enzyme. *Nature, 465*, 891–897.

CHAPTER FIVE

Mechanistic Studies on the Radical SAM Enzyme Tryptophan Lyase (NosL)

Dhananjay M. Bhandari, Dmytro Fedoseyenko, Tadhg P. Begley[1]
Department of Chemistry, Texas A&M University, College Station, TX, United States
[1]Corresponding author: e-mail address: begley@tamu.edu

Contents

Abstract

Tryptophan lyase (NosL) is a radical SAM enzyme that catalyzes the formation of 3-methyl-2-indolic acid from L-tryptophan in the biosynthesis of the antibiotic nosiheptide. NosL is the newest addition to the radical SAM-dependent aromatic amino acid lyase subfamily which includes ThiH, HydG, and CofH. The recently solved crystal structure of NosL challenged the previously accepted mechanistic hypothesis and spurred a renewed interest in investigating the reaction. This led to a series of studies that unraveled several fascinating aspects of the fragmentation–recombination reaction. This chapter describes the various methodologies used for the overexpression of NosL, its purification, in vitro reconstitution, preparation of isotopically labeled substrates, and chemoenzymatic synthesis of substrate analogs. The methods described here can be used to further investigate other aromatic amino acid lyases as well as reactivity of fleeting radicals in enzymology.

Methods in Enzymology, Volume 606
ISSN 0076-6879
https://doi.org/10.1016/bs.mie.2018.06.008

1. INTRODUCTION

Biological systems often catalyze complex transformations via radical mechanisms. In 2001, a subset of enzymes utilizing [4Fe–4S] cluster and *S*-adenosylmethionine (SAM) were classified as a superfamily based on the characteristic CX_3CX_2C motif (Sofia, Chen, Hetzler, Reyes-Spindola, & Miller, 2001). The three-cysteine residues coordinate with three of the four irons in a [4Fe–4S] cluster while SAM binds in a bidentate fashion via the amine and carboxylate groups of the methionine moiety to the fourth iron of the cluster (Frey, Hegeman, & Ruzicka, 2008). Almost all known radical SAM enzymes (except Dph2 and glycerol dehydratase activating enzyme), catalyze the reductive homolytic cleavage of SAM generating the 5′-deoxyadenosyl radical (Ado•) which further abstracts a H-atom either from substrate or an enzyme residue to initiate the desired transformation (Broderick, Duffus, Duschene, & Shepard, 2014).

A subgroup (>2200 annotations by bioinformatics) in the radical SAM superfamily consists of enzymes that generate fragments of aromatic amino acids and utilize them in their respective biosynthetic pathways (Fig. 1) (Gerlt, 2017). These include ThiH from the prokaryotic thiamin biosynthetic pathway, HydG involved in the maturation of the H-cluster of hydrogenase, F_0 synthase or CofG/CofH from F_{420} biosynthesis and NosL from nosiheptide biosynthesis (Mehta et al., 2015). As shown in Fig. 1, ThiH, HydG, and CofH utilize L-tyrosine (**1**) as a substrate while NosL uses L-tryptophan (**10**) as the substrate. ThiH catalyzes the formation of dehydroglycine (**3**), which is then incorporated into thiamin (**4**) (Kriek, Martins, Challand, Croft, & Roach, 2007). HydG generates the cyanide (**5**) and carbon monoxide (**6**) ligands of the [Fe–Fe]-hydrogenase H-cluster (**7**) (Driesener et al., 2010; Shepard et al., 2010). The p-cresol radical (**8**) generated by CofH is incorporated into the cofactor F_{420} (**9**) (Philmus, Decamps, Berteau, & Begley, 2015). NosL is involved in biosynthesis of 3-methyl-2-indolic acid (**11**), which is further incorporated into the thiopeptide antibiotic nosiheptide (**12**) (Yu et al., 2009).

Early reports on the mechanism for the aromatic amino acid lyases proposed that the chemistry is initiated by abstraction of a H-atom (by Ado•) from the phenolic O–H for ThiH, HydG, and F_0 synthase, and from indolic N–H for NosL (Fig. 2A) (Zhang et al., 2011). Recently, the crystal structure of NosL demonstrated that the 5′–C of the adenosyl moiety of bound *S*-adenosyl-L-homocysteine (SAH) is 3.7 Å away from the amino group of

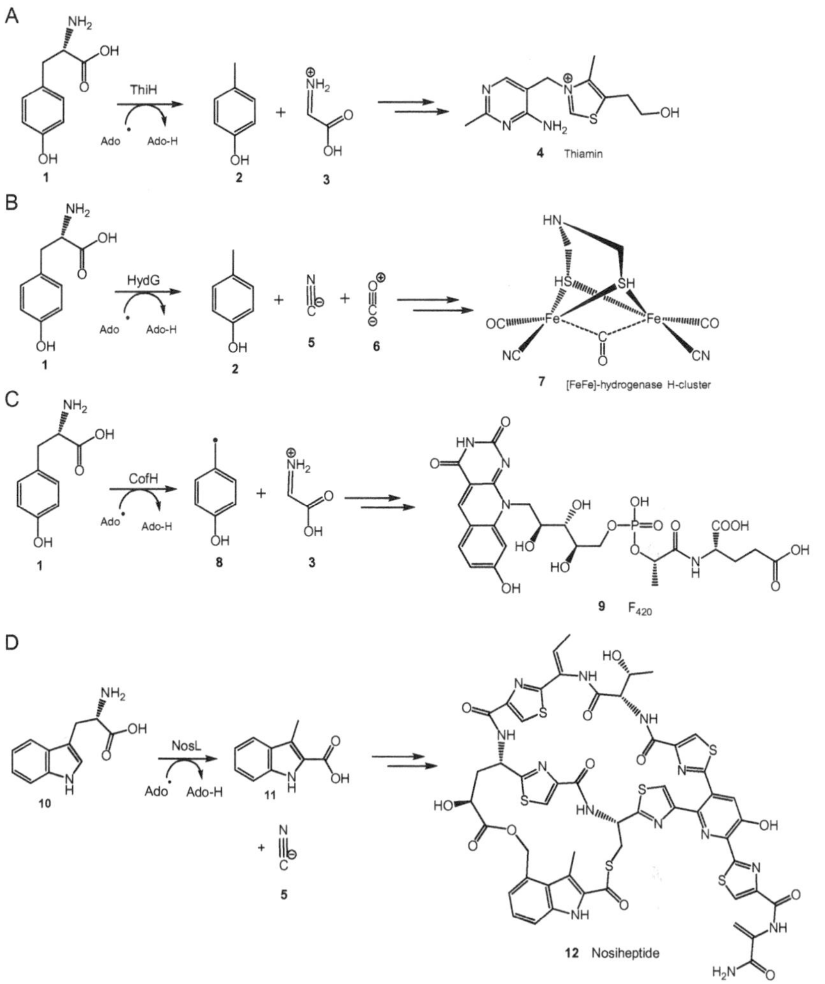

Fig. 1 Examples of the aromatic amino acid lyases in the radical SAM superfamily. (A) ThiH catalyzes formation of dehydroglycine (**3**) which is incorporated into thiamin (**4**). (B) HydG catalyzes formation of carbon monoxide (**6**) and cyanide (**5**) which are then incorporated into [Fe–Fe]-hydrogenase H-cluster (**7**). (C) In CofH, the p-cresol radical (**8**) is incorporated into cofactor F_{420} (**9**). (D) NosL catalyzes the formation of 3-methyl-2-indolic acid (**11**), a component of nosiheptide (**12**).

L-tryptophan (**10**, Fig. 2B) (Nicolet, Zeppieri, Amara, & Fontecilla-Camps, 2014). This indicates that the H-atom is abstracted from the amino group instead of from indole NH of **10** as initially proposed.

Following the insights from the crystal structure of NosL (as shown in Fig. 2B), we undertook a biochemical investigation of the mechanistic

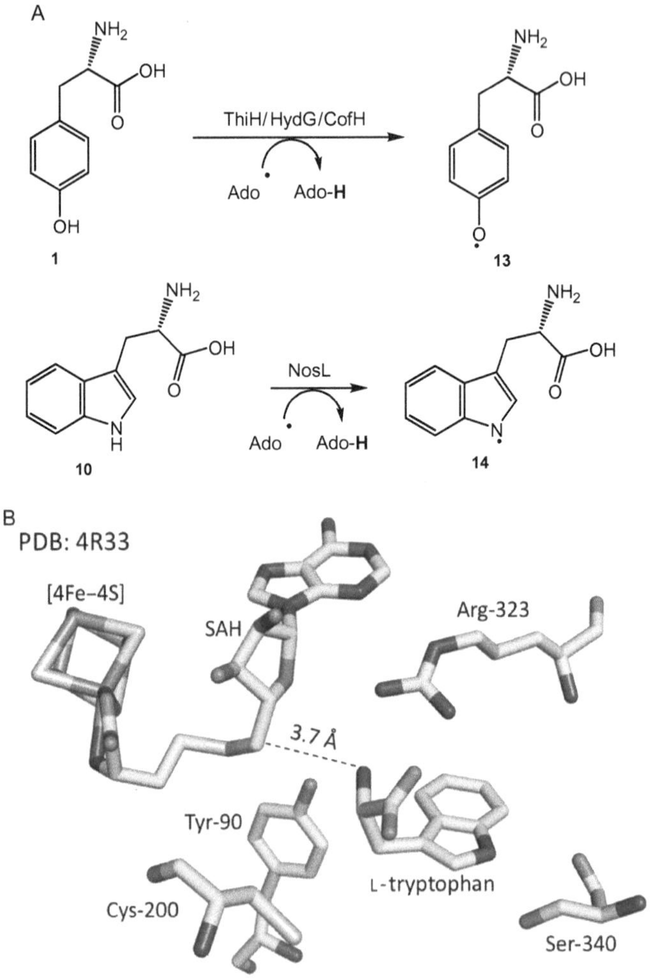

Fig. 2 Proposed H-atom abstraction site in aromatic amino acid lyases and active site architecture of NosL. (A) For tyrosine lyases, it was proposed that the 5′-deoxyadenosyl radical (Ado•) abstracted a H-atom from the phenolic O–H group generating **13**. In the NosL catalyzed reaction, the 5′-deoxyadenosyl radical (Ado•) was proposed to abstract a H-atom from indolic N–H of L-tryptophan generating **14**. (B) Active site architecture of NosL shows that the C'_5 of SAH is close to the amino N–H instead of the indolic N–H of L-tryptophan (**10**).

intricacies of the NosL-catalyzed reaction. Crystal structure of NosL was the first direct evidence of H-atom abstraction from the amino group of a substrate by the 5′-deoxyadenosyl radical (Ado•) in the radical SAM superfamily.

In this chapter, we describe the reconstitution of the NosL-catalyzed reaction which includes identification of cyanide as its byproduct. Additionally, we report on the chemoenzymatic strategies used to obtain isotopically labeled L-tryptophan and stereospecific analogs of L-tryptophan which played a key role in uncovering the mechanistic details of this unique reaction.

2. CLONING, OVEREXPRESSION, AND PURIFICATION OF THE ENZYMES

2.1 Cloning of NosL and Its Active Site Mutants

The *nosl* gene from *Streptomyces actuosus* was PCR amplified and cloned into the pTHT vector (kanamycin resistant, derivative of pET28b with a TEV protease cleavage site after the N-terminal Histag). To study the role of active site residues, the following mutants of NosL were prepared using the QuickChange II site-directed mutagenesis kit (Catalog # 200523, Agilent Technologies): Y90F, Y90A, C200A, S340A, and R323K. To incorporate the [4Fe–4S] cluster in the overexpressed enzyme, WT *nosl* and its variants were overexpressed in the presence of genes involved in iron–sulfur cluster biosynthesis. Comparable yields of NosL were obtained when coexpressed either with the SUF (sulfur mobilization) operon or the ISC (iron sulfur cluster) operon (Mettert & Kiley, 2015). It should be noted that in vitro chemical reconstitution of apo-NosL using Fe^{+2} and sulfide as described elsewhere led to lower activity (20%–30%) in comparison to the in vivo reconstitution of the [4Fe–4S] (Lanz et al., 2012).

Escherichia coli BL21(DE3) was used for the overexpression. The use of this T7 expression host allows for high amounts of coexpression of genes in *suf* operon as well as *nosl*. *E. coli* BL21(DE3) cells were first transformed with the pSuf plasmid which contains the SufABCDSE operon cloned into the pACYC-Duet-1 vector and has a chloramphenicol resistance marker. These cells were further transformed with pTHT-NosL (or its variants) which has a kanamycin resistance marker. Using the protocol described later, the typical yields of NosL were ~35 mg/L of cell culture. The iron content was determined using the Ferene assay while the sulfide content was determined using the methylene blue assay (Beinert, 1983; Van De Bogart & Beinert, 1967). Iron and sulfide content for WT purified NosL were 4.2 ± 0.2 and 3.7 ± 0.1, respectively.

2.2 NosL Overexpression Protocol

1. Sequentially transform *E. coli* BL21(DE3) with pSuf followed by pTHT-NosL (or its variants) using the standard electroporation technique.
2. Pick 3–4 freshly transformed colonies and inoculate in 100 mL LB medium (lennox, EMD Millipore) supplemented with 30 μg/mL chloramphenicol and 40 μg/mL kanamycin.
3. Grow the starter culture at 37°C with shaking (220 rpm) till OD_{600} ~1 (5–6 h).
4. Inoculate 6 L of sterile LB lennox medium (4 × 1.5 L in 2.5 L Ultra-Yield flasks, Thomson Instrument Company), supplemented with 30 μg/mL chloramphenicol and 40 μg/mL kanamycin, with 60 mL of the starter culture.
5. Seal the culture flasks with AirOtop enhanced seals (Thomson Instrument Company), which increases the yield of the cellular wet weight and the purified enzyme.
6. Grow the culture flasks at 37°C with shaking (180 rpm) till OD_{600} ~0.6 (3 h).
7. Incubate the flasks in the cold room (4°C) for ~1 h without shaking.
8. Supplement each culture flask (1.5 L) with 120 mg ferrous ammonium sulfate and 120 mg cysteine.
9. Induce protein expression by adding isopropyl-β-D-thiogalactoside (IPTG) to a final concentration of 200 μ*M*.
10. Incubate the cultures further at 15°C for ~15 h with shaking (110 rpm) to overexpress the enzymes.
11. Harvest the cells by centrifugation in 1 L bottle (9000 rpm) for 10 min. Typical yield is ~4.5 g/L of dark colored wet cell paste.
12. Transfer the wet cell paste to a falcon tube and store under liquid nitrogen in a dewar until required for purification.

2.3 NosL Purification Protocol

All steps for NosL purification were performed in an anaerobic chamber (96% N_2, 4% H_2, $O_2 \leq 2$ ppm, COY laboratories). All buffers described later were autoclaved and immediately transferred to the anaerobic chamber where they were left to equilibrate overnight (to remove dissolved O_2).

1. Thaw the cell pellet (18 g, wet weight) and resuspend in ~65 mL of lysis buffer (100 m*M* Tris–HCl, pH 7.5) at room temperature in the presence of lysozyme (10–12 mg) and benzonase (1000 units).

2. Stir the suspension gently using a magnetic stirrer for ~1.5 h on an ice bath.
3. Lyse the cells using sonication (Qsonica55 with ¼″ tip), 90% amplitude, 2 min. Repeat sonication 4–5 times with 10 min of stirring between each cycle.
4. Transfer the cell lysate to 2 × 40 mL centrifuge tubes with a sealing cap (Nalgene). Further seal the caps with Parafilm.
5. Remove the sealed tubes from the anaerobic chamber and centrifuge at 39,000 × *g*, 25 min, 4°C.
6. In the meantime, equilibrate the 5 mL HisTrap column (GE Healthcare) with five column volumes of lysis buffer (100 m*M* Tris–HCl, pH 7.5) at 4°C in the anaerobic chamber.
7. Transfer the centrifuged cell lysate tubes to the glove box and pour the lysate into a 50 mL syringe. Filter the cell lysate using a 0.45 μm filter. This step will remove most of the cell debris.
8. Load the filtered lysate onto the HisTrap column at 2 mL/min flow rate using a peristaltic pump. As NosL binds to the column, it will appear as a dark band.
9. After loading the lysate, wash the HisTrap column with 20–30 column volumes of wash buffer (100 m*M* Tris–HCl, 30 m*M* imidazole, 300 m*M* NaCl, pH 7.5) at 3 mL/min. Confirm using the Bradford assay that all the nonspecifically bound protein has been washed from the HisTrap column.
10. Elute NosL from the HisTrap column with the elution buffer (100 m*M* Tris–HCl, 250 m*M* imidazole, 300 m*M* NaCl, pH 7.5) at 1–2 mL/min flow rate.
11. Preequilibrate the Econo-Pac® 10DG desalting column (Bio-rad) with five column volumes of water followed by the desalting buffer (50 m*M* potassium phosphate, 10% glycerol, pH 7.5).
12. Pool the dark colored fractions from step 10 and buffer exchange with desalting buffer (50 m*M* potassium phosphate, 10% glycerol, pH 7.5) using 2 × Econo-Pac 10DG desalting column.
13. Prepare 325 μL aliquots of the purified enzyme in Eppendorf tubes having a pinhole on their lid. Remove the aliquots from the anaerobic chamber and flash freeze immediately submerging them in liquid N_2.
14. Store the frozen enzyme aliquots submerged in liquid N_2 in a dewar.
15. Confirm the purity of enzyme using SDS-PAGE and calculate the concentration by measuring the absorbance at 280 nm.

16. Typical protein yields were ~35 mg of NosL per liter of cell culture. The extinction coefficient for WT NosL is calculated as 38,390 $M^{-1} cm^{-1}$ at 280 nm (assuming all cysteines were reduced) using the ProtParam tool of the ExPASy proteomics server.

2.4 Overexpression and Purification of Flavodoxin A/Flavodoxin Reductase

1. *E. coli* flavodoxin A (FldA) and flavodoxin reductase (FldR) were cloned separately in the pTHT vector.
2. Transform the respective plasmids into *E. coli* BL21(DE3) strain using the standard electroporation technique.
3. Pick 2–3 freshly transformed colonies and inoculate in 30 mL LB medium (lennox, EMD Millipore) supplemented with 40 μg/mL kanamycin.
4. Grow the starter culture at 37°C with shaking (220 rpm) till OD_{600} ~1 (~5 h).
5. Inoculate 3 L of sterile LB lennox medium (2 × 1.5 L Ultra-Yield flasks), supplemented with 40 μg/mL kanamycin, with 30 mL of the starter culture for each strain.
6. Seal the culture flasks with AirOtop enhanced seals and grow the cells at 37°C with shaking (210 rpm) till OD_{600} ~0.6 (~3 h).
7. Incubate the flasks in the cold room (4°C) for ~30 min without shaking.
8. Induce protein expression by adding isopropyl-β-D-thiogalactoside (IPTG) to a final concentration of 500 μ*M*.
9. Incubate the cultures at 30°C for ~6 h with shaking (110 rpm) to over-express the enzymes.
10. Harvest the cells by centrifugation for 10 min in 1 L bottles (9000 rpm) and store the resulting cell paste at −80°C until further use. Typical yields are ~4 g/L of wet cell paste. Perform protein purification under aerobic conditions until the last step of desalting.
11. Thaw the cells and resuspend in 50 mL of lysis buffer (50 m*M* Tris–HCl, 10 m*M* imidazole, 150 m*M* NaCl, pH 7.5) in the presence of lysozyme (8 mg) and benzonase (1000 units).
12. Stir the suspension gently using a magnetic stirrer for ~1.5 h on an ice bath.
13. Lyse the cells using 4–5 cycles of sonication (30 s, 50% cycle, 65% power) with an interval of 10 min between each cycle.

14. Transfer the cell lysate to 2 × 40 mL tubes and centrifuge (39,000 × *g*, 25 min, 4°C).
15. In the meantime, equilibrate the 5 mL HisTrap column (GE Healthcare) with 5–10 column volumes of lysis buffer (50 m*M* Tris–HCl, 10 m*M* imidazole, 150 m*M* NaCl, pH 7.5) at 4°C.
16. Filter the cell lysate using a 0.45 μm filter. This step will remove most of the cell debris.
17. Load the filtered lysate onto the HisTrap column at 2 mL/min flow rate using a peristaltic pump.
18. After loading the lysate, wash the HisTrap column with 100 mL of washing buffer (50 m*M* Tris–HCl, 20 m*M* imidazole, 150 m*M* NaCl, pH 7.5) including 50 μ*M* flavin (FMN for FldA and FAD for FldR).
19. Elute FldA/FldR from the HisTrap column with the elution buffer (100 m*M* Tris–HCl, 250 m*M* imidazole, 150 m*M* NaCl, pH 7.5) at 1–2 mL/min flow rate.
20. In the anaerobic chamber, preequilibrate the Econo-Pac® 10DG desalting column (Bio-rad) with five column volumes of water followed by the desalting buffer (50 m*M* potassium phosphate, 150 m*M* NaCl, 1 m*M* DTT, 10% glycerol, pH 7.5).
21. Pool the yellow fractions from step 19, concentrate them to 3 mL and buffer exchange with desalting buffer (50 m*M* potassium phosphate, 150 m*M* NaCl, 1 m*M* DTT, 10% glycerol, pH 7.5) using an Econo-Pac 10DG desalting column.
22. Prepare 100 μL aliquots of the purified FldA/FldR in Eppendorf tubes. Remove the aliquots from the anaerobic chamber and flash freeze immediately by submerging them in liquid N_2.
23. Confirm the purity of the enzyme using SDS-PAGE. Store the frozen enzyme aliquots submerged in liquid N_2 in a dewar.

3. ENZYMATIC REACTIONS OF NosL

3.1 Reconstitution of NosL

The NosL enzymatic reactions were performed in an anaerobic chamber (96% nitrogen, 4% hydrogen, $O_2 \leq 2$ ppm, COY laboratories). A typical enzymatic reaction (total volume 320 μL) was conducted in 100 m*M* phosphate buffer, pH 7.5 in an Eppendorf tube by sequentially adding NosL (75–150 μ*M*), dithionite (750 μ*M*), L-tryptophan, or the substrate analog (400 μ*M*) and SAM (600 μ*M*). The reaction mixture was incubated at RT (25°C) for ~2 h. The samples were quenched by transferring the reaction mixture to a 10 kDa polyethersulfone membrane centrifugal filter (500 μL, VWR) and centrifuging at 13,500 × *g* for 15 min at RT. Alternatively, to investigate formation of hydrophobic products (they bind to the filter membrane), the enzymatic reaction was quenched with 50% trifluoroacetic acid (TFA) solution such that the final TFA concentration was 5%–10%. The contents were centrifuged at 13,500 × *g* for 15 min to remove the precipitated protein. The supernatant was transferred to a fresh Eppendorf tube and adjusted to ~pH 6 using NaOH solution.

A typical HPLC chromatogram of the NosL reaction mixture is shown in Fig. 3. When L-tryptophan (**10**) is used as a substrate with dithionite as a reductant, **10** was entirely consumed generating 3-methyl indole (**15**) as the major product while 3-methyl-2-indolic acid (**11**) is a minor product (ratio 95:5, Fig. 3A) (Bhandari, Xu, Nicolet, Fontecilla-Camps, & Begley, 2015). Lowering the dithionite concentration slightly improved the product ratio but 3-methyl indole (**15**) was still the dominant product (90:10) (Yu et al., 2009). No significant difference in the NosL activity was observed when the Histag was cleaved using TEV protease. To investigate the influence of the reducing agent on the product ratio, the NosL catalyzed reaction was performed in the presence of the *E. coli* flavodoxin/flavodoxin reductase reducing system. The optimized conditions were as follows: glycerol was removed from NosL using a Bio-spin 6 column (Bio-rad) and 100 m*M* phosphate buffer, pH 7.5. The components of the enzymatic reaction were NosL (75–100 μ*M*), FldA (60 μ*M*), FldR (40 μ*M*), NADPH (750 μ*M*), L-tryptophan/substrate analog (500 μ*M*), and SAM (600 μ*M*). The enzymatic reaction was incubated at room temperature for ~3 h. With the FldA/FldR reducing system, 3-methyl-2-indolic acid (**11**) was the major product while 3-methyl indole (**15**) was detected in trace amounts (ratio 95:5, Fig. 3B). However, the turnover was greatly reduced with the *E. coli* FldA/FldR system (maximum activity was ~20% of the NosL concentration) (Bhandari et al., 2015). The effect of various reducing systems on radical SAM enzymes has been well documented and potentially for NosL increased turnovers will be observed in presence of *S. actuosus* FldA/FldR system (Bruender, Young, & Bandarian, 2015).

3.1.1 HPLC Parameters

An Agilent 1260 HPLC equipped with a quaternary pump was used. The system included a diode array UV–vis detector and eluted compounds were detected by absorbance at 254, 260, 280, 288, 320, 340, and 475 nm. The parameters for the fluorescence detector were: excitation at 270 nm and

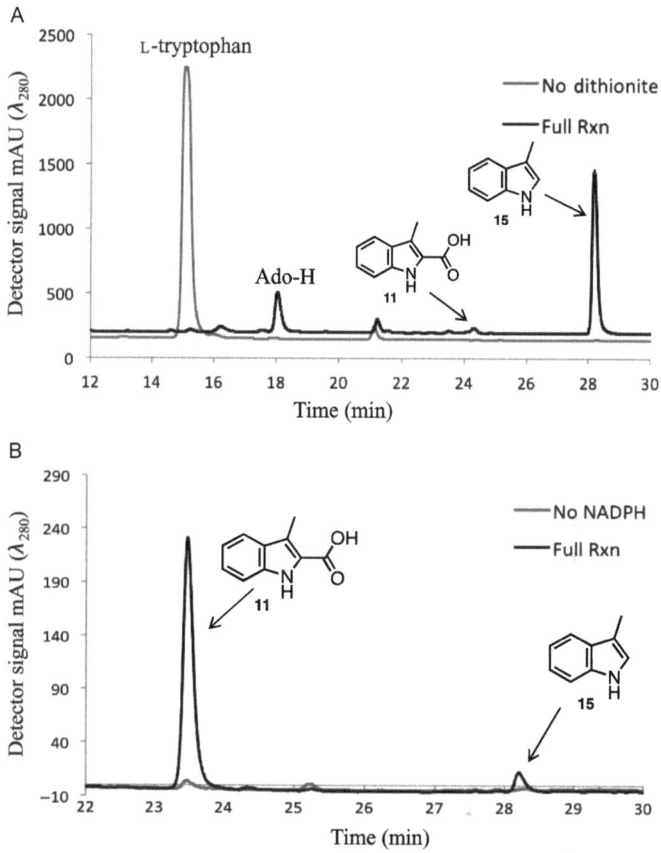

Fig. 3 HPLC chromatogram (280 nm) of the NosL-catalyzed reaction. (A) When dithionite is used as a reductant, L-tryptophan is completely consumed leading to the formation of 3-methylindole (**15**) as a major product when all components of the enzymatic reaction were present. (B) When *E. coli* FldA/FldR is used as the reducing system, 3-methyl-2-indolic acid (**11**) is the major product but the turnover is greatly reduced.

emission at 352 nm (to detect indole-based products) or excitation at 418 nm and emission at 454 nm (to detect derivatized cyanide). For analytical purposes, HPLC analysis was performed on a ZORBAX Eclipse XDB-C18 column (15 cm × 4.6 mm, 5 μm particles, Agilent Technologies, flow rate = 1 mL/min). Typical injection volumes were in the range of 10–80 μL. Data were processed using ChemStation ver. B.04.01 SP1 (Agilent Technologies).

HPLC conditions: (Flow rate 1 mL/min)

A. Water.
B. 100 m*M* potassium phosphate buffer, pH 6.6/10 m*M* ammonium acetate, pH 6.6.
C. Methanol.
HPLC method:
0 min—100% B, 5 min—100% B, 14 min—7% A 70% B 23% C, 25 min—25% A 0% B 75% C, 28 min—25% A 0% B 75% C, 32 min—100% B, 36 min—100% B.

3.1.2 LC–MS Parameters

LC–ESI–TOF–MS was performed using an Agilent 1260 HPLC system equipped with a binary pump and a 1200 series diode array detector followed by a MicroToF-Q II mass spectrometer (Bruker Daltonics) using an ESI source either in negative mode or positive mode. The samples were run either on an LC-18-T column (15 cm × 3 mm, 3 μm particles, Supelco, flow rate 0.4 mL/min) or Poroshell 120 EC-C18 (10 cm × 3 mm, 2.7 μm particles, Agilent, flow rate 0.4 mL/min). Typical injection volumes were in the range of 10–80 μL. The data were processed using DataAnalysis 4.0 SP1 (Bruker Daltonics).

LC conditions: (Flow rate 0.4 mL/min).
A. 5 m*M* Ammonium acetate buffer, pH 6.6.
B. 75% methanol and 25% water.
LC method:
0 min—100% A, 7 min—100% A, 10 min—80% A 20% B, 27 min—100% B, 29 min—100% B, 30 min—100% A, 40 min—100% A.
MS parameters:
Capillary, −4500 V; capillary offset, −500 V; nebulizer gas, 3.0 bar; dry gas, 10 L/min; dry gas temperature, 200°C; funnel 1 RF, 250.0 Vpp; funnel 2 RF, 300.0 Vpp; ISCID, 0.0 eV; hexapole RF, 200 Vpp; quadrapole ion energy, 3.0 eV; collision cell, collision energy, 8.0 eV; collision RF, 150.0 Vpp, transfer time, 80.0 μs; prepulse storage, 5.0 μs.

3.2 Position of H-Atom Abstraction in the NosL-Catalyzed Reaction

When the NosL catalyzed reaction was performed in ~90% D_2O buffer, LC–MS analysis of the 5′-deoxyadenosine (Ado-H) showed incorporation of multiple deuteria from solvent (Bhandari et al., 2015). Similar observations have been made with HydG, CofH, and ThiH (Duffus, Ghose, Peters, & Broderick, 2014; Philmus et al., 2015). This demonstrates that the 5′-deoxyadenosyl radical (Ado•) abstracts a H-atom from a solvent exchangeable position on the substrate. As shown in Fig. 2, the structure of NosL suggested that the 5′-deoxyadenosyl radical (Ado•) abstracts a H-atom from the amino rather than from the indolic N–H of tryptophan. To test this prediction in a biochemical assay, the NosL reaction was run using L-3-benzothienylalalanine (**16**) in which the indole N–H is replaced by sulfur. This reaction generated **18** and **19** as products consistent with hydrogen atom abstraction from the amino group (Fig. 4) (Bhandari et al., 2015). A similar study using the benzofuran analog of tryptophan led to an identical conclusion (Ji, Li, Ding, & Zhang, 2015).

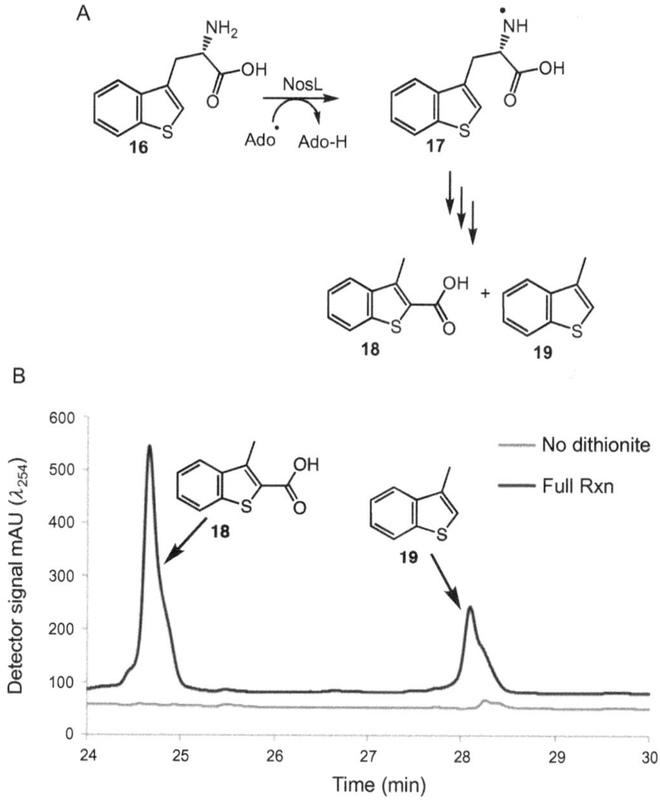

Fig. 4 Analysis of the NosL enzymatic reaction with L-3-benzothienylalanine (**16**). (A) Scheme showing the reaction products formed from **16**. (B) HPLC analysis (254 nm) of the reaction mixture showing formation of compounds **18** and **19**.

3.3 Reaction of NosL With L-Tryptophan Thiocarboxylate

To gain further insights into the β-scission reaction that occurs after the H-atom abstraction from the amino group, we investigated the reaction of NosL in the presence of L-tryptophan thiocarboxylate (**20**). As shown in Fig. 5A, if the reaction proceeds via C_α—C_β bond scission, it will generate 3-methylindole (**15**) while if it proceeds via C_α—C(O) bond scission, it will lead to the formation of indole-3-acetaldehyde (**26**). HPLC analysis of the NosL catalyzed reaction with L-tryptophan thiocarboxylate showed complete substrate consumption and formation of a new product, which was

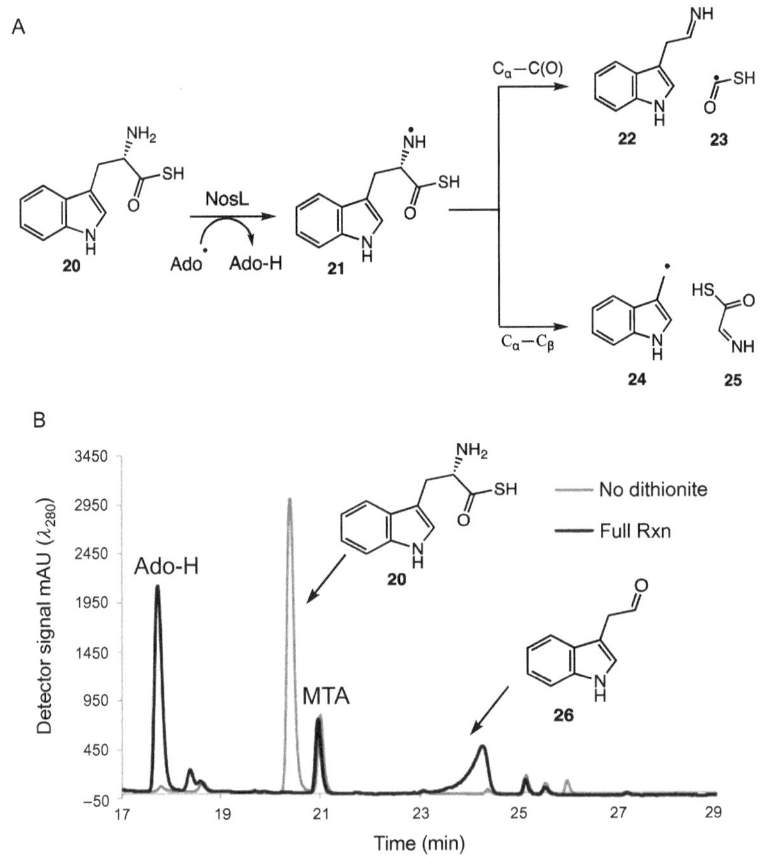

Fig. 5 Analysis of the NosL reaction with L-tryptophan thiocarboxylate (**20**). (A) After H-atom abstraction from the amino group of **20**, two possible β-scission reactions will lead to different products. (B) HPLC analysis shows complete consumption of substrate **20** and formation of the aldehyde product **26** (MTA: methylthioadenosine, impurity in SAM).

identified as aldehyde **26** (Fig. 5B). It was observed that L-tryptophan thiocarboxylate (**20**) underwent slow hydrolysis generating L-tryptophan (**10**) when dissolved in reaction buffer (100 m*M* KH_2PO_4, pH 7.5). To prevent contamination due to L-tryptophan (**10**, native substrate of NosL), a fresh stock of **20** was prepared each time while setting up the enzymatic reaction. During the reaction (~2 h), less than 5% hydrolysis of **20** was observed which ensures that the data are not misinterpreted. The route for the synthesis of L-tryptophan thiocarboxylate (**20**) can be found in the original publication (Bhandari, Fedoseyenko, & Begley, 2018).

3.4 Synthesis of L-Tryptophan Isotopologs Using Tryptophan Synthase

To identify the fate of C_1, C_2, and C_2—N in the NosL-catalyzed reaction of L-tryptophan (**10**, Fig. 3A), by NMR analysis, various isotopologs of L-tryptophan were synthesized using tryptophan synthase. The following procedure was followed:

1. Grow a starter culture of *E. coli* BL21(DE3) containing pSTB7 (ATCC®, 37845™)—the tryptophan synthase overexpression plasmid—in 30 mL LB medium supplemented with 100 μg/mL of ampicillin (Kawasaki, Bauerle, Zon, Ahmed, & Miles, 1987).
2. Inoculate 3 L of sterile LB lennox medium (2 × 1.5 L Ultra-Yield flasks) supplemented with 100 μg/mL of ampicillin, with 30 mL of the starter culture.
3. Seal the culture flasks with AirOtop enhanced seals and grow them at 37°C with shaking (200 rpm) for ~24 h.
4. Harvest the cells by centrifugation in 1 L bottles (9000 rpm) for 10 min. A typical yield is ~7 g/L of wet cell paste.
5. Distribute the cell pellet into six falcon tubes and store them at −80°C until further use.
6. Thaw the contents of one falcon tube and suspend the cells in 30 mL of lysis buffer (50 m*M* KH_2PO_4, pH 7.5) at 4°C.
7. Add 2 mg of pyridoxal 5′-phosphate (PLP) to the suspension and sonicate for four cycles (30 s, 50% cycle, 65% power) with intervals of 10 min between each cycle.
8. Transfer the cell lysate to a 40 mL tube (Nalgene) and centrifuge (39,000 × *g*, 25 min, 4°C).
9. Filter the cell lysate using a 0.45 μm filter to remove most of the cell debris.

10. Store the filtered lysate in the cold room (4°C) for up to 1 month and use whenever required during this period.
11. To synthesize the isotopologs of L-tryptophan (**10**), dissolve 10mg of indole (**27**) and 12mg of the respective isotopolog of L-serine in 12mL of lysis buffer (50m*M* KH_2PO_4, pH 7.5).
12. Add 0.5mg of PLP and 3mL of tryptophan synthase lysate (step 10) to this mixture.
13. Incubate the reaction mixture at 37°C (220rpm) for ~3–4h.
14. Heat denature the reaction mixture at 75°C for 20min and centrifuge at 5000 × *g* for 10min to remove precipitated proteins.
15. Filter the supernatant using a 10kDa cut-off filter (VWR) at 14,000 × *g* for 15min.
16. Purify the L-tryptophan isotopolog by HPLC using a Supelcosil LC-18 (25cm × 10mm, 5μm particles) semiprep column (Bhandari, Fedoseyenko, & Begley, 2016). Typical yield ~8–10mg labeled L-tryptophan (Fig. 6).

3.5 Identification of Byproducts in the NosL Reaction

Enzymatic reactions for NMR analysis were set up as described in Section 3.1. After ~2h incubation, the reaction mixture was transferred to an NMR tube in the anaerobic chamber. The tube was capped and sealed with parafilm. When the NosL reaction was performed with [1,2,3-$^{13}C_3$,2-amino-$^{15}N_1$]-L-tryptophan (**10a**), ^{13}C NMR analysis of the reaction mixture revealed only two new signals (8 and 160ppm, Fig. 7A). The 8ppm signal was assigned

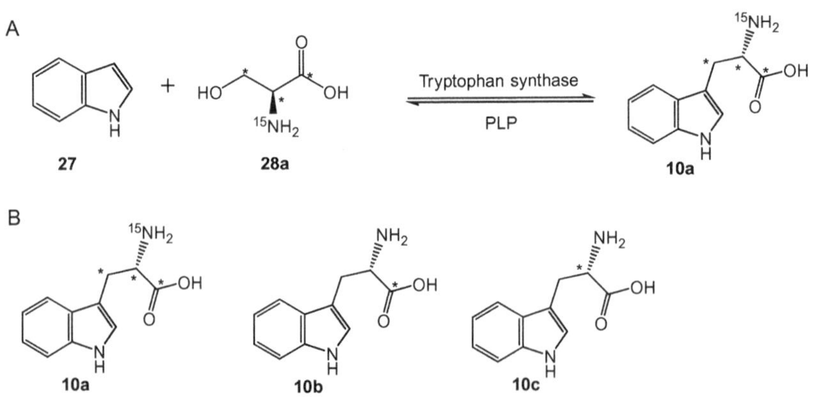

Fig. 6 Isotopologs of L-tryptophan used for product characterization by NMR. (A) Scheme depicting the enzymatic synthesis of the tryptophan isotopologs. (B) Isotopologs synthesized using this methodology (* = ^{13}C).

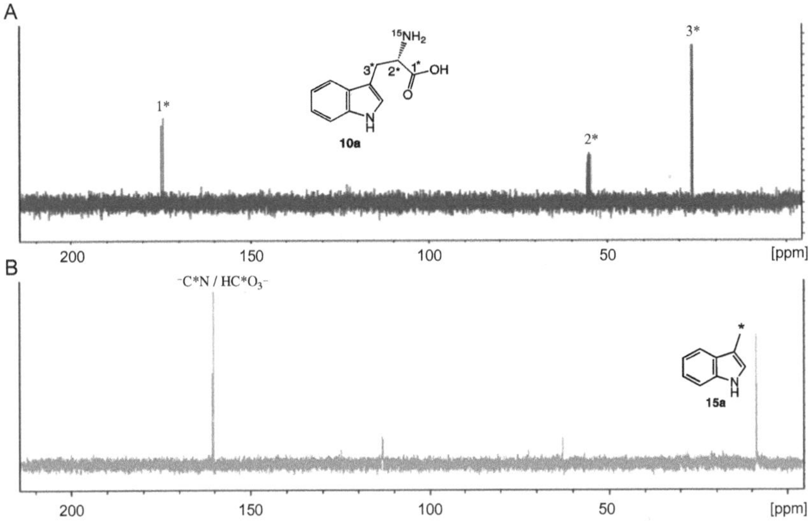

Fig. 7 ^{13}C NMR analysis of the NosL-catalyzed reaction with L-tryptophan isotopolog. (A) ^{13}C NMR of [1,2,3-$^{13}C_3$,2-amino-$^{15}N_1$]-L-tryptophan standard. (B) ^{13}C NMR of the NosL-catalyzed reaction with [1,2,3-$^{13}C_3$,2-amino-$^{15}N_1$]-L-tryptophan.

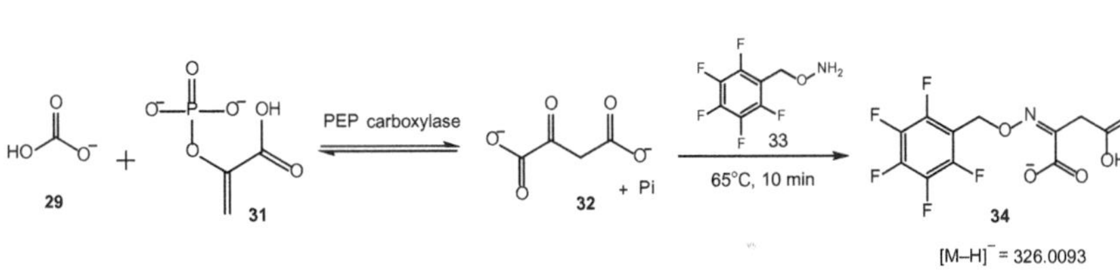

Fig. 8 Trapping of bicarbonate (**29**) using PEP carboxylase and conversion of the resulting oxaloacetate (**32**) to oxime **34** for LC–MS detection.

to the methyl group of 3-methylindole (**15**) and the 160 ppm signal suggested formation of cyanide (**5**) and/or bicarbonate (**29**).

To test for the formation of CO_2/bicarbonate (**29**), the NosL enzymatic reaction was treated with 0.1 unit of phosphoenolpyruvate carboxylase (PEPC, Sigma Aldrich) in the presence of 2 m*M* phosphoenolpyruvate (**31**) to convert any bicarbonate present to oxaloacetate (**32**, Fig. 8) (O'Leary, 1982). To establish the presence of oxaloacetate (**32**), the sample was filtered through a 10 kDa cut-off filter (VWR) and converted to oxime **34** with o-(2,3,4,5,6-pentafluorobenzyl) hydroxylamine.HCl (**33**, 2 m*M*, 65°C for 10 min) which was detected by LC–MS analysis (Cancilla & Hee, 1992). When this experiment was repeated with isotopologs **10a–c**, the CO_2/bicarbonate (**29**) carbon was found to be derived from the carboxy group of L-tryptophan (Bhandari et al., 2018).

To test for cyanide formation, the enzymatic reaction was set up as described in Section 3.1. The reaction was quenched after ~2h with ~50% aqueous TFA solution because cyanide binds to the filter membrane and centrifuged to remove the precipitated protein (13,500 × *g*, 15 min). The supernatant was transferred to a clean Eppendorf tube and the pH was adjusted to ~pH 6 using NaOH solution. A 1 mL stock solution of the cyanide derivatization reagent was prepared by adding 200 μL of 2,3-napthalenedicarboxaldehyde (**35**, 10 m*M* solution in methanol), 200 μL of taurine (**36**, 50 m*M* solution in water), 60 μL of 7 *M* ammonia in methanol, 200 μL of methanol, and 140 μL of water to an Eppendorf tube. This reagent (80 μL) was incubated with 20 μL of the enzymatic reaction mixture at room temperature for 20–30 min (Tracqui, Raul, Geraut, Berthelon, & Ludes, 2002). The derivatization reaction was quenched by adding 100 μL of water and a 10 μL aliquot was analyzed by HPLC (fluorescence detector, E_x = 418 nm, E_m = 454 nm). A new fluorescent signal eluted at 26.6 min (Fig. 9). This compound was identical by HPLC and MS analysis to a synthesized sample of **37**. The ratio of cyanide (**5**) to 3-methylindole (**15**) was ~0.9:1. Repeating this experiment with isotopologs **10a–c** demonstrated that the cyanide (**29**) was derived from C_2 and the amino group of L-tryptophan (Bhandari et al., 2018).

3.6 Chemoenzymatic Synthesis of C_2-Halogenated L-Tryptophan Analogs

Trapping radical intermediates, using exogenous traps, is challenging due to the constrained environment of the enzyme active site and almost all of the successful radical trapping strategies have relied on intramolecular trapping reactions. Halogens substitution of the native substrate can serve as a versatile intramolecular radical trap due to rapid carbon halogen bond fragmentation triggered by radicals beta to the carbon halogen bond (Wagner, Lindstrom, Sedon, & Ward, 1981). Halogen substitution is much smaller than the widely used cyclopropyl radical trap and unlike the cyclopropyl group, halogen atoms are unlikely to inactivate the enzyme. In addition, the β-scission of the carbon halogen bond can be tuned $>10^4$-fold by varying the halogen from iodo to chloro (Joshi, Mahanta, Fedoseyenko, Williams, & Begley, 2017). Based on this, 2′-halo-L-tryptophan (Br, Cl, I) was designed as a C'_3 radical trap. 2′-chloro-L-tryptophan (**41**) was synthesized using the following protocol:

1. In a dry round bottom flask, dissolve 5.1 mmol of L-tryptophan methyl ester.HCl (**38**, 1.3 g) in 50 mL acetic acid and 15 mL formic acid.

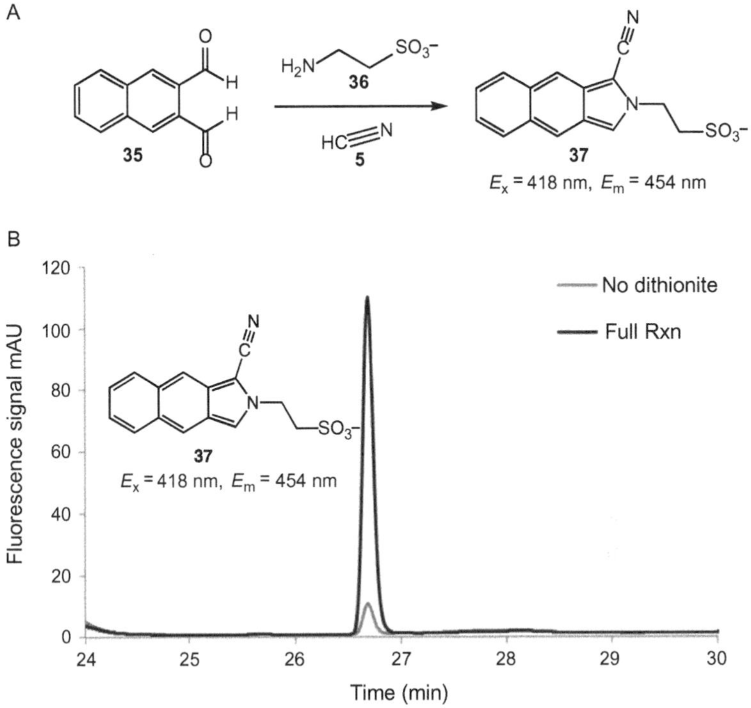

Fig. 9 Detection of cyanide (**5**) as a byproduct of the NosL catalyzed reaction. (A) Reaction used to convert cyanide to the fluorescent derivative **37**. (B) HPLC analysis of the NosL/L-tryptophan (**10**) reaction mixture after derivatization as shown in panel (A).

2. To this solution, add 5.2 mmol of N-chlorosuccinimide (**39**, 700 mg).
3. Stir the reaction mixture at room temperature for 30 min and remove solvents by rotary evaporation.
4. Purify the product **40** by flash chromatography (silica, dichloromethane: methanol = 95:5). Typical yield ~50%.
5. Add a solution of **40** (500 mg in 10 mL dioxane) to a solution containing 50 mg of α-chymotrypsin (Type II, from bovine pancreas, Sigma Aldrich), 45 mL of water, 2 mL of 1 *M* $NaHCO_3$, and 10 mL dioxane. (Since the C_2-chloro group of **40** is susceptible to acid/base hydrolysis, the deprotection of the ester was carried out using α-chymotrypsin.)
6. Stir the reaction mixture for ~6 h at room temperature. Flash freeze in liquid N_2 and lyophilize overnight.
7. Suspend the residual mixture in methanol and filter through celite.
8. Wash the residue with methanol, combine the washings and the filtrate, and remove the solvent by rotary evaporation.

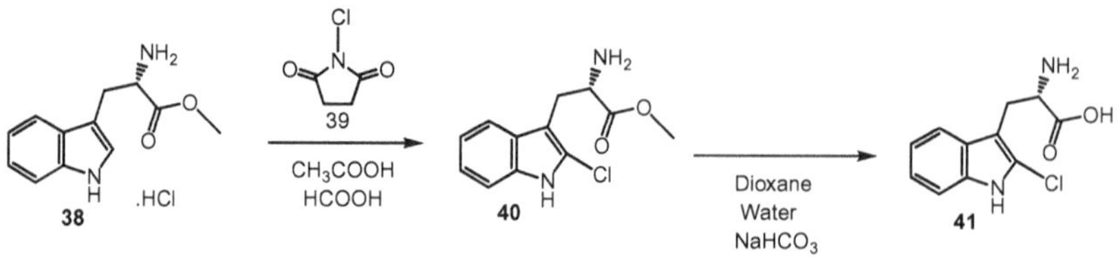

Fig. 10 Route for the chemoenzymatic synthesis of 2′-chloro-L-tryptophan (**41**).

9. Purify compound **41** by HPLC using Supelcosil LC-18 (25 cm × 10 mm, 5 μm particles) semiprep column (Fig. 10).
10. To synthesize 2′-bromo-L-tryptophan (**42**), repeat the earlier procedure using L-tryptophan methyl ester.HBr and N-bromosuccinimide.

When the NosL catalyzed reaction was carried out with 2′-chloro-L-tryptophan (**41**), HPLC analysis showed complete substrate consumption and exclusive formation 3-methylindole-2-carboxylic acid (**11**). Additionally, cyanide (**5**) was the only byproduct generated. The ratio of cyanide (**5**) to 3-methylindole-2-carboxylic acid (**11**) was 0.9:1. Similar results were obtained when the NosL catalyzed reaction was performed in presence of 2′-bromo-L-tryptophan (**42**) (Bhandari et al., 2018). These results are consistent with the NosL-catalyzed formation of **47** (Fig. 11).

4. SUMMARY AND CONCLUSIONS

NosL catalyzes a complex rearrangement reaction that converts L-tryptophan (**10**) to 3-methyl-2-indolic acid (**11**) and generates cyanide as byproduct. However, under in vitro conditions, we predominantly observe formation of 3-methyl indole (**15**), cyanide (**5**), and CO_2/bicarbonate for reasons that are not yet understood.

Our current mechanistic proposal for the NosL catalyzed reaction is shown in Fig. 11. In this proposal, the 5′-deoxyadenosyl radical (Ado•) abstracts a H-atom from the amino group of L-tryptophan (**10**). The amine radical (**43**) further undergoes a β-scission reaction to form **22** and the formate radical (**44**) which then adds to the C'_2 of the indole to form **45**. Loss of H_2CN^+ by a second β-scission reaction gives **46** which can also be described as an aryl radical anion **46′**. Protonation of this radical by the strongly acidic H_2CN^+ followed by electron transfer back to the $[4Fe–4S]^{+2}$ cluster gives the cation **48**. This can lose CO_2 or a proton to give the experimentally observed reaction products **15** and **11**, respectively. Alternatively, loss of a formate radical from **47** could also give **15**. This mechanism is consistent

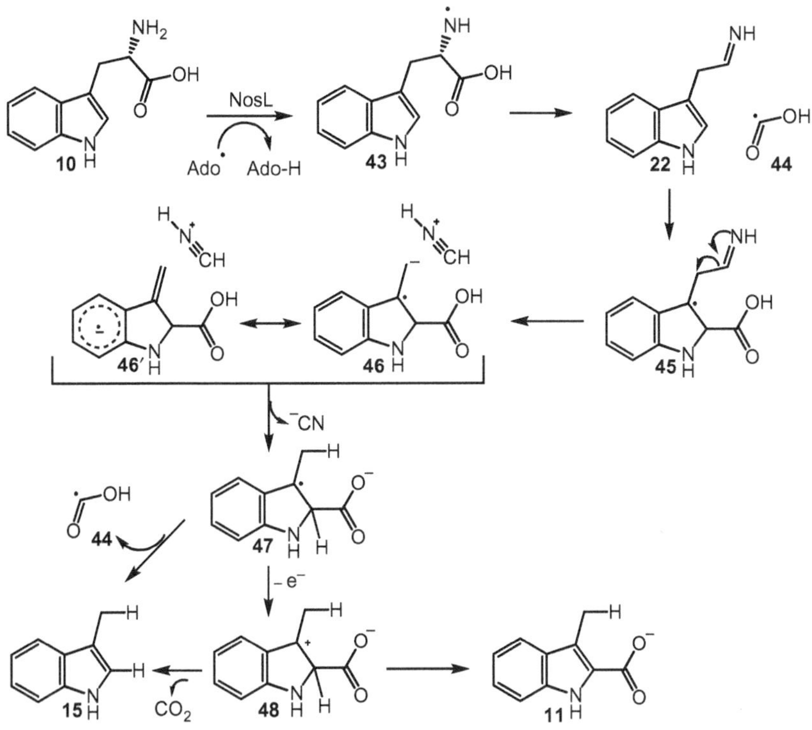

Fig. 11 Current mechanistic proposal for the NosL-catalyzed reaction of L-tryptophan.

with the enzyme structure, with EPR studies that support of the formation of **45** (Sicoli et al., 2016) and with the formation of cyanide as a reaction product. The mechanism is also consistent with numerous other studies using substrate analogs and with trapping studies using 2-chlorotryptophan that support of the formation of **47** and also suggest that β-scission of the C—Cl bond in **45** is slower than loss of protonated HCN.

In this chapter, we have described the methodologies developed and adopted by our lab for the overexpression and purification of NosL, the reconstitution of the enzymatic reaction, and the use of tryptophan analogs to probe the enzyme mechanism. The identification of cyanide as a major NosL byproduct connects NosL chemistry with that of HydG—the radical SAM enzyme that forms cyanide and carbon monoxide from tyrosine during the biosynthesis of the metallo cluster of the [FeFe] hydrogenase (Fig. 1) (Pagnier, Martin, Zeppieri, Nicolet, & Fontecilla-Camps, 2016). The methods described may be generally useful in mechanistic studies on the other aromatic amino acid lyases.

FUNDING SOURCES

This research was supported by the Robert A. Welch Foundation Grant A-0034.

ORCID

Dhananjay M. Bhandari: 0000-0003-4250-3303.
Dmytro Fedoseyenko: 0000-0001-7776-8278.
Tadhg P. Begley: 0000-0001-5134-2623.

REFERENCES

Beinert, H. (1983). Semi-micro methods for analysis of labile sulfide and of labile sulfide plus sulfane sulfur in unusually stable iron–sulfur proteins. *Analytical Biochemistry*, *131*(2), 373–378.

Bhandari, D. M., Fedoseyenko, D., & Begley, T. P. (2016). Tryptophan lyase (NosL): A cornucopia of 5′-deoxyadenosyl radical mediated transformations. *Journal of the American Chemical Society*, *138*(50), 16184–16187.

Bhandari, D. M., Fedoseyenko, D., & Begley, T. P. (2018). Mechanistic studies on tryptophan lyase (NosL): Identification of cyanide as a reaction product. *Journal of the American Chemical Society*, *140*(2), 542–545.

Bhandari, D. M., Xu, H., Nicolet, Y., Fontecilla-Camps, J. C., & Begley, T. P. (2015). Tryptophan lyase (NosL): Mechanistic insights from substrate analogues and mutagenesis. *Biochemistry*, *54*(31), 4767–4769.

Broderick, J. B., Duffus, B. R., Duschene, K. S., & Shepard, E. M. (2014). Radical S-adenosylmethionine enzymes. *Chemical Reviews*, *114*(8), 4229–4317.

Bruender, N. A., Young, A. P., & Bandarian, V. (2015). Chemical and biological reduction of the radical SAM enzyme CPH4 synthase. *Biochemistry*, *54*(18), 2903–2910.

Cancilla, D. A., & Hee, S. S. Q. (1992). O-(2,3,4,5,6-pentafluorophenyl) methylhydroxylamine hydrochloride: A versatile reagent for the determination of carbonyl-containing compounds. *Journal of Chromatography A*, *627*(1–2), 1–16.

Driesener, R. C., Challand, M. R., McGlynn, S. E., Shepard, E. M., Boyd, E. S., Broderick, J. B., et al. (2010). [FeFe]-hydrogenase cyanide ligands derived from S-adenosylmethionine-dependent cleavage of tyrosine. *Angewandte Chemie International Edition*, *49*(9), 1687–1690.

Duffus, B. R., Ghose, S., Peters, J. W., & Broderick, J. B. (2014). Reversible H atom abstraction catalyzed by the radical S-adenosylmethionine enzyme HydG. *Journal of the American Chemical Society*, *136*(38), 13086–13089.

Frey, P. A., Hegeman, A. D., & Ruzicka, F. J. (2008). The radical SAM superfamily. *Critical Reviews in Biochemistry and Molecular Biology*, *43*(1), 63–88.

Gerlt, J. A. (2017). Genomic enzymology: Web tools for leveraging protein family sequence–function space and genome context to discover novel functions. *Biochemistry*, *56*(33), 4293–4308.

Ji, X., Li, Y., Ding, W., & Zhang, Q. (2015). Substrate-tuned catalysis of the radical S-adenosyl-L-methionine enzyme NosL involved in nosiheptide biosynthesis. *Angewandte Chemie International Edition*, *54*(31), 9021–9024.

Joshi, S., Mahanta, N., Fedoseyenko, D., Williams, H., & Begley, T. P. (2017). Aminofutalosine synthase: Evidence for captodative and aryl radical intermediates using β-scission and SRN1 trapping reactions. *Journal of the American Chemical Society*, *139*(32), 10952–10955.

Kawasaki, H., Bauerle, R., Zon, G., Ahmed, S., & Miles, E. (1987). Site-specific mutagenesis of the alpha subunit of tryptophan synthase from Salmonella typhimurium. Changing arginine 179 to leucine alters the reciprocal transmission of substrate-induced conformational changes between the alpha and beta 2 subunits. *Journal of Biological Chemistry*, *262*(22), 10678–10683.

Kriek, M., Martins, F., Challand, M. R., Croft, A., & Roach, P. L. (2007). Thiamine biosynthesis in Escherichia coli: Identification of the intermediate and by-product derived from tyrosine. *Angewandte Chemie International Edition*, *46*(48), 9223–9226.

Lanz, N. D., Grove, T. L., Gogonea, C. B., Lee, K.-H., Krebs, C., & Booker, S. J. (2012). Chapter seven—RlmN and AtsB as models for the overproduction and characterization of radical SAM proteins. In D. A. Hopwood (Ed.), *Methods in Enzymology: Vol. 516* (pp. 125–152). Academic Press.

Mehta, A. P., Abdelwahed, S. H., Mahanta, N., Fedoseyenko, D., Philmus, B., Cooper, L. E., et al. (2015). Radical S-adenosylmethionine (SAM) enzymes in cofactor biosynthesis: A treasure trove of complex organic radical rearrangement reactions. *Journal of Biological Chemistry*, *290*(7), 3980–3986.

Mettert, E. L., & Kiley, P. J. (2015). How is Fe–S cluster formation regulated? *Annual Review of Microbiology*, *69*(1), 505–526.

Nicolet, Y., Zeppieri, L., Amara, P., & Fontecilla-Camps, J. C. (2014). Crystal structure of tryptophan lyase (NosL): Evidence for radical formation at the amino group of tryptophan. *Angewandte Chemie*, *126*(44), 12034–12038.

O'Leary, M. H. (1982). Phosphoenolpyruvate carboxylase: An enzymologist's view. *Annual Review of Plant Physiology*, *33*(1), 297–315.

Pagnier, A., Martin, L., Zeppieri, L., Nicolet, Y., & Fontecilla-Camps, J. C. (2016). CO and CN− syntheses by [FeFe]-hydrogenase maturase HydG are catalytically differentiated events. *Proceedings of the National Academy of Sciences*, *113*(1), 104–109.

Philmus, B., Decamps, L., Berteau, O., & Begley, T. P. (2015). Biosynthetic versatility and coordinated action of 5′-deoxyadenosyl radicals in deazaflavin biosynthesis. *Journal of the American Chemical Society*, *137*(16), 5406–5413.

Shepard, E. M., Duffus, B. R., George, S. J., McGlynn, S. E., Challand, M. R., Swanson, K. D., et al. (2010). [FeFe]-hydrogenase maturation: HydG-catalyzed synthesis of carbon monoxide. *Journal of the American Chemical Society*, *132*(27), 9247–9249.

Sicoli, G., Mouesca, J.-M., Zeppieri, L., Amara, P., Martin, L., Barra, A. L., et al. (2016). Fine-tuning of a radical-based reaction by radical S-adenosyl-L-methionine tryptophan lyase. *Science*, *351*(6279), 1320–1323.

Sofia, H. J., Chen, G., Hetzler, B. G., Reyes-Spindola, J. F., & Miller, N. E. (2001). Radical SAM, a novel protein superfamily linking unresolved steps in familiar biosynthetic pathways with radical mechanisms: Functional characterization using new analysis and information visualization methods. *Nucleic Acids Research*, *29*(5), 1097–1106.

Tracqui, A., Raul, J., Geraut, A., Berthelon, L., & Ludes, B. (2002). Determination of blood cyanide by HPLC–MS. *Journal of Analytical Toxicology*, *26*(3), 144–148.

Van De Bogart, M., & Beinert, H. (1967). Micro methods for the quantitative determination of iron and copper in biological material. *Analytical Biochemistry*, *20*, 325–334.

Wagner, P. J., Lindstrom, M. J., Sedon, J. H., & Ward, D. R. (1981). Photochemistry of Delta-haloketones: Anchimeric assistance in triplet-state. Gamma.-hydrogen abstraction and. Beta.-elimination of halogen atoms from the resulting diradicals. *Journal of the American Chemical Society*, *103*(13), 3842–3849.

Yu, Y., Duan, L., Zhang, Q., Liao, R., Ding, Y., Pan, H., et al. (2009). Nosiheptide biosynthesis featuring a unique indole side ring formation on the characteristic thiopeptide framework. *ACS Chemical Biology*, *4*(10), 855–864.

Zhang, Q., Li, Y., Chen, D., Yu, Y., Duan, L., Shen, B., et al. (2011). Radical-mediated enzymatic carbon chain fragmentation-recombination. *Nature Chemical Biology*, 7(3), 154–160.

CHAPTER SIX

Aminofutalosine Synthase (MqnE): A New Catalytic Motif in Radical SAM Enzymology

Sumedh Joshi, Dmytro Fedoseyenko, Nilkamal Mahanta, Tadhg P. Begley[1]

Department of Chemistry, Texas A&M University, College Station, TX, United States
[1]Corresponding author: e-mail address: begley@tamu.edu

Contents

Abstract

Aminofutalosine synthase (MqnE) is a radical SAM enzyme involved in the futalosine-dependent menaquinone biosynthetic pathway. Its ability to add the 5′-deoxyadenosyl radical to the substrate—rather than abstract a hydrogen atom—and to catalyze radical addition to a stable benzene ring gives it a unique place in the radical SAM superfamily and required the development of new strategies for trapping radical intermediates. This chapter describes the methodologies used for enzyme overexpression, purification, and in vitro reconstitution. We also describe the development of fast, radical triggered, carbon–halogen bond fragmentation reactions for the trapping of intermediates. We anticipate that these methods will be of general use in the study of other transient enzymatic radicals.

Methods in Enzymology, Volume 606
ISSN 0076-6879
https://doi.org/10.1016/bs.mie.2018.05.002

1. INTRODUCTION

Radical SAM enzymes, first classified as a superfamily in 2001 (Sofia, Chen, Hetzler, Reyes-Spindola, & Miller, 2001), utilize a [4Fe−4S] cluster and *S*-adenosylmethionine (SAM, **9**) to initiate a diverse set of radical reactions that are otherwise difficult or impossible to catalyze by polar mechanisms (Broderick, Duffus, Duschene, & Shepard, 2014). Currently, there are >250,000 sequences annotated in this superfamily (Finn et al., 2016; Gerlt, 2017). These enzymes harbor a redox active [4Fe–4S] cluster, which is typically ligated by three conserved cysteine residues constituting a CX_3CX_2C motif (Vey & Drennan, 2011). Radical SAM enzymes catalyze the reductive homolytic cleavage of the C–S bond of SAM **9** to generate the highly reactive 5′-deoxyadenosyl radical **13**. This radical typically abstracts a hydrogen atom from the substrate directly or via a protein residue-based radical. The resulting substrate radical usually undergoes a rearrangement or fragmentation reaction to give the product radical which is typically quenched by an electron transfer or H atom abstraction (Ruszczycky, Zhong, & Liu, 2018). Detailed mechanistic investigations on these enzymes have been limited by poorly understood enzyme inactivation reactions and by the absence of generally applicable strategies for trapping transient radical intermediates (Bhandari, Fedoseyenko, & Begley, 2016; Nicolet, Rohac, Martin, & Fontecilla-Camps, 2013). Radical SAM enzymes are especially prevalent in cofactor biosynthesis and have been shown to catalyze complex radical rearrangement reactions (Mehta et al., 2015). The recently discovered futalosine-dependent menaquinone biosynthetic pathway also features two radical SAM enzymes (MqnE and MqnC) which play a central role in the formation of the naphthoquinone core of menaquinone (Hiratsuka et al., 2008; Mahanta, Fedoseyenko, Dairi, & Begley, 2013).

The current status of this pathway is outlined in Fig. 1. MqnA, a chorismate dehydratase, catalyzes the conversion of chorismate **1**, into VOBA, **2** the first aromatic molecule in the pathway. The radical SAM enzyme MqnE converts VOBA **2**, into aminofutalosine **3** (Mahanta et al., 2013). In the next step, aminofutalosine deaminase (MqnF) deaminates **3** to form futalosine **4** (Goble et al., 2013; Hein, Klimmek, Polly, Kern, & Simon, 2017; Mahanta et al., 2013) which is then converted to dehypoxanthine futalosine **5**, by futalosine hydrolase (MqnB) (Hiratsuka, Itoh, Seto, & Dairi, 2009).

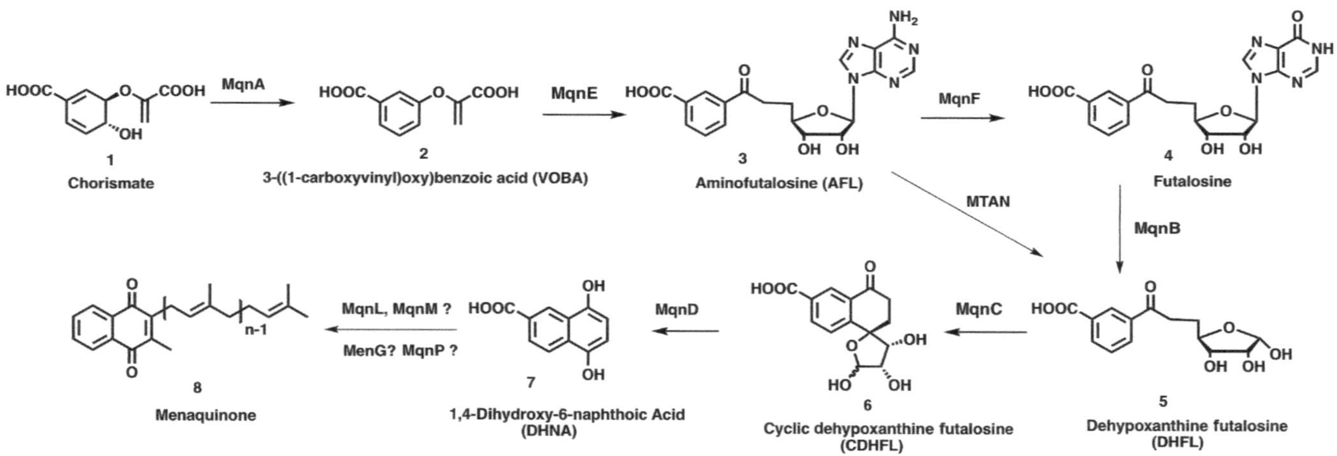

Fig. 1 The futalosine-dependent menaquinone biosynthetic pathway.

In *Campylobacter jejuni* and *Helicobacter pylori*, 5-methylthioadenosine nucleosidase (MTAN) is shown to form **5** directly from **3** (Banco et al., 2016; Li, Apel, Gaynor, & Tanner, 2011). Compound **5** is converted to cyclic dehypoxanthine futalosine **6**, by the radical SAM enzyme MqnC (Cooper et al., 2013; Hiratsuka et al., 2008). MqnD converts **6** to 8-dihydroxy-2-naphthoic acid **7** (Hiratsuka et al., 2008). Based on sequence analysis, the final steps in the pathway are proposed to be catalyzed by MqnP (SCO4491, prenylation), MqnL and MqnM (SCO4490 and SCO4492, decarboxylation), and MenG (SCO4556, methylation) (Cotrim et al., 2017; Hiratsuka et al., 2008; Ravcheev & Thiele, 2016).

The radical SAM enzyme aminofutalosine synthase (MqnE) catalyzes a remarkable reaction involving 5′-deoxyadenosyl radical addition to the vinyl ether of **2**. This utilization of the adenosyl radical for C–C bond formation represents a new catalytic motif in radical SAM enzymology, where all previously characterized enzymes used the adenosyl radical to abstract a hydrogen atom from the cosubstrate or from an active-site glycine (Broderick et al., 2014). Other examples of this efficient catalytic motif have now started to emerge in natural product biosynthetic pathways (Liu et al., 2014).

In this chapter, we describe the reconstitution of the aminofutalosine synthase-catalyzed reaction, and the synthesis and use of halogenated substrate analogs as mechanistic probes.

2. CLONING, OVEREXPRESSION, AND PURIFICATION OF MqnE

2.1 Gene Cloning and Expression Strategy

The catalytic activity of radical SAM enzymes is dependent on the [4Fe–4S] cluster present in the active site and these enzymes expressed on their own typically are cluster deficient. Cluster formation can be enhanced by treating the apo-enzyme with iron and sulfide (Lanz et al., 2012) or in vivo by coexpression with the Fe/S cluster biosynthesis genes. In *Escherichia coli* there are two Fe–S cluster biogenesis pathways: The ISC (Iron Sulfur Cluster) system is utilized for cluster biogenesis under standard growth conditions, whereas the SUF (Sulfur Mobilization) operon is induced under stress conditions (Mettert & Kiley, 2015). Along with ISC (Lanz et al., 2012), the SUF system has also been employed for the efficient in vivo [4Fe–4S] cluster incorporation into overexpressed radical SAM enzymes (Bhandari, Xu, Nicolet, Fontecilla-Camps, & Begley, 2015; Cooper et al., 2013; Hänzelmann & Schindelin, 2004; Joshi, Mahanta, Fedoseyenko, Williams, & Begley, 2017; Wang, Schnell, Baumann, Müller, & Begley, 2017).

MqnE (TTHA0804) from *Thermus thermophilus* HB8 was PCR amplified and cloned into a pTHT vector. This vector is a modified version of pET28b (Novagen) with a TEV protease cleavage site after the N-terminal His_6-tag, a kanamycin resistance marker, and uses a strong T7 promoter compatible with the T7 expression host strain. The use of a T7 expression host also allows the coexpression of the pSuf plasmid required for the in vivo assembly of the [4Fe–4S] cluster. This plasmid encodes the *E. coli* SufABCDSE operon cloned into the pACYC-Duet-1 vector (Hänzelmann & Schindelin, 2004) and has a chloramphenicol resistance marker. This construct resulted in high protein yields of catalytically active enzyme (8 mg/L of culture, >25 turnovers) with good reconstitution of the [4Fe–4S] cluster (4.3 irons and 3.4 sulfides per monomer) (Joshi et al., 2017).

2.2 Protein Overexpression Protocol

1. Insert the pSuf and pTHT-MqnE plasmids into *E. coli* BL 21(DE3) using standard transformation protocols (Green & Rogers, 2013; Lessard, 2013).
2. Use a freshly transformed single colony to inoculate 100 mL of LB medium supplemented with 34 μg/mL chloramphenicol and 40 μg/mL kanamycin.
3. Incubate the seed culture at 37°C overnight with vigorous shaking (220 rpm).
4. Inoculate 9 L of sterile LB medium (6 × 1.5 L flasks), supplemented with 34 μg/mL chloramphenicol and 40 μg/mL kanamycin, with 90 mL of this starter culture. Ultra-Yield Flasks (Thomson), equipped with AirOtop enhanced seals, were used to maximize cell yields and protein overexpression.
5. Grow the culture at 37°C with shaking (220 rpm) till OD_{600} ~0.6.
6. Cool the culture flasks in a cold room (~1 h at 4°C, no shaking).
7. Supplement each flask (1.5 L) with ferrous ammonium sulfate (200 mg) and cysteine (200 mg).
8. Induce protein expression by adding isopropyl-β-D-thiogalactoside (IPTG) to a final concentration of 500 μ*M*.

9. Incubate the culture at 15°C for ~16–18 h with slow shaking (120 rpm).
10. Harvest the cells by centrifugation and store the wet cell paste in liquid nitrogen. Typical overexpression yields are ~2 g/L of dark colored wet cell paste.

2.3 Protein Purification Protocol

All the steps for the purification of MqnE should be carried out in an anaerobic chamber (COY Laboratories). To make oxygen-free buffers, buffers are autoclaved and immediately equilibrated in the anaerobic chamber for 12 h.

1. Thaw the cell pellets (~10 g) in an anaerobic chamber and resuspend in 50–60 mL of lysis buffer at room temperature (100 m*M* Tris–HCl, pH 7.5, 10 mg lysozyme, 1000 U Benzonase) in a stainless-steel cup.
2. Stir the suspension for ~1 h on an ice bath with a magnetic stirrer.
3. Stop the stirring and insert the sonicator probe (Qsonica55 with ¼″ tip).
4. Sonicate for 90–120 s with 90% amplitude. Repeat this step 4–5 times with 8–10 min between each cycle.
5. Pour the dark green-black colored cell lysate into centrifuge tubes with a sealing cap assembly (Thermo Scientific/Nalgene). Tighten the caps on the full tubes and seal with Parafilm M. The sealed tubes can now be removed from the anaerobic chamber.
6. Centrifuge the cell lysate at 39,000 × *g* for 25 min at 4°C.
7. In the anaerobic chamber, wash a HisTrap column (GE Healthcare) with 5 column volumes of column buffer (100 m*M* Tris–HCl, pH 7.5) and wash the Econo-Pac® 10DG prepacked gravity flow Desalting column (Bio-Rad) with 5 column volumes of the desalting buffer (100 m*M* potassium phosphate, 30% glycerol, pH 7.5).
8. Open the centrifuge tubes in the anaerobic chamber and filter the cell lysate through a 0.45 μm filter.
9. Load the lysate onto the HisTrap column at a flow rate of 2 mL/min. When lysate loading is complete, MqnE will be visible as a dark brown band at the top of the column.
10. Wash the HisTrap column with 10–20 column volumes of wash buffer (100 m*M* Tris–HCl, 30 m*M* imidazole, 300 m*M* NaCl, pH 7.5) at a flow rate of 3–4 mL/min.
11. Elute MqnE from the HisTrap column with the elution buffer (100 m*M* Tris–HCl, 250 m*M* imidazole, 300 m*M* NaCl, pH 7.5) at a flow rate of 1–2 mL/min.

12. Collect the dark golden/brown colored fractions and desalt the protein using an Econo-Pac 10DG desalting column eluting with desalting buffer (100 m*M* potassium phosphate, 30% glycerol, pH 7.5). Make aliquots of 250 μL each in the Eppendorf tubes having a pinhole on their lid (pinhole is to avoid the lid breaking open on freeze–thaw in liquid nitrogen). Store the protein submerged in liquid nitrogen in a N_2 dewar. Typical protein yields are ~8 mg MqnE per liter of cell culture.
13. The MqnE concentration was measured using the absorbance at 280 nm. The extinction coefficient was calculated as 61,880 M^{-1} cm^{-1} at 280 n*M* (assuming all cysteines were reduced) using the ProtParam tool of the ExPASy proteomics server. Iron and sulfide content was determined using the ferene and methylene blue assay, respectively (Beinert, 1983; Kennedy et al., 1984; Van De Bogart & Beinert, 1967).

2.4 SDS-PAGE and UV–Visible Spectroscopic Analysis

The purity of the protein was assessed by SDS-PAGE analysis (Brunelle & Green, 2014) which displayed a band at 42 kDa matching the expected molecular mass of TtMqnE (Fig. 2). The UV–visible spectrum of MqnE displays a broad feature at 420 nm corresponding to the cysteinyl thiolate-to-Fe(III) charge transfer transition of the $[4Fe–4S]^{+2}$ cluster (Fig. 2). This feature decreases in intensity on the addition of excess sodium dithionite and completely disappears on the addition of excess Ti(III) citrate indicating conversion of the cluster to the reduced $[4Fe–4S]^{+1}$ state. It is evident

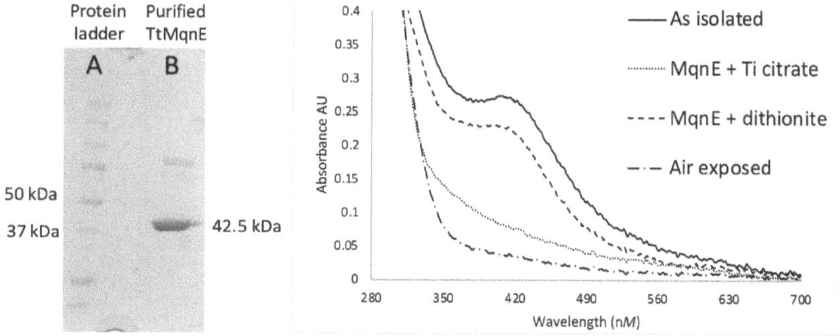

Fig. 2 Characterization of MqnE. I: SDS-PAGE analysis of MqnE, II: UV–visible spectra of as isolated, reduced, and air exposed MqnE.

from the spectra that Ti(III) citrate ($E_M \sim -510$ mV; Holliger, Pierik, Reijerse, & Hagen, 1993; Zehnder & Wuhrmann, 1976) is more efficient at reducing MqnE than sodium dithionite ($E_M \sim -460$ mV at 2 m*M*; Mayhew, 1978).

3. ENZYME ACTIVITY ASSAY

All the assays were conducted in the anaerobic chamber in 100 m*M* potassium phosphate buffer pH 7.5. The reaction components were 2 m*M* sodium dithionite, 100 μ*M* MqnE, 2–4 m*M* substrate (or substrate analog), and 1 m*M* SAM. The reaction mixture was incubated at room temperature (25°C) for 4–6 h and quenched by ultracentrifugation through a 10K polyethersulfone membrane centrifugal filter (VWR) at 13,500 × *g* for 20 min at 4°C. The sodium dithionite reducing agent can be replaced with flavodoxin/flavodoxin reductase/NADPH or Ti(III) citrate depending on the experiment. A typical HPLC chromatogram for the MqnE-catalyzed reaction is shown in Fig. 3.

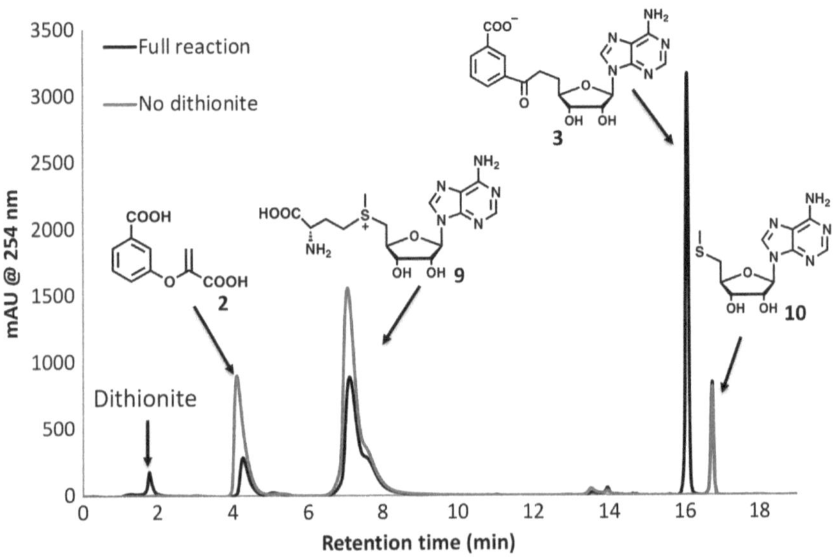

Fig. 3 HPLC analysis of the MqnE-catalyzed reaction.

HPLC parameters

An Agilent 1260 HPLC, equipped with a quaternary pump and a ZORBAX Eclipse XDB-C18 column (15 cm × 4.6 mm, 5 μm particles, Agilent Technologies), was used. Eluted compounds were detected by absorbance at 254, 260, 280, 225, 320, 210, and 475 nm using a diode array UV–visible detector. Typical injection volumes were in the range of 10–50 μL. Data were processed using ChemStation ver. B.04.01 SP1 (Agilent Technologies).

HPLC conditions

A. Water.

B. 100 m*M* potassium phosphate buffer, pH 6.6 or 10 m*M* ammonium acetate pH 6.6.

C. Methanol.

HPLC method

0 min—100% B; 5 min—100% B; 14 min—7% A, 70% B, 23% C; 25 min—25% A, 0% B, 75% C; 28 min—25% A, 0% B, 75% C; 32 min—100% B; 36 min—100% B.

or

0 min—100% B; 5 min—100% B; 12 min—48% A, 40% B, 12% C; 14 min—50% A, 30% B, 20% C; 18 min—30% A, 10% B, 60% C; 20 min—100% B, 25 min—100% B.

4. SYNTHESIS OF $^2H_4$-SAM

The MqnE reaction utilizes SAM as a cosubstrate and incorporates the 5′-deoxyadenosyl moiety into the reaction product. Thus, deuterium-labeled SAM is a useful molecule for the characterization of intermediates by MS and EPR spectroscopy (Wilcoxen, Bruender, Bandarian, & Britt, 2018). The chemical synthesis of SAM is challenging because of its facile epimerization and hydrolysis reactions (Huber et al., 2016; Wu, Huskey, Borchardt, & Schowen, 1983). The method described here utilizes cell-free lysate from a SAM synthetase (MetK) overexpression strain (*E. coli* DM22(pK8), 200-fold overexpression) to enzymatically synthesize SAM which is then purified by cation exchange chromatography (Fig. 4) (Markham, DeParasis, & Gatmaitan, 1984). SAM synthesis by this method is usually associated with severe product inhibition which is overcome by the addition of β-mercaptoethanol or

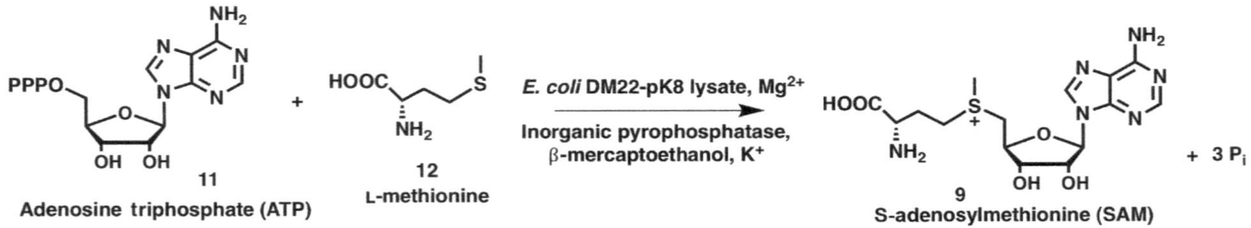

Fig. 4 Enzymatic synthesis of SAM **9** from ATP **11**.

p-toluenesulfonic acid to the reaction mixture (Park, Tai, Roessner, & Scott, 1995).

1. Grow a 100mL seed culture of *E. coli* DM22(pK8) in LB medium supplemented with 30 μg/mL tetracycline.
2. Inoculate 60mL of this culture into 6L of sterile LB medium (4 × 1.5L flasks) supplemented with 30 μg/mL tetracycline.
3. Allow the cultures to grow to stationary phase (12–14h) at 37°C with vigorous shaking (220rpm).
4. Harvest the cells by centrifugation. Typical cell yields are 1.8g/L.
5. Add 30mL lysis buffer (100m*M* Tris–HCl, pH 7.5) to 10g wet cell paste. Also add lysozyme (2mg), Benzonase (200U), and protease inhibitor cocktail (3 "cOmplete" tablets, Roche).
6. Resuspend the cell pellet, sonicate, and remove the cell debris by centrifugation followed by passing through a 0.45 μm filter. This crude lysate can be stored at 4°C (but not at 0°C) for up to 10 days with a minimal loss of activity.
7. The SAM synthesis reaction mixture (3mL) consisted of 600 μL lysate, 6.66m*M* ATP, 8.66m*M* methionine, 100m*M* KCl, 50m*M* $MgCl_2$, 8% (v/v) β-mercaptoethanol, and 1U of inorganic pyrophosphatase.
8. Incubate this reaction mixture at room temperature for 4–5h with monitoring by HPLC.
9. After complete consumption of ATP, acidify to pH 4 (1*M* HCl), centrifuge to remove the precipitated protein, and basify to pH 6 (1*M* NaOH).
10. Load the supernatant dropwise onto a Dowex 50WX8-100 column (7mL bed volume) that had been treated with 6N HCl and thoroughly washed with doubly distilled water.
11. Wash the column with 10–20 column volumes of doubly distilled water until the washings show no absorbance at 260nm.
12. Elute SAM using 15–20mL of 6*M* HCl until the washings show no absorbance at 260nm.
13. Remove the HCl using a rotary evaporator (40°C).
14. Redissolve the SAM in 10m*M* ammonium acetate buffer. Adjust the pH to neutral and lyophilize.
15. Store SAM as a lyophilized powder at −80°C.

LCMS analysis of the MqnE reaction using natural abundance SAM and $^{2}H_4$-SAM is shown in Figs. 5A and B, respectively.

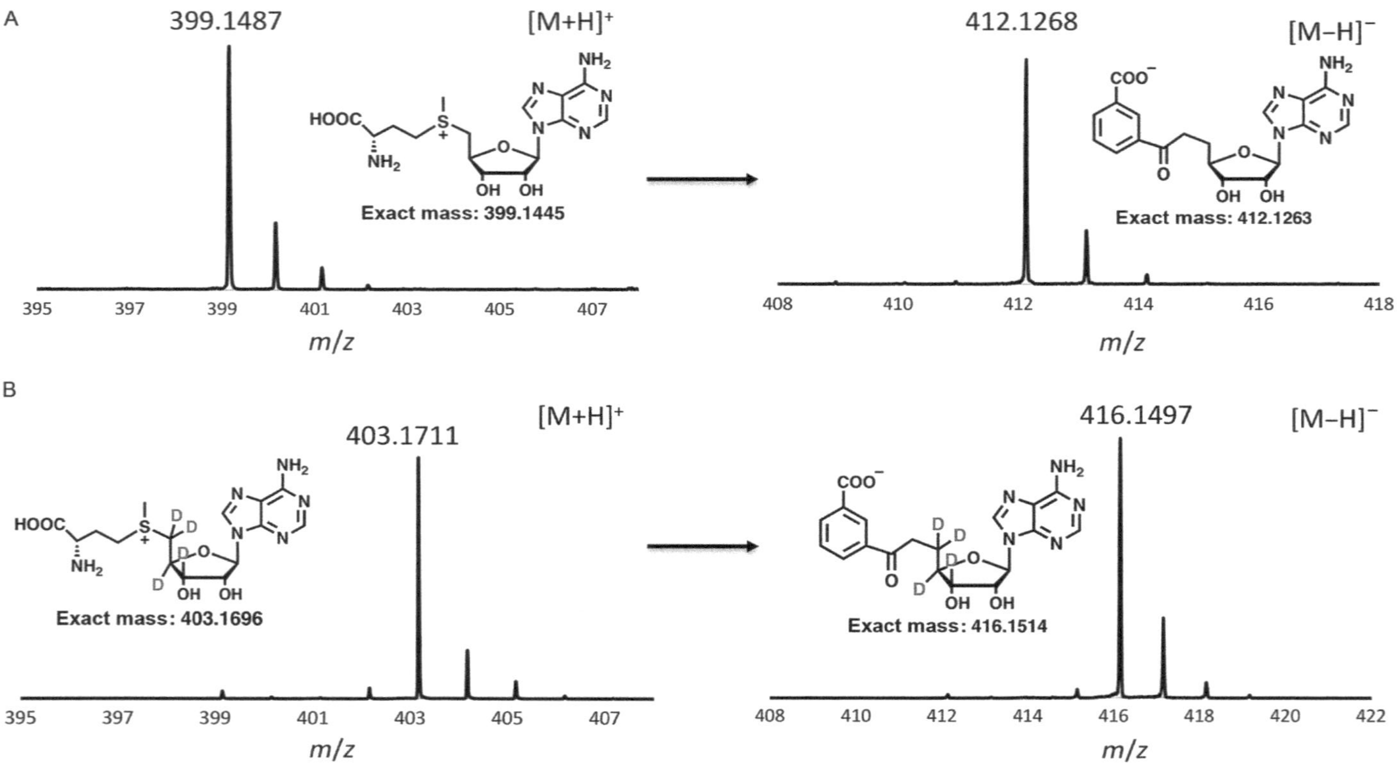

Fig. 5 (A) and (B) Mass spectra showing aminofutalosine and $^{2}H_{4}$-aminofutalosine synthesized from SAM and $^{2}H_{4}$-SAM, respectively.

5. RADICAL TRAPPING STRATEGIES FOR MqnE

Our current mechanistic proposal for the MqnE-catalyzed reaction is outlined in Fig. 6 (Joshi et al., 2017). In this proposal, the 5′-deoxyadenosyl radical **13** adds to the double bond of **2** to form the captodative radical **14**. Rearrangement of **14**, via the spiroepoxide **15**, generates **16**. Decarboxylation followed by deprotonation to form the aryl radical anion **18** and an electron transfer to the $[4Fe–4S]^{+2}$ cluster completes the reaction. The next sections (5.1 and 5.2) describe the strategies we have utilized to trap the captodative and aryl radical anion intermediates (**14** and **18**).

It is generally not possible to trap active-site radical intermediates using exogenous traps because the protein active site effectively shields the radical intermediates from the environment. Thus, intramolecular radical traps, such as the cyclopropyl group, are generally used to trap radical intermediates (Newcomb, 2012). This strategy exploits the very rapid ring opening of a cyclopropyl ring adjacent to a radical (Newcomb, 2012). The main limitations of this approach are: (1) the difficulty in synthesizing the cyclopropyl-substituted substrate analog, (2) the relatively large size of the cyclopropyl group, (3) the stereoelectronic requirements for rapid ring opening (i.e., the SOMO must overlap with the σ^* orbital of the scissile C–C bond), and (4) the radical generated by opening the cyclopropyl group frequently inactivates the enzyme making the isolation

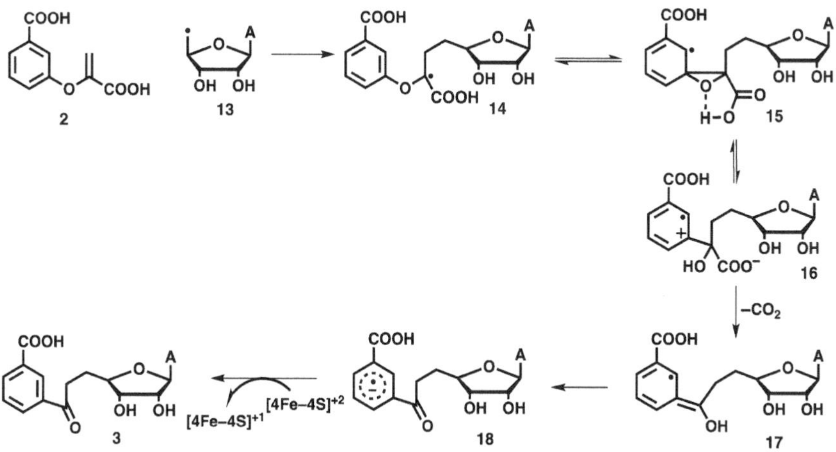

Fig. 6 Current mechanistic proposal for the MqnE-catalyzed reaction.

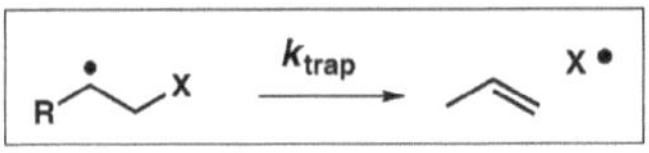

Halogen (X)	k_{trap} (s^{-1})
I	$> 5 \times 10^9$
Br	$\geq 3 \times 10^8$
Cl	4×10^6

Fig. 7 The rapid carbon–halogen β scission reaction as a fast radical probe.

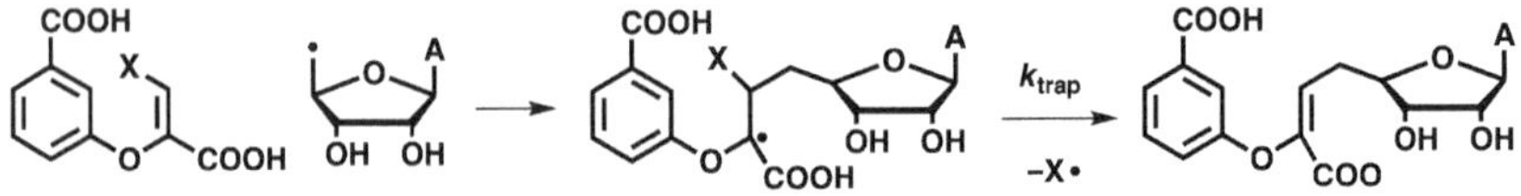

X = Br, 19
X = Cl, 20

13

X = Br, 21
X = Cl, 22

23

Fig. 8 Halogen-based β scission radical trapping in MqnE.

of the products of the trapped radical difficult. To overcome these limitations, we have developed halogen substitution as an alternative small, conformationally mobile, and readily synthesized fast radical trap. This strategy involves design and synthesis of a substrate analog with a halogen beta to the radical. In such systems, carbon–halogen fragmentation is very rapid and tunable by choice of the halogen (Fig. 7) (Wagner, Lindstrom, Sedon, & Ward, 1981). In addition, the halogen radical is small, much less reactive than a primary carbon-centered radical, and likely to diffuse out of the active site. This reduces the likelihood of enzyme inactivation and opens up the possibility of generating the trapped products catalytically.

5.1 Trapping of the Captodative Radical Intermediate

To test these ideas, compound **19** (X = Br) was synthesized as a designed trap for the captodative radical **21**. Analysis of the reaction mixture demonstrated the formation of **23** as the major reaction product. Rearranged products were not detected. Similar results were obtained for compound **20** (X = Cl) suggesting that the rate of the spirocyclization reaction was $<10^4 s^{-1}$. In both cases, the formation of compound **23** was catalytic. We have successfully used similar halogenated substrate analogs to trap radicals for the MoaA- and NosL-catalyzed reactions (Fig. 8) (Bhandari, Fedoseyenko, & Begley, 2018; Mehta, Abdelwahed, Xu, & Begley, 2014).

5.2 Trapping of the Aryl Radical Anion Intermediate

Aryl radical anion **24** undergoes ultrafast substitution. In this reaction, an electron transfer from the SOMO of the aryl radical anion to the C–Br σ* orbital weakens the C–Br bond and results in rapid loss of bromide to give the aryl radical **25**. This radical then reacts with a nucleophile to give **27** (reverse of the **24** to **25** conversion) or abstracts a hydrogen atom from a suitable donor to form **26** (Fig. 9) (Rossi, Pierini, & Peñéñory, 2003; Tanner, Chen, Chen, & Luelo, 1991). This $S_{RN}1$ reaction was ideally suited as a trap for the aryl radical anion intermediate **18** proposed for the MqnE-catalyzed reaction (Fig. 6). Brominated substrate analog **28** was therefore synthesized and tested as a MqnE substrate. HPLC analysis of the resulting reaction mixture demonstrated the formation of **30** and **3** (ratio of the compounds **30**:**3** was ~60:40) consistent with the formation of the aryl radical anion **29** (Fig. 9) (Joshi et al., 2017).

6. SUBSTRATE ANALOG SYNTHESIS

The synthesis of substrate and substrate analogs is a critically important step in any mechanistic investigation. The synthesis of **2**, **19**, and **28** is shown

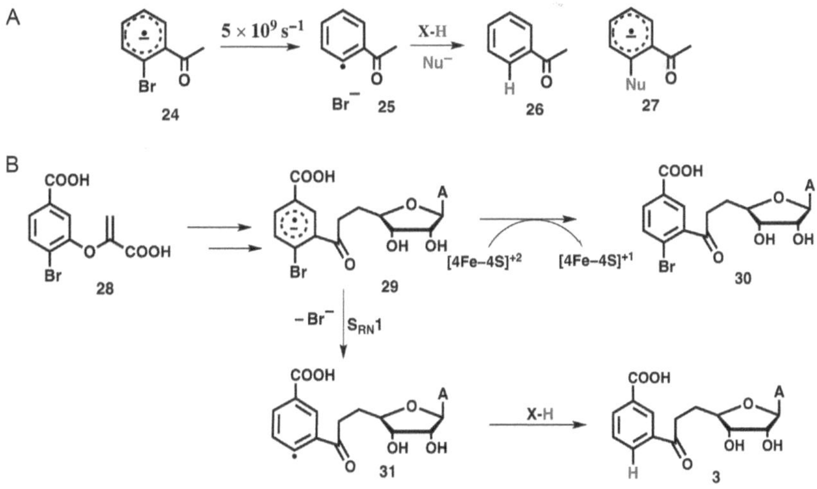

Fig. 9 (A) The $S_{RN}1$ reaction of the bromo-acetophenone radical anion **24**. (B) Mechanistic proposal for the MqnE-catalyzed conversion of **28** to **3**.

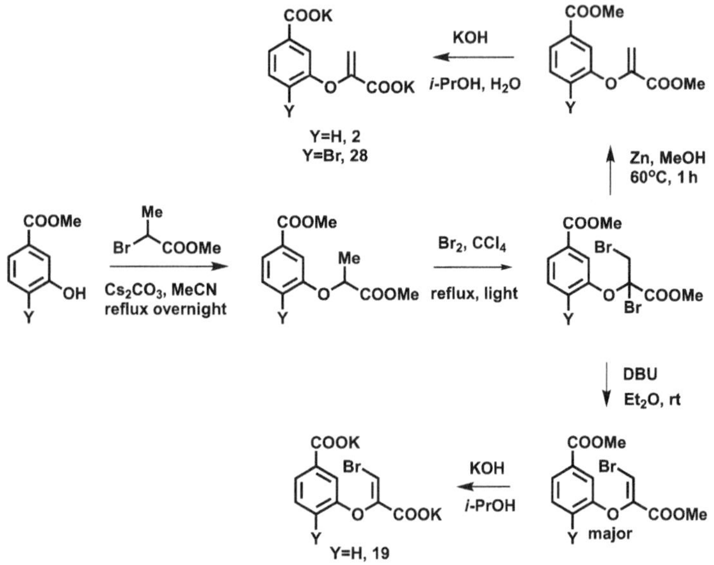

Fig. 10 The synthetic strategy for the synthesis of the MqnE substrate **2** and substrate analogs **28** and **19**.

in Fig. 10. Experimental details can be found in the original publications (Joshi et al., 2017; Mahanta et al., 2013).

7. CONCLUSIONS

MqnE catalyzes the conversion of **2** to **3** and proceeds by the addition of the adenosyl radical to form **14** followed by rearrangement and decarboxylation to give the aryl radical anion **18**. Electron transfer to the iron sulfur cluster completes the reaction (Figs. 6 and 11).

In this chapter, we have described our methodologies for the overexpression and purification of MqnE, the reconstitution of the enzymatic reaction, and the syntheses of deuterated SAM and brominated substrate analogs. These analogs generated experimental support for the formation of the captodative radical **14** and the aryl radical anion **18**. These trapping strategies are likely to be of general utility in the trapping of related radical intermediates in other radical SAM enzyme-catalyzed rearrangement reactions.

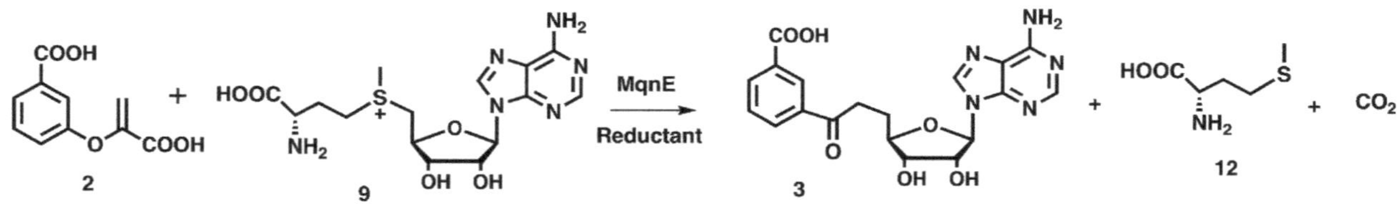

Fig. 11 The MqnE-catalyzed reaction.

ACKNOWLEDGMENTS

This research was supported by a grant from the National Science Foundation (1507191) and by the Robert A. Welch Foundation (A0034). Orcid IDs of authors: Sumedh Joshi: 0000-0002-0407-2920; Dmytro Fedoseyenko: 0000-0001-7776-8278; Nilkamal Mahanta: 0000-0001-8901-2531; Tadhg P. Begley: 0000-0001-5134-2623.

REFERENCES

Banco, M. T., Mishra, V., Ostermann, A., Schrader, T. E., Evans, G. B., Kovalevsky, A., et al. (2016). Neutron structures of the *Helicobacter pylori* 5′-methylthioadenosine nucleosidase highlight proton sharing and protonation states. *Proceedings of the National Academy of Sciences of the United States of America, 113*(48), 13756–13761.

Beinert, H. (1983). Semi-micro methods for analysis of labile sulfide and of labile sulfide plus sulfane sulfur in unusually stable iron-sulfur proteins. *Analytical Biochemistry, 131*(2), 373–378.

Bhandari, D. M., Fedoseyenko, D., & Begley, T. P. (2016). Tryptophan lyase (NosL): A cornucopia of 5′-deoxyadenosyl radical mediated transformations. *Journal of the American Chemical Society, 138*(50), 16184–16187.

Bhandari, D. M., Fedoseyenko, D., & Begley, T. P. (2018). Mechanistic studies on tryptophan lyase (NosL): Identification of cyanide as a reaction product. *Journal of the American Chemical Society, 140*(2), 542–545.

Bhandari, D. M., Xu, H., Nicolet, Y., Fontecilla-Camps, J. C., & Begley, T. P. (2015). Tryptophan lyase (NosL): Mechanistic insights from substrate analogues and mutagenesis. *Biochemistry, 54*(31), 4767–4769.

Broderick, J. B., Duffus, B. R., Duschene, K. S., & Shepard, E. M. (2014). Radical S-adenosylmethionine enzymes. *Chemical Reviews, 114*(8), 4229–4317.

Brunelle, J. L., & Green, R. (2014). One-dimensional SDS-polyacrylamide gel electrophoresis (1D SDS-PAGE). *Methods in Enzymology, 541*, 151–159. Elsevier.

Cooper, L. E., Fedoseyenko, D., Abdelwahed, S. H., Kim, S.-H., Dairi, T., & Begley, T. P. (2013). In vitro reconstitution of the radical S-adenosylmethionine enzyme MqnC involved in the biosynthesis of futalosine-derived menaquinone. *Biochemistry, 52*(27), 4592–4594.

Cotrim, C. A., Weidner, A., Strehmel, N., Bisol, T. B., Meyer, D., Brandt, W., et al. (2017). A distinct aromatic prenyltransferase associated with the futalosine pathway. *ChemistrySelect, 2*(29), 9319–9325.

Finn, R. D., Attwood, T. K., Babbitt, P. C., Bateman, A., Bork, P., Bridge, A. J., et al. (2016). InterPro in 2017—Beyond protein family and domain annotations. *Nucleic Acids Research, 45*(D1), D190–D199.

Gerlt, J. A. (2017). Genomic enzymology: Web tools for leveraging protein family sequence–function space and genome context to discover novel functions. *Biochemistry, 56*(33), 4293–4308.

Goble, A. M., Toro, R., Li, X., Ornelas, A., Fan, H., Eswaramoorthy, S., et al. (2013). Deamination of 6-aminodeoxyfutalosine in menaquinone biosynthesis by distantly related enzymes. *Biochemistry, 52*(37), 6525–6536.

Green, R., & Rogers, E. J. (2013). Transformation of chemically competent E. coli. *Changes, 29*, 329–336.

Hänzelmann, P., & Schindelin, H. (2004). Crystal structure of the S-adenosylmethionine-dependent enzyme MoaA and its implications for molybdenum cofactor deficiency in humans. *Proceedings of the National Academy of Sciences of the United States of America, 101*(35), 12870–12875.

Hein, S., Klimmek, O., Polly, M., Kern, M., & Simon, J. (2017). A class C radical S-adenosylmethionine methyltransferase synthesizes 8-methylmenaquinone. *Molecular Microbiology*, *104*(3), 449–462.

Hiratsuka, T., Furihata, K., Ishikawa, J., Yamashita, H., Itoh, N., Seto, H., et al. (2008). An alternative menaquinone biosynthetic pathway operating in microorganisms. *Science*, *321*(5896), 1670–1673.

Hiratsuka, T., Itoh, N., Seto, H., & Dairi, T. (2009). Enzymatic properties of futalosine hydrolase, an enzyme essential to a newly identified menaquinone biosynthetic pathway. *Bioscience, Biotechnology, and Biochemistry*, *73*(5), 1137–1141.

Holliger, C., Pierik, A. J., Reijerse, E. J., & Hagen, W. R. (1993). A spectroelectrochemical study of factor F430 nickel (II/I) from methanogenic bacteria in aqueous solution. *Journal of the American Chemical Society*, *115*(13), 5651–5656.

Huber, T. D., Wang, F., Singh, S., Johnson, B. R., Zhang, J., Sunkara, M., et al. (2016). Functional AdoMet isosteres resistant to classical AdoMet degradation pathways. *ACS Chemical Biology*, *11*(9), 2484–2491.

Joshi, S., Mahanta, N., Fedoseyenko, D., Williams, H., & Begley, T. P. (2017). Aminofutalosine synthase: Evidence for captodative and aryl radical intermediates using β-scission and SRN1 trapping reactions. *Journal of the American Chemical Society*, *139*(32), 10952–10955.

Kennedy, M. C., Kent, T., Emptage, M., Merkle, H., Beinert, H., & Münck, E. (1984). Evidence for the formation of a linear [3Fe-4S] cluster in partially unfolded aconitase. *Journal of Biological Chemistry*, *259*(23), 14463–14471.

Lanz, N. D., Grove, T. L., Gogonea, C. B., Lee, K.-H., Krebs, C., & Booker, S. J. (2012). RlmN and AtsB as models for the overproduction and characterization of radical SAM proteins. *Methods in Enzymology*, *516*, 125–152. Elsevier.

Lessard, J. C. (2013). Transformation of E. coli via electroporation. *Methods in Enzymology*, *529*, 321–327.

Li, X., Apel, D., Gaynor, E. C., & Tanner, M. E. (2011). 5′-Methylthioadenosine nucleosidase is implicated in playing a key role in a modified futalosine pathway for menaquinone biosynthesis in *Campylobacter jejuni*. *The Journal of Biological Chemistry*, *286*(22), 19392–19398.

Liu, W., Sakr, E., Schaeffer, P., Talbot, H. M., Donisi, J., Härtner, T., et al. (2014). Ribosylhopane, a novel bacterial hopanoid, as precursor of C35 bacteriohopanepolyols in Streptomyces coelicolor A3 (2). *Chembiochem*, *15*(14), 2156–2161.

Mahanta, N., Fedoseyenko, D., Dairi, T., & Begley, T. P. (2013). Menaquinone biosynthesis: Formation of aminofutalosine requires a unique radical SAM enzyme. *Journal of the American Chemical Society*, *135*(41), 15318–15321.

Markham, G. D., DeParasis, J., & Gatmaitan, J. (1984). The sequence of metK, the structural gene for S-adenosylmethionine synthetase in Escherichia coli. *Journal of Biological Chemistry*, *259*(23), 14505–14507.

Mayhew, S. G. (1978). The redox potential of dithionite and SO − 2 from equilibrium reactions with flavodoxins, methyl viologen and hydrogen plus hydrogenase. *The FEBS Journal*, *85*(2), 535–547.

Mehta, A. P., Abdelwahed, S. H., Mahanta, N., Fedoseyenko, D., Philmus, B., Cooper, L. E., et al. (2015). Radical S-adenosylmethionine (SAM) enzymes in cofactor biosynthesis: A treasure trove of complex organic radical rearrangement reactions. *Journal of Biological Chemistry*, *290*(7), 3980–3986.

Mehta, A. P., Abdelwahed, S. H., Xu, H., & Begley, T. P. (2014). Molybdopterin biosynthesis: Trapping of intermediates for the MoaA-catalyzed reaction using 2′-deoxyGTP and 2′-chloroGTP as substrate analogues. *Journal of the American Chemical Society*, *136*(30), 10609–10614.

Mettert, E. L., & Kiley, P. J. (2015). How is Fe-S cluster formation regulated? *Annual Review of Microbiology*, *69*, 505–526.
Newcomb, M. (2012). Radical kinetics and clocks. In *Encyclopedia of radicals in chemistry, biology and materials*. Wiley.
Nicolet, Y., Rohac, R., Martin, L., & Fontecilla-Camps, J. C. (2013). X-ray snapshots of possible intermediates in the time course of synthesis and degradation of protein-bound Fe4S4 clusters. *Proceedings of the National Academy of Sciences of the United States of America*, *110*(18), 7188–7192.
Park, J., Tai, J., Roessner, C. A., & Scott, A. I. (1995). Overcoming product inhibition of S-adenosyl-L-methionine (SAM) synthetase: Preparation of SAM on the 30 mM scale. *Bioorganic & Medicinal Chemistry Letters*, *5*(19), 2203–2206.
Ravcheev, D. A., & Thiele, I. (2016). Genomic analysis of the human gut microbiome suggests novel enzymes involved in quinone biosynthesis. *Frontiers in Microbiology*, 7, 128.
Rossi, R. A., Pierini, A. B., & Peñéñory, A. B. (2003). Nucleophilic substitution reactions by electron transfer. *Chemical Reviews*, *103*(1), 71–168.
Ruszczycky, M. W., Zhong, A., & Liu, H.-w. (2018). Following the electrons: Peculiarities in the catalytic cycles of radical SAM enzymes. *Natural Product Reports*. Epub ahead of print.
Sofia, H. J., Chen, G., Hetzler, B. G., Reyes-Spindola, J. F., & Miller, N. E. (2001). Radical SAM, a novel protein superfamily linking unresolved steps in familiar biosynthetic pathways with radical mechanisms: Functional characterization using new analysis and information visualization methods. *Nucleic Acids Research*, *29*(5), 1097–1106.
Tanner, D. D., Chen, J. J., Chen, L., & Luelo, C. (1991). Fragmentation of substituted acetophenones and halobenzophenone ketyls. Calibration of a mechanistic probe. *Journal of the American Chemical Society*, *113*(21), 8074–8081.
Van De Bogart, M., & Beinert, H. (1967). Micro methods for the quantitative determination of iron and copper in biological material. *Analytical Biochemistry*, *20*, 325–334.
Vey, J. L., & Drennan, C. L. (2011). Structural insights into radical generation by the radical SAM superfamily. *Chemical Reviews*, *111*(4), 2487–2506.
Wagner, P. J., Lindstrom, M. J., Sedon, J. H., & Ward, D. R. (1981). Photochemistry of. delta.-haloketones: Anchimeric assistance in triplet-state. gamma.-hydrogen abstraction and. beta.-elimination of halogen atoms from the resulting diradicals. *Journal of the American Chemical Society*, *103*(13), 3842–3849.
Wang, Y., Schnell, B., Baumann, S., Müller, R., & Begley, T. P. (2017). Biosynthesis of branched alkoxy groups: Iterative methyl group alkylation by a cobalamin-dependent radical SAM enzyme. *Journal of the American Chemical Society*, *139*(5), 1742–1745.
Wilcoxen, J., Bruender, N. A., Bandarian, V., & Britt, R. D. (2018). A radical intermediate in Bacillus subtilis QueE during turnover with the substrate analog 6-carboxypterin. *Journal of the American Chemical Society*, *140*(5), 1753–1759.
Wu, S. E., Huskey, W. P., Borchardt, R. T., & Schowen, R. L. (1983). Chiral instability at sulfur of S-adenosylmethionine. *Biochemistry*, *22*(12), 2828–2832.
Zehnder, A., & Wuhrmann, K. (1976). Titanium (III) citrate as a nontoxic oxidation-reduction buffering system for the culture of obligate anaerobes. *Science*, *194*(4270), 1165–1166.

CHAPTER SEVEN

Iterative Methylations Resulting in the Biosynthesis of the *t*-Butyl Group Catalyzed by a B_{12}-Dependent Radical SAM Enzyme in Cystobactamid Biosynthesis

Yuanyou Wang*, Bastien Schnell†, Rolf Müller†, Tadhg P. Begley*,[1]
*Department of Chemistry, Texas A&M University, College Station, TX, United States
†Department of Microbial Natural Products, Helmholtz Institute for Pharmaceutical Research Saarland, Helmholtz Center for Infection Research, Saarland University, Saarbrücken, Germany
[1]Corresponding author: e-mail address: begley@tamu.edu

Contents

Abstract

B_{12}-dependent radical SAM enzymes that can perform methylations on sp^3 carbon centers are important for functional diversity and regulation of biological activity in several nonribosomal peptides. Detailed studies on these enzymes are hindered by the

Methods in Enzymology, Volume 606
ISSN 0076-6879
https://doi.org/10.1016/bs.mie.2018.05.018

complexity of the substrates and low levels of expression of active enzymes. CysS can catalyze iterative methylations of a methoxybenzene moiety during the biosynthesis of the cystobactamids. Here, we describe the overexpression, purification, substrate identification, and mechanism of this enzyme.

ABBREVIATIONS

ACP	acyl carrier protein
ABB	4-amino-3-butoxybenzoic acid
AMB	4-amino-3-methoxybenzoate acid
APB	4-amino-3-isopropoxybenzoic acid
B_{12}	cobalamin
CoA	coenzyme A
IPTG	isopropyl β-D-1-thiogalactopyranoside
LB	lysogeny broth
MeCbl	methylcobalamin
NADPH	nicotinamide adenine dinucleotide phosphate hydrate
NRP	nonribosomal peptide
PK	polyketides
SAM	*S*-adenosyl-L-methionine
SDS-PAGE	sodium dodecyl sulfate polyacrylamide gel electrophoresis
Tris•HCl	tris(hydroxymethyl)aminomethane hydrochloride

1. INTRODUCTION

Natural products substituted with various alkyl groups are widely distributed in nature, and many of them play an active role in the development of bioactive compounds. While the biosynthesis of these substituents is generally fairly predictable, our understanding of the biosynthesis of the *t*-butyl group is still at an early stage (Bisel, Al-Momani, & Muller, 2008). The only experimentally reconstituted enzymes involved in *t*-butyl biosynthesis are a B_{12}-dependent mutase in pivaloyl-CoA biosynthesis, a B_{12}-dependent radical SAM enzyme in polytheoamide propeptide biosynthesis and a methyltransferase in the Apratoxin A biosynthesis (Freeman, Helf, Bhushan, Morinaka, & Piel, 2017; Kitanishi, Cracan, & Banerjee, 2015; Parent et al., 2016; Skiba et al., 2018). All three of these enzymes catalyze a single methyl transfer to form the *t*-butyl group from an isopropyl group.

The enzymology of methylation of inert sp^3 carbon centers, mediated by B_{12}-dependent radical SAM enzymes, is a difficult problem because of low levels of expression of active enzyme, and the mechanistic details of the methyl transfer from MeCbl remain to be fully elucidated. The few systems that have been successful reconstituted are PhpK, GenK, ThnK, TsrM, PoyC, GenD1, Fom3 (Blaszczyk, Wang, Silakov, Ho, & Booker, 2017; Hu, Werner, Allen, & Wang, 2015; Huang et al., 2015; Kim, Liu, McCarty, & Liu, 2017; Lanz et al., 2018; McLaughlin & van der Donk, 2018; Marous et al., 2015; Parent et al., 2016; Sato, Kudo, Kuzuyama, Hammerschmidt, & Eguchi, 2018; Schweifer & Hammerschmidt, 2018). This chapter will describe the overexpression and characterization of a B_{12}-dependent radical SAM enzyme (CysS) involved in cystobactamid biosynthesis. This enzyme catalyzes iterative methylations on the methoxy group of 4-amino-3-methoxybenzoate thioesters to form the corresponding ethoxy, isopropoxy, *sec*-butoxy, and *t*-butoxy groups (Baumann et al., 2014; Huttel et al., 2017; Wang, Schnell, Baumann, Muller, & Begley, 2017).

2. OVEREXPRESSION AND PURIFICATION OF CysS

CysS was cloned into a pET28b vector and coexpressed with a plasmid encoding the *suf* operon ([4Fe–4S] biosynthesis) in *Escherichia coli* BL21 (λDE3) (Bhandari, Fedoseyenko, & Begley, 2018; Hanzelmann et al., 2004; Joshi, Mahanta, Fedoseyenko, Williams, & Begley, 2017). The protein was then purified, under anaerobic conditions, by Ni-NTA affinity chromatography. Cobalamin was not required in the growth medium for the production of soluble protein. The UV–visible spectrum of purified CysS revealed a 420 nm shoulder, typical of a bound Fe/S cluster. Under optimized conditions, isolated CysS contained 3.3 irons and 3.0 sulfides per monomer. After chemical reconstitution with Fe(II) and sulfide, the iron content increased to 3.7 per monomer, while the sulfur content and the enzyme activity remained unchanged.

2.1 Protocol for Overexpression and Purification of CysS

A 15 mL overnight culture was grown in LB medium in the presence of 40 μg/mL kanamycin and 30 μg/mL chloramphenicol. This was then added to 1.5 L of LB medium containing 40 μg/mL kanamycin and 30 μg/mL chloramphenicol. The cultures were incubated at 37°C with shaking (180 rpm) until the OD600 reached 0.45. The culture was then incubated at 4°C without shaking for 1 h. Expression was induced by the addition of 100 μ*M* IPTG followed by ferrous ammonium sulfate (120 mg) and cysteine (120 mg), and the culture was grown at 15°C with shaking (110 rpm) overnight. The cells were harvested by centrifugation and stored in liquid nitrogen until further use. All steps for protein purification were carried out in an anaerobic chamber (COY laboratories). Cell pellets (~12 g) were thawed and resuspended in 30 mL lysis buffer (100 m*M* Tris–HCl, pH 7.5) in the presence of lysozyme (0.2 mg/mL) and benzonase (100 units). This mixture was cooled in an ice-bath with continuous stirring for 1–2 h. The suspension was then sonicated to complete cell lysis for 5 × 120 s using a Misonix Sonicator XL-2000 (setting = 9). Cell debris was removed by centrifugation, and the lysate was loaded onto a HisTrap column preequilibrated with lysis buffer. The column was washed with 5–10 column volumes of wash buffer (100 m*M* Tris–HCl, 30 m*M* imidazole, pH 7.5). The CysS protein was then eluted using elution buffer (100 m*M* Tris–HCl, 300 m*M* NaCl, 250 m*M* imidazole, pH 7.5). The brown-colored fractions, containing the desired protein, were pooled and buffer exchanged to final buffer (100 m*M* Tris–HCl, pH 7.5, 30% glycerol) using Econo-Pac 10DG size exclusion chromatography. The purified CysS was divided into aliquots in Eppendorf tubes, frozen in liquid nitrogen, and stored at −80°C, submerged in liquid nitrogen, until future use. Each tube has a pin hole in the cap to allow for gas release during thawing. Protein concentration was measured by the absorbance at 280 nm (A280) when the protein was exposed to air or the $[4Fe–4S]^{2+}$ cluster was reduced to $[4Fe–4S]^{1+}$ by dithionite (prevent the interference from absorbance from $[4Fe–4S]^{2+}$) with an extinction coefficient calculated using the ProtParam tool of the ExPASy proteomics server. Typical yields were approximately 5 mg/L.

2.2 Iron and Sulfide Content of CysS

Iron content was analyzed using the ferene assay (Cooper et al., 2013). The following reagents were prepared: A, 1.35 g SDS, 30 mL water, 450 μL saturated sodium acetate; B, 540 mg L-ascorbic acid, 18 mg sodium metabisulfite, 11.2 mL water, 800 μL saturated sodium acetate; C, 18 mg ferene in 1 mL water. Guanidine hydrochloride (8 *M*, 300 μL), Reagent A (300 μL), and Reagent B (300 μL) were added to 300 μL of a CysS solution. After mixing by inversion, the samples were incubated for 15 min at 37°C and then 15 μL of Reagent C was added. The samples were incubated for 5 min at 25°C and centrifuged for 5 min at 25°C. The absorbance at 593 nm of the supernatant was recorded using a Varian Cary Bio 300 UV–Visible Spectrophotometer. The iron content was finally determined by comparing the reading to a standard curve that was generated under identical conditions using $FeCl_3$ in 2% HNO_3 with a concentration range from 0 to 200 μ*M*.

Sulfide content was determined using the methylene blue assay (Mahanta, Fedoseyenko, Dairi, & Begley, 2013). To 300 μL of the assay solution containing CysS, 1 mL of 1% (w/v) zinc acetate was added followed by 50 μL of 3 *M* NaOH. This mixture was agitated gently and 250 μL of 0.1% *N*,*N*-dimethyl-*p*-phenylenediamine (DMPD) monohydrochloride in 5 *M* HCl and 50 μL of 23 m*M* $FeCl_3$ in 1.2 *M* HCl were added. The resulting solution was mixed vigorously for 30 min. The samples were then centrifuged for 10 min at 25°C. The supernatant was collected, and the absorbance at 670 nm was recorded using a Varian Cary Bio 300 UV–Visible Spectrophotometer. The sulfide content was determined by comparing this reading to a standard curve that was generated under identical conditions using a fresh solution of sodium sulfide (Na_2S) in 0.1 *M* NaOH with a concentration range from 0 to 100 μ*M*.

3. IDENTIFICATION OF THE CysS SUBSTRATE

The most logical substrate for CysS is the 4-amino-3-methoxybenzoate thioester with the peptidyl carrier protein of the polypeptide synthase (CysG; Baumann et al., 2014). This was synthesized as shown in Fig. 1. Condensation of **1** with coenzyme A yielded the CoA ester **2**. The phosphopantetheinyl thioester moiety of **2** was then transferred to CysG1 in a reaction catalyzed by phosphopantetheinyl transferase (Mori et al., 2016; Sanchez, Du, Edwards, Toney, & Shen, 2001; Sunbul, Zhang, & Yin, 2009).

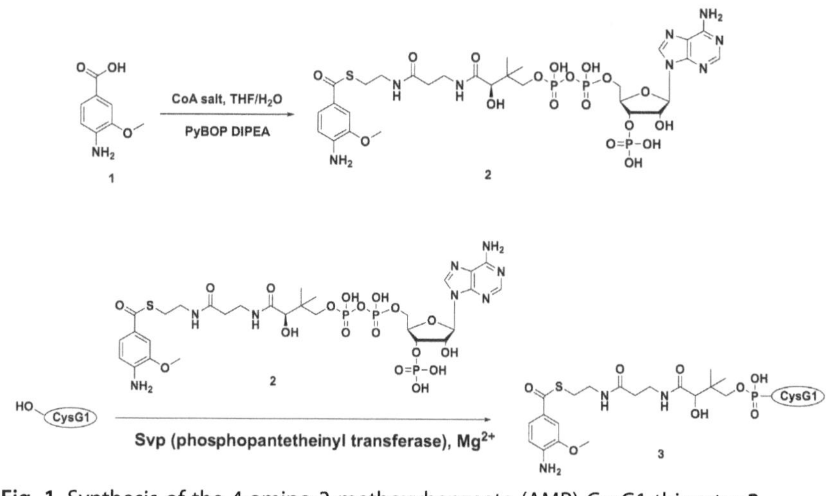

Fig. 1 Synthesis of the 4-amino-3-methoxybenzoate (AMB) CysG1 thioester **3**.

3.1 Synthesis of AMB-CoA Analog 2

4-Amino-3-methoxybenzoic acid **1** (1.4 mg, 8.7 μmol), coenzyme A (sodium salt hydrate, 9 mg, 13 μmol), and benzotriazol-1-yl-oxytripyrrolidinophosphonium hexafluorophosphate (PyBOP, 6.7 mg, 13 μmol) were dissolved in 500 μL of a tetrahydrofuran (THF)/ddH_2O (1:1, v/v). *N,N*-Diisopropylethylamine (DIPEA, 15 μL, 87 μmol) was added, and the reaction mixture was stirred at 25°C for 2 h and then lyophilized. The dried product was resuspended in 10% aqueous acetonitrile and purified by preparative HPLC. The product-containing fractions were combined and lyophilized yielding AMB-CoA **2** as a white solid. ^{1}H NMR (400 MHz, D_2O) δ 8.45 (s, 1H), 8.10 (s, 1H), 7.37 (dd, J=8.3, 1.6 Hz, 1H), 7.21 (d, J=1.6 Hz, 1H), 6.70 (d, J=8.3 Hz, 1H), 6.06 (d, J=5.6 Hz, 1H), 4.75, 4.55 (m, 1H), 4.23 (m, 2H), 4.00 (s, 1H), 3.85 (s, 3H), 3.82–3.43 (m, 6H), 3.18 (m, 2H), 2.43 (m, 2H), 0.89 (s, 3H), 0.75 (s, 3H). ^{13}C NMR (101 MHz, D_2O) δ 192.92, 174.78, 174.09, 155.40, 152.65, 146.42, 143.10, 139.66, 126.07, 122.86, 113.91, 108.95, 86.60, 74.01, 71.97, 55.84, 38.96, 38.43, 35.44, 28.04, 23.28, 20.94, 18.25.

3.2 Overexpression and Purification of CysG1

CysG1 (adenylation domain and peptidyl-carrier-protein of CysG module 5; Baumann et al., 2014) was cloned into a pHisTEV plasmid (derived from pET30a). Single colony transformants of *E. coli* BL21 (DE3) carrying pHisTEV-CysG1 were grown aerobically for 12–15 h at 37°C in 10 mL of LB medium supplemented with 50 μg/mL kanamycin (LB-kan). Cells from the starter culture were added to 1 L of LB-kan, and the culture was grown aerobically at 37°C until the A600 was 0.6–0.8. Expression was induced by the addition of 1 m*M* IPTG, and the culture was grown aerobically at 15°C for an additional ~20 h. Cells were harvested by centrifugation. The medium was discarded, and the cell paste was stored at −80°C until use. Cells were thawed, resuspended in 20 mL of lysis buffer (50 m*M* potassium phosphate, pH 7.8, 150 m*M* NaCl), and then lysed by sonication on ice for 6 × 20 s using a Misonix Sonicator 3000 (24 W, 1.5 s pulse, 1.5 s pause). Cell debris was removed by centrifugation at 4°C, and the supernatant was filtered through a 0.45-μm syringe filter. His_6-CysG1 was purified by nickel affinity chromatography eluting with a step gradient of increasing imidazole concentration with a 5-mL HiTrap chelating HP column. The resin was charged with 2 × 5 mL of 0.1 *M* $NiSO_4$, washed with 5 × 5 mL of filtered water to remove excess nickel, and equilibrated with 2 × 5 mL of lysis buffer. After loading the filtered protein sample, the resin was washed with 10 × 5 mL of wash buffers (50 m*M* potassium phosphate, 50 m*M* imidazole,

150 m*M* NaCl, pH 7.8) and then eluted with elution buffer (50 m*M* potassium phosphate, 200 m*M* imidazole, 150 m*M* NaCl, pH 7.8). Fractions were collected and analyzed for purity by 12% SDS-PAGE. Column fractions containing the desired pure protein were pooled and concentrated to ~3 mL using a centrifugal filtering device with a 10 kDa MWCO membrane and the buffer was exchanged to 50 m*M* potassium phosphate, pH 7.5 containing 150 m*M* KCl and 15% glycerol via Econo-Pac 10DG size exclusion chromatography. The desired protein has a molecular weight of ~72 kDa. Protein concentrations were determined using A280 measurements. Theoretical extinction coefficients for A280 calculations were obtained from the ProtParam function on the ExPASy Proteomics Server.

3.3 Chemoenzymatic Synthesis of the 4-Amino-3-Methoxybenzoate (AMB) CysG1 Thioester 3

A typical enzymatic reaction for synthesizing the AMB-CysG1 thioester was performed in 100 m*M* phosphate buffer, pH 7.5 containing the Svp phosphopantetheinyl transferase (20 μ*M*), CysG1 (43 μ*M*), AMB-CoA (0.3 m*M*), and $MgCl_2$ (10 m*M*). The reaction was allowed to proceed aerobically at room temperature for about 6 h and was quenched by desalting using a Bio-spin 6 column (Bio-Rad). Then the sample was injected onto a reverse phase HPLC Zorbax 300SB-C8 3.5 μ, 3 × 150 mm column operating at room temperature coupled to an ESI-TOF mass spectrometer operating in positive ionization mode. ESI-TOF analysis of the CysG1 protein mass after reaction showed mass increases of around 490 Da, which corresponding to the mass of the attached phosphopantetheinyl-AMB thioester group (Fig. 2).

3.4 Reconstitution of the CysS-Catalyzed Methyl Transfer Reaction

AMB-CysG1 thioester (53 μ*M*), prepared as described earlier, was buffer exchanged into anaerobic 100 m*M* phosphate buffer, pH 7.5 using a Bio-spin 6 column in a glove box and then treated with 0.28 m*M* MeCbl, 0.56 m*M* SAM and 0.6 m*M* methyl viologen, 1.8 m*M* NADPH, and 22 μ*M* CysS or 44 μ*M* CysS (the formation of trimethylated products required higher enzyme concentrations). The reaction was allowed to proceed anaerobically at room temperature for about 14 h. The protein in the reaction mixture was desalted using 30K Amicon Ultra-0.5-mL centrifugal filters. LC-MS analysis of the AMB-CysG1-derived products demonstrated the conversion of the substrate methoxy to *i*-propoxy and butoxy

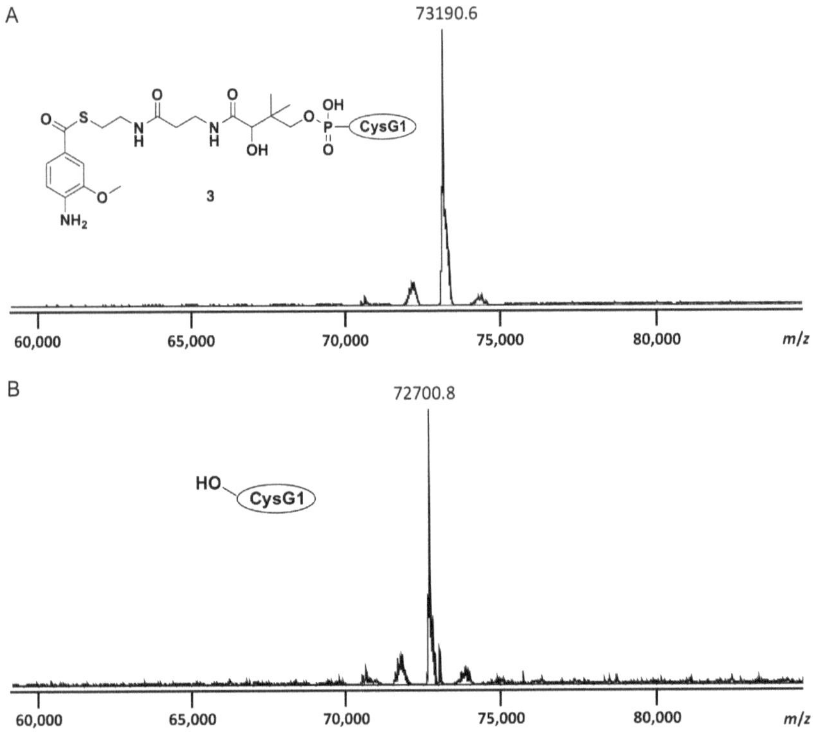

Fig. 2 MS analysis of the peptidyl carrier protein CysG1. (A) CysG1 thioester **3** (calculated monoisotopic mass: 73189.7 Da, experimental mass: 73190.6 Da) and (B) CysG1 without modification (calculated monoisotopic mass: 72700.5 Da, experimental mass: 72700.8 Da).

groups (Fig. 3) as well as the formation of 5′-deoxyadenosine (5′-dA) and *S*-adenosylhomocysteine.

LC-ESI-TOF-MS on protein samples was performed using an Agilent 1260 HPLC system equipped with a binary pump and a 1200 series diode array detector followed by a MicroToF-Q II mass spectrometer (Bruker Daltonics) using an ESI source in positive mode. The analysis was performed on a Zorbax 300SB-C8 column (3 × 150 mm column, 3.5 μm particles, Agilent).

LC conditions: A—0.1% formic acid in H_2O, B—0.1% formic acid in acetonitrile.

LC method: 0 min—90% A, 10 min—80% A, 25 min—10% A 90% B, 27 min—90% A, 30 min—90% A.

The mass spectrometer parameters for protein analysis: Capillary, 4500 V; end plate offset, 500 V; nebulizer gas, 3.0 bar; dry gas, 10.0 L/min; dry gas temperature, 220°C; funnel 1 RF, 400.0 Vpp; funnel 2 RF,

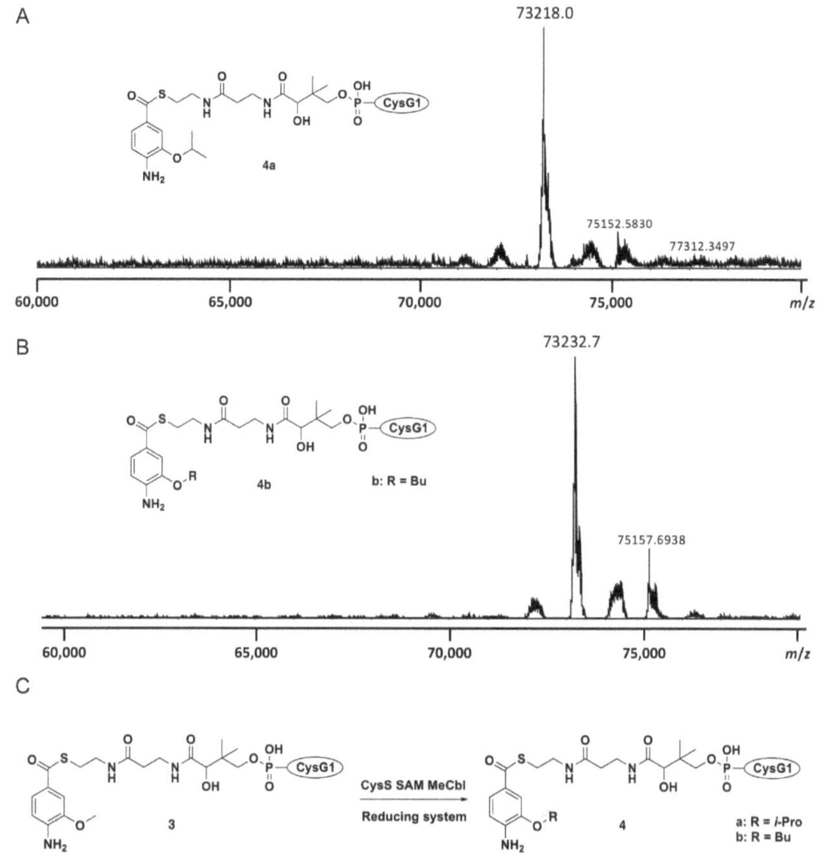

Fig. 3 The CysS-catalyzed reaction of the AMB-CysG1 thioester. (A) MS of the dimethylated product (APB-CysG1 thioester **4a**, calculated monoisotopic mass: 73217.8 Da, experimental mass: 73218.0 Da), (B) MS of the trimethylated product (ABB-CysG1 thioester **4b**, calculated monoisotopic mass: 73231.7 Da, experimental mass: 73232.7 Da), (C) The reaction catalyzed by CysS.

400.0 Vpp; ISCID, 0.0 eV; hexapole RF, 250 Vpp; quadrupole, Ion energy, 4.0 eV; low mass, 200 *m/z*; collision cell, collision energy, 8.0 eV; collision RF, 1000.0 Vpp; transfer time, 120.0 μs; prepulse storage, 10.0 μs. Data were processed with DataAnalysis ver. 4.1 (Bruker Daltonics)

4. IDENTIFICATION OF THE MINIMAL CysS SUBSTRATE

AMB-CysG1 thioester is a complex molecule and not ideal for mechanistic studies. We therefore synthesized and tested the truncated analogs of AMB-CysG1 thioester shown in Table 1.

Table 1 Identification of the Minimal CysS Substrate

Substrate	Product	Relative Reactivity[a]
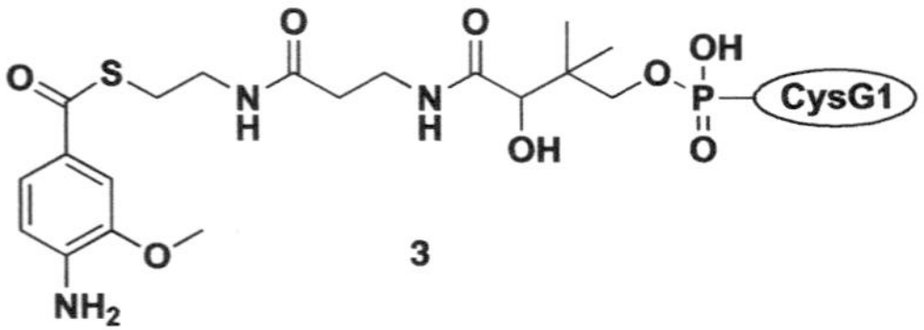 3	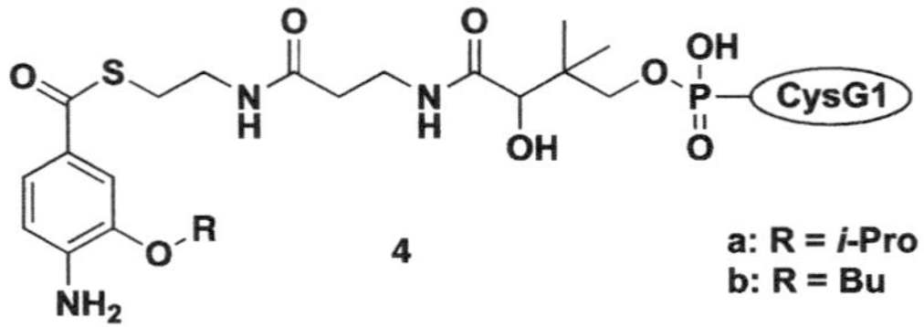 4 a: R = *i*-Pro b: R = Bu	++++
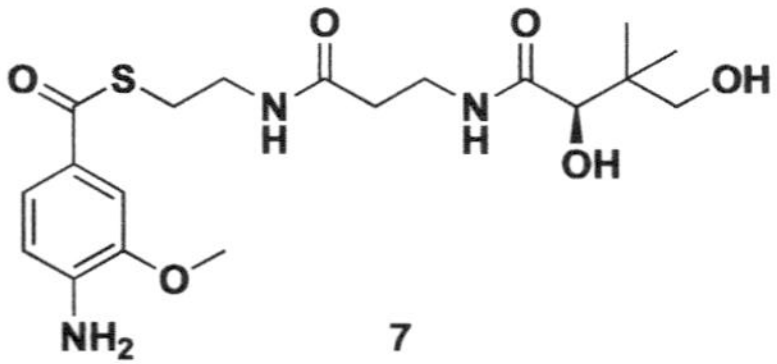 7	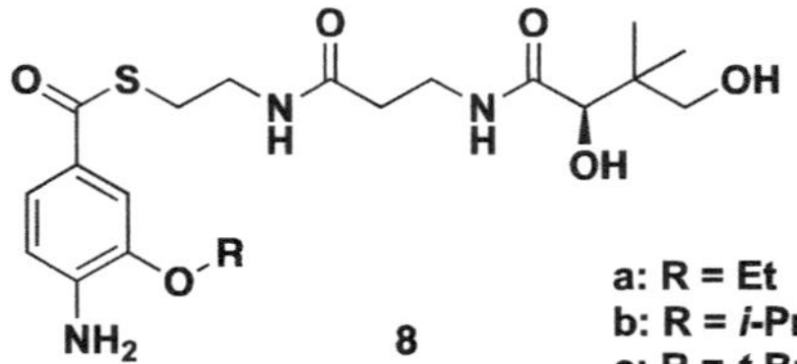 8 a: R = Et b: R = *i*-Pro c: R = *t*-Bu d: R = *sec*-Bu	+++
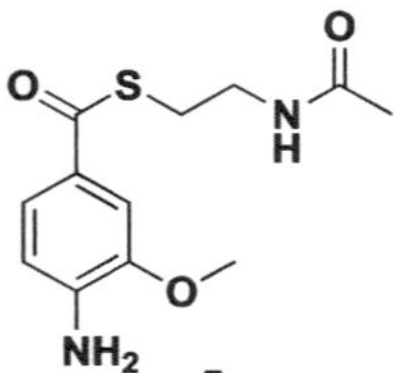 5	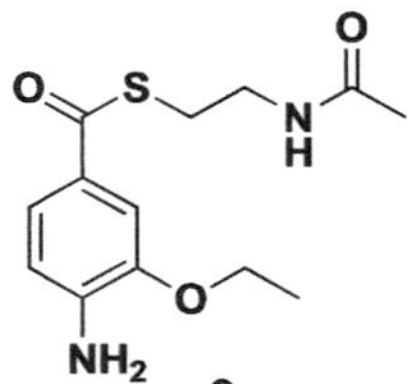 9	++

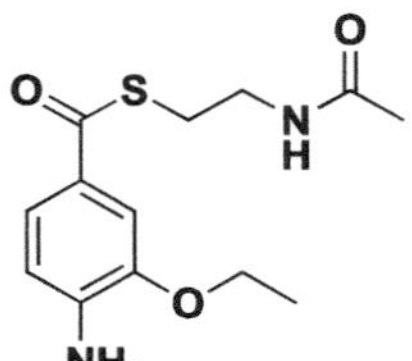 9	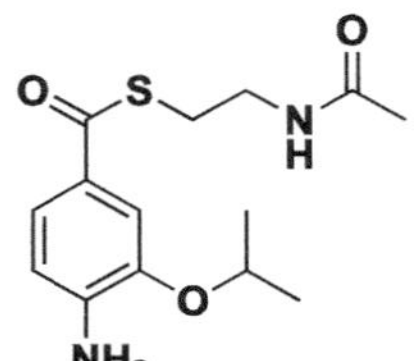 10	+

[a]CysS undergoes only five turnovers. It was therefore not possible to determine the steady-state kinetic parameters for the reaction.

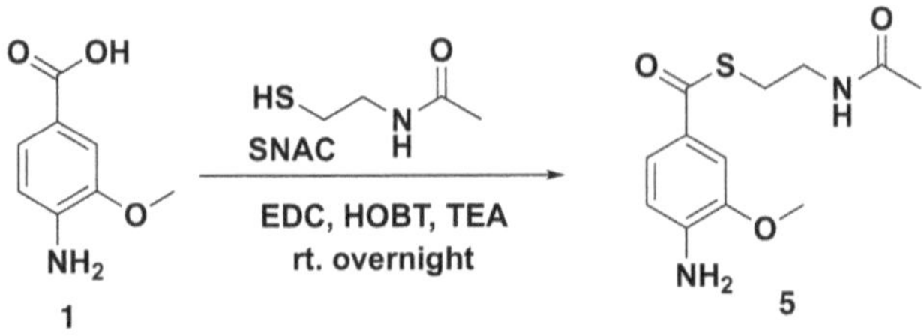

Fig. 4 The synthesis scheme of AMB *N*-acetylcysteamine thioester **5**.

4.1 Synthesis of the AMB *N*-Acetylcysteamine Thioester 5

To a solution of triethylamine (TEA, 0.39 mL, 2.80 mmol) in dichloromethane (10 mL) was added 4-amino-3-methoxybenzoic acid (AMB, **1**, 234 mg, 1.40 mmol), (3-dimethylaminopropyl)-3-ethylcarbodiimide hydrochloride (EDC, 268 mg 1.40 mmol), 1-hydroxybenzotriazole (HOBt, 214 mg, 1.40 mmol), and *N*-acetylcysteamine (SNAC, 166 mg, 1.4 mmol) (Fig. 4). The reaction mixture was stirred overnight under argon. The organic layer was washed, first with saturated aqueous $NaHCO_3$, then with 0.1 N HCl solution, and finally with brine. It was then dried over anhydrous sodium sulfate and concentrated under vacuum. Purification of the crude mixture by flash chromatography (silica gel, 1:1–2:1 hexanes:EtOAc) afforded **5** in 20% isolated yield. ^{1}H NMR (400 MHz, $CDCl_3$) δ 7.53 (d, J=8.2, 1H), 7.36 (s, 1H), 6.63 (d, J=8.2 Hz, 1H), 3.89 (s, 3H), 3.51 (dd, J=12.2, 5.9 Hz, 2H), 3.18 (t, J=6.3 Hz, 2H), 1.95 (s, 3H). ^{13}C NMR (101 MHz, $CDCl_3$) δ 190.29, 170.29, 146.21, 142.34, 126.60, 122.85, 112.78, 108.72, 55.63, 40.08, 28.31, 23.21.

4.2 Synthesis of the AMB-Pantetheinyl Thioester 7

To a stirred solution of 4-amino-3-methoxybenzoic acid (AMB, **1**, 306 mg, 2.0 mmol) in 10 mL of dry THF was added *N*-hydroxysuccinimide (NHS, 230 mg, 2.0 mmol), followed by *N,N'*-dicyclohexylcarbodiimide (DCC, 413 mg 2.0 mmol) at room temperature. The resulting mixture was stirred overnight. The white precipitate was filtered and washed with EtOAc. The filtrate and washings were concentrated under reduced pressure. The resulting residue was triturated with CH_2Cl_2 to give the NHS ester **6** as a pale brown powder. Pantethine (41 mg, 0.075 mmol), sodium bicarbonate (159 mg, 1.9 mmol), and 1,4-dithiothreitol (DTT, 12 mg, 0.08 mmol) were dissolved in 3 mL of water and left to stand for 10 min to allow reduction of the pantethine to pantetheine to take place. The NHS ester (**6**, 143 mg, 0.54 mmol) dissolved in 7 mL of THF was added, the reaction mixture was stirred overnight, and the solvent was evaporated under reduced pressure (Fig.5). The crude product was dissolved in water, and the insoluble

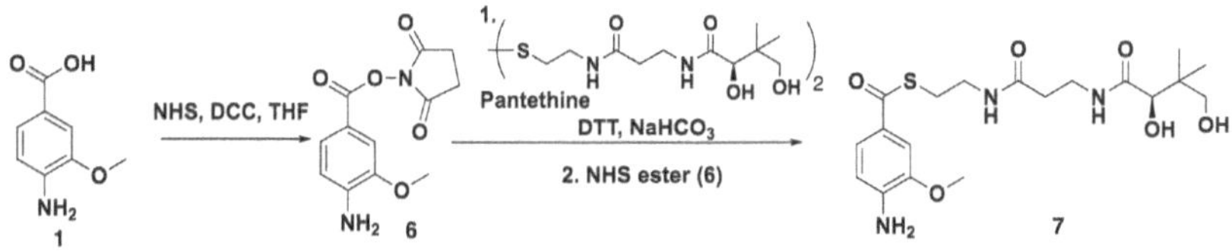

Fig. 5 The synthesis scheme for AMB-pantetheinyl thioester **7**.

residue was removed by filtration. The resulting solution was lyophilized and purified by flash chromatography (silica gel, 20:1 to 10:1 CH_2Cl_2: MeOH). The resulting crude product was further purified by Prep HPLC using an SPLC-18DB column (10 × 250 mm, 5 μm, Supelco) at a flow rate of 2 mL/min on an Agilent 100 HPLC system with a quaternary pump and an automatic injector. ^{1}H NMR (400 MHz, D_2O) δ 7.52 (dd, J=8.3, 1.9 Hz, 1H), 7.35 (d, J=1.9 Hz, 1H), 6.79 (d, J=8.3 Hz, 1H), 3.91 (s, 1H), 3.88 (s, 3H), 3.47–3.29 (m, 6H), 3.17 (t, J=6.3 Hz, 2H), 2.44 (t, J=6.5 Hz, 2H), 0.83 (d, J=13.9 Hz, 6H). ^{13}C NMR (101 MHz, D_2O) δ 193.17, 175.04, 174.02, 146.63, 126.36, 123.04, 114.04, 109.43, 75.86, 68.49, 55.92, 38.93, 38.58, 35.50, 35.33, 28.03, 22.30, 20.46, 19.11.

4.3 Assay of Alternative CysS Substrates

A typical reaction was performed in 100 m*M* phosphate buffer, pH 7.5 containing 55 μ*M* of CysS with 1 m*M* MeCbl, 1 m*M* SAM, 1 m*M* substrate, and the flavodoxin/flavodoxin reductase/NADPH-reducing system. The reaction was run for 12–15 h in an anaerobic chamber at room temperature. The enzymes were removed by ultrafiltration using 10 kDa cutoff filters (VWR), and the crude reaction mixture was further analyzed by LC-MS/MS. The relative activities of the truncated substrate analogs are shown in Table 1. These experiments demonstrate that it was possible to remove all of the CysG protein and to reduce the phosphopantetheinyl-AMB thioester to a simple *N*-acetylcysteamine thioester **5**. AMB **1** and AMB-CoA **2** were not substrates for CysS.

Various other [4Fe–4S] cluster-reducing agents were tested in addition to the flavodoxin/flavodoxin reductase/NADPH system. NADPH/methyl viologen, a commonly used electron source for cobalamin-dependent radical SAM enzymes, gave similar activity. However, dithionite, titanium citrate, and the combination of methyl viologen and dithionite gave lower activity. Buffer thiols inactivate the substrate by *trans*-thioesterification and should be avoided.

LC-ESI-TOF-MS on small molecules was performed using an Agilent 1260 HPLC system equipped with a binary pump and a 1200 series diode array detector followed by a MicroToF-Q II mass spectrometer (Bruker Daltonics) using an ESI source in positive mode. The analysis was performed on a ZORBAX Eclipse XDB-C18 column (15 cm × 4.6 mm, 5 μm particles, Agilent).

LC conditions: A—5 m*M* ammonium acetate buffer, pH 6.6, B—75% methanol and 25% water.

LC method: 0 min—100% A, 2 min—100% A, 4 min—80% A 20% B, 27 min—100% B, 29 min—100% B, 30 min—100% A, 40 min—100% A.

The mass spectrometer parameters for small molecules: Capillary, 4500 V; end plate offset, 500 V; nebulizer gas, 3.0 bar; dry gas, 10.0 L/min; dry gas temperature, 200°C; funnel 1 RF, 250.0 Vpp; funnel 2 RF, 300.0 Vpp; isCID, 0.0 eV; hexapole RF, 200 Vpp; quadrupole, Ion energy, 3.0 eV; low mass, 150 *m/z*; collision cell, collision energy, 8.0 eV; collision RF, 150.0 Vpp; transfer time, 80.0 μs; prepulse storage, 5.0 μs. Data were processed with DataAnalysis ver. 4.0 (Bruker Daltonics).

5. CONCLUSIONS

Our current mechanistic proposal for the CysS-catalyzed methylation reactions is shown in Fig. 6. This mechanism is based in part on the proposed mechanisms for GenK (Kim et al., 2017), ThnK (Marous et al., 2015), and on our unpublished studies. The reaction requires two molecules of SAM per methylation reaction. The first SAM methylates the cobalamin, and the second SAM undergoes reductive cleavage by the $[4Fe–4S]^{+1}$ cluster to give the 5′-deoxyadenosyl radical and the oxidized cluster ($[4Fe–4S]^{+2}$). The 5′-deoxyadenosyl radical abstracts a hydrogen atom from the methyl group of the substrate **7** to give radical **11**, which then undergoes a radical substitution with methyl cobalamin to form **8a** and cobalamin(II). The starting redox state of the enzyme is then regenerated by reduction of cobalamin(II) to cobalamin(I), either directly by the exogenous reducing system or via the $[4Fe–4S]^{+2}$ cluster. Repetition of this sequence leads to the formation of the isopropyl, *t*-butyl, and *sec*-butyl ethers of **7**. So far, there is no published crystal structure available for any B_{12}-dependent radical SAM methyltransferase, and it is unknown how the $[4Fe–4S]^{+1}$ cluster, SAM, cobalamin, and the substrate interact in the enzyme active site to catalyze complex iterative methylations on an sp^3 carbon center that are still without precedent in synthetic organic chemistry.

The B_{12}-dependent radical SAM methyltransferases in NRP and NRP-PK hybrids biosynthesis catalyze novel chemistry and have potential applications in the engineering of unique functional groups on bioactive

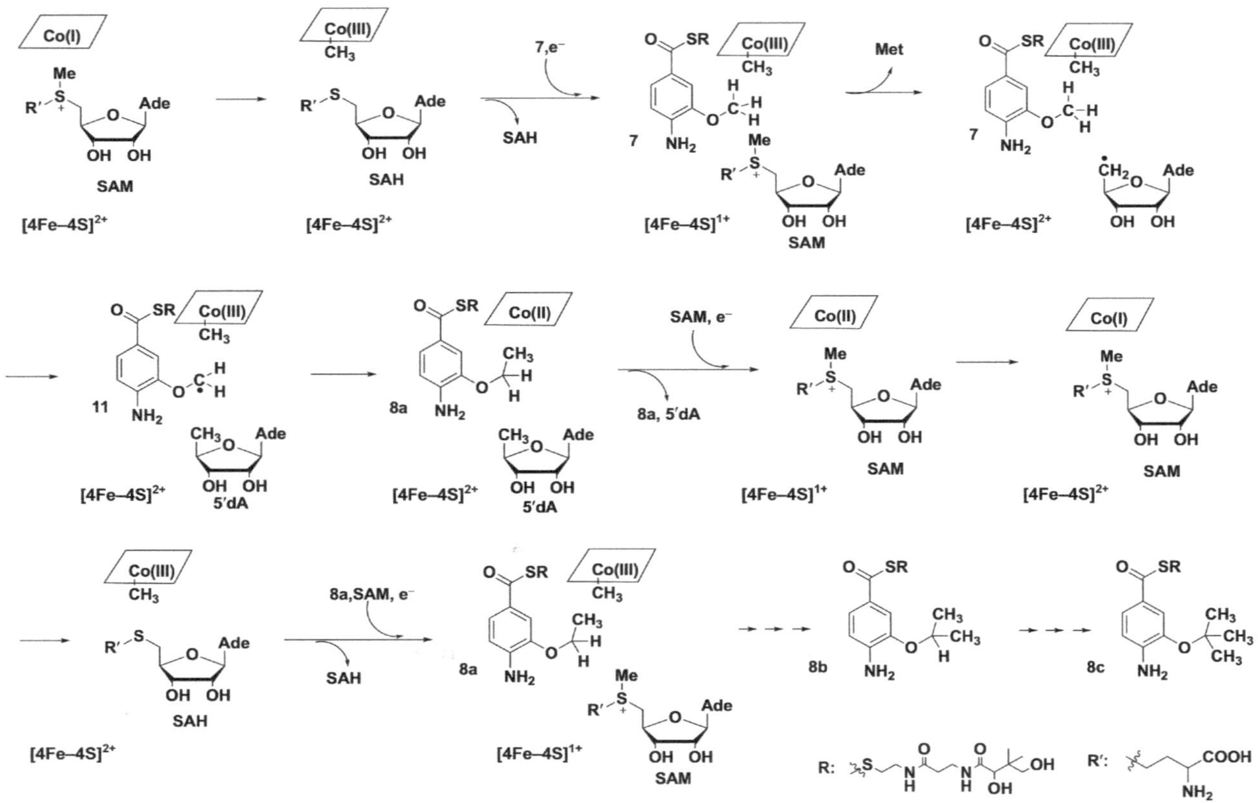

Fig. 6 Proposed mechanism for the CysS-catalyzed iterative methylations of the AMB-CysG1 thioester.

natural products. For example, the iterative methylation chemistry catalyzed by CysS makes possible the formation of a 25-member combinatorial library of cystobactamid antibiotics. Analogous methyl-based combinatorial libraries could potentially be generated in natural products such as kedarcidin (Lohman et al., 2013) and quinomycin SW-163s (Watanabe et al., 2009), which have ethyl, isopropyl, and *sec*-butyl substituents on a thiomethyl group.

ACKNOWLEDGMENTS

This work was supported by NIH (DK44083) and by the Robert A. Welch Foundation (A-0034) as well as the German Centre for Infection Research (DZIF).

REFERENCES

Baumann, S., Herrmann, J., Raju, R., Steinmetz, H., Mohr, K. I., Huttel, S., et al. (2014). Cystobactamids: Myxobacterial topoisomerase inhibitors exhibiting potent antibacterial activity. *Angewandte Chemie (International Ed. in English)*, *53*(52), 14605–14609.

Bhandari, D. M., Fedoseyenko, D., & Begley, T. P. (2018). Mechanistic studies on tryptophan lyase (NosL): Identification of cyanide as a reaction product. *Journal of the American Chemical Society*, *140*(2), 542–545.

Bisel, P., Al-Momani, L., & Muller, M. (2008). The tert-butyl group in chemistry and biology. *Organic & Biomolecular Chemistry*, *6*(15), 2655–2665.

Blaszczyk, A. J., Wang, B., Silakov, A., Ho, J. V., & Booker, S. J. (2017). Efficient methylation of C2 in L-tryptophan by the cobalamin-dependent radical S-adenosylmethionine methylase TsrM requires an unmodified N1 amine. *Journal of Biological Chemistry*, *292*(37), 15456–15467.

Cooper, L. E., Fedoseyenko, D., Abdelwahed, S. H., Kim, S. H., Dairi, T., & Begley, T. P. (2013). In vitro reconstitution of the radical S-adenosylmethionine enzyme MqnC involved in the biosynthesis of futalosine-derived menaquinone. *Biochemistry*, *52*(27), 4592–4594.

Freeman, M. F., Helf, M. J., Bhushan, A., Morinaka, B. I., & Piel, J. (2017). Seven enzymes create extraordinary molecular complexity in an uncultivated bacterium. *Nature Chemistry*, *9*(4), 387–395.

Hanzelmann, P., Hernandez, H. L., Menzel, C., Garcia-Serres, R., Huynh, B. H., Johnson, M. K., et al. (2004). Characterization of MOCS1A, an oxygen-sensitive iron-sulfur protein involved in human molybdenum cofactor biosynthesis. *Journal of Biological Chemistry*, *279*(33), 34721–34732.

Hu, K. F., Werner, W. J., Allen, K. D., & Wang, S. C. (2015). Investigation of enzymatic C-P bond formation using multiple quantum HCP nuclear magnetic resonance spectroscopy. *Magnetic Resonance in Chemistry*, *53*(4), 267–272.

Huang, C., Huang, F., Moison, E., Guo, J., Jian, X., Duan, X., et al. (2015). Delineating the biosynthesis of gentamicin x2, the common precursor of the gentamicin C antibiotic complex. *Chemistry & Biology*, *22*(2), 251–261.

Huttel, S., Testolin, G., Herrmann, J., Planke, T., Gille, F., Moreno, M., et al. (2017). Discovery and total synthesis of natural cystobactamid derivatives with superior activity against Gram-negative pathogens. *Angewandte Chemie (International Ed. in English)*, *56*(41), 12760–12764.

Joshi, S., Mahanta, N., Fedoseyenko, D., Williams, H., & Begley, T. P. (2017). Aminofutalosine synthase: Evidence for captodative and aryl radical intermediates using beta-scission and SRN1 trapping reactions. *Journal of the American Chemical Society, 139*(32), 10952–10955.

Kim, H. J., Liu, Y. N., McCarty, R. M., & Liu, H. W. (2017). Reaction catalyzed by GenK, a cobalamin-dependent radical S-adenosyl-l-methionine methyltransferase in the biosynthetic pathway of gentamicin, proceeds with retention of configuration. *Journal of the American Chemical Society, 139*(45), 16084–16087.

Kitanishi, K., Cracan, V., & Banerjee, R. (2015). Engineered and native coenzyme B12-dependent isovaleryl-CoA/pivalyl-CoA mutase. *Journal of Biological Chemistry, 290*(33), 20466–20476.

Lanz, N. D., Blaszczyk, A. J., McCarthy, E. L., Wang, B., Wang, R. X., Jones, B. S., et al. (2018). Enhanced solubilization of class B radical S-adenosylmethionine methylases by improved cobalamin uptake in Escherichia coli. *Biochemistry, 57*(9), 1475–1490.

Lohman, J. R., Huang, S. X., Horsman, G. P., Dilfer, P. E., Huang, T. T., Chen, Y. H., et al. (2013). Cloning and sequencing of the kedarcidin biosynthetic gene cluster from Streptoalloteichus sp. ATCC 53650 revealing new insights into biosynthesis of the enediyne family of antitumor antibiotics. *Molecular BioSystems, 9*(3), 478–491.

Mahanta, N., Fedoseyenko, D., Dairi, T., & Begley, T. P. (2013). Menaquinone biosynthesis: Formation of aminofutalosine requires a unique radical SAM enzyme. *Journal of the American Chemical Society, 135*(41), 15318–15321.

Marous, D. R., Lloyd, E. P., Buller, A. R., Moshos, K. A., Grove, T. L., Blaszczyk, A. J., et al. (2015). Consecutive radical S-adenosylmethionine methylations form the ethyl side chain in thienamycin biosynthesis. *Proceedings of the National Academy of Sciences of the United States of America, 112*(33), 10354–10358. https://doi.org/10.1073/pnas.1508615112.

McLaughlin, M. I., & van der Donk, W. A. (2018). Stereospecific radical-mediated B12-dependent methyl transfer by the fosfomycin biosynthesis enzyme Fom3. *Biochemistry*. https://doi.org/10.1021/acs.biochem.8b00616.

Mori, S., Simkhada, D., Zhang, H., Erb, M. S., Zhang, Y., Williams, H., et al. (2016). Polyketide ring expansion mediated by a thioesterase, chain elongation and cyclization domain, in azinomycin biosynthesis: Characterization of AziB and AziG. *Biochemistry, 55*(4), 704–714. https://doi.org/10.1021/acs.biochem.5b01050.

Parent, A., Guillot, A., Benjdia, A., Chartier, G., Leprince, J., & Berteau, O. (2016). The B12-radical SAM enzyme PoyC catalyzes valine Cbeta-methylation during polytheonamide biosynthesis. *Journal of the American Chemical Society, 138*(48), 15515–15518.

Sanchez, C., Du, L. C., Edwards, D. J., Toney, M. D., & Shen, B. (2001). Cloning and characterization of a phosphopantetheinyl transferase from Streptomyces verticillus ATCC15003, the producer of the hybrid peptide-polyketide antitumor drug bleomycin. *Chemistry & Biology, 8*(7), 725–738.

Sato, S., Kudo, F., Kuzuyama, T., Hammerschmidt, F., & Eguchi, T. (2018). C-methylation catalyzed by Fom3, a cobalamin-dependent radical SAM enzyme in fosfomycin biosynthesis, proceeds with inversion of configuration. *Biochemistry*. https://doi.org/10.1021/acs.biochem.8b00614.

Schweifer, A., & Hammerschmidt, F. (2018). Stereochemical course of methyl transfer by cobalamin-dependent radical SAM methyltransferase in Fosfomycin biosynthesis. *Biochemistry, 57*(14), 2069–2073.

Skiba, M. A., Sikkema, A. P., Moss, N. A., Lowell, A. N., Su, M., Sturgis, R. M., et al. (2018). Biosynthesis of t-butyl in apratoxin A: Functional analysis and architecture of a PKS loading module. *ACS Chemical Biology, 13*(6), 1640–1650.

Sunbul, M., Zhang, K., & Yin, J. (2009). Using phosphopantetheinyl transferases for enzyme posttranslational activation, site specific protein labeling and identification of natural product biosynthetic gene clusters from bacterial genomes. *Methods in Enzymology, 458*, 255.

Wang, Y., Schnell, B., Baumann, S., Muller, R., & Begley, T. P. (2017). Biosynthesis of branched alkoxy groups: Iterative methyl group alkylation by a cobalamin-dependent radical SAM enzyme. *Journal of the American Chemical Society*, *139*(5), 1742–1745.

Watanabe, K., Hotta, K., Nakaya, M., Praseuth, A. P., Wang, C. C., Inada, D., et al. (2009). Escherichia coli allows efficient modular incorporation of newly isolated quinomycin biosynthetic enzyme into echinomycin biosynthetic pathway for rational design and synthesis of potent antibiotic unnatural natural product. *Journal of the American Chemical Society*, *131*(26), 9347–9353.

CHAPTER EIGHT

Biochemical Approaches for Understanding Iron–Sulfur Cluster Regeneration in *Escherichia coli* Lipoyl Synthase During Catalysis

Erin L. McCarthy*, Squire J. Booker*,†,‡,1

*Department of Biochemistry and Molecular Biology, The Pennsylvania State University, University Park, PA, United States

†Department of Chemistry, The Pennsylvania State University, University Park, PA, United States

‡The Howard Hughes Medical Institute, The Pennsylvania State University, University Park, PA, United States

[1]Corresponding author: e-mail address: squire@psu.edu

Contents

Methods in Enzymology, Volume 606
ISSN 0076-6879
https://doi.org/10.1016/bs.mie.2018.06.006

Abstract

Lipoyl synthase (LipA in bacteria) is a radical *S*-adenosylmethionine (SAM) enzyme that catalyzes the second step of the de novo biosynthesis of the lipoyl cofactor: the insertion of sulfur at C6 and C8 of a pendant octanoyl chain. In addition to the [4Fe—4S] cluster that is characteristic of the radical SAM (RS) enzymes, LipA contains a second [4Fe—4S] cluster that, though controversial, has been proposed to be degraded during turnover to supply the inserted sulfur atoms. A consequence of this proposed role is that the destruction of its iron–sulfur cluster renders the enzyme in an inactive state. Recently, it was shown that *Escherichia coli* proteins NfuA or IscU can confer catalytic properties to *E. coli* LipA in vitro. In this chapter, we present methods for characterizing LipA and analyzing its activity in vitro, and provide strategies to monitor the pathway for the regeneration of LipA's auxiliary cluster by *E. coli* iron–sulfur carrier protein NfuA.

1. INTRODUCTION

Lipoic acid is an eight carbon straight-chain fatty acid containing sulfur atoms at C6 and C8 (Reed, Busk, Gunsalus, & Schnakenberg, 1951; Reed et al., 1953). It plays a central role in multienzyme complexes involved in the oxidative decarboxylation of α-keto acids, the breakdown of glycine, and the degradation of acetoin in certain types of bacteria (Reed, 1974). In this capacity, it functions as a redox-active cofactor due to its ability to convert between states containing free thiol groups and a dithiolane ring and as a shuttle to move reaction intermediates between successive active sites in multienzyme complexes. In its role as a cofactor, lipoic acid is covalently attached through an amide linkage to a specific lysyl residue of a lipoyl carrier protein (LCP), one of the subunits of the multienzyme complexes (Cronan, 2016).

Early studies identified the pathway for the de novo biosynthesis of the lipoyl cofactor in *Escherichia coli* and outlined the roles of two dedicated enzymes: an octanoyltransferase, LipB, and a radical *S*-adenosylmethionine (SAM) enzyme, LipA (Cronan, 2016; Miller et al., 2000; Zhao, Miller, Jiang, Marletta, & Cronan, 2003). In an offshoot from fatty acid biosynthesis, LipB transfers an octanoyl group from an acyl carrier protein to an LCP, which generates the substrate for LipA. LipA subsequently inserts sulfur atoms at C6 and C8 to generate the lipoyl cofactor. The inability to synthesize the lipoyl cofactor is embryonic lethal in mice, and defects in its biosynthesis result in disastrous biochemical outcomes in humans, such as seizures, developmental delays, and death (Landgraf, McCarthy, & Booker, 2016; Yi & Maeda, 2005). Further, lipoic acid supplementation has been shown to be therapeutic for a wide range of clinical ailments associated with oxidative stress, including diabetes and Alzheimer's disease (Borcea et al., 1999; Holmquist et al., 2007).

There are several challenges associated with the biosynthesis of the lipoyl cofactor, despite its structural simplicity. The first obstacle is removal of hydrogens from C6 and C8 of the octanoyl chain, which LipA overcomes through the use of radical chemistry. As a radical SAM (RS) enzyme, LipA coordinates a $[4Fe{-}4S]^{2+}$ cluster that is absolutely required for its function. In its reduced $[4Fe{-}4S]^{+}$ form, the iron–sulfur (Fe/S) cluster can inject an electron into the sulfonium of SAM, inducing its cleavage to L-methionine, and a 5′-deoxyadenosyl 5′-radical (5′-dA•). This high energy radical species is a potent oxidant capable of cleaving C—H bonds displaying homolytic bond-dissociation energies of 98 kcal/mol or greater.

The second outstanding question is the source of the sulfur used for the sulfur insertion. In addition to the [4Fe—4S] cluster common to all RS enzymes, LipA coordinates a second "auxiliary" cluster that appears to be the source of the inserted sulfur atoms (Lanz & Booker, 2015). As such, the [4Fe—4S] auxiliary cluster is destroyed during catalysis, which is supported by an overwhelming amount of biochemical evidence (Cicchillo & Booker, 2005; Cicchillo, Iwig, et al., 2004; Cicchillo, Lee, et al., 2004; Douglas, Kriek, Bryant, & Roach, 2006; Harmer et al., 2014; Lanz et al., 2014; McLaughlin et al., 2016). For example, LipA can only undergo up to one turnover in vitro in the absence of exogenously added sulfur. Further, when LipA is isolated from *E. coli* cultures using ^{34}S as the sole sulfur source and then subsequently used in in vitro assays, the lipoyl product contains almost exclusively ^{34}S at C6 and C8 (Cicchillo & Booker, 2005).

Due to the sacrificial role of its auxiliary cluster, LipA typically catalyzes less than one equivalent of product formation in in vitro assays. Recently, the *E. coli* [4Fe–4S] cluster carrier protein NfuA was shown to regenerate LipA's auxiliary cluster to support catalysis in vitro (McCarthy & Booker, 2017). Remarkably, it accomplishes this feat at a rate that does not limit the rate of catalysis. In the absence of NfuA, the scaffold protein IscU can fulfill a similar role. Herein, we describe methods for isolating and characterizing LipA and monitoring the regeneration of its auxiliary cluster in vitro. This chapter aims to complement other previously published *Methods in Enzymology* papers that highlight the characterization of RS enzymes that contain multiple [Fe—S] clusters (Lanz et al., 2012) and RS enzymes that contain cobalamin cofactors (Blaszczyk, Wang, & Booker, 2017), with additional insight into analysis of Fe—S cluster regeneration in vitro in RS enzymes that consume their clusters during catalysis.

2. OVERPRODUCTION AND ISOLATION OF *E. COLI* LipA AND *E. COLI* NfuA

The generation of LipA protein of high purity and activity has been notoriously difficult due to issues of solubility and the air sensitive nature of its two [4Fe–4S] clusters. Initial characterizations resulted in protein that had poor activity most likely due to incomplete cofactor incorporation which is essential for the enzyme's activity. With regard to NfuA, the *E. coli* protein had been typically isolated aerobically and then subjected to a chemical reconstitution procedure to insert its iron–sulfur (Fe/S) cluster, leaving it unclear as to what the in vivo cluster configuration is (Angelini et al., 2008; Boutigny et al., 2013; Py et al., 2012). Following the methods detailed herein, we have generated robust amounts of LipA and NfuA proteins that are >90% homogenous (Figs. 1 and 2). Further, the proteins contain intact [4Fe—4S] clusters in its As-isolated state without the need for chemical reconstitution allowing for extensive biochemical analysis

2.1 Gene Cloning and Expression Strategy

The genes encoding LipA and NfuA were amplified from *E. coli* genomic DNA (strain W3110) using the polymerase chain reaction (PCR) and inserted into the pET28a vector using the *Nde*I/*Eco*RI restriction sites and *Hin*dIII/*Xho*I restriction sites, respectively (Cicchillo, Iwig, et al., 2004; Cicchillo, Lee, et al., 2004; McCarthy & Booker, 2017). The resulting constructs allowed for overexpression of the *lipA* and *nfuA* genes under the control of the strong T7 promoter. Further, these constructs conferred kanamycin resistance for selection in vivo and encoded an N-terminal hexahistidine (His_6) tag, which provided a simple purification strategy. A second plasmid, pDB1282, encodes the *isc* operon from *Azotobacter vinelandii* and confers ampicillin resistance, making it compatible for coexpression with the pET28a-*lipA* and pET28a-*nfuA* constructs. *E. coli* BL21(DE3) cells, which are commonly used with the pET systems, were cotransformed with either the pET28a-*lipA* *or* pET28a-*nfuA* construct and pDB1282 for gene expression.

E. coli LipA and *E. coli* NfuA are overproduced in *E. coli* cultured in M9 minimal medium supplemented with glucose as a carbon source and ammonium chloride as a nitrogen source, and cultures are grown with limited aeration. During the overproduction, iron is not typically added to the cultures until gene expression is induced at $\sim OD_{600} = 0.6$. We have found that in some cases, iron starvation can induce cells to release siderophores, which have an extremely high affinity for iron and can increase iron–sulfur cluster incorporation in some proteins, such as LipA.

2.2 Protocol for Expression of the *lipA* and *nfuA* Genes

1. Cotransform *E. coli* BL21(DE3) cells with pDB1282 and either pET28a-*lipA* or pET28a-*nfuA* using standard protocols. Use the antibiotic markers within the plasmids to select for colonies containing both pDB1282 and pET28a-*lipA* or pET28a-*nfuA* constructs by plating transformants on Luria-Bertani (LB) plates containing kanamycin (50 μg/mL) and ampicillin (100 μg/mL).
2. Select a single colony and inoculate 200 mL LB media containing kanamycin (50 μg/mL) and ampicillin (100 μg/mL). Incubate at 37°C with shaking at 250 rpm for ~14 h. Simultaneously, place four 6 L flasks containing 4 L M9 minimal media in a shaker overnight at 37°C in order to warm culture media.
3. Immediately prior to inoculation of large flasks, add kanamycin (50 μg/mL) and ampicillin (100 μg/mL).
4. Inoculate each 6 L flask with 10 mL of the 200 mL overnight LB culture and incubate the cultures at 37°C with shaking at 180 rpm.
5. Once the cells reach an $OD_{600}=0.3$ (~5 h), add 0.2% (w/) L-arabinose to induce expression of genes on pDB1282. Take a 1 mL sample of each flask directly before arabinose induction (preinduction sample) for SDS-PAGE analysis.
6. At an $OD_{600}=0.6$, place the flasks in an ice bath for ~1 h and add $FeCl_3$ to a final concentration of 50 μ*M*. Adjust the temperature of the incubator to 18°C. Take a sample of each flask (postarabinose sample) for SDS-PAGE analysis, normalizing the number of cells to the preinduction sample.
7. Induce gene expression by adding 200 μ*M* isopropyl-β-D-thiogalactopyranoside (IPTG) and allow expression to proceed for ~18 h at 18°C with shaking at 180 rpm.
8. The following day, take a sample of each flask (post-IPTG sample) for SDS-PAGE analysis, normalizing the number of cells to the postarabinose sample. Harvest the bacterial cells by centrifugation at 7500 × *g* for 12 min. Flash-freeze the resulting cellular pellet (~3 g/L) in liquid nitrogen and store in liquid nitrogen until further use.
9. Analyze the overproduction of the desired protein by SDS-PAGE (12.5%) using the samples collected during the growth (preinduction, postarabinose induction, post-IPTG induction).

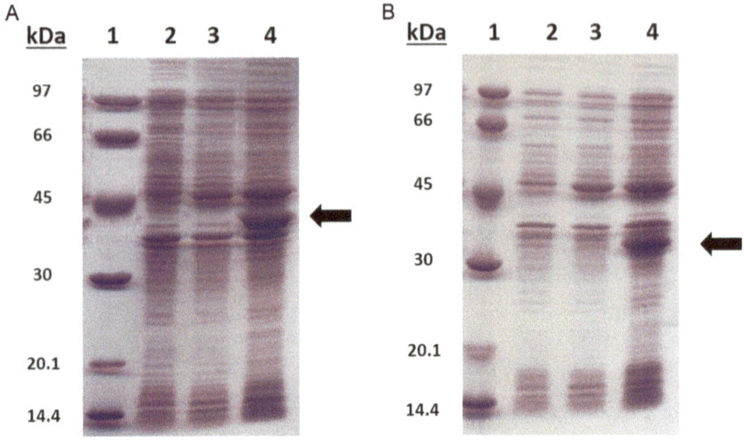

Fig. 1 SDS-PAGE analysis of (A) *E. coli* LipA overproduction and (B) *E. coli* NfuA overproduction. The lanes are as follows: (1) Molecular Weight Ladder, (2) Uninduced cells, (3) Arabinose-induced cells, (4) IPTG-induced cells. The *arrows* point to the protein of interest.

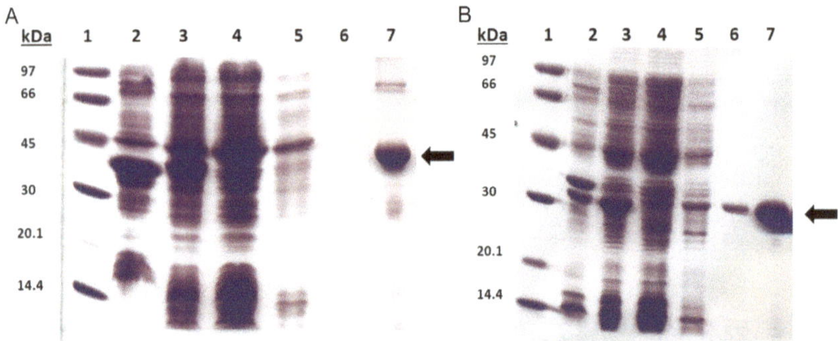

Fig. 2 SDS-PAGE analysis of (A) *E. coli* LipA isolation and (B) *E. coli* NfuA isolation. The lanes are as follows: (1) Molecular Weight Ladder, (2) Pellet, (3) Supernatant, (4) Flow-through, (5) Wash 1, (6) Wash 2, (7) Pure protein. The *arrows* point to the protein of interest.

2.3 Purification of *E. coli* LipA and NfuA

Isolating proteins that contain Fe/S clusters are challenging due to their oxygen-sensitive nature. Oxygen contamination can lead to the degradation of the clusters, resulting in proteins with poor stability and enzymatic activity. As such, careful handling of the protein during all purification steps is vital to ensure that the quality of the protein does not suffer because of oxidative damage. Ideally, the isolation and characterization of Fe/S cluster-containing proteins should be performed in an anaerobic chamber.

The Fe/S clusters coordinated by LipA and NfuA are especially sensitive to oxygen, so maintaining an anaerobic environment is essential. LipA is also sensitive to repeated freeze–thaw cycles, so the purification should be carried out to completion without interruption using chilled buffers, and the protein should be stored in small aliquots. The general procedure for isolating Fe—S cluster-containing proteins has been previously described in detail (Lanz et al., 2012).

1. Autoclave plastic centrifuge tubes and cycle into anaerobic chamber while hot. Allow the tubes to equilibrate in the chamber several days before purification.
2. In the anaerobic chamber, prepare 250 mL of each of the buffers needed for the purification using oxygen-free water: lysis buffer (50 m*M* HEPES, pH 7.5, 300 m*M* KCl, 20 m*M* imidazole, 10 m*M* 2-mercaptoethanol), wash buffer (50 m*M* HEPES, pH 7.5, 300 m*M* KCl, 45 m*M* imidazole, 10 m*M* 2-mercaptoethanol, 10% glycerol), elution buffer (50 m*M* HEPES, pH 7.5, 300 m*M* KCl, 500 m*M* imidazole, 10 m*M* 2-mercaptoethanol, 10% glycerol), and storage buffer (50 m*M* HEPES, pH 7.5, 300 m*M* KCl, 5 m*M* dithiothreitol, 20% glycerol).
3. Resuspend the cell paste (~60 g) in 150 mL lysis buffer supplemented with lysozyme (1 mg/mL) and DNase I (0.1 mg/mL) in a metal sonication cup. Once the solution is homogeneous, allow it to stir for 30 min at room temperature and then add one EDTA-free Sigmafast protease inhibitor tablet.
4. Place the sonication cup in an ice-water bath on a stir-plate and slowly stir.
5. Lyse the cells by sonic disruption using a Fisher 550 Sonic Dismembrator by sonicating for 1 min at 30% output (setting of 7) followed by resting for 8 min to allow for temperature reequilibration. Repeat for a total of five cycles. The cell lysate should become noticeably darker as the cells burst, releasing Fe—S containing proteins into solution.
6. Following sonication, pour the cell lysate into the plastic centrifuge tubes that were previously autoclaved and allowed to equilibrate in the anaerobic chamber. Seal the juncture between the cap and tube tightly with vinyl tape to avoid oxygen contamination.
7. Remove the tightly sealed tubes from the glove box and centrifuge at 45,000 × *g* for 1 h at 4°C.
8. Cycle the centrifuge tubes back into the anaerobic glove box and pour the supernatant into a 250 mL flask. Take a 50 μL sample of the

supernatant and scrape a small amount of the pellet into an Eppendorf tube for SDS-PAGE analysis.

9. Equilibrate a column containing 50 mL Ni-NTA resin with 100 mL chilled lysis buffer.
10. After equilibration, load the supernatant onto the column. Collect the flow-through in a separate container and take a 50 μL sample for SDS-PAGE analysis.
11. Wash the column with 100 mL chilled wash buffer and take a 50 μL sample for SDS-PAGE analysis. Repeat the wash step and collect another 50 μL sample for SDS-PAGE analysis.
12. Elute the desired protein using chilled elution buffer. Allow ~15 mL of elution buffer to flow over the column before collecting the protein, which is when the dark brown colored band will begin to elute. Stop collecting once the color of the eluate returns to clear.
13. Concentrate the protein in an Amicon stir-cell fitted with an YM-10 membrane (10 kDa molecular weight cutoff) in the anaerobic glove box. It is vital to keep the stirring apparatus on ice throughout the entire concentration step.
14. Once the protein has been concentrated to ~2.5 mL, exchange it into chilled storage buffer using a PD-10 column. Collect 3.5 mL and aliquot into Eppendorf tubes.
15. Take a 5 μL aliquot of the final purified protein for SDS-PAGE analysis.
16. Remove the protein aliquots from the anaerobic chamber and immediately flash-freeze in liquid nitrogen. Store the protein aliquots in liquid nitrogen until use, avoiding freezing and thawing, which leads to activity loss.
17. Analyze the purification by SDS-PAGE (12.5%) using the samples collected of the cellular pellet, supernatant, flow-through, wash, and final protein.

3. INITIAL CHARACTERIZATION OF AS-ISOLATED PROTEIN

Following the isolation, routine characterizations should be performed to determine the yield and to obtain a qualitative estimate of the cofactor incorporation in the protein. These routine analyses provide a straightforward method to monitor consistency among enzyme preparations. UV–visible spectroscopy can be used to analyze [4Fe–4S] cluster-containing proteins which have distinct features at 330 and 400 nm (Fig. 3). Protein

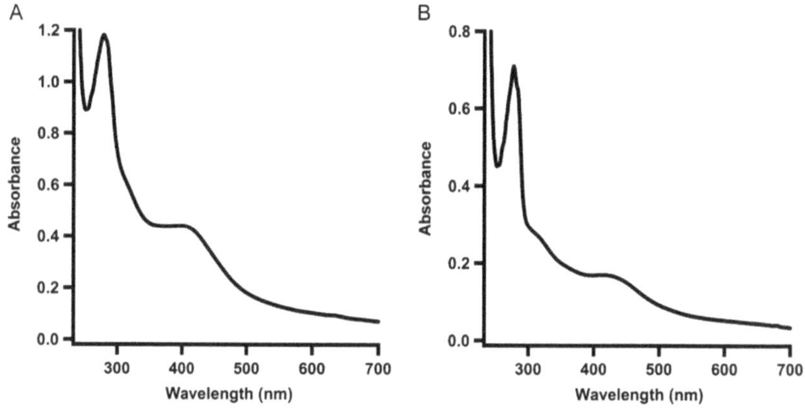

Fig. 3 UV–visible spectrum of (A) As-isolated *E. coli* LipA and (B) As-isolated *E. coli* NfuA.

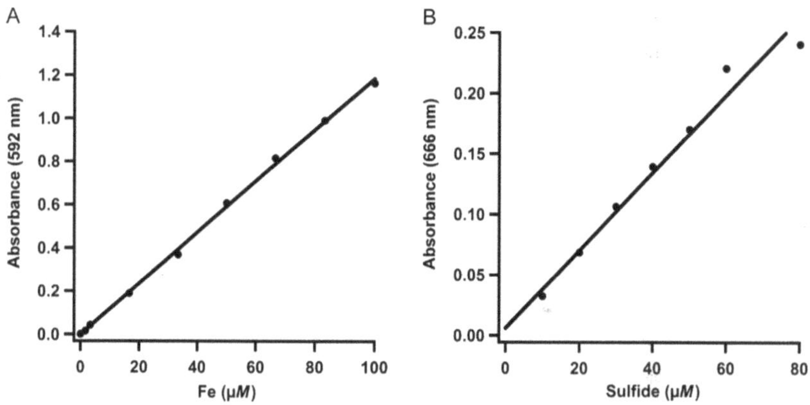

Fig. 4 (A) Iron analysis standard curve. (B) Sulfide analysis standard curve.

concentration is determined via the Bradford method (Bradford, 1976) using bovine serum albumin (fraction V) as a standard. However, because the Bradford method is biased by amino acid composition, a correction factor should be obtained through amino acid analysis to account for differences in amino acid sequence. This step was especially important in the analysis of LipA, which has a correction factor of nearly 1.5. In this case, its amino acid composition significantly changes the protein concentration empirically determined by the Bradford method, which has further implications in determining its activity and the stoichiometry of its Fe—S cluster cofactors. In addition to protein concentration, colorimetric iron and sulfide analyses should be performed (Beinert, 1978, 1983) (Fig. 4). It is important to note that these methods simply estimate the amount of iron and sulfide in solution; they do not provide evidence for the biologically relevant form of the cofactor or its coordination environment. For a more advanced understanding, Mössbauer and EPR spectroscopies should be used, as detailed in other methods papers (Lanz et al., 2012; Pandelia, Lanz, Booker, & Krebs, 2015).

3.1 Protocol for Obtaining Amino Acid Correction Factor

1. Contact a facility capable of performing amino acid analysis, such as the University of California-Davis proteomics facility.
2. Ensure that the protein you wish to analyze is >95% homogenous.
3. Using a NICK column (GE healthcare) exchange 100 μL of the protein (~40 mg/mL) into 10 m*M* NaOH in order to remove the Fe—S cluster.
4. Perform three Bradford assays each with an independent standard curve to determine the protein concentration.
5. Average the three values together and record as the concentration determined via the Bradford method.
6. Carefully place 100 μL of the protein in three Corning Pyrex test tubes (10 × 75 m*M*) and seal the top with parafilm.
7. Use a needle to add a hole to the top of each of the glass tubes and lyophilize the protein samples in vacuo.
8. Ship to the proteomics facility in order to determine the true concentration of the sample based on its amino acid composition.
9. Average the three values of the concentration determined by amino acid analysis.
10. Generate the amino acid correction factor by dividing the concentration determined by amino acid analysis by the concentration determined by the Bradford method.
11. Apply this ratio in every subsequent Bradford analysis performed on the protein to ensure an exact determination of the concentration.

3.2 Colorimetric Iron Analysis Protocol

1. Prepare the following solutions: Reagent A (1.35 g SDS, 450 μL saturated iron-free ammonium acetate, in a final volume of 30 mL dH_20); Reagent B (270 mg ascorbic acid, 9 mg sodium metabisulfite, 450 μL saturated iron-free ammonium acetate, in a final volume of 6 mL dH_20); and Reagent C (18 mg ferene in a final volume of 1 mL dH_20).

2. In the anaerobic chamber, dilute the protein to 10 μ*M* in 50 m*M* NaOH.
3. Aliquot into four 300 μL samples and remove the samples from the anaerobic chamber.
4. Prepare a 100 μ*M* iron standard stock using Claritas PPT iron standard (Fisher Scientific).
5. Dilute the 100 μ*M* iron standard to the following concentrations in a final volume of 300 μL: 0, 1.66, 3.33, 16.6, 33.3, 50, 66.6, 83.3, and 100 μ*M*.
6. Add 300 μL of Reagent A to each sample and standard and mix.
7. Add 300 μL Reagent B to each sample and standard and mix.
8. Incubate the samples and standards in a 37°C water bath for 15 min.
9. Add 15 μL of Reagent C to each sample and standard. Mix and then centrifuge at 14,000 × *g* for 5 min.
10. Measure the absorbances of the standards at 592 nm using the first standard (0 μ*M*) as a blank.
11. Measure the absorbances of each of the four samples.
12. Plot the concentration vs absorbance of each of the standards and use the linear equation to calculate the concentrations of each of the four samples (Fig. 4).
13. Average the concentration of iron in each of the four samples and normalize to the protein concentration in order to determine the amount of iron per protein.

3.3 Colorimetric Sulfide Analysis Protocol

1. Prepare the following solutions anaerobically: 1% zinc acetate (w/v%); 3 *M* NaOH; 0.2% dimethyl-4-phenylenediamine (DMPD) in 5 N HCl; 23 m*M* $FeCl_3$ in 1.2 N HCl; and 200 μ*M* Na_2S.
2. Complete steps 3–10 anaerobically.
3. Dilute the protein to 10 μ*M* in 50 m*M* NaOH (prepared in deoxygenated water) to a final volume of 1000 μL.
4. Aliquot 100 μL of the 10 μ*M* protein solution into five Eppendorf tubes.
5. Create a standard curve using the 200 μ*M* Na_2S solution by making 100 μL standards at the following concentrations: 0, 10, 20, 30, 40, 50, 60, and 80 μ*M*.
6. Add 300 μL of the 1% zinc acetate solution to each of the standards and samples.
7. Add 15 μL of 3 *M* NaOH to each of the standards and samples. Mix by inversion.

8. Let stand for 2 h to allow zinc sulfide to form and settle to the bottom of the tubes.
9. Underlay the standards and samples with 75 μL of the 0.2% DMPD solution.
10. Quickly add 2 μL of 23 m*M* $FeCl_3$ to each sample and mix by inversion.
11. Remove standards and samples from the anaerobic chamber and centrifuge at 14,000 × *g* for 15 min.
12. Remove 400 μL of each standard and sample and place in unused Eppendorf tubes.
13. Add 600 μL 2 m*M* EPPS, pH 8.0, to each standard and sample.
14. Let stand for 30 min at room temperature and then measure the absorbance at 666 nm.
15. Plot the concentration vs the absorbance from each standard. Use the linear equation to determine the concentration of sulfide in each sample and normalize the sulfide to the protein concentration to determine the amount of sulfide per protein (Fig. 4).

4. METHODS FOR ISOTOPICALLY LABELING IRON–SULFUR CLUSTERS TO MONITOR CLUSTER TRANSFER

4.1 Protocol for the Expression of the *nfuA* Gene in Apo Form

1. Introduce the pET28a-*nfuA* construct into BL21(DE3) cells using standard transformation procedures. Spread cells on an LB plate supplemented with 50 μg/mL kanamycin.
2. The following day, use a single colony to inoculate 200 mL LB media in a 250-mL flask supplemented with 50 μg/mL kanamycin. Incubate the starter culture at 37°C with shaking at 250 rpm for ~14 h. Place four 6-L flasks containing 4 L M9 minimal media in the shaker overnight at 37°C to warm the culture media.
3. Prior to inoculation of the large flasks, add 50 μg/mL kanamycin and 400 μL 10,000 × micronutrient mix (3 n*M* $(NH_4)_2Mo_7O_{24}$, 30 n*M* $CoCl_2·6H_20$, 400 n*M* H_3BO_3, 10 n*M* $CuSO_4·5H_2O$, 80 n*M* $MnCl_2·4H_20$, and 10 n*M* $ZnSO_4·7H_20$).
4. Use 20 mL of the overnight starter culture to inoculate each of the four large flasks.
5. At an OD_{600} = 0.3, supplement cultures with 200 mL amino acid mixture (alanine (800 μ*M*), arginine (400 μ*M*), asparagine (400 μ*M*),

aspartic acid (400 μ*M*), cysteine (100 μ*M*), glutamic acid (600 μ*M*), glutamine (600 μ*M*), glycine (800 μ*M*), histidine (200 μ*M*), isoleucine (400 μ*M*), leucine (800 μ*M*), lysine (400 μ*M*), methionine (200 μ*M*), phenylalanine (400 μ*M*), proline (400 μ*M*), serine (10 m*M*), threonine (400 μ*M*), tryptophan (100 μ*M*), tyrosine (200 μ*M*), and valine (600 μ*M*). Take a 1 mL sample (uninduced sample) for SDS-PAGE analysis.

6. At an $OD_{600}=0.6$, add 1,10-phenanthroline to each of the flasks to a final concentration of 75 μ*M* and allow the flasks to shake for an additional 15 min.
7. Place the flasks in an ice-water bath for ~1 h. Turn the temperature of the incubator to 18°C.
8. Induce overexpression of the desired gene by adding 200 μ*M* IPTG and allow expression to take place at 18°C with shaking at 180 rpm for ~18 h.
9. On the following day, take a sample (post-IPTG induction) normalized to the number of cells in the uninduced sample for SDS-PAGE analysis.
10. Harvest the cells by centrifugation at 7500 × *g* for 12 min. Flash-freeze the resulting cell pellet (~2 g/L) in liquid nitrogen and store in liquid nitrogen until further use.
11. Isolate the apo protein using the exact method described for the holo protein.
12. Analyze the overproduction of the protein by SDS-PAGE analysis (12.5%) using the samples collected.

4.2 Synthesis of ^{34}S-Labeled Sulfide From ^{34}S-Labeled Elemental Sulfur

1. Perform all steps under a constant flow of nitrogen.
2. Cool a two-neck round-bottomed flask containing a magnetic stir bar to −78°C in a dry ice/acetone bath.
3. Add 0.5 g elemental ^{34}S-labeled elemental sulfur to the cooled flask and pass ammonia gas into the flask until approximately 75 mL liquid ammonia forms.
4. Carefully and slowly add a twofold molar excess of sodium metal to the reaction and allow it to stir for 30 min.
5. Remove the reaction from the dry ice/acetone bath and allow it to warm to room temperature.
6. Evaporate the ammonia under a stream of nitrogen gas to yield a white precipitate.
7. Add isopropanol until the blue color dissipates, indicating that the excess sodium metal is quenched.

8. Dissolve the white precipitate in 15 mL H_2O.
9. Remove the solvent using a rotary evaporator to yield a solid containing both NaOH and ^{34}S-labeled sodium sulfide ($Na_2^{34}S$).
10. Dissolve the solid in a minimum amount of water and quantify the sulfide concentration using the colorimetric sulfide analysis protocol detailed above.

4.3 Reconstitution of Apo NfuA Using $Na_2^{34}S$

1. Perform all reconstitution steps under strictly anaerobic conditions.
2. Dilute Apo NfuA to 100 μ*M* in reconstitution buffer (50 m*M* HEPES, pH 7.5, 300 m*M* KCl, 5 m*M* DTT, and 20% glycerol) in a total volume of 20 mL.
3. Add DTT (5 m*M* final concentration) in three increments every 20 min. Incubate the solution on ice for 1 h.
4. Slowly, add $FeCl_3$ (2 m*M* final concentration) every 5 min a total of five times. Incubate the solution for an additional 30 min.
5. Add the $Na_2^{34}S$ (2 m*M* final concentration) in five increments every 30 min. Check the pH of the $Na_2^{34}S$ solution before adding and neutralize the NaOH left from the synthesis procedure if necessary. Incubate the reaction on ice overnight.
6. The following day, remove protein precipitation by centrifuging the reconstituted protein at 14,000 × *g* for 15 min.
7. Concentrate the solution to ~2 mL.
8. Remove aggregates from the reconstitution procedure by size-exclusion chromatography using a HiPrep 16/60 Sephacryl HR S-200 column (GE Health Sciences) equilibrated in storage buffer (flow rate of 0.5 mL/min) and connected to an ÅKTA protein liquid chromatography system (GE Health Sciences) housed in an anaerobic chamber.
9. Collect the fractions containing the reconstituted protein and concentrate to ~2 mL. Flash-freeze in liquid nitrogen until further use. Redetermine the protein concentration using the Bradford method and perform an iron and sulfide analyses on the reconstituted protein.

5. STRATEGIES FOR ANALYZING PROTEIN COMPLEX FORMATION

Various methods exist to measure protein–protein interactions; however, it is often difficult to characterize interactions that are transient in nature. A nonexhaustive list of methods includes coimmunoprecipitation, isothermal titration calorimetry (ITC), surface plasmon resonance (SPR), and analytical molecular-sieve chromatography. The benefit of using coimmunoprecipitation is the ability to pull down protein complexes in vivo that are biologically relevant. However, while it is possible to pull protein complexes out of the cell, the use of a cross-linking reagent is often needed to capture weak interactors. Both ITC and SPR are enticing options because they can quantitatively determine the K_d associated with the binding event in protein–protein interactions or with protein–small molecule interactions. One downfall associated with ITC is the need for large amounts of material for analysis.

Qualitatively, molecular-sieve chromatography can be used to determine if in vitro mixtures of proteins coelute as a complex rather than as individual components. A standard curve can be generated using a suite of standards in order to estimate the molecular weight of the complexes (Fig. 5). This method is inferior to ITC and SPR in the sense that it only provides a qualitative assessment of whether the proteins form a complex. However, it takes very small amounts of material and can be performed in a single day. Similar to coimmunoprecipitation, this method does not capture transient interactions. Following its completion, other methods can be used for a more quantitative assessment of the binding event.

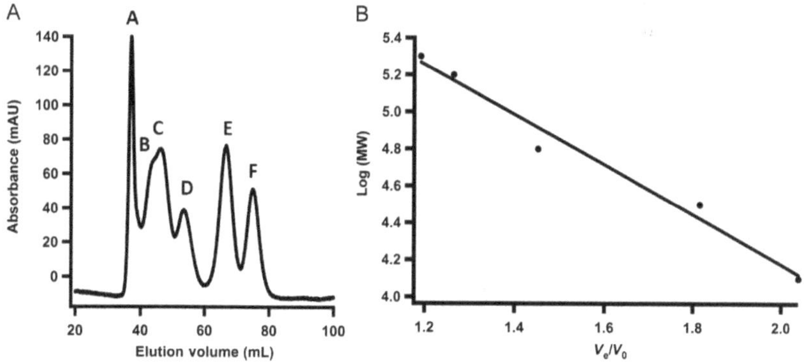

Fig. 5 Molecular sieve chromatography. (A) Chromatogram of the suite of standards used to generate the standard curve as follows: (A) Blue dextran, 2000 kDa; (B) β-Amylose, 200 kDa; (C) Alcohol dehydrogenase, 150 kDa; (D) Bovine serum albumin, 66 kDa; (E) Carbonic anhydrase, 29 kDa; (F) Cytochrome C, 12.4 kDa. (B) Standard curve generated from the elution volume of each of the standards corrected for the void volume of the column vs the log molecular weight of each of the standards.

5.1 Protocol for Performing Analytical Molecular-Sieve Chromatography

1. Equilibrate a HiPrep 16/60 Sephacryl HR S-200 column (GE Healthcare) connected to an ÅKTA protein liquid chromatography system (GE Health Sciences) housed in an anaerobic chamber with buffer (50 m*M* HEPES, pH 7.5, 300 m*M* KCl, 5 m*M* DTT, and 15% glycerol).
2. It is essential to ensure that the glove box is fully devoid of oxygen contamination (<1 ppm O_2) and to add fresh reductant (5 m*M* DTT) to buffer directly before use to prevent disulfide formation between proteins.
3. Prepare a 500 μL sample in buffer (50 m*M* HEPES, pH 7.5, 300 m*M* KCl, 5 m*M* DTT, and 15% glycerol) containing the following proteins as standards: cytochrome *C* (0.3 mg/mL), carbonic anhydrase (0.3 mg/mL), bovine serum albumin (0.80 mg/mL), alcohol dehydrogenase (0.65 mg/mL), β-amylase (0.75 mg/mL), and Blue Dextran (0.40 mg/mL).
4. Load the mixture containing the standards onto the column and flow at a rate of 0.5 mL/min until all of the standards elute. Record the elution volume of the Blue Dextran, which indicates the experimentally determined void volume (V_0), and the elution volumes (V_e) of each of the additional standards, which will be used to generate a standard curve.
5. Dilute the LipA protein to make a 500 μL sample at a concentration of 100 μ*M*. Load the sample onto the column and elute at a rate of 0.5 mL/min. Record its elution volume.
6. Dilute the NfuA protein to make a 500 μL sample with a concentration of 100 μ*M*. Load the sample onto the column and elute at a rate of 0.5 mL/min. Record its elution volume.
7. Make a 500 μL sample mixture containing 100 μ*M* LipA protein and 100 μ*M* NfuA protein. Incubate at room temperature for 5 min and then load onto the column and elute at a rate of 0.5 mL/min. Collect protein containing fractions and record the elution volumes of each peak observed in the chromatogram.
8. Concentrate the fractions corresponding to each protein peak observed in the chromatogram. Analyze by SDS-PAGE (12.5%) to confirm the identity of the proteins in each fraction.
9. Generate a standard curve of the Log(molecular weight) vs V_e/V_0 for each of the standards.
10. Divide the experimentally determined elution volume of each of the peaks in the chromatograms of each sample by the void volume of the column (V_e/V_0).

11. Use the linear equation from the standard curve to solve for the Log(molecular weight) of the individual proteins as well as the protein complexes.

6. MONITORING IRON–SULFUR CLUSTER TRANSFER USING QUANTITATIVE ANALYSIS

A strategy for accurately monitoring lipoyl cofactor production in vitro was not established until the early 2000s (Miller et al., 2000). This strategy was based on an indirect assay in which activation of apo-pyruvate dehydrogenase by lipoylation was coupled to the reduction of NAD^+, which could be monitored spectrophotometrically. Through this method, the rate of NADH formation was directly dependent on the amount of lipoylated pyruvate dehydrogenase. Since that time, more direct methods have been developed to quantitatively monitor the formation of the monothiolated intermediate and lipoyl product using liquid chromatography-mass spectrometry (LC-MS), which is described below. This method allows the activity of LipA to be quantified and the effect of NfuA on LipA's activity to be assessed (Fig. 6). Additionally, by using NfuA with isotopically labeled clusters, as detailed above, the transfer of an iron–sulfur cluster can be directly monitored. Further, including a competitive chelator such as citrate in the reaction can serve to ensure that the cluster transfer from

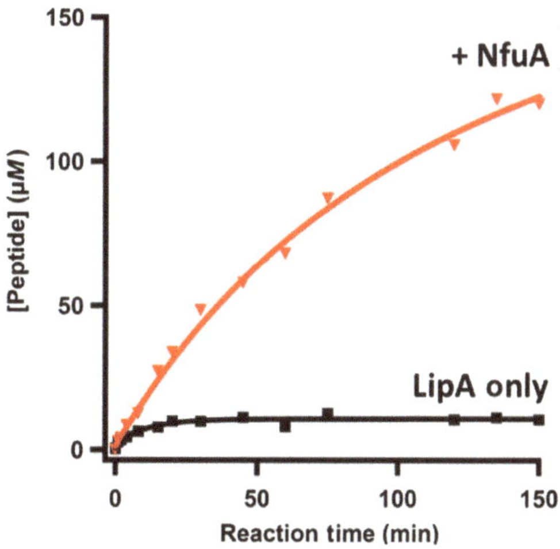

Fig. 6 Time-dependent formation of the monothiolated intermediate and lipoyl product in the absence (15 μ*M* LipA, shown in *black*) and presence (400 μ*M* NfuA, shown in *red*) of NfuA.

NfuA to LipA is direct, as opposed to a diffusive mechanism wherein LipA obtains a cluster from solution (Ding, Clark, & Ding, 2004).

6.1 Sample Preparation for Analysis of Reaction Products by LC-MS to Assess the Effect of NfuA on LipA Activity

1. Reactions are performed anaerobically. Reactions are typically prepared in a total volume of 200 μL and contain the following: buffer (50 m*M* HEPES, pH 7.5, 300 m*M* KCl, 10% glycerol), 600 μ*M* octanoyl peptide substrate (Glu-Ser-Val-[N^6-octanoyl]Lys-Ala-Ala-Ser-Asp), 25 μ*M* LipA, 1 μ*M* *S*-adenosylhomocysteine (SAH) nucleosidase (to prevent product inhibition), and 2 m*M* SAM. An additional and identical reaction, but containing 400 μ*M* NfuA, is also prepared. Importantly, DTT has been shown to be capable of nonphysiological iron–sulfur cluster transfer and should not be used in assay buffers (Vranish et al., 2015).
2. A 500 μL quench solution containing 100 m*M* H_2SO_4 and 20 μ*M* AtsA (peptide internal standard, Pro-Met-Ser-Ala-Pro-Ala-Arg-Ser-Met) is prepared, and 15 μL is aliquoted into 11 Eppendorf tubes.
3. Initiate the reaction by adding the low-potential reductant. In LipA/NfuA assays, 2 m*M* dithionite is typically used. The initiation of the two reactions (with and without NfuA) can be staggered by 30 s in order to perform them simultaneously.
4. At the specified time points ($t=0$, 1, 2, 4, 8, 15, 30, 60, 90, 120, and 150 min), add 15 μL of the reaction into 15 μL of the quench solution.
5. After the reaction is complete, remove the reaction tubes from the anaerobic chamber and centrifuge for 30 min at 14,000 × *g*.
6. Avoiding any precipitation, pipette 25 μL of each time point into an appropriate autosampler vial for LC-MS analysis.

6.2 Sample Preparation for Analysis of Reaction Products in the Presence of a Competitive Chelator or Exogenous ^{34}S-Labeled Sulfide to Ensure Direct Cluster Transfer From NfuA to LipA

1. Prior to performing activity assays with LipA and NfuA, incubate each protein individually with a fivefold excess sodium citrate. Assess whether the citrate has a significant effect on the Fe—S clusters of LipA and NfuA by obtaining a UV–visible spectrum of the protein before and after a 1.5 h incubation with citrate.
2. Perform a LipA only control and a LipA/NfuA activity assay exactly as described previously.

3. Set up two additional reactions containing both LipA and NfuA. In one reaction, add a fivefold excess of citrate. In the other reaction, add $Na_2^{34}S$ to a final concentration of 1 m*M*.
4. When analyzing data from the LipA/NfuA reaction with citrate, assess whether the addition of this chelator inhibits product formation by LipA. Citrate can chelate metals from solution and will inhibit multiple turnover of LipA if LipA's auxiliary cluster is regenerated chemically from iron and sulfide in solution after each turnover. The absence of an effect is consistent with a direct cluster transfer from NfuA through protein–protein interactions.
5. As a second method (or if citrate chelates the cluster out of the protein of interest), analyze whether ^{34}S from exogenous $Na_2^{34}S$ included in the reaction mixture is observed in the lipoyl product. The observation of product that contains sulfur at natural abundance suggests that the attached sulfur atom did not come from solution.

6.3 Protocol for the Quantification of Reaction Products by LC-MS

1. Prepare 50 μL of a 500 μ*M* standard in deoxygenated water that contains buffer (50 m*M* HEPES, pH 7.5, 300 m*M* KCl, and 10% glycerol), 50 m*M* H_2SO_4, 2 m*M* TCEP, 20 μ*M* AtsA internal standard, 500 μ*M* octanoyl peptide, 500 μ*M* 6-thio-octanoyl peptide, and 500 μ*M* lipoyl peptide. The peptides used for quantification (octanoyl, 6-thio-octanoyl, lipoyl, AtsA) are >98% pure and can be purchased from ProImmune Custom Peptides.
2. Prepare 300 μL of a dilution mix containing the following: buffer (50 m*M* HEPES, pH 7.5, 300 m*M* KCl, and 10% glycerol), 50 m*M* H_2SO_4, 2 m*M* TCEP, and 20 μ*M* AtsA internal standard.
3. Aliquot 25 μL of the dilution mix into 11 Eppendorf tubes.
4. Perform a 1:2 serial dilution of the 500 μ*M* standard by mixing 25 μL of the standard with 25 μL of the dilution mix.
5. Continue serial diluting the standards by a factor of 2 to generate standards for quantification (500, 250, 125, 62.5, 31, 16, 8, 4, 2, 1, 0.5, and 0.25 μ*M*).
6. Analyze the standards and reaction mixtures by LC-MS (Agilent Technologies 1200 system coupled to an Agilent Technologies 6410 QQQ mass spectrometer) using an Agilent Technologies Zorbax Extend-C18 column Rapid Resolution HT (4.6 mm × 50 mm, 1.8 μm particle size) equilibrated in 98% Solvent A (0.1% formic acid, pH 2.6) and 2% Solvent B (100% acetonitrile). Detect the compounds using

electrospray ionization in positive mode (ESI^+) with the following parameters: 340°C nitrogen gas temperature, 9.0L/min flow rate, 40 PSI nebulizer pressure, and 4000V capillary pressure. Apply a gradient of 2%–23% Solvent B from 0.8 to 3.5min and maintain at 23% Solvent B until 8min. Return to 2% Solvent B from 8 to 10min and then allow the column to reequilibrate for 2min before the next injection. Maintain a flow rate of 0.4mL/min throughout the entire method.

7. Use the associated MassHunter program to quantify the decay of the octanoyl peptide substrate and formation of the 6-thio-octanoyl peptide intermediate and lipoyl product (Tables 1 and 2).

Table 1 Parameters for LC-MS/MS Analysis of Unlabeled Reaction Intermediate and Product

Compound	Parent Ion	Fragmentator Voltage	Product Ion 1[a]	Product Ion 2[a]
Octanoyl peptide	932.5	240	552.3 (37)	210.1 (51)
6-Thio-octanoyl peptide	964.5	240	584.3 (37)	242.1 (51)
Lipoyl peptide	996.5	240	616.4 (37)	274.1 (51)
AtsA peptide (IS)	474.4	135	229.1 (12)	153.1 (26)
Tryptophan (IS)	188.0	130	146.1 (10)	118.0 (21)

[a]Respective collision energies are in parentheses (McCarthy & Booker, 2017).

Table 2 Parameters for LC-MS/MS Analysis of ^{34}S-Labeled Reaction Intermediate and Product

Compound	Parent Ion	Fragmentator Voltage	Product Ion 1[a]	Product Ion 2[a]
Octanoyl peptide	932.5	240	552.3 (37)	210.1 (51)
^{32}S-6-Thio-octanoyl peptide	964.5	240	584.3 (37)	242.1 (51)
^{34}S-6-Thio-octanoyl peptide	966.5	240	586.3 (37)	244.1 (51)
$^{32}S/^{32}S$ Lipoyl peptide	996.5	240	616.4 (37)	274.1 (51)
$^{32}S/^{34}S$ Lipoyl peptide	998.5	240	618.4 (37)	276.1 (51)
$^{34}S/^{34}S$-Lipoyl peptide	1000.5	240	620.4 (37)	278.1 (51)
AtsA peptide (IS)	474.4	135	229.1 (12)	153.1 (26)
Tryptophan (IS)	188	130	146.1 (10)	118.0 (21)

[a]Respective collision energies are in parentheses (McCarthy & Booker, 2017).

7. CONCLUSIONS

LipA catalyzes the attachment of sulfur atoms to an unactivated octanoyl chain in the biosynthesis of the lipoyl cofactor, an essential compound used in various multienzyme complexes in Nature and implicated in human health and disease. Refined techniques for the overproduction and isolation of Fe—S containing proteins allowed for the isolation of homogenous LipA with high activity for robust mechanistic study, although the enzyme only catalyzes a single turnover due its inactivation upon the destruction of its auxiliary cluster. Recent work has identified a pathway for reforming the auxiliary cluster to allow for multiple turnovers. Biochemical methods to study the molecular details of cofactor repair and/or reinsertion in proteins that sacrifice their Fe—S clusters will help fill outstanding gaps in our understanding of the mechanism of protein maturation.

ACKNOWLEDGMENTS

This work was supported by the National Institutes of Health Grant GM122595 (S.J.B.) and NSF Grant MCB-1158486 (S.J.B.). S.J.B. is an investigator of the Howard Hughes Medical Institute.

REFERENCES

Angelini, S., Gerez, C., Ollagnier-de Choudens, S., Sanakis, Y., Fontecave, M., Barras, F., et al. (2008). NfuA, a new factor required for maturing Fe/S proteins in Escherichia coli under oxidative stress and iron starvation conditions. *The Journal of Biological Chemistry*, *283*(20), 14084–14091.

Beinert, H. (1978). Micro methods for the quantitative determination of iron and copper in biological material. *Methods in Enzymology*, *54*, 435–445.

Beinert, H. (1983). Semi-micro methods for analysis of labile sulfide and of labile sulfide plus sulfane sulfur in unusually stable iron-sulfur proteins. *Analytical Biochemistry*, *131*(2), 373–378.

Blaszczyk, A. J., Wang, R. X., & Booker, S. J. (2017). TsrM as a model for purifying and characterizing cobalamin-dependent radical S-adenosylmethionine methylases. *Methods in Enzymology*, *595*, 303–329.

Borcea, V., Nourooz-Zadeh, J., Wolff, S. P., Klevesath, M., Hofmann, M., Urich, H., et al. (1999). Alpha-lipoic acid decreases oxidative stress even in diabetic patients with poor glycemic control and albuminuria. *Free Radical Biology & Medicine*, *26*, 1495–1500.

Boutigny, S., Saini, A., Baidoo, E. E., Yeung, N., Keasling, J. D., & Butland, G. (2013). Physical and functional interactions of a monothiol glutaredoxin and an iron sulfur cluster carrier protein with the sulfur-donating radical S-adenosyl-L-methionine enzyme MiaB. *The Journal of Biological Chemistry*, *288*(20), 14200–14211.

Bradford, M. M. (1976). A rapid and sensitive method for the quantitation of microgram quantities of protein utilizing the principle of protein-dye binding. *Analytical Biochemistry*, *72*, 248–254.

Cicchillo, R. M., & Booker, S. J. (2005). Mechanistic investigations of lipoic acid biosynthesis in *Escherichia coli*: Both sulfur atoms in lipoic acid are contributed by the same lipoyl synthase polypeptide. *Journal of the American Chemical Society*, *127*, 2860–2861.

Cicchillo, R. M., Iwig, D. F., Jones, A. D., Nesbitt, N. M., Baleanu-Gogonea, C., Souder, M. G., et al. (2004). Lipoyl synthase requires two equivalents of S-adenosyl-L-methionine to synthesize one equivalent of lipoic acid. *Biochemistry*, *43*, 6378–6386.

Cicchillo, R. M., Lee, K. H., Baleanu-Gogonea, C., Nesbitt, N. M., Krebs, C., & Booker, S. J. (2004). Escherichia coli lipoyl synthase binds two distinct [4Fe–4S] clusters per polypeptide. *Biochemistry*, *43*, 11770–11781.

Cronan, J. E. (2016). Assembly of lipoic acid on its cognate enzymes: An extraordinary and essential biosynthetic pathway. *Microbiology and Molecular Biology Reviews*, *80*, 429–450.

Ding, H., Clark, R. J., & Ding, B. (2004). IscA mediates iron delivery for assembly of iron-sulfur clusters in IscU under the limited accessible free iron conditions. *The Journal of Biological Chemistry*, *279*, 37499–37504.

Douglas, P., Kriek, M., Bryant, P., & Roach, P. L. (2006). Lipoyl synthase inserts sulfur atoms into an octanoyl substrate in a stepwise manner. *Angewandte Chemie*, *118*, 5321–5323.

Harmer, J. E., Hiscox, M. J., Dinis, P. C., Fox, S. J., Iliopoulos, A., Hussey, J. E., et al. (2014). Structures of lipoyl synthase reveal a compact active site for controlling sequential sulfur insertion reactions. *The Biochemical Journal*, *464*, 123–133.

Holmquist, L., Stuchbury, G., Berbaum, K., Muscat, S., Young, S., Hager, K., et al. (2007). Lipoic acid as a novel treatment for Alzheimer's disease and related dementias. *Pharmacology & Therapeutics*, *113*(1), 154–164.

Landgraf, B. J., McCarthy, E. L., & Booker, S. J. (2016). Radical S-adenosylmethionine enzymes in human health and disease. *Annual Review of Biochemistry*, *85*(1), 485–514.

Lanz, N. D., & Booker, S. J. (2015). Auxiliary iron-sulfur cofactors in radical SAM enzymes. *Biochimica et Biophysica Acta*, *1853*(6), 1316–1334.

Lanz, N. D., Grove, T. L., Gogonea, C. B., Lee, K. H., Krebs, C., & Booker, S. J. (2012). RlmN and AtsB as models for the overproduction and characterization of radical SAM proteins. *Methods in Enzymology*, *516*, 125–152.

Lanz, N. D., Pandelia, M. E., Kakar, E. S., Lee, K. H., Krebs, C., & Booker, S. J. (2014). Evidence for a catalytically and kinetically competent enzyme-substrate cross-linked intermediate in catalysis by lipoyl synthase. *Biochemistry*, *53*, 4557–4572.

McCarthy, E. L., & Booker, S. J. (2017). Destruction and reformation of an iron-sulfur cluster during catalysis by lipoyl synthase. *Science*, *358*(6361), 373–377.

McLaughlin, M. I., Lanz, N. D., Goldman, P. J., Lee, K. H., Booker, S. J., & Drennan, C. L. (2016). Crystallographic snapshots of sulfur insertion by lipoyl synthase. *Proceedings of the National Academy of Sciences of the United States of America*, *113*, 9446–9450.

Miller, J. R., Busby, R. W., Jordan, S. W., Cheek, J., Henshaw, T. F., Ashley, G. W., et al. (2000). *Escherichia coli* LipA is a lipoyl synthase: In vitro biosynthesis of lipoylated pyruvate dehydrogenase complex from octanoyl-acyl carrier protein. *Biochemistry*, *39*, 15166–15178.

Pandelia, M. E., Lanz, N. D., Booker, S. J., & Krebs, C. (2015). Mossbauer spectroscopy of Fe/S proteins. *Biochimica et Biophysica Acta*, *1853*(6), 1395–1405.

Py, B., Gerez, C., Angelini, S., Planel, R., Vinella, D., Loiseau, L., et al. (2012). Molecular organization, biochemical function, cellular role and evolution of NfuA, an atypical Fe-S carrier. *Molecular Microbiology*, *86*(1), 155–171.

Reed, L. (1974). Multienzyme complexes. *Accounts of Chemical Research*, 7, 40–46.

Reed, L. J., Busk, B. G. D., Gunsalus, I. C., & Schnakenberg, G. H. F. (1951). Crystalline a-lipoic acid: A catalytic agent associated with pyruvate dehydrogenase. *Science*, *114*, 93.

Reed, L. J., Gunsalus, I. C., Schnakenberg, G. H. F., Soper, Q. F., Boaz, H. E., Kern, S. F., et al. (1953). Isolation, characterization and structure of α-lipoic acid. *Journal of the American Chemical Society*, *75*, 1267–1270.

Vranish, J. N., Russell, W. K., Yu, L. E., Cox, R. M., Russell, D. H., & Barondeau, D. P. (2015). Fluorescent probes for tracking the transfer of iron-sulfur cluster and other metal cofactors in biosynthetic reaction pathways. *Journal of the American Chemical Society, 137*(1), 390–398.

Yi, X., & Maeda, N. (2005). Endogenous production of lipoic acid is essential for mouse development. *Molecular and Cellular Biology, 25*, 8387–8392.

Zhao, S., Miller, J. R., Jiang, Y., Marletta, M. A., & Cronan, J. E., Jr. (2003). Assembly of the covalent linkage between lipoic acid and its cognate enzymes. *Chemistry & Biology, 10*, 1293–1302.

FURTHER READING

Frey, P. A., & Booker, S. J. (2001). Radical mechanisms of S-adenosylmethionine-dependent enzymes. *Advances in Protein Chemistry, 58*, 1–45.

Snell, E. E. (1992). Microorganisms in vitamin and biofactor research. *Journal of Nutritional Science and Vitaminology, 38*, 34–39. Spec No:34–9.

Snell, E. E. (1993). From bacterial nutrition to enzyme structure: A personal odyssey. *Annual Review of Biochemistry, 62*, 1–27.

CHAPTER NINE

Using Peptide Mimics to Study the Biosynthesis of the Side-Ring System of Nosiheptide

Bo Wang*, Joseph W. LaMattina*, Edward D. Badding*, Lauren K. Gadsby†, Tyler L. Grove*, Squire J. Booker*,†,‡,1

*Department of Chemistry, The Pennsylvania State University, University Park, PA, United States
†Department of Biochemistry and Molecular Biology, The Pennsylvania State University, University Park, PA, United States
‡The Howard Hughes Medical Institute, The Pennsylvania State University, University Park, PA, United States
[1]Corresponding author: e-mail address: squire@psu.edu

Contents

Abstract

Thiopeptide natural products have gained interest recently for their diverse pharmacological properties, including antibacterial, antifungal, anticancer, and antimalarial activities. Due to their inherent poor solubility and uptake, there is interest in developing new thiopeptides that mimic these unique structures, but which exhibit better pharmacokinetic properties. One strategy is to exploit the biosynthetic pathways using a

Methods in Enzymology, Volume 606
ISSN 0076-6879
https://doi.org/10.1016/bs.mie.2018.06.005

chemoenzymatic approach to make analogs. However, a complete understanding of thiopeptide biosynthesis is not available, especially for those molecules that contain a large number of modifications to the thiopeptide core. This gap in knowledge and the lack of a facile method for generating a variety of thiopeptide intermediates makes studying particular enzymatic steps difficult. We developed a method to produce thiopeptide mimics based on established synthetic procedures to study the reaction catalyzed by NosN, the class C radical *S*-adenosylmethionine methylase involved in carbon transfer to C4 of 3-methylindolic acid and completion of the side-ring system in nosiheptide. Herein, we detail strategies for overproducing and isolating NosN, as well as procedures for synthesizing substrate mimics to study the formation of the side-ring system of nosiheptide.

1. INTRODUCTION

Thiopeptide natural products are ribosomally synthesized and post-translationally modified peptides (RiPPs) that exhibit a variety of biologically activities (Ortega & van der Donk, 2016). Most notably, these activities include inhibition of protein synthesis either by binding to the 50S ribosome or by interacting with elongation factor thermal unstable (EF-tu), and thus can produce antibiotic effects (Bagley, Dale, Merritt, & Xiong, 2005). These natural products originate from a precursor peptide before undergoing a multistep transformation to achieve, among other modifications, multiple thiazole and/or oxazole rings, dehydrated alanine (Dha) and dehydrated threonine (Dht) side chains, and a central nitrogen-containing six-membered ring of different oxidation states (Ortega & van der Donk, 2016; Zheng, Fang, & Liu, 2017). The core macrocycle comprises a 26- or 29-membered ring, which is related to the specific biological target. The larger thiopeptides include thiostrepton and nosiheptide (NOS), which have been cocrystallized with the large subunit of the ribosome (Harms et al., 2008). Such thiopeptides interact with H44 and L11 in the A-site, thus confirming the observation that thiostrepton prevents elongation factor G (EF-G) crosslinking (Cameron, Thompson, March, & Dahlberg, 2002).

NOS from *Streptomyces actuosus* 40,037 (NRRL 2954) was one of the first thiopeptides to be identified and characterized and has been used as a feed additive to promote the growth of poultry livestock (Benazet & Cartier, 1980). NOS exhibits potent in vitro activity against bacteria of urgent clinical relevance, such as methicillin-resistant *Staphylococcus aureus*, penicillin-resistant *Streptococcus pneumoniae*, and vancomycin-resistant-enterococci (Benazet et al., 1980). Although it exhibits a minimum inhibitory concentration in the range of 10^{-3} μg/mL against a list of Gram-positive bacteria, NOS is inactive when used on experimentally infected mice (Benazet et al., 1980). The inability to act as an antibiotic in vivo is due to a large extent to its poor solubility in aqueous solution and gastrointestinal absorption (Just-Baringo, Albericio, & Alvarez, 2014). Remarkably, the total synthesis of NOS has been achieved, but with poor yields (Wojtas et al., 2016). One strategy for optimizing the yields and for generating analogs that are better suited as therapeutic agents is to use a chemoenzymatic approach (Just-Baringo et al., 2014; Lin, He, & Liu, 2017).

Prior to determining the complete structure of NOS by X-ray crystallography, NMR spectroscopy and chemical hydrolysis helped to identify key fragments of the molecule (Depaire, Thomas, Brun, Hull, et al., 1977; Depaire, Thomas, Brun, & Lukacs, 1977; Depaire, Thomas, Brun, Olesker, & Lukacs, 1977; Pascard, Ducruix, Lunel, & Prange, 1977; Prange, Ducruix, Pascard, & Lunel, 1977). Notably, NOS carries a 3,4-dimethyl-2-indolic acid (DMIA) bridge that connects the 4-methyl carbon of DMIA to Glu6 of the peptide framework and the carboxylic acid group of DMIA to Cys8 by ester and thioester linkages, respectively, to form a second macrocycle, termed the side-ring (Fig. 1). The presence of the DMIA moiety can be confirmed by hydrolysis, which generates (4-hydroxymethyl)-3-methyl-2-indolic acid (Depaire, Thomas, Brun, & Lukacs, 1977). The biosynthesis of the side-ring system in NOS has been of general interest because of the complex and unique chemistry that it entails.

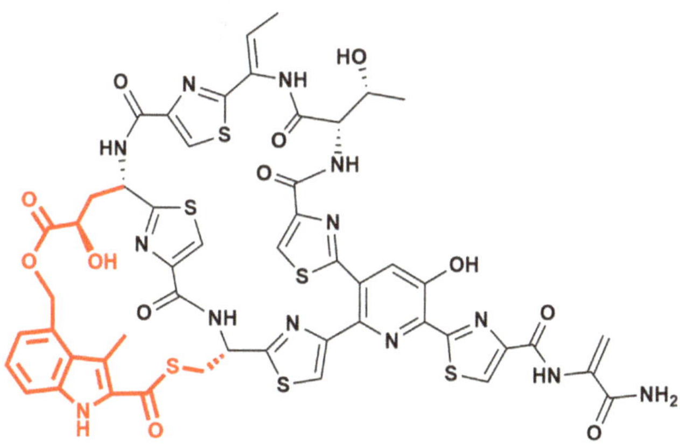

Fig. 1 Chemical structure of nosiheptide. The side-ring system formed by DMIA was highlighted in *red color*.

Pioneering studies by Wen Liu and coworkers led to the identification of the nosiheptide biosynthetic operon (*nosA–nosP*), and encoded gene products were assigned preliminary functions based on amino acid sequence similarities to proteins of known function and gene knockout studies in *S. actuosus* (Yu et al., 2009). A working model was proposed wherein NosA, B, and C are responsible for terminal posttranslational modifications to the peptide core (Liu et al., 2015; Wang et al., 2015), and NosD, E, F, G, H, M, O are responsible for generating the core thiopeptide (Yu et al., 2009). NosG contains a YcaO-like domain, which is found in Ser/Thr cyclodehydratases involved in natural product biosynthesis (Dunbar, Melby, & Mitchell, 2012). Recent work by Wen Liu and coworkers showed that NosGHF modify Cys39, 42, 44, 46, and 48 to form thiazole rings on the linear precursor peptide while selectively leaving Cys47 for incorporation of 3-methylindolic acid (MIA) by NosK (Qiu et al., 2017). Dehydrated amino acids are presumably formed by NosD and NosE, which together are homologous to NisB, a dehydratase involved in nisin biosynthesis. This dehydration requires activation of the Ser or Thr residue by glutamylation using $tRNA^{Glu}$, and subsequent elimination to form Dha and Dht, respectively (Garg, Salazar-Ocampo, & van der Donk, 2013). Despite our understanding of these enzymes, the timing in which these modifications are installed is unclear. In thiomuracin biosynthesis, thiazole rings are generated before dehydrated amino acids (Hudson, Zhang, Tietz, Mitchell, & van der Donk, 2015), but the sequence of events may be pathway specific. NosO is annotated as a Diels–Alderase that catalyzes a [4+2] *aza*-cycloaddition and two subsequent eliminations to remove water and the leader peptide to afford the central pyridine ring of NOS (Wever et al., 2015).

NosI, J, K, L, and N are involved in the preparation and installation of the 3,4-dimethylindolic acid moiety to form the side-ring of NOS. Biosynthesis of the side-ring begins with NosL, which catalyzes the C_α–C_β fragmentation–recombination of L-Trp to afford MIA (Bhandari, Fedoseyenko, & Begley, 2016; Sicoli et al., 2016; Zhang et al., 2011). AMP is then attached to the carboxyl group of MIA by NosI to allow for transfer of MIA to the 4′-phosphopantetheine prosthetic group on NosJ (Badding et al., 2017; Ding, Ji, et al., 2017; Qiu et al., 2017). NosJ then shuttles MIA to NosK, an α/β hydrolase that attaches MIA onto a seryl residue within its active site. The structure of NosK revealed that the active site contains a cleft that is large enough to accommodate the structural peptide (Badding et al., 2017). Further studies showed that NosK transfers MIA to Cys45 of NosM at some stage during NOS maturation after the installation of the thiazole rings (Qiu et al., 2017).

NosN is annotated as a class C radical SAM methylase, a subgroup of the radical SAM superfamily that is largely understudied (Bauerle, Schwalm, & Booker, 2015; Zhang, van der Donk, & Liu, 2012). Other reported class C radical SAM methylases appear to be involved in the biosynthesis of complex natural products or the degradation of heme in pathogenic bacteria (Hein, Klimmek, Polly, Kern, & Simon, 2017; LaMattina, Nix, & Lanzilotta, 2016). We proposed that upon attachment of MIA to the precursor peptide, NosN attaches a carbon unit at C4 of MIA to form an electrophilic exocyclic methylene intermediate. A glutamyl (Glu6) residue on NosM then adds to this electrophilic species, forming the ester linkage in NOS and completing the side-ring (Fig. 2). Previous studies of NosN showed that MIA containing an *N*-acetylcysteamine group (MIA-SNAC) is recognized as a substrate to form DMIA. However, considering the lack of additional gene products that could transform DMIA into a closed side-ring system, we posited that DMIA is formed as a side product when NosN cannot complete its catalytic cycle (Ding, Li, et al., 2017; Ding, Wu, et al., 2017; LaMattina et al., 2017). Only when using substrate analogs that better mimic the predicted physiological one could we probe the full reaction catalyzed by NosN. With the increased interest in thiopeptides and other RiPPs, we thought that providing a synthetic strategy to generate good substrate mimics might be useful to the community.

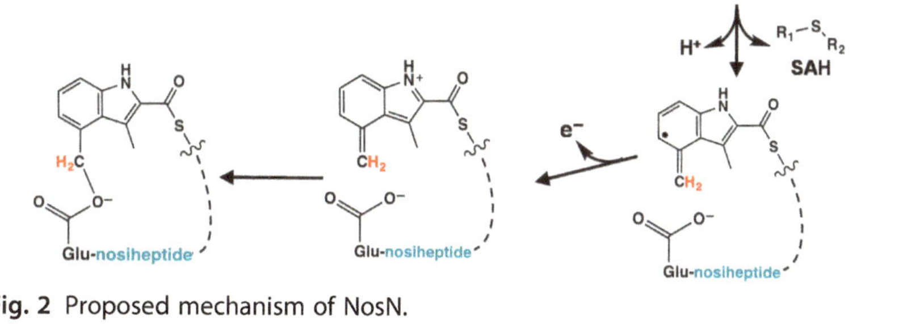

Fig. 2 Proposed mechanism of NosN.

2. OVERPRODUCTION AND PURIFICATION OF NosN

Our strategy to purify NosN focused on using similar overproduction and purification methods previously shown to be successful for other radical SAM enzymes (Lanz et al., 2012). Below is a detailed description of the methods used to overproduce NosN and purify it to homogeneity.

2.1 Gene Cloning and Expression Strategy

The *nosN* gene from *S. actuosus* was synthesized by Geneart (Thermo Fisher) with codon optimization for *Escherichia coli* to aid in optimal protein production. The synthesized gene incorporated *Nhe*I and *Xho*I restriction sites at its N- and C-termini, respectively, and was delivered in a pMK DNA production plasmid. The plasmid is digested with *Nhe*I and *Xho*I, and the fragment containing the *nosN* gene is ligated into a pET28a expression plasmid digested with the same restriction enzymes. The resulting construct encodes NosN with an N-terminal hexahistidine (His_6) tag linked by a thrombin cleavage site for removal of the tag (LaMattina et al., 2017). The pET-based cloning system, coupled with the strong T7 promoter, allows for facile cloning and robust overproduction of proteins (Panavas, Sanders, & Butt, 2009). The pET28a vector also contains a kanamycin (Kan) resistance marker for in vivo selection, which, in addition, allows for coexpression of genes contained on plasmid pDB1282, because this plasmid confers ampicillin (Amp) resistance. pDB1282 harbors the *isc* operon from *Azotobacter vinelandii*, which encodes proteins involved in Fe—S cluster assembly (Cicchillo et al., 2004; Zheng, Cash, Flint, & Dean, 1998). The BL21(DE3) strain of *E. coli* is transformed both with pDB1282 and with pET28a-*nosN*, giving rise to a strain that confers resistance to Kan and Amp.

2.2 Protocol for Expression of the *nosN* Gene

1. Transform *E. coli* BL21 (DE3) with pDB1282 and pET28a-NosN using standard procedures and at least 1 μg of each plasmid. Select transformants on Luria–Bertani (LB) plates containing Kan and Amp at final concentrations of 50 and 100 μg/mL, respectively.
2. Use a single colony to inoculate 5 mL of LB medium containing Kan and Amp at 50 and 100 μg/mL, respectively. Incubate at 37°C and

Table 1 Composition of M9 Minimal Medium

1 L of 20 × M9 Salts[a]	**g**
Na_2HPO_4 (Anhydrous)	136
KH_2PO_4	60
NaCl	10
NH_4Cl	20
Volumes for 4 L of M9 Minimal Medium	
20 × M9 Salts[b]	200 mL
40% w/v Glucose[c]	40 mL
1 M $CaCl_2$[c]	0.4 mL
1 M $MgSO_4$[c]	8 mL
ddH_2O[b]	3700 mL

[a]After salts dissolve, adjust the pH of the medium to 7.4 with HCl and KOH; adjust volume to 4 L with water.
[b]Medium should be autoclaved after addition to ddH_2O in a 6-L Erlenmeyer flask.
[c]Solutions should be sterilized by filtration and added just before inoculation.

shake at 250 rpm for ~12 h. In addition, place four 6-L flasks, each containing 4 L of M9 media (Table 1), in an incubator-shaker to allow them to equilibrate at 37°C.

3. Add 50 μg/mL Kan and 100 μg/mL Amp to each flask just prior to inoculation.
4. Inoculate each 6-L flask with ~200 μL of the 5 mL starter culture grown to an OD_{600} of 2.0.
5. At an OD_{600} of 0.3 (~15 h), add L-arabinose to a final concentration of 0.1% (w/v) to induce the expression of the *isc* operon.
6. At an OD_{600} of 0.6, cool flasks to ~20°C in an ice bath and add ferric ammonium citrate to a final concentration of 25 μ*M* before adjusting the temperature in the incubator to 18°C. Once the medium is equilibrated (about 30 min), return the flasks to the shaker and induce expression of the *nosN* gene by adding isopropyl-β-D-thiogalactopyranoside (IPTG) to a final concentration of 200 μ*M*.
7. Allow cultures to incubate for ~16 h at 18°C while shaking at 180 rpm.
8. Harvest the bacteria by centrifugation and freeze the resulting cell paste (~1.5 g/L) in liquid nitrogen and store in a liquid nitrogen dewar.

2.3 Purification of NosN

One of the greatest challenges of isolating radical SAM enzymes is their instability upon exposure to oxygen. To address this issue, purification of NosN and all subsequent manipulations of the protein are performed in a Coy anaerobic chamber under an atmosphere of 95% nitrogen and 5% hydrogen, as has been previously described (Lanz et al., 2012). Additionally, the purification process is completed without interruption while keeping the protein chilled during each step. The following steps detail the protocol that is used to isolate homogeneous NosN with an N-terminal hexahistidine (His_6)-tag.

1. In an anaerobic chamber, prepare 500 mL each of lysis buffer (50 m*M* HEPES, pH 7.5, 400 m*M* KCl, 10% glycerol, 10 m*M* BME, 10 m*M* imidazole), wash buffer (50 m*M* HEPES, pH 7.5, 400 m*M* KCl, 10% glycerol, 10 m*M* BME, 20 m*M* imidazole), elution buffer (50 m*M* HEPES, pH 7.5, 400 m*M* KCl, 10% glycerol, 10 m*M* BME, 250 m*M* imidazole), and storage buffer (50 m*M* HEPES, pH 7.5, 400 m*M* KCl, 10% glycerol, 2 m*M* DTT) with water that is oxygen free. Except for centrifugation, all subsequent steps are performed in an anaerobic chamber.
2. In the anaerobic chamber, resuspend 40 g of frozen cell paste in 120 mL lysis buffer containing lysozyme (1 mg/mL) and DNase I (0.1 mg/mL) by stirring in a 250 mL metal sonication cup inserted into a crystallizing dish packed with ice.
3. Once the cell paste is resuspended, sonicate the mixture for 45 s at 60% output using a Fisher 1550 sonic dismembrator. Repeat for at least six cycles with an 8 min interval between each cycle to allow for temperature reequilibration (~6°C).
4. Pour the lysate into centrifuge tubes. Cap the tubes tightly and seal the edges with vinyl tape before removing the tubes from the anaerobic chamber.
5. Centrifuge the lysate at 50,000 × *g* for 1 h at 4°C. During centrifugation, equilibrate a column of Ni-NTA resin with the chilled lysis buffer.
6. After centrifugation, reintroduce the centrifuge tubes into the anaerobic chamber and carefully decant the supernatant into a clean 250 mL bottle. Remove a 100 μL sample of the supernatant and a sample of the pellet for SDS-PAGE analysis.
7. Load the supernatant onto the column. Proceed to wash with 150 mL of prechilled wash buffer. Take 100 μL samples of the flow-through both from the loaded lysate and from the wash for SDS-PAGE analysis.

8. Elute the protein from the column with prechilled elution buffer. Once the colored band begins to elute, collect fractions until all of the protein (brown/black band) elutes from the column. Concentrate the protein to a final volume of ≤2.5 mL by centrifugal ultrafiltration at 3500 × *g* at 4°C using an Amicon centrifugal filter device (10 kDa molecular weight cutoff). While the protein is concentrating, equilibrate a PD-10 column with prechilled storage buffer.
9. Collect a 20 μL sample of the concentrated protein for SDS-PAGE analysis. Exchange the protein into prechilled cleavage buffer using the preequilibrated PD-10 column. Collect the flow-through and concentrate the protein by centrifugal ultrafiltration if desired. Aliquot and flash-freeze the protein in liquid N_2 before storing in a liquid N_2dewar.
10. Typical expression and purification gels for NosN are shown in Fig. 3A and B, respectively.

2.4 Chemical Reconstitution of as-Isolated NosN

Although the overproduction and purification methods described above result in NosN that contains 4.6 and 3.2 mol of iron and sulfide, respectively, reconstitution can enhance Fe—S cluster incorporation.

1. Introduce the following into the anaerobic chamber: 27 mg of $FeCl_3 \cdot 6H_2O$ (270.3 g/mol), 24 mg of $Na_2S \cdot 9H_2O$ (240.2 g/mol), 17.5 μL of stock BME (14.3 M), and a 50-mL Erlenmeyer flask.
2. Add NosN and 40 mL of Storage Buffer to the Erlenmeyer flask and let incubate on ice for 10 min.

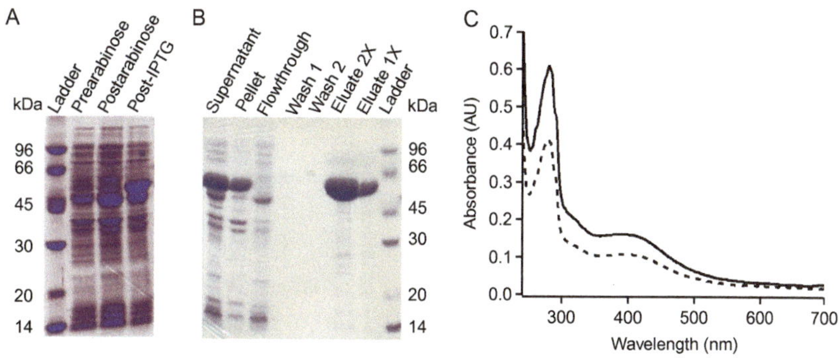

Fig. 3 (A) NosN expression on SDS-page gel. (B) NosN purification on SDS-page gel. (C) UV–visible spectra for as-isolated (*dashed line*) and reconstituted (*solid line*) NosN.

3. Dissolve the $FeCl_3{\cdot}6H_2O$ and $Na_2S{\cdot}9H_2O$ in 1 mL of water. Add a 4:1 *M* equivalent of the $FeCl_3$ solution to NosN and place mixture on ice for 30 min.
4. While incubating on ice, add 1 *M* equivalent of the Na_2S solution to the reconstitution mixture every 20 min until reaching a 4:1 *M* ratio of sulfide to NosN. Let the mixture sit on ice overnight.
5. Concentrate the reconstitution mixture to $\leq$2.5 mL by centrifugal ultrafiltration as described above.
6. Exchange the concentrated protein into Gel-Filtration Buffer using a PD-10 column.
7. Concentrate the flow-through to $\leq$2 mL by centrifugal ultrafiltration as described above.
8. Load the protein onto an S-200 column equilibrated in Gel-Filtration Buffer and connected to an ÅKTA fast-protein liquid chromatography system (housed in an anaerobic chamber) for further purification by size exclusion chromatography.
9. UV–visible spectra for as-isolated (dashed line) and reconstituted (solid line) NosN are shown in Fig. 3C.

2.5 Determining the Stoichiometry and Forms of Iron–Sulfur Clusters

Determining the correct number of Fe—S clusters per polypeptide is critical for accurate characterization of radical SAM enzymes, which is typically performed in concert with Mössbauer and electron paramagnetic resonance (EPR) spectroscopies (Pandelia, Lanz, Booker, & Krebs, 2015). We follow protocols that have been previously described for determining the amount of iron and sulfide per protein in other RS enzymes (Lanz et al., 2012). A brief description of these methods follows.

1. Following purification, the concentration of enzyme is determined by the Bradford assay using bovine serum albumin, fraction V (BSA) as a standard (Bradford, 1976).
2. Amino acid analysis is conducted to determine a correction factor for NosN when BSA is used as a standard for Bradford analysis.
3. Quantification of the amount of iron and acid-labile sulfide per protein is done using established procedures (Beinert, 1978, 1983).

Although determining the amount of iron and sulfide per protein is useful, further characterization is required to determine the types of Fe—S clusters bound to an enzyme, which is done using EPR and ^{57}Fe-Mössbauer spectroscopies. Preparation of samples for Mössbauer spectroscopy is carried out using methods that we have previously described (Lanz et al., 2012).

3. PREPARATION OF PEPTIDE SUBSTRATES

3.1 Design and Synthesis of Peptide Mimics to Study NosN

In our proposed mechanism, NosN transfers a carbon unit to the C-4 position of MIA attached to the precursor peptide through a thioester bond. The resulting electrophilic methide intermediate is attacked by the γ-carboxylate from a glutamyl residue on the thiazolyl peptide to form a lactone bond (Fig. 2). Therefore, we designed the peptide mimic, t_3NosM, as shown in Fig. 4A to validate the proposed mechanism of NosN. In our design, we maintained the thiazolylcysteine residue (structure in red) to carry the MIA (structure in black), as well as the thiazolyl glutamic acid residue (structure in blue) to act as the nucleophile for ring closure. The C-terminus of the peptide was capped with an amide to mimic a peptide bond and to avoid potential competition with the glutamyl residue. Capping a free carboxylic acid with an amide increases the stability of the entire molecule and provides a facile method for synthesis and purification. The N-terminus was capped with a thiazole-4-carboxylic acid (structure in pink) to extend the length of the peptide mimic.

Our strategy involves connecting individual amino acid building blocks through amide coupling reactions to create the desired substrate. The synthetic route is illustrated briefly in Fig. 4B. Due to the presence of a labile thioester bond, the protecting group should be chosen carefully. The thioester linkage is more sensitive than its oxygen counterpart (ester). Therefore, harsh conditions, such as the use of strong acids/bases and intense heat should be avoided to prevent hydrolysis of the MIA thioester bond. Benzyl ester/ether and Cbz protection groups, which are usually removed by catalytic hydrogenation, are also not compatible, because the molecule is sulfur rich, which is poisonous to the palladium catalyst used for catalytic hydrogenation. Hence, we use an allyl ester as the protecting group on the carboxylic acid, because it can be removed by $(PPh_3)_4Pd(0)$ under very mild conditions following MIA incorporation.

3.2 Protocol to Synthesize Thiazolyl Amino Acid Building Blocks

For the synthesis of the NOS mimic, two thiazolyl amino acid building blocks are needed, a thiazolyl cysteine residue (S1) and a thiazolyl glutamic acid residue (S2); however, the methods we describe can be applied to the synthesis of any other thiazolyl amino acid analogs. The method starts from

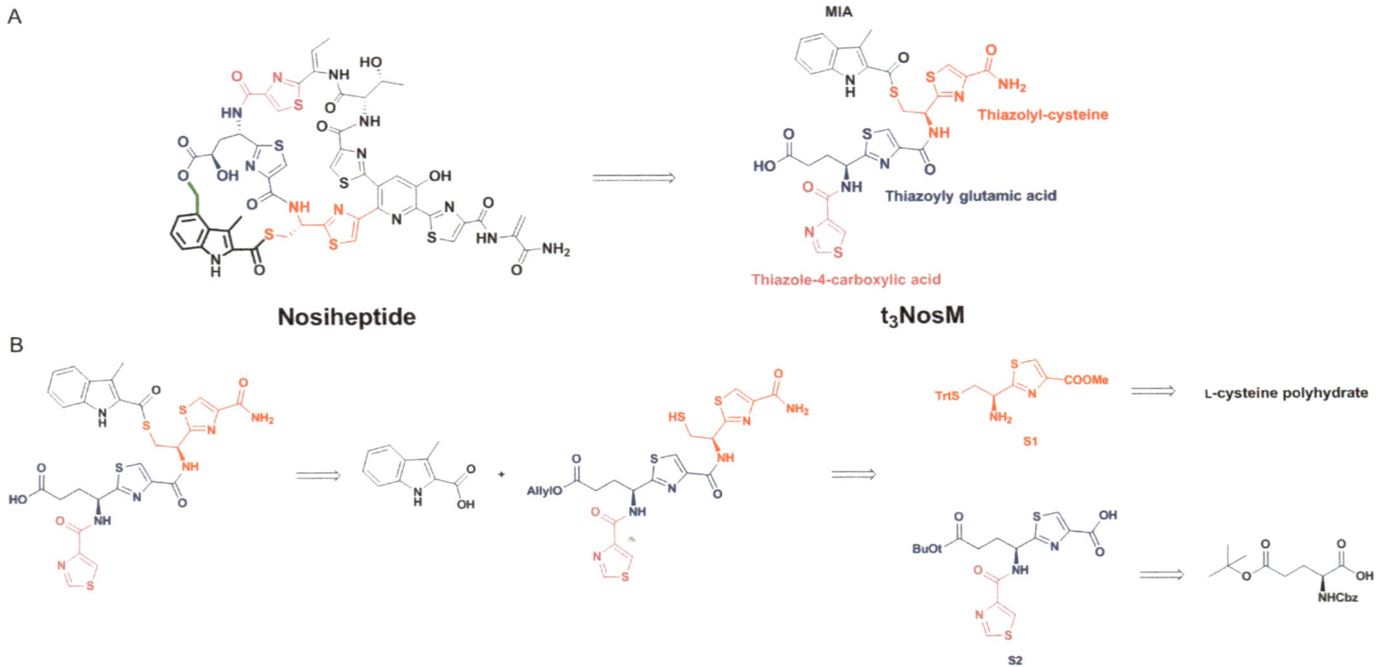

Fig. 4 (A) Chemical structure of nosiheptide mimic (t_3NosM) used for mechanistic studies of NosN. (B) Retro-synthetic analysis of nosiheptide mimic.

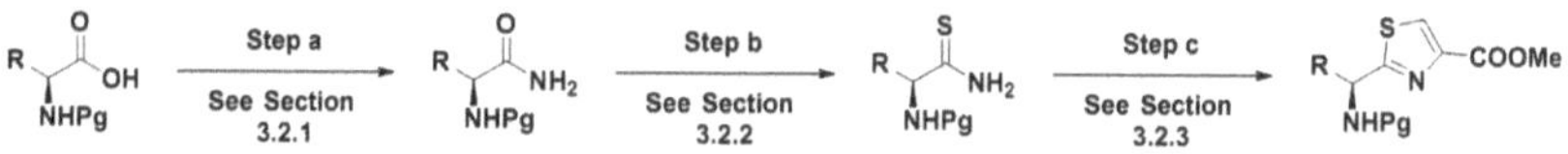

Fig. 5 General procedures for the synthesis of thiazolyl amino acid building blocks.

amino acids with proper protecting groups on their side chains and amino groups, most of which are commercially available from Chem-Impex International Inc. Our three-step method includes amidation of the carboxylic acid (step a), thioamide synthesis (step b), and cyclization/dehydration (step c), as shown in Fig. 5.

3.2.1 Amidation of the Carboxylic Acid

1. In a round-bottom flask, add the substrate, DMF (5 mL/1 mmol), and a magnetic stir bar. Stir the solution and add CDI (1.5 mmol/1 mmol substrate) in three portions. Let react for 1 h.
2. Bubble ammonia gas into the reaction slowly and monitor the reaction by silica TLC until disappearance of the starting material. For separation, use chloroform:methanol (5:1 v/v) as the mobile phase.
3. After completion, dilute the reaction mixture with ethyl acetate (20 mL/1 mmol substrate) and transfer to a properly sized separatory funnel.
4. Wash the organic phase thoroughly with 0.5 *M* HCl, aqueous saturated $NaHCO_3$, and brine. Dry the organic layer with anhydrous sodium sulfate and concentrate by rotary evaporation with the bath water set no higher than 45°C. The resulting residue is used for the next step without further purification.

Caution: The reaction conditions of step a tolerates most of the protecting groups for commercially available starting material, including Fmoc, Boc, Cbz, Alloc for the amino group, as well as *t*Bu and benzyl for side chains of amino acids. However, excessive ammonia gas can cleave Fmoc. Therefore, when using such a protecting group, silica TLC monitoring should be done frequently and, once the starting material is gone, the bubbling of ammonia gas should be ceased immediately.

3.2.2 Thioamide Synthesis

1. Weigh the resulting material from Section 3.2.1 in a round-bottom flask with a magnetic stir bar. Add 50 mL of THF per 1 mmol of the substrate and stir before the addition of Lawesson's reagent (MW = 404, 0.5 mmol per 1 mmol substrate). Stir the resulting reaction mixture at room temperature overnight.

2. After completion, concentrate the reaction by rotary evaporation in a fume hood and resuspend the resulting residue in ethyl acetate (20 mL/1 mmol of substrate). Wash the organic phase thoroughly with 0.5 *M* HCl, aqueous saturated $NaHCO_3$, and brine.
3. Dry the organic layer with anhydrous sodium sulfate and concentrate by rotary evaporation. The resulting residue is used for the next step without further purification.

Caution: Lawesson's reagent gives a strong unpleasant odor. Use in a fume hood that functions well. Deodorization of glassware used for this step can be done by rinsing with 5% potassium permanganate in water.

3.2.3 Cyclization/Dehydration

1. Weigh the resulting material from Section 3.2.2 in a round-bottom flask with a magnetic stir bar and add THF (30 mL/1 mmol substrate).
2. While stirring, add potassium bicarbonate (MW = 100, 10 mmol per 1 mmol substrate) as well as methyl bromopyruvate (MW = 181, 3 mmol per 1 mmol substrate). Stir the reaction at room temperature for 3 h or until disappearance of the starting material as determined by silica TLC.
3. After completion, dilute the reaction with ethyl acetate (30 mL/1 mmol substrate) and wash thoroughly with 0.5 M HCl, aqueous saturated $NaHCO_3$, and brine. Dry the organic layer over anhydrous sodium sulfate and concentrate by rotary evaporation.
4. Dissolve the resulting residue in ethyl acetate (30 mL/1 mmol substrate) and pyridine (MW = 79, 10 mmol per 1 mmol substrate). After cooling thoroughly in an ice-water bath, add trifluoroacetic anhydride (MW = 210, 3 mmol per 1 mmol substrate) dropwise using a dropping funnel.
5. After completion, as indicated by TLC, quench the reaction by adding methanol (MW = 32, 2 mmol per 1 mmol trifluoroacetic anhydride).
6. Wash the reaction mixture thoroughly with 0.5 *M* HCl, aqueous saturated $NaHCO_3$, and brine.
7. Dry the organic layer over anhydrous sodium sulfate and concentrate by rotary evaporation.
8. Purify the resulting residue using a silica gel column with hexanes and ethyl acetate as the eluent.

3.3 General Method for Protection/Deprotection and Amide Coupling

The synthesis of S1 (Fig. 6, step d), S2 (Fig. **7**, step e), and the final product (Fig. 8) requires several protection/deprotection and amide coupling steps.

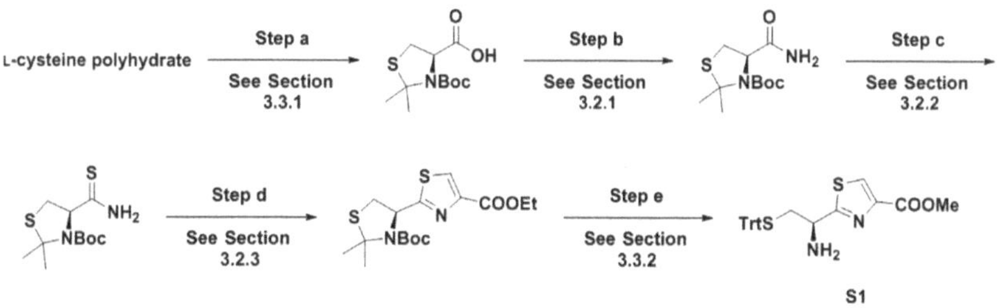

Fig. 6 Synthesis of S1.

Here, we describe the detailed procedure. To avoid repetition, each subsection corresponds to a particular reaction step during the synthesis and can be referred to at different stages in the synthesis. Each of the figures has section references for each of the reaction steps to make it easier to follow.

3.3.1 Protection of the Amino Group and Side Chain Thiol Group With Boc and Acetonide

1. Suspend L-cysteine in polyhydratein, a mixture of acetone (20 mL/1 g substrate) and 2,2-dimethoxypropane (4 mL/1 g substrate), and reflux the resulting suspension for 48 h.
2. After concentrating by rotary evaporation, dissolve the remaining residue in pyridine (12 mL/1 g substrate) and Boc_2O (1.5 mmol per 1 mmol substrate).
3. Stir the reaction mixture overnight at room temperature.
4. After completion, dilute the mixture with EtOAc (50 mL/1 g substrate) and follow with a thorough washing of the organic layer with hydrochloric acid (0.5 *M*), saturated aqueous sodium bicarbonate, and then brine.
5. Concentrate the organic layer by rotary evaporation and use the resulting viscous residue for the next step without further purification.

3.3.2 Deprotection of Boc and Acetonide and Protection of the Thiol Group With Trt

1. Weigh the round-bottom flask to obtain the amount of starting material. Add TFA (10 mL per 1 g substrate) and water (0.35 mL per 1 g substrate). Stir the solution at 50°C overnight and then concentrate by rotary evaporation.
2. Resuspend the resulting residue in TFA (5 mL per 1 mmol substrate) and trityl chloride (1 mmol per 1 mmol substrate). Stir the reaction mixture at room temperature for 2 h and then concentrate by rotary evaporation.

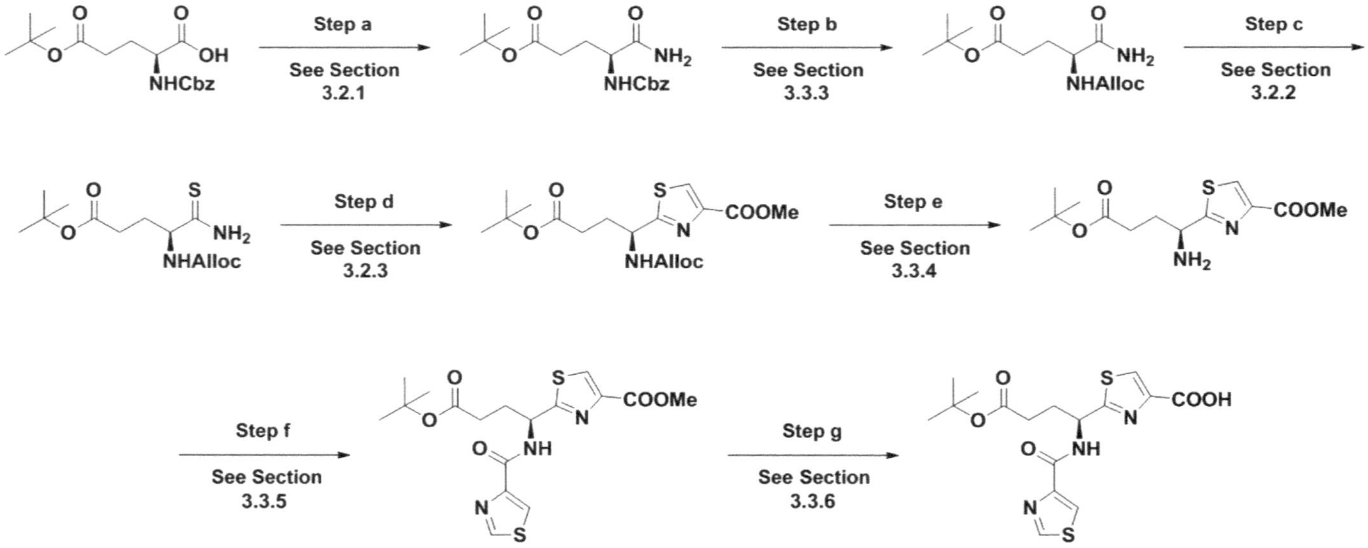

S2

Fig. 7 Synthesis of S2.

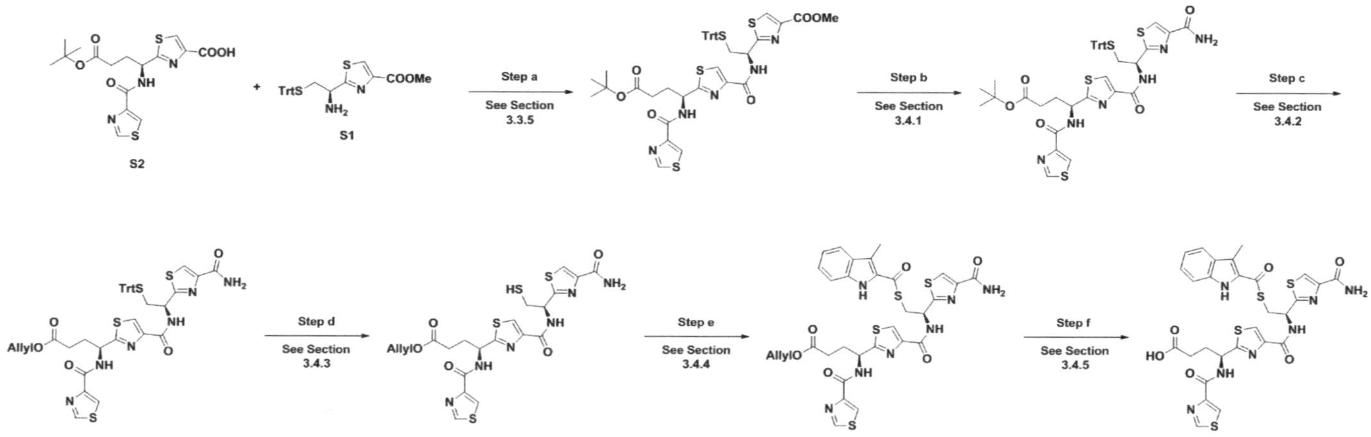

Fig. 8 Assembly of S1 and S2, incorporation of MIA, and deprotection.

3. Resuspend the resulting residue in EtOAc (75 mL/1 g substrate) and wash the organic phase thoroughly with aqueous saturated $NaHCO_3$, and then brine.
4. Dry the organic layer over anhydrous Na_2SO_4 and concentrate by rotary evaporation. Purify the resulting residue by silica-gel flash-chromatography (hexanes:ethyl acetate = 1:1 as the eluent).

3.3.3 Deprotection of Cbz and Reprotection With Alloc of Amino Group

1. Dissolve the starting material in methanol (20 mL per 1 g of substrate) and add palladium on carbon (10% palladium on carbon, 200 mg per 1 g substrate). Stir the reaction under a hydrogen atmosphere overnight.
2. Filter the palladium on carbon through a pad of Celite before concentrating the filtrate by rotary evaporation. The resulting residue is used for the next step without further purification.
3. Add THF (20 mL per 1 g of crude product from previous step). Place the round-bottom flask in an ice-water bath and stir the solution to chill.
4. Once chilled, add aqueous saturated $NaHCO_3$ (20 mL/1 g substrate).
5. Add allyl chloroformate (1.5 mmol per 1 mmol substrate) dropwise to the reaction while stirring.
6. Continue stirring the reaction at 0°C for 2 h before diluting with EtOAc (20 mL/1 g substrate). Wash the organic phase thoroughly with water and then brine.
7. Dry the organic layer over anhydrous Na_2SO_4 and concentrate by rotary evaporation. Purify the resulting residue by silica-gel flash-chromatography (hexanes:ethyl acetate = 2:1 as the eluent).
8. Follow steps outlined in Sections 3.2.2 and 3.2.3 to prepare the substrate for the next step

3.3.4 Removal of Alloc

1. In a round-bottom flask containing the starting material, add dioxane (10 mmol per 1 mmol substrate) and water (1 mL per 1 mmol substrate).
2. Add $(PPh_3)_4Pd$ (MW = 1155, 0.05 mmol per 1 mmol substrate) and 1,3-dimethylbarbituric acid (MW = 156, 1 mmol per 1 mmol substrate) to the reaction and stir at room temperature for 3 h. Dilute the reaction mixture with EtOAc (30 mL per 1 g substrate) and wash the organic phase thoroughly with aqueous saturated $NaHCO_3$, and brine.
3. Dry the organic layer over anhydrous Na_2SO_4 and concentrate by rotary evaporation. The resulting residue is used in the next step without further purification.

3.3.5 Amide Coupling Reaction

1. Dissolve acid (1 mmol) and amine (1 mmol) in DCM (20 mL) and chill the reaction in an ice-water bath while stirring.
2. Add DIPEA (MW = 129, 3 mmol per 1 mmol acid) and HATU (MW = 380, 1.5 mmol per 1 mmol acid). Let the reaction gradually warm up to room temperature and continue stirring overnight.
3. After completion, transfer the reaction mixture to a separatory funnel and wash the reaction mixture thoroughly with 0.5 M HCl, aqueous saturated $NaHCO_3$, and brine.
4. Dry the organic layer over anhydrous Na_2SO_4 and concentrate by rotary evaporation. Purify the resulting residue by silica-gel flash-chromatography using hexanes and acetone as the eluent.

3.3.6 Hydrolysis of the Methyl Ester

1. In a round-bottom flask containing the starting material, add a cosolvent of THF, methanol, and water (5 mL of each solvent per 1 mmol substrate).
2. Add lithium hydroxide monohydrate (MW = 42, 2 mmol per 1 mmol substrate) and stir at room temperature for 5 h.
3. After completion, as indicated by silica TLC, dilute the reaction with EtOAc (30 mL per 1 mmol substrate) and wash thoroughly with 0.5 M HCl and then brine.
4. Dry the organic layer over anhydrous Na_2SO_4 and concentrate by rotary evaporation. The resulting residue is used for the next step with no further purification.

3.4 Incorporation of MIA and Final Deprotection

As shown in Figs. 6 and 7, precursors S1 and S2 are prepared from the previous synthetic protocols described above. The two precursors are first coupled using the amide coupling procedure in Section 3.3.5. After replacement of the *t*-butyl protecting group with an allyl group on the carboxylic acid of the glutamate, the Trt can be removed selectively to allow for the incorporation of MIA on the thiol group via an acyl chloride intermediate of MIA. Below is the detailed protocol to make the final peptide mimic used to study NosN turnover.

3.4.1 Amidation of Carboxylic Ester

1. Resuspend the starting material in 7 N ammonia in methanol (200 mL/1 g substrate) and stir at room temperature overnight.

2. After completion, as indicated by silica TLC, concentrate the reaction mixture by rotary evaporation.
3. Purify the resulting residue by silica-gel flash-chromatography using hexanes and acetone as the eluent.

3.4.2 Deprotection of t-Butyl Ester and Reprotection With Allyl Ester

1. Add 20 mL of DCM per 1 mmol substrate in a round-bottom flask.
2. While stirring, add TFA (1/4 in volume of DCM used) and continue stirring for 1 h at room temperature.
3. Concentrate the reaction mixture by rotary evaporation and use the resulting residue for the next step without further purification.
4. Add DMF (10 mL per 1 mmol substrate) to the residue while stirring.
5. Add K_2CO_3 (MW = 138, 5 mmol per 1 mmol substrate) in one portion and stir for 10 min.
6. Add allyl bromide (MW = 121, 2 mmol per 1 mmol substrate) dropwise to the stirred reaction.
7. Stir the reaction mixture at room temperature for 2 h before diluting with EtOAc (30 mL per 1 mmol substrate).
8. Wash the organic phase thoroughly with 0.5 M HCl, aqueous saturated $NaHCO_3$, and brine.
9. Dry the organic layer over Na_2SO_4 and concentrate by rotary evaporation.
10. Purify the resulting residue by silica-gel flash-chromatography using chloroform and methanol as the eluent.

3.4.3 Deprotection of Trt

1. Add DCM (20 mL per 1 mmol substrate) in a round-bottom flask that contains the substrate from the previous step.
2. Add triethylsilane (4 mL per 1 mmol substrate) while stirring with a magnetic stir bar. Continue stirring for 5 min before adding TFA (4 mL per 1 mmol substrate).
3. Stir the resulting clear solution for 30 min at room temperature.
4. Concentrate the reaction mixture by rotary evaporation and use the resulting thiol intermediate for the next step without further purification.

3.4.4 Synthesis of MIA Acyl Chloride and Thioesterification

1. Prepare MIA acyl chloride (2 mmol per 1 mmol of crude product from previous step) for the subsequent thioesterification as described below.
2. Add DCM (10 mL per 1 mmol of MIA) to a round-bottom flask with MIA and a magnetic stir bar.

3. While stirring, add oxalyl chloride (MW=127, 3 mmol per 1 mmol MIA) and DMF (1 drop).
4. Stir the reaction for 1 h at room temperature.
5. After completion, concentrate the reaction mixture by rotary evaporation. The resulting acyl chloride is used in subsequent steps without further purification.
6. In the round-bottom flask with the thiol intermediate from Section 3.4.3, add THF (10 mL per 1 mmol of thiol intermediate).
7. While stirring, add DIPEA (MW=129, 5 mmol per 1 mmol thiol intermediate).
8. Dissolve the MIA acyl chloride intermediate in THF (5 mL per 1 mmol MIA acyl chloride).
9. While stirring, add MIA acyl chloride in THF from the previous step dropwise to the round-bottom flask containing the thiol intermediate and DIPEA.
10. Stir the resulting reaction mixture for 30 min at room temperature.
11. After completion, dilute the reaction with EtOAc (50 mL/1 mmol thiol intermediate). Wash the resulting mixture thoroughly with 0.5 M HCl, 0.1 M aqueous NaOH, and brine.
12. Dry the organic layer over anhydrous Na_2SO_4 and concentrate by rotary evaporation.
13. Purify the resulting residue by silica-gel flash-chromatography using chloroform and methanol as the eluent.

3.4.5 Deprotection of Allyl Ester

1. Add dioxane (10 mL per 0.1 mmol substrate) and water (0.5 mL per 0.1 mmol substrate) in the round-bottom flask with the allyl ester starting material.
2. While stirring, add $(PPh_3)_4Pd$ (MW=1155, 0.005 mmol per 0.1 mmol substrate) and 1,3-dimethylbarbituric acid (MW=156, 0.1 mmol per 0.1 mmol substrate).
3. Stir the reaction for 2 h at room temperature.
4. Once complete, dilute the reaction mixture with EtOAc (50 mL per 0.1 mmol substrate).
5. Wash the organic layer thoroughly with brine before drying the organic layer over anhydrous Na_2SO_4 and concentrating by rotary evaporation.
6. Purify the resulting residue by silica-gel flash-chromatography using chloroform and methanol as the eluent.

4. QUANTITATIVE ANALYSIS OF NosN TURNOVER

NosN has been stated to produce 5′-deoxyadenosine (5′-dA) and either *S*-adenosylhomocysteine (SAH) or thioadenosine (tA), depending on whether *S*-adenosylmethionine (SAM) or methylthioadenosine (MTA) acts as a methyl donor, respectively (Ding, Li, et al., 2017; LaMattina et al., 2017). SAM typically degrades to MTA during long periods of storage, which is heavily dependent on the pH of the storage solution and the temperature. When using unlabeled MTA and (^{13}C-methyl) SAM in NosN assays, we only observe formation of labeled product, indicating that MTA is not a direct methyl donor in our reactions. These products can be quantified by LC–MS/MS using multiple reaction monitoring, as described below for a NosN-catalyzed reaction. The substrate for NosN is more difficult to quantify due to a lack of synthetic protocols to make a peptide mimic that contains the intact side-ring product. One alternative is to use MIA-SNAC as a substrate analog for NosN assays and quench with sodium hydroxide to form a hydroxymethyl adduct of MIA (LaMattina et al., 2017). However, such methods result in a mixture of products formed due to a lack of specificity with the poor substrate, making the quantification less meaningful.

The NosN peptide reaction products can be analyzed structurally by MS, as well as by MS/MS. For the addition of a $-CH_2$ group, the mass will only change 12 units (684–696 as shown in Fig. 9B and C), which can be a result of either the predicted exocyclic methylene intermediate or the closed side-ring. For this reason, MS/MS is used to detect the fragmented molecule and determine which product is produced. For NosN turnover, the side-ring closed product (t_3NosM_{SR}) will fragment to an *m/z* of 188 as shown in Fig. 9D, which corresponds to the hydroxyl adduct from the glutamate residue. This fragment is only observed when the side ring is formed and can be verified by ^{18}O-labeling of the carboxylate oxygens of Glu6 of the peptide mimic (LaMattina et al., 2017).

4.1 NosN Enzyme Assays Using Synthesized Peptide Mimics

1. Assays should contain final concentrations of the following in a total volume of 200 μL: 50 m*M* HEPES, pH 7.5, 100 m*M* KCl, 5% glycerol, 1 m*M* SAM, 1 m*M* substrate, 10–50 μ*M* NosN, 100 μ*M* Trp (internal standard), and 2 m*M* sodium dithionite as a reductant. *Note: We have not observed NosN activity with the E. coli flavodoxin/flavodoxin reductase reducing system.*

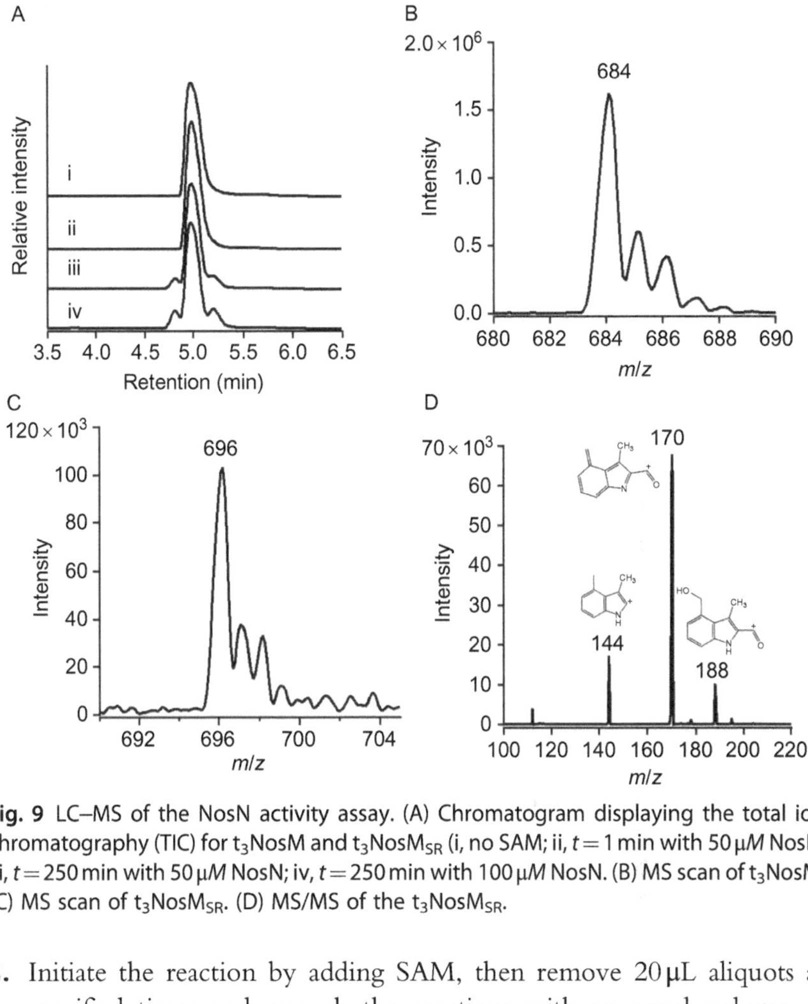

Fig. 9 LC–MS of the NosN activity assay. (A) Chromatogram displaying the total ion chromatography (TIC) for t_3NosM and t_3NosM$_{SR}$ (i, no SAM; ii, $t=1$ min with 50 μ*M* NosN; iii, $t=250$ min with 50 μ*M* NosN; iv, $t=250$ min with 100 μ*M* NosN. (B) MS scan of t_3NosM. (C) MS scan of t_3NosM$_{SR}$. (D) MS/MS of the t_3NosM$_{SR}$.

2. Initiate the reaction by adding SAM, then remove 20 μL aliquots at specified times and quench the reactions with an equal volume of 100% MeOH.
3. Centrifuge the samples for 15 min at 14,000 × *g* before removing supernatants and placing them into MS vials.

4.2 Protocol for Quantification of Reaction Products by HPLC Coupled to Electrospray-Ionization Mass Spectrometry

1. Prepare standards of SAH, 5′-dA, and Trp using extinction coefficients for each molecule. SAH and 5′-dA display absorption features at 260 nm with extinction coefficients of 16,000 M^{-1} cm^{-1} and 15,400 M^{-1} cm^{-1}, respectively, while Trp displays an absorption feature at 278 nm with an

extinction coefficient of 5560 M^{-1} cm^{-1} (Wood, 1987). Using these standards, generate a standard curve containing 0.5–250 μ*M* of each molecule and 100 μ*M* Trp (internal standard).

2. Analyze samples on an Agilent Technologies Zorbax Extend-C18 RRHD column (4.6 mm × 50 mm, 1.8 μm particle size) equilibrated in 98% solvent A (0.1% formic acid, pH 2.6) and 2% solvent B (methanol). Apply a linear gradient of 2%–10% solvent B from 0 to 0.5 min, and then a linear gradient of 10%–50% solvent B from 0.5 to 2.5 min. Next, increase solvent B to 100% from 2.5 to 3.5 min and hold constant from 3.5 to 4.5 min before returning to 2% solvent B from 4.5 to 5 min. Allow the column to reequilibrate for 2.5 min under initial conditions before subsequent sample injections.
3. Detect products using MS/MS with electrospray ionization in positive mode. All products are detected in their +1 charge states. Table 2 displays the *m/z* values (parent and daughter ions), fragmentor voltages, collision energies, and retention times of all molecules that are monitored in a NosN reaction.
4. Quantify product formation. Shown in Fig. 10 is the time-dependent formation of SAH, 5′-dA, MTA, and $NosM_{SR}$ in the NosN-catalyzed reaction.

Table 2 Parameters for LC–MS/MS Analysis of Reaction Products

Compound	Precursor Ion	Product Ion	Fragmentor	Collision Energy
SAH	385	136	92	17
		87.9	92	61
5′-dA	252	136	63	29
		119	92	13
MTA	298	136	106	18
		119	106	60
MIA-t_3NosM	684	158	100	50
		112	100	50
$NosM_{SR}$	696	188	100	50
		170	100	50
L-Tryptophan	205	188	63	9
		118	63	29

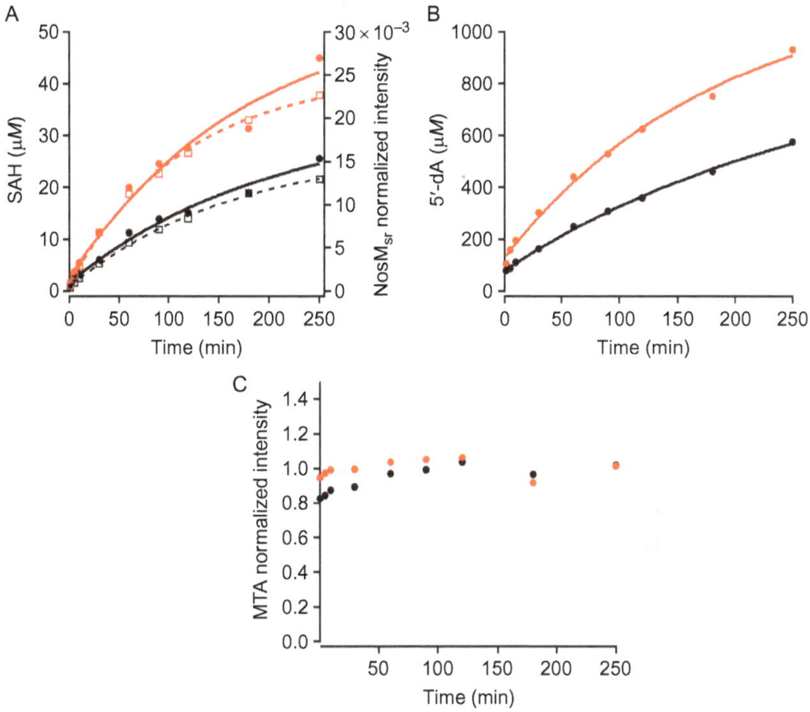

Fig. 10 (A) Detection of SAH (*closed circles*) and NosM$_{SR}$ (*open squares*) production when 50 μ*M* NosN (*black*) or 100 μ*M* NosN (*red*) is used in the assay. *Solid curves* are the fit for SAH production, while *dashed curves* are for NosM$_{SR}$. (B and C) *Progress curves* for 5′-dA and MTA production, respectively.

5. CONCLUSIONS

Class C radical SAM methylases catalyze the appendage of a carbon unit to sp^2-hybridized carbon centers, resulting in the transfer of either a methyl group ($-CH_3$), as in the example of TbtI for thiazole methylation, or a methylene group ($-CH_2$), which appears to be the case for NosN and YtkT (Bauerle et al., 2015; LaMattina et al., 2017). With the role that these enzymes play in the biosynthesis of medically relevant natural products, determining their mechanisms of catalysis can lead to producing RiPP analogs to combat antibiotic resistant bacteria. However, the previous understanding of these biosynthetic pathways and the lack of substrate syntheses have hindered the study of the class C radical SAM methylaes. This method provides a facile synthetic approach to generate similar thiopeptides as a means to probe the reactions catalyzed by some of the unique enzymes within the respective pathway.

ACKNOWLEDGMENTS

This work was supported by the National Institutes of Health (GM-122595 and AI-133318). S.J.B. is an investigator of the Howard Hughes Medical Institute.

REFERENCES

Badding, E. D., Grove, T. L., Gadsby, L. K., LaMattina, J. W., Boal, A. K., & Booker, S. J. (2017). Rerouting the pathway for the biosynthesis of the side ring system of nosiheptide: The roles of NosI, NosJ, and NosK. *Journal of the American Chemical Society*, *139*(16), 5896–5905. https://doi.org/10.1021/jacs.7b01497.

Bagley, M. C., Dale, J. W., Merritt, E. A., & Xiong, X. (2005). Thiopeptide antibiotics. *Chemical Reviews*, *105*(2), 685–714. https://doi.org/10.1021/cr0300441.

Bauerle, M. R., Schwalm, E. L., & Booker, S. J. (2015). Mechanistic diversity of radical S-adenosylmethionine (SAM)-dependent methylation. *The Journal of Biological Chemistry*, *290*(7), 3995–4002. https://doi.org/10.1074/jbc.R114.607044.

Beinert, H. (1978). Micro methods for the quantitative determination of iron and copper in biological material. *Methods in Enzymology*, *54*, 435–445.

Beinert, H. (1983). Semi-micro methods for analysis of labile sulfide and of labile sulfide plus sulfane sulfur in unusually stable iron-sulfur proteins. *Analytical Biochemistry*, *131*(2), 373–378.

Benazet, F., & Cartier, J. R. (1980). Effect of nosiheptide as a feed additive in chicks on the quantity, duration, prevalence of excretion, and resistance to antibacterial agents of Salmonella typhimurium; on the proportion of Escherichia coli and other coliforms resistant to antibacterial agents; and on their degree and spectrum of resistance. *Poultry Science*, *59*(7), 1405–1415.

Benazet, F., Cartier, M., Florent, J., Godard, C., Jung, G., Lunel, J., et al. (1980). Nosiheptide, a sulfur-containing peptide antibiotic isolated from Streptomyces actuosus 40037. *Experientia*, *36*(4), 414–416.

Bhandari, D. M., Fedoseyenko, D., & Begley, T. P. (2016). Tryptophan Lyase (NosL): A cornucopia of 5′-deoxyadenosyl radical mediated transformations. *Journal of the American Chemical Society*, *138*(50), 16184–16187. https://doi.org/10.1021/jacs.6b06139.

Bradford, M. M. (1976). A rapid and sensitive method for the quantitation of microgram quantities of protein utilizing the principle of protein-dye binding. *Analytical Biochemistry*, *72*, 248–254.

Cameron, D. M., Thompson, J., March, P. E., & Dahlberg, A. E. (2002). Initiation factor IF2, thiostrepton and micrococcin prevent the binding of elongation factor G to the Escherichia coli ribosome. *Journal of Molecular Biology*, *319*(1), 27–35. https://doi.org/10.1016/S0022-2836(02)00235-8.

Cicchillo, R. M., Iwig, D. F., Jones, A. D., Nesbitt, N. M., Baleanu-Gogonea, C., Souder, M. G., et al. (2004). Lipoyl synthase requires two equivalents of S-adenosyl-L-methionine to synthesize one equivalent of lipoic acid. *Biochemistry*, *43*(21), 6378–6386. https://doi.org/10.1021/bi049528x.

Depaire, H., Thomas, J. P., Brun, A., Hull, W. E., Olesker, A., & Lukacs, G. (1977). N-15 NMR-spectroscopy of nosiheptide—Determination of elemental formula and molecular-weight of antibiotic. *Tetrahedron Letters*, *16*, 1401–1402.

Depaire, H., Thomas, J. P., Brun, A., & Lukacs, G. (1977). Acid and alkaline-hydrolysis of antibiotic nosiheptide—Structure elucidation of 5 fragments. *Tetrahedron Letters*, *16*, 1395–1396.

Depaire, H., Thomas, J. P., Brun, A., Olesker, A., & Lukacs, G. (1977). C-13 NMR-spectroscopy of nosiheptide. *Tetrahedron Letters*, *16*, 1397–1400.

Ding, W., Ji, W., Wu, Y., Wu, R., Liu, W. Q., Mo, T., et al. (2017). Biosynthesis of the nosiheptide indole side ring centers on a cryptic carrier protein NosJ. *Nature Communications*, *8*(1), 437. https://doi.org/10.1038/s41467-017-00439-1.

Ding, W., Li, Y., Zhao, J., Ji, X., Mo, T., Qianzhu, H., et al. (2017). The catalytic mechanism of the class C radical S-Adenosylmethionine methyltransferase NosN. *Angewandte Chemie (International Ed. in English)*, *56*(14), 3857–3861. https://doi.org/10.1002/anie.201609948.

Ding, W., Wu, Y., Ji, X., Qianzhu, H., Chen, F., Deng, Z., et al. (2017). Nucleoside-linked shunt products in the reaction catalyzed by the class C radical S-adenosylmethionine methyltransferase NosN. *Chemical communications (Cambridge)*, *53*(37), 5235–5238. https://doi.org/10.1039/c7cc02162c.

Dunbar, K. L., Melby, J. O., & Mitchell, D. A. (2012). YcaO domains use ATP to activate amide backbones during peptide cyclodehydrations. *Nature Chemical Biology*, *8*(6), 569–575. https://doi.org/10.1038/nchembio.944.

Garg, N., Salazar-Ocampo, L. M., & van der Donk, W. A. (2013). In vitro activity of the nisin dehydratase NisB. *Proceedings of the National Academy of Sciences of the United States of America*, *110*(18), 7258–7263. https://doi.org/10.1073/pnas.1222488110.

Harms, J. M., Wilson, D. N., Schluenzen, F., Connell, S. R., Stachelhaus, T., Zaborowska, Z., et al. (2008). Translational regulation via L11: Molecular switches on the ribosome turned on and off by thiostrepton and micrococcin. *Molecular Cell*, *30*(1), 26–38. https://doi.org/10.1016/j.molcel.2008.01.009.

Hein, S., Klimmek, O., Polly, M., Kern, M., & Simon, J. (2017). A class C radical S-adenosylmethionine methyltransferase synthesizes 8-methylmenaquinone. *Molecular Microbiology*, *104*(3), 449–462. https://doi.org/10.1111/mmi.13638.

Hudson, G. A., Zhang, Z., Tietz, J. I., Mitchell, D. A., & van der Donk, W. A. (2015). In vitro biosynthesis of the core scaffold of the thiopeptide thiomuracin. *Journal of the American Chemical Society*, *137*(51), 16012–16015. https://doi.org/10.1021/jacs.5b10194.

Just-Baringo, X., Albericio, F., & Alvarez, M. (2014). Thiopeptide engineering: A multidisciplinary effort towards future drugs. *Angewandte Chemie (International Ed. in English)*, *53*(26), 6602–6616. https://doi.org/10.1002/anie.201307288.

LaMattina, J. W., Nix, D. B., & Lanzilotta, W. N. (2016). Radical new paradigm for heme degradation in Escherichia coli O157:H7. *Proceedings of the National Academy of Sciences of the United States of America*, *113*(43), 12138–12143. https://doi.org/10.1073/pnas.1603209113.

LaMattina, J. W., Wang, B., Badding, E. D., Gadsby, L. K., Grove, T. L., & Booker, S. J. (2017). NosN, a radical S-adenosylmethionine methylase, catalyzes both C1 transfer and formation of the ester linkage of the side-ring system during the biosynthesis of nosiheptide. *Journal of the American Chemical Society*, *139*(48), 17438–17445. https://doi.org/10.1021/jacs.7b08492.

Lanz, N. D., Grove, T. L., Gogonea, C. B., Lee, K. H., Krebs, C., & Booker, S. J. (2012). RlmN and AtsB as models for the overproduction and characterization of radical SAM proteins. *Methods in Enzymology*, *516*, 125–152. https://doi.org/10.1016/B978-0-12-394291-3.00030-7.

Lin, Z., He, Q., & Liu, W. (2017). Bio-inspired engineering of thiopeptide antibiotics advances the expansion of molecular diversity and utility. *Current Opinion in Biotechnology*, *48*, 210–219. https://doi.org/10.1016/j.copbio.2017.06.008.

Liu, S., Guo, H., Zhang, T., Han, L., Yao, P., Zhang, Y., et al. (2015). Structure-based mechanistic insights into terminal amide synthase in nosiheptide-represented thiopeptides biosynthesis. *Scientific Reports*, *5*, 12744.

Ortega, M. A., & van der Donk, W. A. (2016). New insights into the biosynthetic logic of ribosomally synthesized and post-translationally modified peptide natural products. *Cell Chemical Biology*, *23*(1), 31–44. https://doi.org/10.1016/j.chembiol.2015.11.012.

Panavas, T., Sanders, C., & Butt, T. R. (2009). SUMO fusion technology for enhanced protein production in prokaryotic and eukaryotic expression systems. In H. D. Ulrich (Ed.), *SUMO Protocols* (pp. 303–317). Totowa, NJ: Humana Press.

Pandelia, M. E., Lanz, N. D., Booker, S. J., & Krebs, C. (2015). Mossbauer spectroscopy of Fe/S proteins. *Biochimica et Biophysica Acta, 1853*(6), 1395–1405. https://doi.org/10.1016/j.bbamcr.2014.12.005.

Pascard, C., Ducruix, A., Lunel, J., & Prange, T. (1977). Highly modified cysteine-containing antibiotics. Chemical structure and configuration of nosiheptide. *Journal of the American Chemical Society, 99*(19), 6418–6423.

Prange, T., Ducruix, A., Pascard, C., & Lunel, J. (1977). Structure of nosipeptide, a polythiazole-containing antibiotic. *Nature, 265*(5590), 189–190.

Qiu, Y., Du, Y., Zhang, F., Liao, R., Zhou, S., Peng, C., et al. (2017). Thiolation protein-based transfer of indolyl to a ribosomally synthesized polythiazolyl peptide intermediate during the biosynthesis of the side-ring system of nosiheptide. *Journal of the American Chemical Society*. https://doi.org/10.1021/jacs.7b11367.

Sicoli, G., Mouesca, J. M., Zeppieri, L., Amara, P., Martin, L., Barra, A. L., et al. (2016). Fine-tuning of a radical-based reaction by radical S-adenosyl-L-methionine tryptophan lyase. *Science, 351*(6279), 1320–1323. https://doi.org/10.1126/science.aad8995.

Wang, Y., Liu, S., Yao, P., Yu, Y., Zhang, Y., Lan, W., et al. (2015). Crystallographic analysis of NosA, which catalyzes terminal amide formation in the biosynthesis of nosiheptide. *Acta Crystallographica Section F, 71*, 1033–1037.

Wever, W. J., Bogart, J. W., Baccile, J. A., Chan, A. N., Schroeder, F. C., & Bowers, A. A. (2015). Chemoenzymatic synthesis of thiazolyl peptide natural products featuring an enzyme-catalyzed formal [4 + 2] cycloaddition. *Journal of the American Chemical Society, 137*(10), 3494–3497. https://doi.org/10.1021/jacs.5b00940.

Wojtas, K. P., Riedrich, M., Lu, J. Y., Winter, P., Winkler, T., Walter, S., et al. (2016). Total synthesis of nosiheptide. *Angewandte Chemie (International Ed. in English), 55*(33), 9772–9776. https://doi.org/10.1002/anie.201603140.

Wood, E. J. (1987). Data for biochemical research (3rd ed.) by R. M C Dawson, D C Elliott, W H Elliott and K M Jones, pp 580. Oxford Science Publications, OUP, Oxford, 1986. £35/$59. ISBN 0-19-855358-7. Biochemical Education, 15(2), 97. https://doi.org/10.1016/0307-4412(87)90110-5

Yu, Y., Duan, L., Zhang, Q., Liao, R., Ding, Y., Pan, H., et al. (2009). Nosiheptide biosynthesis featuring a unique indole side ring formation on the characteristic thiopeptide framework. *ACS Chemical Biology, 4*(10), 855–864. https://doi.org/10.1021/cb900133x.

Zhang, Q., Li, Y., Chen, D., Yu, Y., Duan, L., Shen, B., et al. (2011). Radical-mediated enzymatic carbon chain fragmentation-recombination. *Nature Chemical Biology*, 7(3), 154–160. https://doi.org/10.1038/nchembio.512.

Zhang, Q., van der Donk, W. A., & Liu, W. (2012). Radical-mediated enzymatic methylation: A tale of two SAMS. *Accounts of Chemical Research, 45*(4), 555–564. https://doi.org/10.1021/ar200202c.

Zheng, L., Cash, V. L., Flint, D. H., & Dean, D. R. (1998). Assembly of iron-sulfur clusters. Identification of an iscSUA-hscBA-fdx gene cluster from Azotobacter vinelandii. *The Journal of Biological Chemistry, 273*(21), 13264–13272.

Zheng, Q., Fang, H., & Liu, W. (2017). Post-translational modifications involved in the biosynthesis of thiopeptide antibiotics. *Organic & Biomolecular Chemistry, 15*(16), 3376–3390. https://doi.org/10.1039/c7ob00466d.

CHAPTER TEN

Mechanistic Studies of Radical SAM Enzymes: Pyruvate Formate-Lyase Activating Enzyme and Lysine 2,3-Aminomutase Case Studies

Amanda S. Byer, Elizabeth C. McDaniel, Stella Impano, William E. Broderick, Joan B. Broderick[1]

Department of Chemistry and Biochemistry, Montana State University, Bozeman, MT, United States
[1]Corresponding author: e-mail address: jbroderick@chemistry.montana.edu

Contents

Abstract

The radical SAM enzyme superfamily is large and diverse, with ever-increasing numbers of examples of characterized reactions. This chapter focuses on the methodology we have developed over the last 25 years for working with these enzymes, with the specific examples discussed being the pyruvate formate-lyase activating enzyme (PFL-AE) and lysine 2,3-aminomutase (LAM). Both enzymes are purified from overexpressing *Escherichia coli*, but differ in that PFL-AE is expressed without an affinity tag and does not require iron–sulfur cluster reconstitution, while LAM purification is carried out

Methods in Enzymology, Volume 606
ISSN 0076-6879
https://doi.org/10.1016/bs.mie.2018.04.013

through use of a His_6 affinity tag and the enzyme benefits from cluster reconstitution. Because of radical SAM enzymes' catalytic need for a [4Fe–4S] cluster, we present methods for characterization and incorporation of a full [4Fe–4S] cluster in addition to enzyme activity assay protocols. Synthesis of SAM (S-adenosyl-L-methionine) and its analogs have played an important role in our mechanistic studies of radical SAM enzymes, and their synthetic methods are also presented in detail.

ABBREVIATIONS

A_{426nm}	UV–vis absorbance at 426 nm
AdoK	adenosine kinase
ADP	adenosine diphosphate
AK	adenylate kinase
AMP	adenosine monophosphate
amp	ampicillin
anAdo	anhydroadenosine
anADP	anhydroadenosine diphosphate
anAMP	anhydroadenosine monophosphate
anATP	anhydroadenosine triphosphate
anSAM	anhydroadenosyl-L-methionine
ATP	adenosine triphosphate
C. subterminale	*Clostridium subterminale*
CoA	coenzyme A
CV	column volume
dAdo	5′-deoxyadenosine
DEAE	diethylaminoethyl
DT	dithionite
DTT	dithiothreitol
ENDOR	electron nuclear double resonance
EPPS	3-[4-(2-hydroxyethyl)piperazin-1-yl]propane-1-sulfonic acid
EPR	electron paramagnetic resonance
E. coli	*Escherichia coli*
FAS	ferrous ammonium sulfate
FeS	iron–sulfur
FPLC	fast protein liquid chromatography
GRE-AE	glycyl radical enzyme activating enzyme
GTP	guanosine triphosphate
HCl	hydrochloric acid
HEPES	2-[4-(2-hydroxyethyl)piperazin-1-yl]ethanesulfonic acid
His	histidine
His_6	6-histidine affinity purification tag
HPLC	high-performance liquid chromatography
IPTG	isopropyl ß-D-1-thiogalactopyranoside
ISC	iron–sulfur cluster
kDa	kiloDalton
KOH	potassium hydroxide
LAM	lysine 2,3-aminomutase
LB	lysogeny broth

LMCT	ligand-to-metal charge transfer
MOPS	3-(*N*-morpholino)propanesulfonic acid
MWCO	molecular weight cut off
NaCl	sodium chloride
NADH	nicotinamide adenine dinucleotide
NH_4OH	ammonium hydroxide
NMR	nuclear magnetic resonance
OD_{600}	optical density at 600 nm absorbance
PCK	phosphocreatine kinase
PFL	pyruvate formate-lyase
PFL_a	active PFL-AE
PFL-AE	pyruvate formate-lyase activating enzyme
PIPES	piperazine-*N*,*N*′-bis(2-ethanesulfonic acid)
PITC	phenylisothiocyanate
PLP	pyridoxal 5′-phosphate
PMSF	phenylmethylsulfonyl fluoride
SAM	*S*-adenosyl-L-methionine
SDS-PAGE	sodium dodecyl sulfate polyacrylamide gel electrophoresis
SPL	spore photoproduct lyase
TIM	triose-phosphate isomerase
TLC	thin-layer chromatography
Tris	tris(hydroxymethyl)aminomethane
UV–vis	ultraviolet–visible

1. INTRODUCTION

Enzymes in the radical SAM superfamily use an iron–sulfur cluster (ISC) in conjunction with *S*-adenosyl-L-methionine (SAM) to catalyze a diverse range of radical reactions (Broderick, Duffus, Duschene, & Shepard, 2014). The radical SAM iron–sulfur (FeS) cluster is a site-differentiated [4Fe–4S] cluster coordinated at three of the irons by cysteines in a conserved cysteine motif (CX_3CX_2C). The iron lacking a cysteine ligand, referred to as the unique iron, is coordinated by the amino and carboxylate moieties of SAM; this SAM–cluster interaction likely plays a central role in catalysis (Krebs, Broderick, Henshaw, Broderick, & Huynh, 2002; Walsby, Ortillo, Broderick, Broderick, & Hoffman, 2002). Radical SAM catalysis requires that the [4Fe–4S] cluster be in its reduced (1+) state, which is achieved in vivo by partner proteins such as flavodoxin (Crain & Broderick, 2013). Because the cluster reduction potentials are generally in the −400 to −600 mV range, strong reductants such as dithionite are often required to achieve this reduced state in vitro (Hinckley & Frey, 2006;

Pierrel, Hernandez, Johnson, Fontecave, & Atta, 2003; Ugulava, Gibney, & Jarrett, 2001; Wang & Frey, 2007a). The [4Fe–4S]$^+$ has been proposed to transfer an electron to the bound SAM to promote the homolytic S—C (5′) bond cleavage that generates methionine and a 5′-deoxyadenosyl radical (Fig. 1) (Henshaw, Cheek, & Broderick, 2000). This 5′-deoxyadenosyl (dAdo) radical is thought to be a central intermediate in radical SAM enzyme mechanism and to abstract a hydrogen atom from substrate initiating subsequent mechanistic steps that diverge widely depending on the specific enzyme and substrate in question (Broderick et al., 2014). Recently, a novel organometallic intermediate (Ω) was discovered during pyruvate formate-lyase activating enzyme (PFL-AE) catalysis; Ω involves a direct C—Fe bond between the 5′-carbon of a deoxyadenosyl moiety and the unique iron of the [4Fe–4S] cluster (Fig. 2) (Horitani et al., 2016). Whether Ω is a central feature of radical SAM mechanisms more broadly, or is unique to PFL-AE, remains to be determined.

Radical SAM enzymes can be divided into two broad categories based on their mode of SAM utilization: some use SAM as a substrate and catalyze a net oxidative reaction on substrate, while others use SAM as a cofactor and catalyze a rearrangement reaction that involves no net redox chemistry on the substrate (Fig. 3) (Wang & Frey, 2007b). Most characterized radical SAM enzymes, including PFL-AE, fall into the former category, cleaving SAM stoichiometrically with substrate turnover, to produce methionine and 5′-deoxyadenosine (dAdo) as products (the orange pathway in Fig. 3). A few radical SAM enzymes, including lysine 2,3-aminomutase (LAM) and spore photoproduct lyase (SPL), use SAM as a cofactor that is not consumed during substrate turnover (the green pathway in Fig. 3).

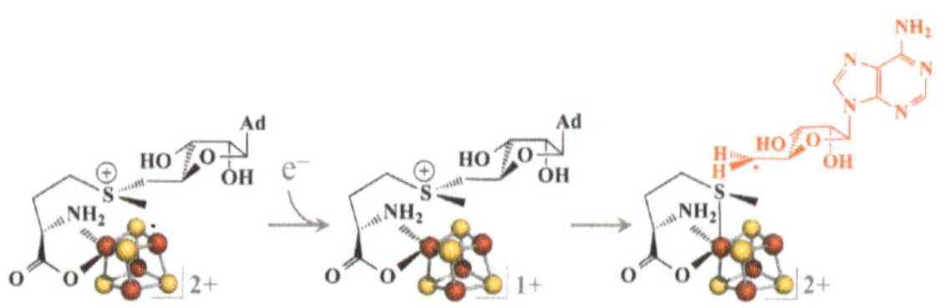

Fig. 1 Generation of the dAdo$^•$ radical by radical SAM enzymes. The resting enzyme with a [4Fe–4S]$^{2+}$ cluster (*left*: sulfur, *yellow spheres*; iron, *rust-colored spheres*) is reduced by one electron to generate the catalytically active state containing a [4Fe–4S]$^{1+}$ cluster (*center*). An electron is thought to undergo innersphere electron transfer to the sulfonium of SAM, promoting S—C(5′) bond cleavage and generation of a 5′-dAdo$^•$ radical intermediate (red) and methionine (*right*).

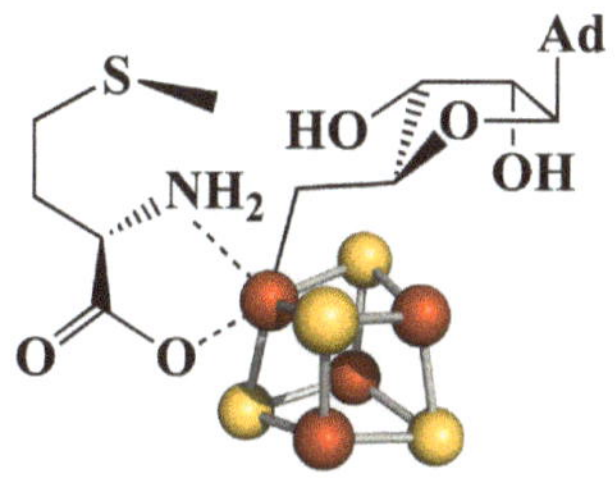

Fig. 2 The organometallic intermediate Ω observed in PFL-AE. This Ω intermediate was captured through rapid freeze-quench preparations of spectroscopy samples; electron paramagnetic resonance and electron-nuclear double resonance spectroscopy characterization illustrated a direct bond between the 5′-C of the dAdo moiety and the unique iron of the cluster (sulfur, *yellow spheres*; iron, *rust-color spheres*).

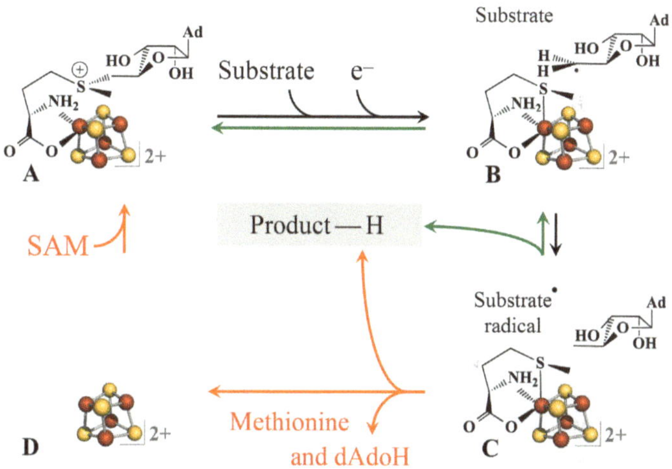

Fig. 3 Radical SAM enzyme catalysis scheme. This scheme represents a simplification of the current understanding of how radical SAM enzymes use SAM as a cofactor (*green pathway*) or a cosubstrate (*orange pathway*). Prior to catalysis, radical SAM enzymes house a [4Fe–4S] cluster in the 2+ state (A: sulfur, *yellow spheres*; iron, *rust-color spheres*). One electron reduction of this cluster generates the catalytically active $[4Fe–4S]^+$ state (see Fig. 1); in this cluster state, an electron can be transferred to the sulfonium of SAM to promote homolytic bond cleavage of the S—C(5′) bond to generate a 5′-dAdo• radical intermediate (B). This dAdo• radical abstracts a hydrogen from substrate to generate a substrate-based radical and dAdoH (C). From here, the scheme branches depending on whether SAM is used as a cofactor (*green arrows*) or as a cosubstrate (*orange arrows*). If SAM is used as a cofactor, a product radical abstracts a hydrogen from dAdoH to regenerate the dAdo radical (C → B) and the product leaves the active site. If SAM is used as a cosubstrate, the substrate-based radical reacts to form the product, while methionine and dAdoH leave the active site (C → D) allowing a new SAM molecule to enter the enzyme active site (D → A) and catalysis to begin again.

This remarkable chemistry catalyzed by radical SAM enzymes is harbored within the ancient triose-phosphate isomerase (TIM) barrel fold, with the ISC at one end of the barrel (Figs. 4 and 5). While some radical SAM enzymes comprise a full TIM barrel $(\beta\alpha)_8$ (Fig. 4A), others function with only partial TIM barrels (Fig. 4B and C). The completeness of the TIM barrel appears to inversely correlate with substrate size in that the radical SAM enzymes catalyzing reactions with large substrates tend to have partial TIM barrels. Many radical SAM enzymes also have additional domains beyond the TIM barrel (Fig. 4B), and some harbor accessory ISCs that can serve a variety of catalytic functions. Beyond these ISCs, additional metal centers can be found in radical SAM enzymes. For example, some radical SAM enzymes also use B_{12} cofactors, and recently PFL-AE has been shown to require a monovalent cation interacting directly with SAM in the active site for optimal activity (Fig. 5B) (Shisler et al., 2017).

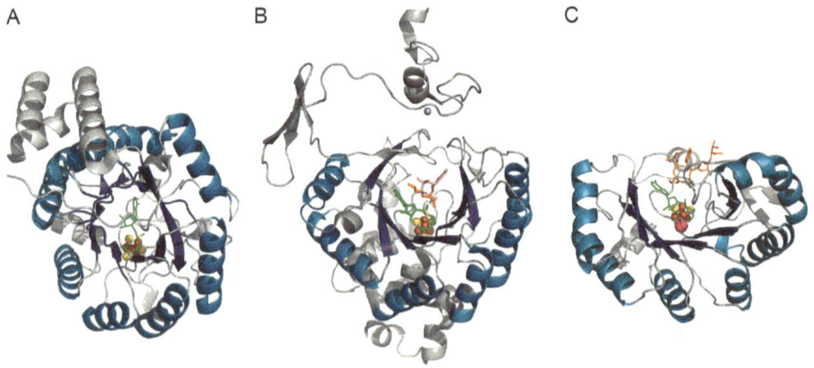

Fig. 4 Radical SAM enzymes full or partial TIM barrels. In all structures, the alpha-helices of the TIM barrel are shown in *light blue* and the beta strands are displayed in *dark blue* and the enzyme structure remainder is shown in *light grey*; in the center of the TIM barrel, essential components of radical SAM enzymes are displayed: the [4Fe–4S] (sulfur, *yellow spheres*; iron, *rust-colored spheres*), SAM (*green sticks*), and substrate (*orange sticks*, when present). (A) The full TIM barrel structure $(ß\alpha)_8$ is [FeFe]-hydrogenase maturase, HydE; no substrate is present in this structure (PDB: 3IIZ). (B) In LAM's partial TIM barrel $(ß\alpha)_6$, the substrate (lysine, *orange sticks*) and LAM's second cofactor PLP (pyridoxyl 5′-phosphate, *pink sticks*) are visible above the SAM bound [4Fe–4S] cluster; a zinc ion (*blue–grey sphere*) resides at the subunit interface, however the additional subunits are not displayed (PDB: 2A5H). (C) PFL-AE's partial TIM barrel $(ß\alpha)_6$ houses the cluster, SAM, and a peptide (*orange sticks*) representing the arm of PFL that sits in the PFL-AE active site; additionally in this active site, a monovalent cation (Na^+, *magenta*) is positioned in close proximity to SAM and the cluster (PDB: 3CB8).

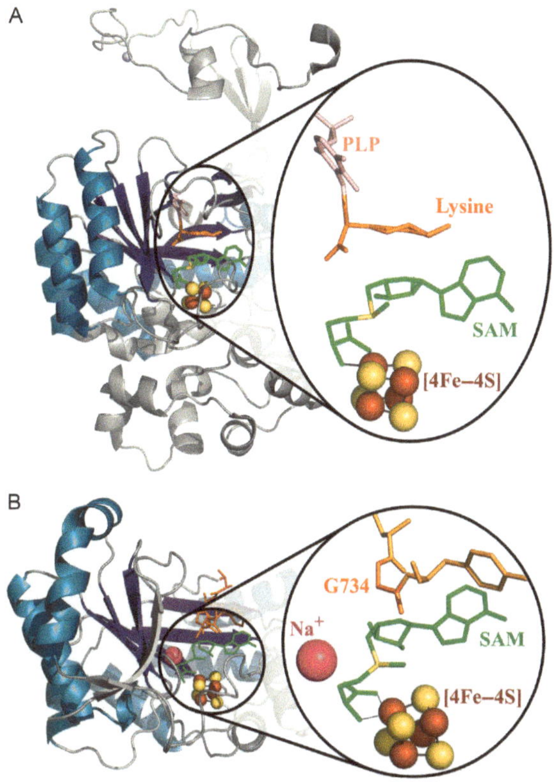

Fig. 5 Radical SAM enzyme active site. Both structures of LAM (A) and PFL-AE (B) illustrate similar approximate positioning (from *bottom* to *top*) of the [4Fe–4S] cluster (sulfur, *yellow spheres*; iron, *rust-colored spheres*), SAM (*green sticks*), and substrate (*orange sticks*). (A) The LAM X-ray crystal structure (PDB: 2A5H) shows an additional cofactor PLP (*pink sticks*). (B) PFL-AE's active site (PDB: 3CB8) depicts a monovalent cation (Na^+, *magenta sphere*) and only part of the PFL substrate (*orange sticks*; for clarity only three residues are shown, S733–G734–Y735, with the G734 highlighted in a *darker orange*).

Our lab has been studying radical SAM enzymes for the past 25 years, with a particular focus on understanding the fundamental chemistry of radical initiation (Broderick, 2003; Broderick et al., 2014; Cheek & Broderick, 2001; Shepard & Broderick, 2010; Walsby et al., 2005). Our early efforts centered on PFL-AE (Broderick et al., 1997), and this enzyme has remained an important focus of our work, because it is well behaved, relatively easy to isolate in large quantities, and is amenable to a range of spectroscopic and biochemical approaches. We also study a number of other radical SAM enzymes including LAM, SPL, and the [FeFe]-hydrogenase maturase enzymes, HydE and HydG. In this chapter, we present our methods for preparing and characterizing PFL-AE, its substrate protein PFL, and LAM (Section 2). We also describe the methodology we have developed for synthesizing SAM and SAM analogs (Section 3), which are useful in a wide range of spectroscopic and biochemical approaches to investigating these radical SAM enzyme mechanisms.

2. PREPARATION AND CHARACTERIZATION OF RADICAL SAM ENZYMES

Owing to the ISC content, radical SAM enzymes usually require protection from oxygen exposure to minimize cluster oxidation and degradation. Without anaerobic conditions, enzyme purification is often accompanied by partial or complete loss of the ISC. We have developed approaches for preparation of a number of different radical SAM enzymes such that more FeS clusters remain intact. Here we describe these methods for two representative examples, PFL-AE and LAM. In the case of PFL-AE, no affinity tag is present on the protein, and purification involves multiple runs over a size exclusion column with no need for ISC reconstitution; our LAM construct introduces a His_6 tag for protein purification on an affinity column, and the enzyme benefits from cluster reconstitution. Techniques for ISC reconstitution, characterization of protein and ISCs, and enzyme activity assays are presented in this section.

2.1 Preparation of PFL-AE and Its Substrate Protein PFL

PFL-AE is a glycyl radical enzyme activating enzyme, or GRE-AE, that converts pyruvate formate-lyase (PFL) to its catalytically active state by generating a stable glycyl radical in the active site at G734. Once activated, PFL catalyzes the conversion of pyruvate and CoA to formate and acetyl-CoA (Fig. 6) during anaerobic glucose metabolism.

PFL was the first enzyme discovered to require a stable glycyl radical for catalytic activity (Knappe, Neugebauer, Blaschkowski, & Gänzler, 1984; Wagner, Frey, Neugebauer, Schäfer, & Knappe, 1992), and although many other glycyl radical enzymes (GREs) are now known, PFL remains the best characterized (Shisler & Broderick, 2014). During the early studies of PFL, PFL-AE was discovered (Conradt, Hohmann-Berger, Hohmann, Blaschkowski, & Knappe, 1984; Knappe & Blaschkowski, 1975; Knappe et al., 1969). Though PFL-AE's cofactor requirements were not immediately clear, absorption features in the visible region of the spectrum pointed to a possible metal cofactor (Conradt et al., 1984). This metal cofactor was

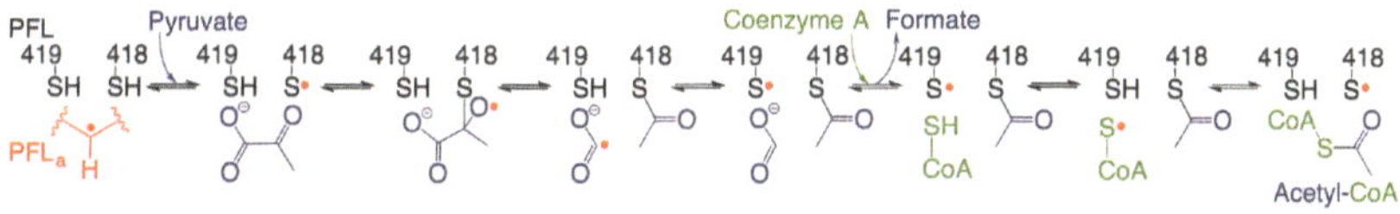

Fig. 6 PFL reaction scheme. Through this proposed ping pong mechanism pathway, PFL converts pyruvate into formate, and CoA into acetyl-CoA. Activated PFL (PFL_a) has a glycyl radical at residue G734 which abstracts a hydrogen from residue C418, generating a cysteinyl radical which initiates a nucleophilic attack on the 2′C of pyruvate to generate an oxygen-centered pyruvate-based radical. Homolytic cleavage between the C1 and C2 position on pyruvate produces a carbonyl radical, which quickly reacts to form a different cysteinyl radical, this time on C419, while formate is released from the active site. This C419-based radical initiates a nucleophilic attack on the incoming CoA, generating a CoA sulfur-based radical. Finally, this CoA-based radical initiates a nucleophilic attack on the C418 bound carbonyl to form acetyl-CoA and reestablish the C418 sulfur-based radical so the catalytic cycle can repeat.

ultimately shown to be an ISC (Broderick et al., 1997); and this finding, together with the dependence of PFL activation on SAM (Frey, Rothe, Wagner, & Knappe, 1994; Knappe & Blaschkowski, 1975; Knappe, Bohnert, & Brummer, 1965; Knappe & Schmitt, 1976), revealed a striking similarity to LAM, which also requires SAM and an ISC to carry out radical catalysis (Baraniak, Moss, & Frey, 1989; Chirpich, Zappia, Costilow, & Barker, 1970; Moss & Frey, 1987; Moss & Frey, 1990; Petrovich, Ruzicka, Reed, & Frey, 1991, 1992). Together, these discoveries in PFL-AE and LAM helped to set the stage for the identification of the radical SAM superfamily (Cheek & Broderick, 2001; Sofia, Chen, Hetzler, Reyes-Spindola, & Miller, 2001).

The critical discovery of an ISC in PFL-AE arose from early purification trials under partially anaerobic conditions, which revealed a red-brown color that was easily lost upon exposure of the protein to oxygen. This observation led to our development of methods to reliably purify this protein with an intact [4Fe–4S] cluster (Broderick et al., 1997, 2000; Krebs, Henshaw, Cheek, Huynh, & Broderick, 2000; Walsby et al., 2005).

In contrast, the substrate protein PFL does not house a metallocofactor and thus a stringent O_2-free preparation is not initially necessary; however, the activated PFL enzyme harbors a glycyl radical which is extremely reactive with oxygen. In order to produce effective PFL-AE substrate, therefore, we combine aerobic overexpression and purification, with subsequent Schlenk techniques to make the enzyme anaerobic prior to activation experiments. The methods described below for PFL purification were adapted by us from the early work published by Knappe and coworkers (Conradt et al., 1984).

2.1.1 Overexpression of PFL-AE

PFL-AE can be overexpressed to high levels in the *Escherichia coli* cell line BL21(DE3)pLysS. The expression vector we use has the *pflA* gene, encoding PFL-AE, inserted at the NdeI/HindIII sites of pCAL-n-EK (Broderick et al., 2000), thereby putting the gene under the control of the T7 promoter and allowing for induction with isopropyl ß-D-1-thiogalactopyranoside (IPTG). Using this system, PFL-AE induces to such a high level that we are able to use Mössbauer spectroscopy to characterize the cluster states in the induced cells, prior to protein purification (Yang et al., 2009). In an effort to minimalize the protein's contact with oxygen, the lysis, purification, and concentration of PFL-AE are conducted in an anaerobic chamber using anaerobic techniques to minimize oxygen contamination, as is detailed later.

To prepare for the growth, the freshly transformed BL21(DE3)pLysS cells (~25 μL) or cells from a small amount of glycerol stock are spread on a LB agar plate containing 50 μg/mL ampicillin (*amp*) and incubated at 37°C overnight (14–16 h). After incubation, colonies should be small and scattered across the plate, and can be used to inoculate media for about a week if handled properly (plate sealed with parafilm and stored at 4°C). The day before the large-scale growth, two 250 mL Erlenmeyer flasks of media (50 mL LB media, 50 μg/mL *amp*) are inoculated, each with a different cell colony; these cultures are grown overnight (14–16 h) at 37°C with shaking (220 rpm), and then either used directly to inoculate the large-scale growth or stored at 4°C until use.

For the large-scale growth, LB media (9.0 L divided between 6 baffled 2.8 L Fernbach flasks) are sterilized and brought to 37°C. Prior to inoculation with the overnight culture (10 mL overnight culture per liter of growth media), *amp* is added to a final concentration of 50 μg/mL. After inoculation, the cells are grown in an incubator shaker (New Brunswick Scientific I 26) at 37°C with shaking at 200 rpm. The pH is monitored every 30 min and adjusted as necessary with 6 *M* NH_4OH and 6 *M* HCl to maintain pH between 7.2 and 7.5.

Once the cells have reached an optical density (OD_{600}) of 0.3, glucose is added (final concentration, 5 mg/mL); and when the OD_{600} reaches 0.8, a 1 mL aliquot of the culture is removed and then the remaining culture is induced by adding IPTG to a final concentration of 0.25 m*M*. At induction, ferrous ammonium sulfate (FAS) and L-cysteine are added to a final concentration of 0.20 m*M* each, and the temperature is lowered to 30°C. The culture is then incubated with shaking until the pH no longer undergoes a significant change in 30 min; often this requires ~5 h.

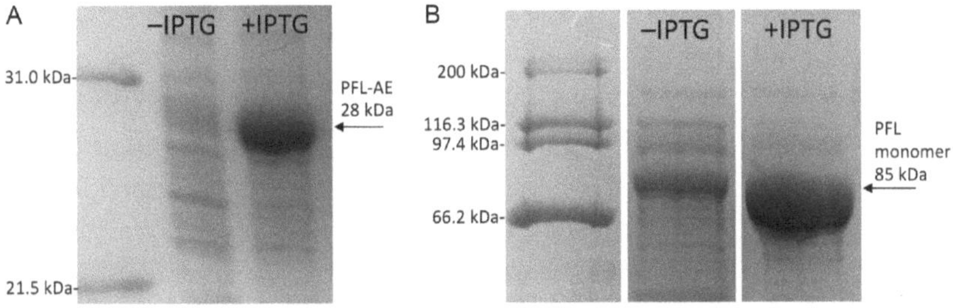

Fig. 7 PFL-AE and PFL overexpression. (A) SDS-PAGE of PFL-AE shows cellular protein content before induction (−IPTG) and after (+IPTG, 2 × dilute) induction. (B) SDS-PAGE of PFL showing cellular protein content before induction (− IPTG) and after (+IPTG, 2 × dilute) induction. All samples are run against a BioRad Broad Range Molecular Marker with molecular weights indicated; in (B), the gel fragments are juxtaposed and aligned appropriately, but remain separated to illustrate gel lanes were not adjacent.

At this point, additional FAS and L-cysteine are added (bringing the final concentrations to 0.40 m*M*) before removing flasks to a 4°C refrigerator and sparging cultures with nitrogen overnight (15–16 h). After the ~15 h anaerobic phase, another 1 mL aliquot of the culture is removed and the rest of culture is transferred to centrifuge bottles for harvesting; to minimize exposure to oxygen, $N_{2(g)}$ is bubbled into the centrifuge bottle while adding the sparged culture. The cell cultures are then centrifuged (12,000 × *g*, 10 min, 4°C). The supernatant is gently poured off, and the pellets are flash frozen in liquid nitrogen immediately; these pellets are stored at −80°C until ready to begin purification.

To assess overexpression, the two 1 mL aliquots of culture (pre- and postinduction) are centrifuged, the supernatant is discarded, and sodium dodecyl sulfate polyacrylamide gel electrophoresis (SDS-PAGE) sample buffer (50 μL for preinduction and 100 μL for postinduction) is used to resuspend each cell pellet. This suspension is boiled for 5 min, centrifuged for 1 min at 14,500 × *g*, and approximately 2 μL of this supernatant is loaded onto a 12% SDS-PAGE gel to check expression levels (Fig. 7A).

2.1.2 Purification of PFL-AE

All lysis and purification steps are carried out in a vinyl anaerobic chamber (Coy Laboratory Products, Inc.), unless otherwise indicated. Prior to lysis, a Superdex-75 prep grade gel filtration column (volume ~1000 mL, 5 cm × 55 cm) is equilibrated with two column volumes (CVs) of anaerobic gel filtration buffer (50 m*M* Tris, 100 m*M* sodium chloride (NaCl), 1 m*M* dithiothreitol (DTT), and pH 7.5) in an anaerobic chamber. If there are any concerns about potential oxygen contamination in the column, the column should be rinsed with 1.5 CVs of a 1 m*M* dithionite solution, prior to equilibration with buffer. To the anaerobic lysis buffer (50 m*M* Tris, 100 m*M* NaCl, 1% (w/v) Triton X-100, 5% (w/v) glycerol, 10 m*M* magnesium chloride ($MgCl_2$), and pH 7.5), the following components are added to the indicated final concentrations: phenylmethylsulfonyl fluoride (PMSF, in ~200 µL methanol, 1.0 m*M*), lysozyme (0.20 mg/mL), DTT (1.0 m*M*), and RNAse and DNAse (10 µg/mL each). The final lysis volume depends on the cell mass, such that for every 1.0 g of cell pellet, 1.0 mL of buffer is used. The PFL-AE cell pellet (typically ~15–20 g) is added to the lysis mixture and gently broken up with a spatula. The lysate is placed on ice and agitation is continued with an 18-gauge needle and 30 mL syringe until the cell lysate consistency becomes watery (~45–60 min). The lysate is transferred to centrifuge bottles (using o-ring caps) and then brought out of the anaerobic chamber for centrifugation (38,000 × *g*, 1 h, 4°C). The centrifuged lysates are then brought back into the anaerobic chamber, and the supernatant is immediately and gently decanted into a Superloop (50 mL). This clarified lysate is loaded onto the Superdex 75 column at 3 mL/min using a fast protein liquid chromatography (FPLC) system, and is eluted with the same buffer used for the initial column equilibration. PFL-AE will begin to elute a little after 0.5 CVs. Fraction (10 mL) collection begins as PFL-AE first elutes off the column, as judged by the position of the brown band on the column, and the darker color of eluate; PFL-AE should be a clearly visible dark brown band on the column. The darkest fractions are collected (Fig. 8A), combined, and concentrated to ~10 mL. If the remainder of purification cannot be immediately completed, the protein is flash frozen in liquid nitrogen and stored at −80°C.

The pooled and concentrated protein from the first column purification is reapplied to the same column, after it has been reequilibrated. The second gel filtration utilizes the same procedures as for the first column run, except that 5 mL fractions are collected once the colored protein begins to elute from the column (Fig. 8B). To determine which fractions will be kept, aliquots (100 µL) of the darkest fractions are diluted to 1 mL with gel filtration buffer, and UV–vis absorbance is measured at 426 and 280 nm, to quantify the ISC and protein content, respectively. Fractions with the highest A_{426}/A_{280} ratios, typically in the 0.18–0.19 range, are pooled, concentrated, and designated as the "A-fractions." Fractions with the second highest A_{426}/A_{280} ratio, typically in the 0.17–0.18 range, are pooled, concentrated separately, and designated the "B-fractions." To minimize protein

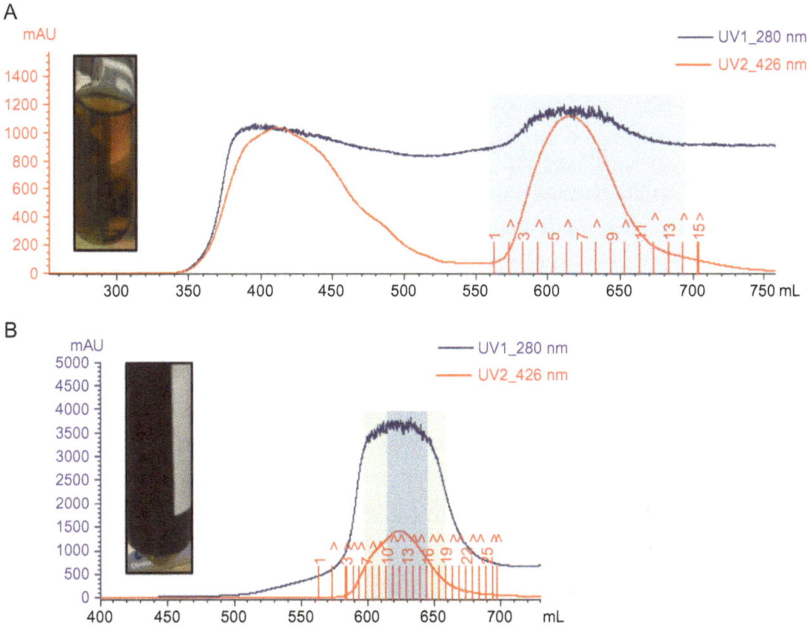

Fig. 8 PFL-AE purifications. PFL-AE is purified twice on a 1 L (5 cm × 55 cm) Superdex 75 prep grade gel filtration column, while UV–vis traces at 280 and 426 nm monitor protein and iron content, respectively. (A) First PFL-AE purification. The elution chromatogram shows two peaks, wherein the first peak is an impurity and the second is the PFL-AE protein; in this chromatogram, the 0–1600 mAU scale applies only to the 426 nm trace, as the 280 nm trace is on a scale of approximately 0–4500 mAU. A representative fraction (*inset*) illustrates the *brown color* of the PFL-AE solution isolated from a first purification; select fractions (*highlighted in blue*) are run over the column a second time. (B) Second PFL-AE purification. This second purification chromatogram shows PFL-AE eluting at the same volume as the first run, but without the impurity; both UV–vis traces are on the same scale. The *brown color* of PFL-AE isolated from this second purification (*inset*) is often darker than that of the first purification fractions. To select enzyme with the best cluster content, fractions are partitioned into A (*highlighted in blue*) and B (*highlighted in green*) sets, based on A_{426nm}:A_{280nm} ratios.

freeze/thaw cycles, the A-fraction protein is aliquoted (250 μL) and flash frozen in liquid nitrogen before storage at −80°C. B-fractions are combined, concentrated, and run over the gel filtration column a third time to collect more A-fractions.

2.1.3 Manipulation of the Monovalent Cation Site in PFL-AE

Recently, we reported the identification of a monovalent cation site in PFL-AE (Shisler et al., 2017). Intriguingly, the presence and identity of this cation have a strong effect on PFL-AE's activity, with potassium giving rise to the highest activity, while other cations of larger or smaller radii providing lower activity, and divalent cations eliminating activity completely (Shisler et al., 2017).

To control the cation present in PFL-AE's active site, PFL-AE is prepared similar to the procedures described (Sections 2.1.1 and 2.1.2); however, all alkali metal cations are omitted from lysis and purification buffers. To provide buffer lacking alkali metal cations, the buffer is made from Tris base and the pH is adjusted with HCl. Owing to the lack of simple salts in the lysis buffer (50 m*M* Tris pH 7.4, 5% (w/v) glycerol, 1% (w/v) Triton X-100), lysis requires additional time, up to 1–1½ h total lysis time. After purification using alkali metal-free gel filtration buffer (50 m*M* Tris, pH 7.5), this PFL-AE is diluted into 100 m*M* Tris, pH 7.6 buffer containing 100 m*M* of the desired alkali metal cation. If a divalent cation is needed, the final salt concentration is 1 m*M*. Before initiating a reaction of PFL-AE with PFL, excess metal cations must also be removed from the PFL solution by buffer exchange. Typically this salt removal from PFL is performed by buffer exchange into 20 m*M* Tris pH 7.2 buffer (made from Tris base with pH adjusted using HCl) and prior to Schlenk line gas exchange (Section 2.1.5).

2.1.4 Overexpression of PFL

Overnight cultures and media for large-scale growth are prepared as described in the PFL-AE protocol. Fernbach flasks (2.8 L) containing LB/*amp* are inoculated with the overnight culture, and then placed in an incubator shaker (New Brunswick Scientific I 26) at 37°C, with shaking at 200 rpm. Once cells have reached an OD_{600} of 0.8, a 1 mL aliquot is removed for SDS-PAGE analysis, and then IPTG is added to a final concentration of 0.20 m*M*. The incubation temperature is lowered to 30°C and the culture is allowed to grow overnight (~16 h) with shaking at 200 rpm. The following morning, another 1 mL aliquot is removed for SDS-PAGE postinduction analysis, and then the culture is transferred to centrifuge bottles and centrifuged to harvest the cells (12,000 × *g*, 10 min, 4°C). The supernatant is decanted, and the cell pellet is flash frozen in liquid nitrogen. While cultures may produce foam growing overnight, antifoam (200 μL) can be added to each flask before allotting into centrifuge bottles to make culture transfer easier. Cells are stored at −80°C until lysis and purification. Induction is verified by SDS-PAGE using aliquots taken before and after induction, as described earlier for PFL-AE. Strong overexpression of PFL should be observed (Fig. 7B).

2.1.5 Purification of PFL

To prepare PFL cell lysis solution, the following components are combined in lysis buffer (20 m*M* Tris, 1% (w/v) Triton X-100, 5% (w/v) glycerol, 10 m*M* $MgCl_2$, pH 7.2) to the specified final concentrations: PMSF (dissolved in methanol, 1.0 m*M*), lysozyme (0.20 mg/mL), and RNAse and DNAse (10 μg/mL). Using 2 mL buffer for every 1 g of cell pellet, PFL cell pellets (typically 15–25 g) are added to lysis buffer and gently broken up with a spatula. Using an 18-gauge needle and 30 mL syringe, with care to avoid bubbles, agitation is continued until the cell lysate has a watery consistency (30–45 min). The lysate is centrifuged (38,000 × *g*, 1 h, 4°C), and the supernatant (~50 mL) is decanted and immediately purified or stored at −80°C.

Initial chromatographic separation of PFL occurs on an Accell Plus quaternary methylamine (QMA) (300 Å) AP5-ion exchange column (600 mL, 5.0 cm × 30 cm). After the column is equilibrated with two CVs of no salt buffer (20 m*M* Tris, pH adjusted to 7.2 with HCl), ~50 mL lysate is loaded onto the column at 3 mL/min, and then a three-step program is run at 5 mL/min using an FPLC system. The first step is a wash with 0.5 CVs of no salt buffer. Next, a 1.5 CV linear gradient is run starting with 100% no salt buffer and ending in 100% high salt buffer (20 m*M* Tris, 500 m*M* NaCl, 1 m*M* DTT, pH 7.2). Finally, the column is washed with 0.5 CVs of high salt buffer.

Beginning halfway through the linear gradient, fractions (10 mL) are collected (Fig. 9A). SDS-PAGE is used to determine which PFL fractions are sufficiently pure to be saved: 10 μL of each fraction is mixed with an equal volume of SDS-PAGE sample buffer, boiled for 3 min, centrifuged, and then 2–5 μL is loaded into the well of a 12% SDS-PAGE gel (Fig. 9A, inset). PFL monomer will run around 85 kDa. The main criteria for saving PFL fractions from this first column require those with a strong band at ~30 kDa be avoided, as this impurity will not be removed by subsequent purification steps. The pooled fractions are concentrated to ~10 mL using centrifuge filters with a molecular weight cut off (MWCO) of 10 kDa. The pooled concentrated protein is buffer exchanged into Buffer B (20 m*M* Tris, 1 *M* ammonium sulfate, 1 m*M* DTT, pH 7.2) by bringing the 10 mL of protein up to 50 mL with Buffer B then concentrating back to 10 mL and repeating this two additional times for a total of three exchanges.

The final purification step employs a HighLoad 16/10 phenyl Sepharose high-performance column (20 mL, 1.6 cm × 10 cm) equilibrated with two CVs of Buffer B. The protein is loaded onto the column and eluted via a

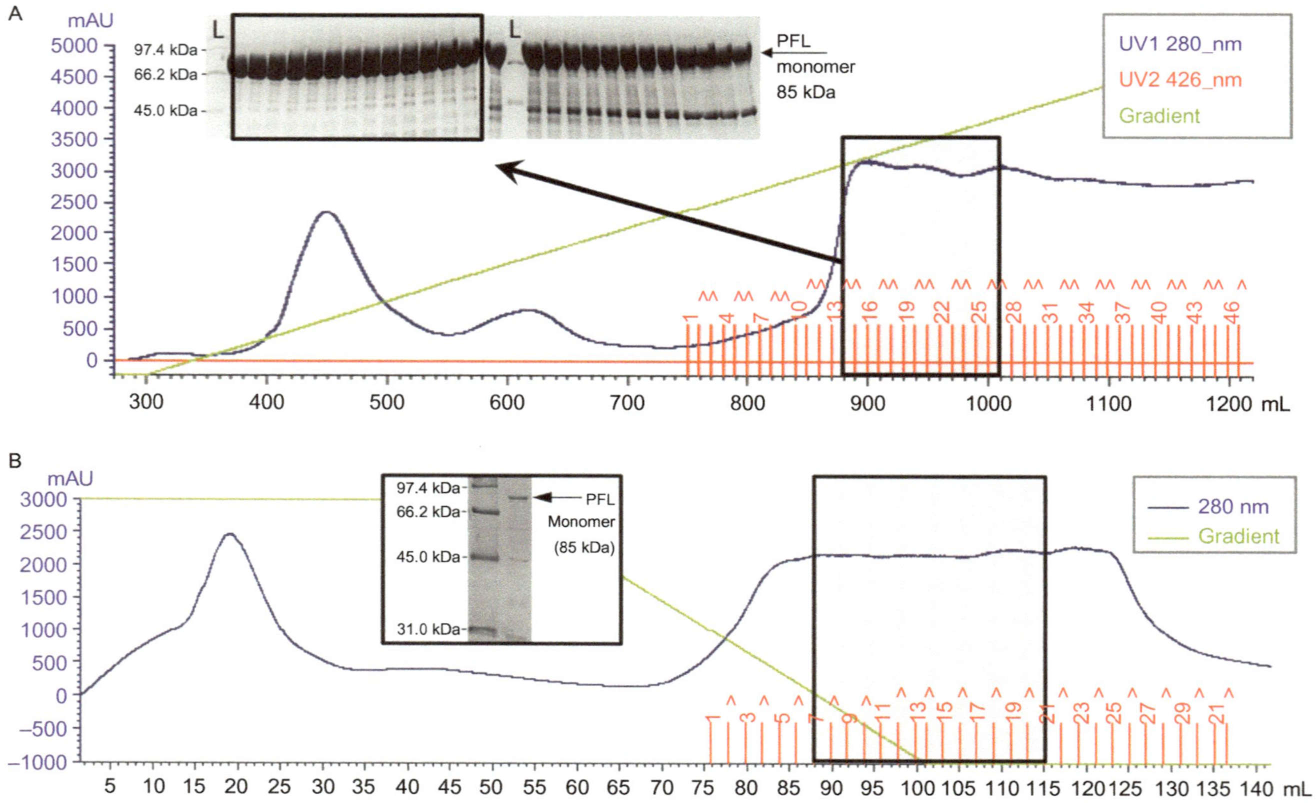

Fig. 9 PFL protein purifications. (A) PFL elution off the Accell Plus™ quaternary methylamine anion exchange column will show a trailing peak beginning roughly halfway through the linear gradient. The SDS-PAGE (*inset* and *boxed*) illustrates only those fractions without the lower molecular weight impurity are kept for buffer exchange and further purification. (B) PFL elution off of the HR 16/10 Highload phenyl Sepharose high-performance column commences around 70 mL, halfway through the linear gradient into the no salt buffer. The collected fractions (*boxed*) are predominantly PFL protein (*inset*).

three-step program: (1) wash with 2.5 CVs of Buffer B; (2) run a linear gradient from 100% Buffer B to 100% no salt buffer over 2.5 CVs; and (3) wash with 2.5 CVs of no salt buffer. Beginning halfway through the linear gradient, fractions (2 mL) are collected (Fig. 9B) and SDS-PAGE analysis (as described for the previous step) determines which fractions to save. Fractions of acceptable purity are collected (Fig. 9B, inset), concentrated, and flash frozen in liquid nitrogen; protein is stored at −80°C.

Because PFL is purified under aerobic conditions, it is necessary to remove O_2 prior to experiments. To make it anaerobic, PFL is placed in a Schlenk flask and three pump purge cycles are applied. First the protein solution is frozen in the Schlenk flask by swirling while slowly dipping in liquid nitrogen to achieve a thin layer of frozen protein along the bottom. With the flask submerged in liquid nitrogen, three vacuum and nitrogen cycles are applied and then the flask is transferred to an anaerobic chamber ($O_2 < 1$ ppm) before thawing. To minimize further oxygen exposure, PFL is aliquoted into anaerobic vials with compressible o-rings before flash freezing in liquid nitrogen and storage at −80°C.

2.2 Preparation of LAM

LAM, another well-characterized radical SAM enzyme, has been the subject of a multitude of studies examining cluster content and mechanism. Through several radical intermediates, this exquisite isomerase converts alpha-lysine to the antibiotic precursor ß-lysine using both SAM and PLP as cofactors (Ballinger, Frey, & Reed, 1992; Chang, Ballinger, Reed, & Frey, 1996; Wu, Lieder, Reed, & Frey, 1995). As LAM is a representative member of the group of radical SAM enzymes that use SAM as a cofactor, SAM is regenerated in the active site during catalysis.

The crystal structure of LAM reveals a homohexameric protein, wherein each subunit has a $(\beta\alpha)_6$ TIM barrel (49,200 kDa) that houses an active site with the canonical radical SAM machinery, the [4Fe–4S] cluster ligated by a tricysteine motif and SAM, in addition to the alpha-lysine substrate bound to PLP (Fig. 4B and 5A). Zinc ions at the subunit interface are expected to play a structural role in the tertiary and quaternary assembly (Lepore, Ruzicka, Frey, & Ringe, 2005).

Using substrate and SAM surrogates, many intermediate steps of the LAM mechanism have been spectroscopically identified (Ballinger, Frey, & Reed, 1992; Chang et al., 1996; Horitani et al., 2015; Magnusson, Reed, & Frey, 1999; Wu et al., 2000, 1995). Detailed reduction potential studies found LAM's [4Fe–4S] cluster bound without (−470 mV) and with (−430 mV) SAM had higher reduction potentials than when lysine and PLP were also bound (−600 mV) addressing longstanding energetic questions in the radical SAM mechanism (Wang & Frey, 2007a). The methods for LAM described later are adaptions from these studies.

2.2.1 Overexpression of LAM

Genetic constructs with a C-terminal His_6 tag (such as pET14-b) containing the LAM gene from *Clostridium subterminale* SB4 can be transformed into *E. coli*—BL21(DE3)pLysS (Stratagene) cells for overexpression. The transformed cells are plated on LB agar plates containing 50 μg/mL *amp*. A single colony is used to inoculate 50 mL of LB media containing *amp* (50 μg/mL) in a 250 mL flask, and this culture is grown in a 37°C incubator shaker at 250 rpm overnight (~15 h). The overnight culture is used to initiate the large-scale growth in minimal media. To prepare minimal media for a 10 L growth, casamino acids (110 g), 3-(*N*-morpholino)propanesulfonic acid (MOPS) buffer (84 g, 40.1 m*M*), tricine (8 g, 4.5 m*M*), NaCl (29.3 g, 50.1 m*M*), potassium hydroxide (KOH) (16 g, 28.5 m*M*), and ammonium chloride (5.1 g, 9.5 m*M*) are dissolved in water to make 9.8 L and autoclaved in a 14 L fermenter flask. A sterile filtered 200 mL solution containing *amp* (2.5 m*M*), glucose (50 g), $CaCl_2$ (31.25 m*M*), K_2SO_4 (17.25 m*M*), KH_2PO_4 (125 m*M*), and "O solution" (25 mL) is used to wash vitamins (pyridoxine, niacinamide, riboflavin, folic acid, thiotic acid, thiamine, vitamin B_{12}, pantothenic acid, and biotin at ~10 mg each) into the autoclaved, cooled media in the 14 L fermenter flask. The "O solution" contains $FeCl_2$ (0.1 g) dissolved in 10 mL concentrated HCl, mixed with 10 mL H_2O, 2.68 g $MgCl_2 \cdot 6H_2O$, and "T solution" (1 mL) brought to a final volume of 50 mL. The "T solution" is $CaCl_2 \cdot 2H_2O$ (0.0184 g), H_3BO_4 (0.064 g), $MnCl_2 \cdot 4H_2O$ (0.04 g), $CoCl_2 \cdot 6H_2O$ (0.018 g), $CuCl_2 \cdot 2H_2O$ (0.004 g), $ZnCl_2$ (0.340 g), and $Na_2MoO_4 \cdot 2H_2O$ (0.605 g) dissolved in 8 mL concentrated HCl and diluted to 100 mL in filtered water. The 10 L of minimal media is inoculated with the overnight culture, and then incubated at 37°C with shaking (250 rpm) to an OD_{600nm} of ~0.5. IPTG is then added to 0.5 m*M* final concentration and the growth is supplemented with FAS (0.75 g) and zinc sulfate (0.09 g). The culture is incubated at 37°C with shaking for an additional 2 h, before being cooled to approximately 30°C and supplemented with additional FAS (0.75 g). The culture is then cooled to 4°C and purged with nitrogen for an additional ~15 h; a drop of antifoam may be added prior to initiation of this anaerobic step. The cells are then centrifuged with minimal cell pellet exposure to oxygen as described in Section 2.1.1 and the resulting cell pellets are stored at −80°C until use.

2.2.2 Purification of LAM

Cell lysis and protein purification are performed under anaerobic conditions, typically in a vinyl anaerobic chamber (Coy Laboratory Products, Inc.). Cell pellets (~15 g) are thawed and resuspended in a lysis buffer (2 mL per 1 g of cell paste) containing 50 m*M* 3-[4-(2-hydroxyethyl)piperazin-1-yl] propane-1-sulfonic acid (EPPS), 300 m*M* KCl, 0.1 m*M* lysine, 10 μ*M* PLP, pH 8.0, 1% Triton X-100 (w/v), 0.2 mg/mL PMSF, 0.18 mg/mL lysozyme, and ~0.02 mg/mL each RNAse A and DNAse I. The lysis mixture (~50 mL) is gently stirred for ~1 h, though mechanical lysis via sonication (5 min total sonication, 60% amplitude, 5 s on, and 15 s off) or passage through an 18-gauge needle may supplement the chemical lysis. The lysate is removed from the anaerobic chamber in gas tight bottles, centrifuged at 38,000 × *g* for 30 min, and brought back into the anaerobic chamber prior to decanting the supernatant. This resulting clarified lysate is loaded onto a 5 mL HisTrap Ni^{2+} affinity column (GE Healthcare) equilibrated with 5 CV of Buffer A (50 m*M* EPPS, 300 m*M* KCl, 0.1 m*M* L-lysine, 10 μ*M* PLP, 10 m*M* imidazole, and pH 8.0). By increasing the imidazole concentration in a stepwise manner from 10% to 20% to 50% to 75% to 100% using Buffer B (50 m*M* EPPS, 300 m*M* KCl, 0.1 m*M* L-lysine, 10 μ*M* PLP, pH 8.0, and 500 m*M* imidazole) and allowing absorbance (280 nm) to return to baseline prior to step initiation (~50–100 mL per step), the purest LAM fractions elute at 20% and 50% Buffer B. This eluate is concentrated using a Millipore (13–14 kDa MWCO) spin concentrator. After purification and concentration, the protein is flash frozen in liquid N_2 and stored at −80°C or in liquid N_2 until further use. The concentrated protein is dialyzed or buffer exchanged into 50 m*M* EPPS, 1 m*M* DTT, 0.1 m*M* L-lysine, 10 μ*M* PLP, pH 8.0 prior to ISC reconstitution. Typically, LAM purifies with about 3.2 irons/subunit, which is lower than an optimal four irons per LAM subunit and thus an iron–sulfur reconstitution is performed.

Prior to reconstitution, the swiftly thawed LAM enzyme is centrifuged (14,500 × *g*, 5 min), diluted to about 100–150 μ*M*, and incubated with DTT (5 m*M*) for ~5 min, similar to that described in Section 2.5. From 10 m*M* stock solutions, sodium sulfide nonahydrate and FAS (to a final concentration that is 6 × the protein concentration) are added incrementally over 1–2 h, with gentle stirring. This reconstitution solution is incubated for

6–12 h at 4°C before being centrifuged (14,500 × g, 5 min), concentrated (10 kDa MWCO filter), passed over a desalting G25 column in Buffer C (50 mM EPPS, 1 mM DTT, 10 μM PLP, pH 8.0), and finally, reconcentrated. To minimize freeze/thaw cycles, LAM is aliquoted into several anaerobic vials (with compressible o-rings and screw-tops) and flash frozen in liquid nitrogen prior to storage at −80°C. Concentration can be assessed by Bradford assay or UV–vis absorbance, wherein $\varepsilon_{280} = 7.63 \times 10^4$ subunit M^{-1} cm^{-1} (Wang & Frey, 2007a).

2.3 Enzyme Activity Assays

2.3.1 Coupled Enzyme Activity Assay for PFL-AE

We can probe PFL-AE activity through a coupled assay that mimics a small segment of the citric acid cycle; using minimal amounts of PFL and PFL-AE, this assay indirectly quantifies PFL-AE activity by measuring malic dehydrogenase catalyzed NADH formation via UV–vis spectroscopy (Brush, Lipsett, & Kozarich, 1988; Conradt et al., 1984). For this coupling assay, two solutions are prepared, a coupling mixture containing coupling enzymes and substrates, and an activation mixture containing PFL-AE and PFL; once combined, activated PFL (PFL_a) converts pyruvate and CoA to formate and acetyl-CoA (Fig. 10). To couple the PFL-catalyzed reaction to the spectroscopically observable production of NADH, two citric acid cycle enzymes are included; malic dehydrogenase replenishes oxaloacetate from malate while reducing NAD^+, and citrate synthase catalyzes the conversion of acetyl-CoA and oxaloacetate to citrate.

As the PFL glycyl radical and the PFL-AE cluster are extremely susceptible to oxygen, the entire assay is carried out under anaerobic conditions ($O_2 < 1$ ppm). All solutions are made anaerobic before entering the anaerobic chamber, and the quartz spectroscopy cuvette is fitted with a septum (Tuf-bond Teflon disks 72) in a screw cap to keep the contents anaerobic when outside the anaerobic chamber.

The coupling mix is prepared in anaerobic coupling buffer (100 mM Tris, pH 8.1) and contains NAD^+ (3 mM), CoA (55 μM), bovine serum albumin (0.05 mg/mL), pyruvate (10 mM), malate (10 mM), DTT (10 mM), citrate synthase (40 U), and malic dehydrogenase (600 U). The activation mix is prepared in anaerobic activation buffer (100 mM Tris, 100 mM KCl, pH 7.5) and contains oxamate (10 mM), DTT (8 mM), PFL-AE (0.05 μM), PFL (5.0 μM), SAM (0.1 mM), and 5-deazariboflavin (25 μM). To guard against unintended reduction, the 5-deazariboflavin is

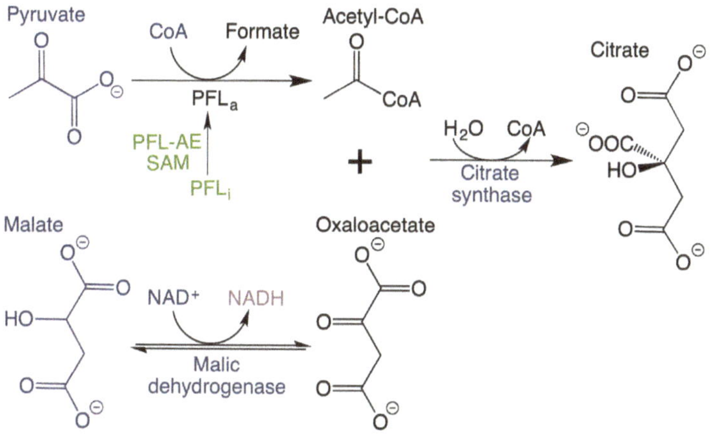

Fig. 10 PFL-AE coupling assay scheme. Inactive PFL (PFL_i) is activated by PFL-AE and SAM to generate active PFL (PFL_a) and enable conversion of pyruvate and CoA to formate and acetyl-CoA. Catalyzed by citrate synthase, acetyl-CoA reacts with oxaloacetate (replenished by the reversible reaction catalyzed by malic dehydrogenase) to produce CoA and citrate. Monitored by UV–vis absorbance (A_{340}), PFL-AE activity is measured by malic dehydrogenase catalyzed NADH (*purple*) production. As the assay requires mixing of two solutions, this scheme illustrates select components of the coupling solution (*blue*) and activation solution (*green*).

added last and the activation mix is covered with foil until ready to initiate the reaction.

To start the assay, the activation mix is placed in a glass or quartz tube (shortened EPR or NMR tube) that is submerged in an ice bath with the tube ~5 cm from a 500 W halogen lamp; multiple activation mixes can be photoreduced simultaneously. In order to minimize air pockets that would introduce O_2 into the chamber, the ice for the ice bath is bathed in liquid nitrogen, and then crushed prior to bringing it in to the anaerobic chamber. The reaction is initiated by turning on the lamp to start deazariboflavin-mediated photoreduction. At desired times (starting as early as 1 min and going as late as 60 min), an aliquot (generally 5 μL) is removed from the activation mix and carefully placed on the lid of a cuvette that contains 895 μL of the coupling mix. The cuvette is carefully sealed and placed in a 30°C water bath to equilibrate for 30 s. To initiate the analysis of active PFL, the cuvette is inverted several times and a kinetic scan at 340 nm (0.1 s scan, 1.5 min cycle) is run to measure the rate of NADH production. A control reaction is also run in the same way by adding 5.0 μL of unreduced activation mix to 895 μL of coupling mix.

To determine the specific activity of PFL-AE, the units of PFL activity must first be calculated, and then converted to AE-specific activity. To determine the units of active PFL (PFL_a) in the activation assay, Eq. (1) incorporates the activity assay slope (AA slope), the path length of the cuvette (cm), the extinction coefficient for NADH ($\varepsilon_{340} = 6.2\,\text{m}M^{-1}\,\text{cm}^{-1}$), the assay volume in the cuvette (V_c) of the combined coupling and activation mixes, the total volume of activation mix that was photoreduced ($V_{\text{AM Total}}$), the volume of the activation mix added to the cuvette lid ($V_{\text{AM Lid}}$), and a few conversion factors. PFL_a units are reported in μmol/min.

$$\frac{\text{AA slope}\left(\frac{\text{AU}}{\text{s} \times \text{cm}}\right)}{\varepsilon_{\text{NADH}}\left(\frac{6.2\text{AU}}{\text{m}M \times \text{cm}}\right)} \times V_{\text{C}}(L) \times \left(\frac{60\text{s}}{\text{min}}\right) \times \left(\frac{10^3\ \mu\text{mol}}{1\ \text{mmol}}\right) \times \left(\frac{V_{\text{AM Total}}(L)}{V_{\text{AM Lid}}(L)}\right) = \text{PFL}_a\text{U}\left(\frac{\mu\text{mol}}{\text{min}}\right) \quad (1)$$

The specific activity of PFL-AE (SA_{AE}) is calculated as shown in Eq. (2), using the relationship that 35 $PFL_aU = 1$ nmol PFL active sites (PFL_{AS}) (Frey et al., 1994) and 1 U AE is the amount of enzyme activating 1 nmol PFL per min. The illumination time (*hν* time) and the total milligram of AE in the activation mix (AE_{AM}) also appear in this calculation.

$$(\text{PFL}_a\text{U})\left(\frac{1\ \text{nmol PFL}_{\text{AS}}}{35\text{PFL}_a\text{U}}\right)\left(\frac{AEU(\mu\text{mol} \times \text{min}^{-1})}{1\ \text{nmol PFL} \times \text{min}^{-1}}\right)\left(\frac{1}{h\nu\ \text{time}(\text{min})}\right)\left(\frac{1}{\text{AE}_{\text{AM}}(\text{mg})}\right) = \text{SA}_{\text{AE}}\left(\frac{U}{\text{mg}}\right) \quad (2)$$

The PFL-AE assay is challenging due to the complexity and the multiple components involved. Care must be taken to ensure that the coupling enzymes have the needed specific activity and that fresh DTT is used. Unexpectedly, low activities can also be due to oxygen contamination, as even minimal O_2 contamination can quench the glycyl radical of activated PFL. Other causes of low measured activity could be partial [4Fe–4S] cluster content of PFL-AE, or poor photoreduction due to overheating or improper illumination geometry. Finally, for accurate activity measurements, it is important to perform the reaction analysis as soon as possible after removing an aliquot from the activation mix.

2.3.2 EPR-Based Activity Assay for PFL-AE

An alternative approach to measuring PFL-AE activity is a direct assay monitoring the production of glycyl radical over time using EPR spectroscopy; this assay is effective and simpler, however requires more enzyme if experiments are performed with X-band EPR spectroscopy. To carry out this assay, a mixture containing PFL (50–200 μ*M*), PFL-AE (1–10 μ*M*), 20 m*M* oxamate, 2 m*M* SAM, and 100 μ*M* 5-deazariboflavin is prepared in Tris (100 m*M*), KCl (100 m*M*), pH 7.5 buffer. This mixture (~180 μL) is illuminated in an EPR tube as in a typical photoreduction (Section 2.6) for times ranging from 5 min to 1 h. After photoreduction, the solution is immediately frozen. EPR spectra are recorded in an X-band instrument at 60 K, 20 μW, and 5.054 G modulation amplitude. Under these conditions, the glycyl radical is centered at $g = 2.007$ with a splitting of 17 G. In order to quantify the glycyl radical, signal intensity is compared to a $K_2(SO_3)_2NO$ (1.04 m*M*) standard (Aasa & Vänngård, 1975; Murib & Ritter, 1952). Plotting the glycyl radical spin concentration as a function of illumination time will give the PFL-AE activity directly.

2.3.3 Activity Assay for LAM

As LAM activity assays generate beta-lysine from alpha-lysine, analysis requires distinguishing between these isomers; thus, detection involves either radioactive assessment of these molecules (Ballinger, Reed, & Frey, 1992; Chirpich et al., 1970) or a derivatization based on the Edman degradation method of polypeptides and high-performance liquid chromatography (HPLC) analysis (Miller, Bandarian, Reed, & Frey, 2001; Ruzicka & Frey, 2010; Wu et al., 1995). For both approaches, an initial incubation is followed by the activity assay; and as with other radical SAM enzymes, the assay is carried out under strict anaerobic conditions.

For activity assessment via radiolabeled substrate (Ballinger, Reed, & Frey, 1992; Chirpich et al., 1970; Wu et al., 1995), LAM is reductively incubated, transferred to the appropriate buffer, and assayed for activity. The initial reductive incubation mix is composed of ferrous ammonium citrate (0–4.5 m*M*) or FAS (0–2 m*M*), dihydrolipoate (87–91 μ*M*), PLP (2.1–2.2 m*M*), cysteine (0–10 m*M*), lysine (80 μ*M*), DTT (0.8 m*M*), sodium dithionite (6 m*M*), EPPS (20–200 m*M*), pH 8.0, and ~14–270 μ*M* LAM, incubated for 4 h at 37°C. After this activation, the solution can be passed over a Sephacryl S200 column (e.g., 2.5 cm × 15 cm), dialyzed, or buffer exchanged into 30–42 m*M* EPPS, 1 m*M* DTT, 10 μ*M* PLP, 0–20 μ*M*

FAS, pH 8.0, whereafter it is concentrated (via Amicon YM10, 10 kDa MWCO centrifugation filters) to ∼1 m*M* and can be frozen and stored at −80°C. This LAM solution (prepared as described earlier, 3 μL, ∼200 μ*M*) is combined with a cofactor-radiolabeled substrate mixture (12 μL) to achieve the following final concentrations: L-U^{14}C-lysine (0–180 m*M*, 0–0.045 μCi/μmol), cysteine (0–1.2 m*M*), SAM (1.2 m*M*), pH 8.0, and sodium dithionite (20 m*M*). To assess the formation of beta-lysine, radiolabeled assays can also be performed in Tris buffer (60 m*M*) pH 8.0. The reaction is quenched after 30 s by adding 25 μL of 0.2 *M* formic acid. Alpha- and beta-lysine are separated by paper electrophoresis. Activity can be observed to be 28–38 IU mg^{-1} (Ballinger, Reed, & Frey, 1992; Chirpich et al., 1970; Ruzicka & Frey, 2010; Wu et al., 1995).

An alternate method for distinguishing alpha- and beta-lysine requires derivatization using phenylisothiocyanate (PITC) and separation via reverse-phase HPLC (Heinrikson & Meredith, 1984; Miller et al., 2001; Ruzicka & Frey, 2010; Wu et al., 1995). The initial reductive incubation mix is composed of ferrous ammonium citrate (0–4.5 m*M*) or FAS (0–2 m*M*), dihydrolipoate (0.08–91 m*M*), PLP (2.1–2.2 m*M*), cysteine (0–10 m*M*), lysine (80 μ*M*), DTT (0.8 m*M*), EPPS (20–200 m*M*), pH 8.0, and ∼14–270 μ*M* LAM, incubated for 3–4 h at 37°C. Under anaerobic conditions, an assay solution includes LAM (85 μ*M*), EPPS (80–340 m*M*), SAM (0.16–1.3 m*M*), sodium dithionite (0.7–2.3 m*M*), L-lysine (1–200 m*M*), pH 8.0, 37°C; at specific intervals under 30 min, aliquots (30 μL) of the reaction are quenched with perchloric acid (2 N, 12 μL) and centrifuged. To derivatize samples, assay aliquots (10 μL, 100 nmol lysine) are dried to a solid and incubated at room temperature with a coupling buffer (acetonitrile:pyridine:triethylamine:water (10:5:2:3), 100 μL) and PITC (5 μL); the solution is then redried and resuspended in 250 μL of ammonium acetate (50 m*M*), water, or water:acetonitrile (7:2). To separate derivatized samples via HPLC, samples (25–250 pmol lysine) are applied to an analytical C8 column under a linear 1.25% min^{-1} gradient (70% A to 70% B) at 1 mL min^{-1}, wherein Buffer A is ammonium acetate (50 m*M*, pH 6.8) and Buffer B is ammonium acetate (100 m*M*, pH 6.8) in a water:acetonitrile:methanol (23:22:5) solution. Recording absorbance at 254 nm, beta-lysine elutes at 11.5 min, and alpha-lysine elutes at 13 min (Miller et al., 2001).

2.4 Assessing Cluster Content via Iron and Protein Assays

As [4Fe–4S] cluster content is necessary for functional radical SAM enzyme catalysis, determining this cluster content is critical and relies on identifying both protein concentration as well as iron number per protein. To gauge protein concentration, Bradford assays are performed (Bradford, 1976); this time sensitive assay requires the standards and samples be run at the same time. To understand "loading" or ISC content of the protein, iron assays are performed via the colorimetric method (Fish, 1988) or via an atomic absorption spectrometer; a value of four irons per protein, for an enzyme that is expected to house one [4Fe–4S] cluster, suggests the majority of the clusters are [4Fe–4S] clusters without further spectroscopic characterization. This iron content assessment is performed after radical SAM enzyme purification to identify if an in vitro ISC reconstitution (Section 2.5) is needed and is particularly helpful as EPR spectroscopy cannot detect diamagnetic $[4Fe–4S]^{2+}$ cluster species.

2.5 In Vitro Reconstitution of ISCs

As discussed in Section 1, radical SAM enzymes require the presence of [4Fe–4S] clusters to accomplish catalysis. In vivo, these clusters are assembled by the help of ISC and sulfur assimilation (SUF) multicomponent ISC systems (Wollers et al., 2010) and though organisms can use either of these systems, the ISC machinery encoded by the *isc* operon is primarily responsible for cluster assembly; however, in conditions of insufficient iron or when cells are under oxidative stress, the SUF machinery is recruited to generate and repair clusters (Py & Barras, 2010). Cysteine desulfurase enzymes, which catalyze the conversion of cysteine to alanine and sulfide, generate the inorganic sulfur for ISCs; these FeS clusters are built on a scaffold protein prior to their transfer to target proteins such as radical SAM enzymes. How well the cluster assembly machinery systems work during *E. coli* growth and gene overexpression significantly affects in the cluster content of the radical SAM enzyme after purification. Although enzymes such as PFL-AE can be isolated with near-full cluster occupancy, in cases where the iron loading is lower than 4.0 irons per protein subunit, the [4Fe–4S] clusters are chemically reconstituted in vitro.

To increase the number of these catalytically relevant [4Fe–4S] clusters, chemical reconstitution is accomplished by addition of exogenous sulfide and iron. During reconstitution, a color change (light brown to dark brown) can be observed as iron and sulfide are being incorporated into the protein active site; after removal of excess iron and sulfide, UV–vis spectroscopy can quantify the increase in cluster content as judged by the ligand-to-metal charge transfer (LMCT) bands characteristic of ISCs (Fig. 11A). Additionally, EPR spectroscopy may be employed to further identify clusters

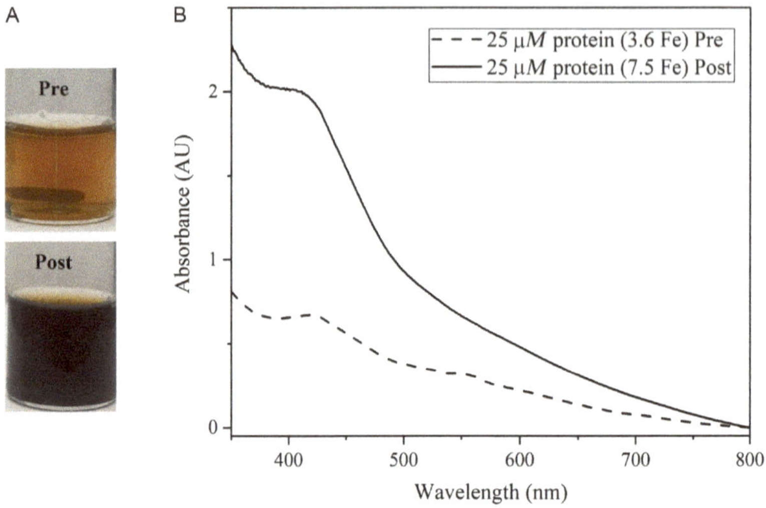

Fig. 11 UV–vis spectroscopy of a radical SAM enzyme pre- and postreconstitution. (A) Radical SAM enzyme solutions pre- and postreconstitution demonstrate iron–sulfur cluster content changes can be observed with the human eye. (B) The UV–vis absorbance spectrum of an as-purified radical SAM enzyme (*dashed line*) illustrates iron–sulfur cluster content; however, after reconstitution, this same radical SAM enzyme has significantly increased iron–sulfur cluster content (*solid line*) as demonstrated by an increase in intensity of the LMCT bands between 400 and 500nm.

types (Section 2.7.2). The following protocol describes a generic reconstitution for radical SAM enzymes. Owing to FeS cluster susceptibility to degradation upon exposure to oxygen, all steps should be performed in an anaerobic chamber, and are generally carried out on ice to maintain protein integrity.

Prior to reconstitution, the protein is adjusted to a concentration generally between 100 and 150 μ*M* using a buffer appropriate to the enzyme, and then DTT is added from a freshly made stock solution to a final concentration of 5 m*M* DTT. The protein is allowed to sit for 5 min, and then iron is added from a 10 m*M* stock of Fe^{3+} (typically ferric chloride) via 10–20 aliquots dispensed at intervals over an hour to a final concentration that is 6 × more than the protein concentration. This slow addition maximizes distribution of the iron and minimizes protein precipitation. Following iron addition, sulfide is added from a stock solution of 50 m*M* $Na_2S{\cdot}9H_2O$ in a similar manner and to the same final concentration as the iron. This mixture is incubated on ice with gentle stirring for 3–12 h, depending on the radical SAM enzyme. After incubation, the mixture should look visibly darker brown than before reconstitution (Fig. 11B); however, a solution that appears black often has significant FeS particulates and/or adventitiously bound iron that can be deleterious. After incubation, the protein solution is centrifuged (14,500 × *g*, 10 min) to remove precipitated protein and FeS particulates, and the supernatant is concentrated (10 kDa MWCO Amicon spin filters, 4000 × *g*) and run over a desalting column to remove free iron, sulfide, and FeS particulates. After this gel filtration step, the enzyme solution is concentrated again, aliquoted, and flash frozen in liquid N_2, prior to stored at −80°C.

To evaluate the reconstitution success, pre- and postreconstitution samples are examined via UV–vis spectroscopy (350–700 nm) to check for increase in absorbance at wavelengths associated with LMCT bands typical for ISCs (Fig. 11; Section 2.7.1). Further characterization can be achieved through low temperature (10–50 K) EPR (Fig. 13; Section 2.7.2) and/or Mössbauer spectroscopies.

2.6 Reduction of ISCs

In radical SAM enzymes, the [4Fe–4S] cluster must be reduced from the resting 2+ state to the catalytically active 1+ state prior to SAM cleavage (Fig. 1). The reductant in vivo is one of the several electron transfer proteins, however in vitro chemical reducing agents are generally used. This section addresses two different approaches to reducing the [4Fe–4S] cluster: chemical reduction via dithionite and photoreduction via 5-deazariboflavin. Each approach has advantages, and the choice of reducing method depends on the particular enzyme, experiment, and objectives.

The advantage of dithionite reduction is the convenience and ease with which this chemical can reduce clusters. The addition of dithionite, in several fold excess over the [4Fe–4S] cluster content, quickly reduces these clusters requiring only brief incubation and therefore allowing swift sample preparation. The disadvantage of dithionite reduction is that this small molecule remains in solution, providing a strongly reducing environment that is not always compatible with particular experiments. In the specific case of radical SAM enzymes, this reducing environment can result in the sometimes problematic nonproductive SAM cleavage, which is the reductive cleavage of SAM to produce methionine and dAdo in the absence of substrate. Furthermore, in certain instances dithionite can react with radical species, and therefore lead to artifactual results (Dong et al., 2016).

The benefit of 5-deazariboflavin-mediated photoreduction is that reducing equivalents are present only during illumination and are removed by simply putting the sample in the dark. Under light and in the presence 5-deazariboflavin, the $[4Fe–4S]^{2+}$ clusters can be reduced to the 1+ state, however once the light is removed and catalysis is initiated, no exogenous sources of reduction exist near the active site. The control inherent in this technique minimizes off-pathway reactions and experimental artifacts.

While this diminished potential for artifacts is invaluable, the set-up for photoreductions requires more investment than a dithionite reduction. The most substantial initial work is in the synthesis of 5-deazariboflavin, which is described elsewhere (Ashton, Brown, & Tolman, 1978; Janda & Hemmerich, 1976; Smit, 1986). Once the 5-deazariboflavin is in hand, photoreduction can be carried out in a strict anaerobic environment (<1 ppm O_2). Typically, photoreductions are performed for about an hour and use a protein concentration between 50 and 500 μ*M*, and 5-deazariboflavin concentrations typically ranging between 10 and 50 μM. Photoreduction time may vary with each enzyme, and photoreductions at higher enzyme concentrations can be hampered by self-absorption and therefore may require longer illumination times. If the protein is not in Tris buffer, then some Tris buffer must be added as a sacrificial electron donor. To carry out the photoreduction, the enzyme solution is placed in an EPR or NMR tube in an ice water bath, and photoreduction is initiated by sample illumination with a standard 500 W halogen (Xe) lamp. To minimize heat and IR radiation damage during photolysis, the EPR or NMR tubes containing the enzyme solutions are placed in the ice water bath in a glass beaker kept ~5 cm from the lamp, with ice being replenished as needed. The ice for this ice water bath is brought into the anaerobic chamber after bathing it in liquid nitrogen and breaking it apart with a metal spatula; when the ice appears more opaque than translucent, it can be brought into the anaerobic chamber while maintaining O_2 levels below 1 ppm.

After photoreduction, several details must be considered in order to obtain optimal results. As the $[4Fe–4S]^{+}$ cluster may begin to reoxidize once the photoreduction ends, the sample must be swiftly frozen or assayed. If the exact photoreduction time is critical, as in activity assays, the sample must be isolated from light by foil or a secondary container; though this reduction will be minimal in comparison with the halogen lamp, the 5-deazariboflavin in the sample can continue to reduce in the presence of ambient light. Congruently, to maintain integrity of the 5-deazariboflavin stock, it should remain protected from light.

Regardless of the reduction technique (chemical or photoreduced), the intensity of the dark brown enzyme coloration should diminish to a paler

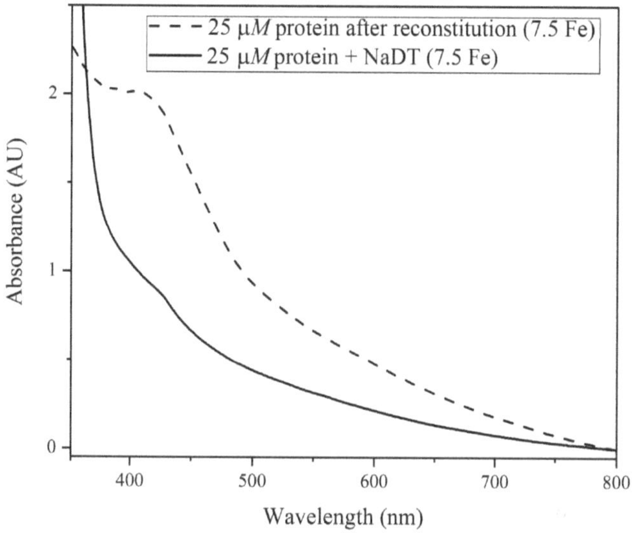

Fig. 12 UV–vis spectroscopy of FeS clusters pre- and postreduction. The UV–vis absorbance spectrum of a reconstituted radical SAM enzyme (*dashed line*) displays iron–sulfur cluster content that can be associated with a $[4Fe–4S]^{2+}$ cluster state; after partial reduction, the LMCT bands diminish, illustrating the oxidation state change in clusters to the $[4Fe–4S]^{+}$ state (*solid line*).

brown during the course of the reduction. UV–vis and EPR spectroscopies (Figs. 12 and 13; Sections 2.7.1 and 2.7.2) can inform on the success of either reduction process.

2.7 Spectroscopic Characterization of Radical SAM Enzymes

As radical SAM enzymes need a [4Fe–4S] cluster to function properly, spectroscopic tools to identify and characterize these clusters are critical. Fortunately, these clusters are spectroscopically rich, with EPR, UV–vis, CD, and Mössbauer spectroscopies providing important insights into cluster type, oxidation state, and coordination environment. However, a key preliminary tool is simply our eyes; the LMCT bands in the visual region lead to the characteristic brown or tan coloration that can visually report on FeS cluster content. UV–vis absorption spectroscopy (Figs. 11 and 12) and EPR spectroscopy (Fig. 13) provide more detailed and quantitative information as to cluster content and type. Additionally, CD spectroscopy in the visible region can inform on cluster content (Mapolelo, Zhang, Naik, Huynh, & Johnson, 2012), and Mössbauer spectroscopy has the advantage of reporting on all types of cluster species. This section will focus primarily on UV–vis absorbance and EPR spectroscopies as tools for further cluster analysis.

2.7.1 UV–Visible Spectroscopy of ISCs

Owing to the sulfur-to-iron LMCT bands inherent to ISCs, and the changes that occur to these bands with oxidation, reduction, or change in cluster type, UV–vis spectroscopy is a useful tool for characterizing proteins with ISC content. As these protein solutions are generally tan to reddish-brown in color, broad absorption features in the 350–700 nm region can be easily observed (Figs. 11 and 12). Due to the breadth of these features and the similarities between different cluster types, UV–vis spectroscopy is not generally considered as a means to unequivocally determine all cluster types present; however, it can provide valuable insight regarding presence of certain cluster types and oxidation states.

In general, $[4Fe–4S]^{2+}$ clusters have high absorbance around 420 nm, and the majority of spectroscopic analyses of radical SAM enzymes report a λ_{max} around 420 nm, although λ_{max} values at 410 or 400 nm are reported in some cases (Broderick et al., 2014). While [2Fe–2S] clusters also absorb in this 400–500 nm region, they can also have additional shoulders at lower energy (550–600 nm) where [4Fe–4S] clusters typically have little absorption.

UV–vis spectroscopy can be useful for measuring changes in these LMCT bands upon cluster oxidation and reduction. For example, reduction of $[4Fe–4S]^{2+}$ clusters results in a significant decrease in intensity of the visible absorption features, such that even to the naked eye the color of the protein appears to "bleach" with reduction. Thus this decrease in visible absorption upon reduction is a useful indicator of reduction of the cluster (Fig. 12).

2.7.2 EPR Spectroscopy of ISCs

EPR spectroscopy can be a more definitive spectroscopic tool for determining cluster content; however, the disadvantage of EPR spectroscopy is that it only reports on paramagnetic species, such as $[4Fe–4S]^{+}$, $[3Fe–4S]^{+}$, and $[2Fe–2S]^{+}$ clusters. Regardless, this technique is quite useful for radical SAM enzymes as the $[4Fe–4S]^{+}$ cluster is the catalytically relevant state. Typically, continuous wave EPR spectroscopy of FeS clusters is carried out in the X- or Q-band (~9.5 and 35 GHz, respectively), at temperatures below 30 K due to the relaxation properties of ISCs. The predominant signals for $[4Fe–4S]^{+}$, $[3Fe–4S]^{+}$, and $[2Fe–2S]^{+}$ clusters appear in the same region of the spectrum near $g=2$; however, these cluster species can often be distinguished based on lineshape (Fig. 13) and relaxation properties, as illustrated by temperature and power dependences of the signal intensity.

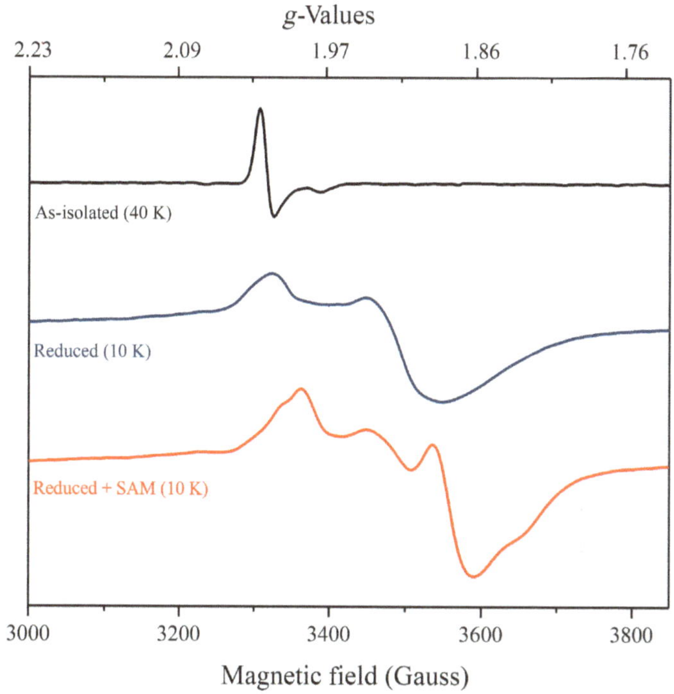

Fig. 13 EPR spectroscopy of FeS clusters in radical SAM enzymes. The *top spectrum* (*black*) is of an as-purified enzyme and illustrates an isotropic $[2Fe–2S]^+$ cluster signal shown at 40K, T_{opt}. The *middle spectrum* (*blue*) is of a reduced enzyme and displays an axial $[4Fe–4S]^+$ cluster signal. The *lower spectrum* (*red*) is of a reduced enzyme with exogenous SAM added and displays two overlapping axial $[4Fe–4S]^+$ cluster signals, suggesting cluster content with and without SAM bound. The lower two spectra are shown at 10K, as this temperature is the T_{opt} for these $[4Fe–4S]^+$ cluster signals.

To examine the temperature dependence of EPR signals, the power and microwave frequency are held constant, and EPR spectra are recorded at several temperatures. Often [2Fe–2S] clusters relax more slowly than [4Fe–4S] and [3Fe–4S] clusters due to the fewer accessible electronic states; thus, [2Fe–2S] clusters can be observed at higher temperatures, and often the signal reaches maximal intensity between 35 and 50K. In contrast, a $[4Fe–4S]^+$ cluster signal displays maximum signal intensity between 12 and 20K, while the maximum signal intensity of $[3Fe–4S]^+$ clusters is in a temperature region between that of $[2Fe–2S]^+$ and $[4Fe–4S]^+$ clusters. Thus, collecting EPR data at a range of temperatures can help to distinguish between these three cluster types.

Power-dependence studies are carried out by collecting EPR data over a range of powers while the temperature is held constant, generally at T_{opt} (the temperature with the maximum signal intensity). Among many advantages, power-dependence studies may distinguish between two clusters with nearly identical temperature profiles and similar or overlapping *g*-values, such as two different $[2Fe–2S]^+$ clusters.

In order to accurately assign *g*-values and to help deconvolute overlapping signals, simulations of the EPR spectra are often performed using EasySpin software (http://www.easyspin.org/) (Stoll & Schweiger, 2006). This graphical software allows the user to set several constraints, such as *g*-values, to simulate the observed signal and identify the multiple cluster signals present in a given spectrum. For example, these simulations can be particularly helpful for distinguishing the $[4Fe–4S]^+$ cluster signal from the $[4Fe–4S]^+$ cluster with SAM bound, if both happen to be present in one spectrum.

EPR spectroscopy of radical SAM enzymes in the as-purified state often reveals the presence of $[2Fe–2S]^+$ (*g*-values of ~ 2.01, 2.00, 1.96 at 30 K) and/or $[3Fe–4S]^+$ (*g*-values of ~ 2.03, 2.01, 1.99 at ~15 K) clusters (Fig. 13, top), although these are frequently minor components in a protein that contains primarily EPR-silent $[4Fe–4S]^{2+}$ clusters. The more the enzyme is exposed to trace oxygen during purification and/or reconstitution, the more of these [3Fe–4S] and [2Fe–2S] cluster states are likely to be observed in the EPR spectrum, as oxidation of $[4Fe–4S]^{2+}$ clusters, particularly those that are site-differentiated like the radical SAM clusters, can lead to oxidative degradation to these cluster states. The reduced $[4Fe–4S]^+$ state is unlikely to be observed in as-isolated protein, as the low reduction potential for radical SAM clusters means that strong reductants are required to achieve the 1+ state. Upon reduction with dithionite or photoreduction with 5-deazariboflavin, typically a $[4Fe–4S]^+$ cluster signal appears (*g*-values of ~ 2.03, 1.92, 1.90 at ~10 K, Fig. 13, middle). Addition of SAM to a reduced sample often perturbs this EPR spectrum lineshape as SAM binding slightly alters the electronics of the $[4Fe–4S]^+$ cluster (*g*-values of ~ 2.00, 1.88, 1.84 at ~10 K, Fig. 13, bottom). EPR spectroscopy can thus provide evidence for SAM binding to the enzyme-bound [4Fe–4S] cluster.

Knowing the concentration of $[4Fe–4S]^+$ spins per protein is valuable because it reflects the amount of catalytically active enzyme. To quantify the percent of catalytically active $[4Fe–4S]^+$ state in a particular sample, the sample is compared to that of a standard of known spin quantity, often a copper standard (Cu(II)-ethylenediaminetetraacetic acid (EDTA)).

Specifically, a standard calibration curve is constructed by taking the double integration of the standard's derivative shaped EPR signals at multiple concentrations with the applied correction factor to account for how the spread of *g*-values affects the signal intensity (Aasa & Vänngård, 1975); the double integration of the derivative shaped EPR signal of the sample is then compared to this standard curve to provide a number that reflects the quantity of spins. Furthermore, the sensitivity of EPR spectroscopy means that $[4Fe–4S]^+$ cluster can be detected even if it is present in low quantities; for certain experiments, it is essential to know whether 10% of the protein, or 100%, is present in this catalytically active state.

It may be helpful to keep in mind a few practical considerations regarding the handling of EPR samples. Owing to the expansion of water while freezing, protein samples in EPR tubes must be slowly lowered into liquid nitrogen, or liquid isopentane, over ~10 s. If the tube is lowered too quickly, the top of the sample may freeze before the bottom of the sample, causing the tube to rupture as the remainder expands during freezing. Furthermore, due to increased brittleness of the quartz tubes at low temperature, EPR tubes should be handled carefully when placing them in dewars or inserting them into the EPR cavity. Finally, to minimize oxygen contamination, EPR samples are capped with a rubber septum prior to removing them from the anaerobic chamber, and samples are stored in liquid N_2. Removal of the rubber septum after freezing can sometimes introduce a broad signal associated with oxygen, which can be eliminated by returning the EPR sample to a minimal oxygen environment, recapping, and refreezing.

EPR spectroscopy can be an incredibly valuable tool for FeS cluster analysis, particularly for radical SAM enzymes. Knowing a radical SAM enzyme has high $[4Fe–4S]^+$ cluster content can yield more productive investigations into the function of these magnificent enzymes.

3. SYNTHESIS OF SAM AND SAM ANALOGS

As SAM is essential to the radical SAM enzyme mechanism, securing optimal preparations of this small molecule minimizes experimental artifact. SAM, the progenitor to the deoxyadenosyl (dAdo) radical (Broderick et al., 2014), comprises the nucleotide adenosine and the amino acid methionine and can be enzymatically synthesized (Section 3.2.1) from adenosine 5′-triphosphate (ATP) and methionine. Though SAM can be obtained commercially, synthesis is preferable as this approach generally yields

SAM with fewer racemic contaminants and impurities (Gross, Geresh, & Whitesides, 1983). While SAM can be made in vivo by supplementing yeast cultures with methionine (Tsuchida, Yoshinaga, & Okumura, 1974), we prefer to generate SAM in vitro while monitoring the reaction progress to achieve optimal yields. The relatively simple synthesis of SAM provides not only high quality SAM for a range of experiments but also the flexibility to incorporate isotopic labels from methionine or ATP. The following sections present methods associated with the enzymatic synthesis of SAM, including preparation of isotopically labeled SAM and SAM analogs.

3.1 SAM Synthetase

SAM synthesis requires the enzyme SAM synthetase (EC 2.5.1.6), which is found in *E. coli*. Our lab received a SAM synthetase strain as a generous gift from the Markham lab, and the methods for preparation of SAM sythetase arise from this lab (Markham, DeParasis, & Gatmaitan, 1984; Markham, Hafner, Tabor, & Tabor, 1980). This SAM synthetase gene is on the pk8 plasmid, a pBR322 plasmid derivative containing the *metK* (SAM synthesis) gene. This pk8 plasmid is in the DM22 *E. coli* strain, itself a derivative of *E. coli* EWH154 wherein the entire region of speC, a protogenic exotoxin, has been deleted (Wright & Boyle, 1984). SAM synthetase is generated in abundance from these cells without the necessity for induction or any special treatments. The clarified lysate of SAM synthetase is sufficient for use in the SAM synthesis reaction and thus purification of SAM synthetase is not required.

3.1.1 Growth of the SAM Synthetase Strain

Using sterile technique, a LB agar plate (oxytetracyclin 30 μg/mL final) is streaked from a glycerol stock of the strain described and is incubated at 37°C for ~15 h. A single colony is used to inoculate LB media (50 mL, oxytetracyclin 30 μg/mL) in a 250 mL Erlenmeyer flask and incubated at 37°C overnight (~15 h) with shaking (240 rpm). A small amount of this culture (~3.5 mL per 700 mL) is used to inoculate the large-scale growth (five 2.8 L beveled Fernbach flasks with 700 mL media and antibiotic in each). This culture is grown at 37°C for 12–14 h with shaking (240 rpm). The cells are aerobically harvested by centrifugation (4000 × *g*, 12 min, 4°C); the supernatant is discarded and the resulting cell pellets are flash frozen in liquid nitrogen and stored at −80°C.

3.1.2 Preparation of the SAM Synthetase Lysate

As the SAM synthesis reaction is performed aerobically, lysis of SAM synthetase cells can be done in aerobic conditions. The lysis buffer contains 100 m*M* Tris–HCl, pH 8.0, 50 μg/mL lysozyme, 0.1 m*M* PMSF (dissolved in 1 mL methanol), 1 m*M* EDTA, and trace DNase and RNase. Lysis buffer is added to the cell pellet at a ratio of 3 mL buffer per 1 g of cell paste. The lysis mixture is incubated at ~25°C with gentle stirring for 30 min, followed by sonication (5 min total sonication, 60% amplitude, 15 s on, and 60 s off) on ice. After chemical and mechanical lysis, the lysate is centrifuged (27,000 × *g*, 30 min, 4°C) and the clarified lysate is aliquoted into 1 mL microcentrifuge tubes and frozen immediately in liquid nitrogen. The lysate aliquots are stored at −80°C, and should only be thawed immediately prior to use in the SAM synthesis reaction.

3.2 SAM Synthesis and Purification

Using SAM synthetase, SAM can be synthesized from ATP and methionine, and subsequently purified on a cation exchange column. Since SAM can decompose into methylthioadenosine and other derivatives, efforts are made to minimize the duration between synthesis initiation and final freezing. The following section details how SAM is synthesized and purified.

3.2.1 SAM Synthesis Reaction

The assay components (50 m*M* KCl, 26 m*M* $MgCl_2$, 0.9 m*M* EDTA, 14.4 m*M* ATP, 12.0 m*M* L-methionine, 0.114 m*M* 2-mercaptoethanol, ~0.25 U inorganic pyrophosphatase, and 1 mL SAM synthetase clarified lysate) are mixed in 100 m*M* Tris–HCl (pH 8.0) to a total volume of 10 mL and incubated with stirring for ~18 h at room temperature. The reaction progress is monitored by thin-layer chromatography (TLC) using a silica TLC plate and a mobile phase containing an 8:2:2:1 ratio of butanol, glacial acetic acid, water, and 2 *M* formic acid. Upon completion (Fig. 14A), the reaction is quenched using 1 *M* HCl and centrifuged (38,000 × *g*, 30 min, 4°C). The supernatant is saved to a clean tube and the pellet discarded.

3.2.2 SAM Purification

Prior to SAM purification, the 15S cation exchange column (HR 16/10) is primed with the following wash steps: (1) water (40 mL, 4 mL/min), (2) 1 *M* HCl (20 mL, 4 mL/min), and (3) water (40 mL, 4 mL/min). The supernatant from the centrifuged SAM synthesis reaction is loaded via a Superloop

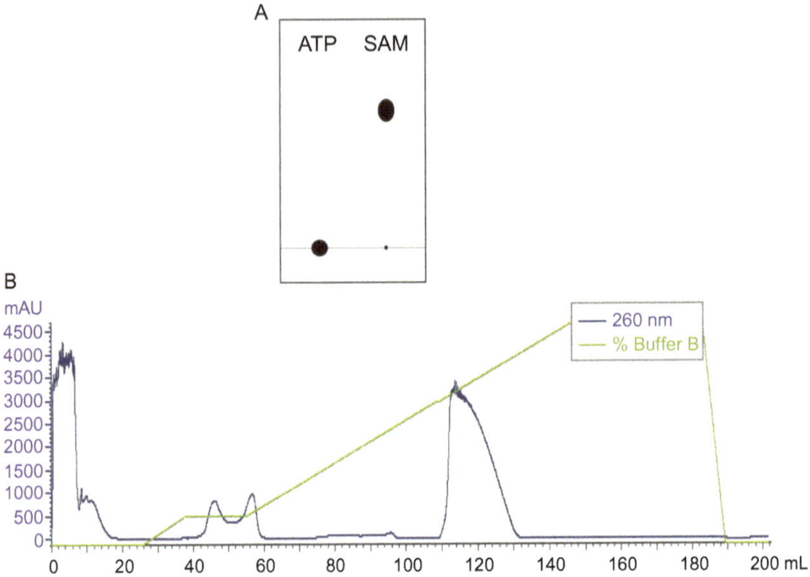

Fig. 14 SAM purification monitors. (A) TLC can illustrate formation of SAM, using a solvent system of butanol:acetic acid:water:2 M formic acid (8:2:2:1) on a silica solid phase. (B) While the SAM elution profile is reproducible, monitoring the UV–vis absorbance during the described purification can be incredibly useful for SAM analogs that can have a slightly altered elution profile.

(10–50 mL) onto the column and SAM is eluted (Fig. 14B) with a combination of isocratic flow and linear gradients (Table 1) using water (solution A) and 1 *M* HCl (solution B). At 2 mL/min, an isocratic elution of water (100% A) is run for 13 min (26 mL) followed by a linear gradient over 6 min to 10% B; another isocratic elution at 10% B is maintained for 8 min, until 54 mL total elution volume, at which point a linear gradient over 60 min to 100% B is applied. While impurities wash off at lower HCl concentration, SAM elutes between 0.35 and 0.55 *M* HCl, corresponding to about 50–60 min, based on the method described (Table 1).

To finalize SAM preparation, SAM is dried to a solid and resuspended in buffer as delineated here. The SAM eluate is concentrated to ~2 mL using a rotary evaporator with a water bath temperature set between 28 and 30°C; to minimize chances of degradation, the duration for which SAM is kept above 4°C is limited as much as possible. After transfer to either a 10-mL Schlenk flask, or a 4-mL round bottom tube (covered with perforated foil) within a Schlenk flask, this concentrated SAM solution is frozen and ready for lyophilization; this vessel should have a sufficiently large

Table 1 SAM purification method. This SAM purification method applies to purification on a Source 15S cation exchange resin in GE, HR 10/10 column with water (Buffer A) and HCl (1 M, Buffer B) and a flow rate of 2 mL/min. After loading the centrifuged SAM synthesis reaction onto the column, an isocratic elution at 0% B is run for 13 min, or 26 mL, followed by a linear gradient over 6 min to 10% B. This isocratic elution at 10% B is maintained for 8 min, until 54 mL total elution volume, at which point a linear gradient over 60 min to 100% B is applied. If this procedure is maintained, SAM will elute around 110 mL.

Elution	Concentration (% solvent B)	End volume (mL)	Time (min)
Isocratic	0%	26	13
Linear gradient	0%–10%	38	6
Isocratic	10%	54	8
Linear gradient	10%–100%	174	60
Isocratic	100%	194	10
Linear gradient	100%–0	202	4
Isocratic	0%	204	20

surface area to lyophilize SAM quickly, but be small enough to minimize SAM loss during resuspension in 200 μL buffer. SAM is lyophilized to a solid overnight and the colorless SAM solid is resuspended in ~200 μL of 100 m*M* Tris at pH 7.0 (or an appropriate assay buffer), in an anaerobic chamber at <1 ppm O_2. Since SAM elutes in HCl, the pH is adjusted to 7.5 with KOH (6 *M*). To minimize thaw/freeze cycles and oxygen exposure, SAM is aliquoted (~25 μL), flash frozen, and immediately stored at −80°C. The final SAM concentration is calculated from the absorbance at 260 nm based on the molar extinction coefficient (ε_{260}) of 15,400 M^{-1} cm^{-1}. Gross et al. (1983) and Shapiro and Ehninger (1966) and typical yields are ~200 μL of a 25–35 m*M* SAM solution. The purity of SAM can be evaluated via mass spectrometry and NMR spectroscopy as needed.

3.3 Isotopically Labeled SAMs

SAM synthesis (Section 3.2) provides facile access to isotopically labeled SAMs by using ^{2}H-, ^{13}C-, and/or ^{15}N-labeled methionine or ATP. The resulting labeled SAMs can be used to trace the fate of specific atoms or probe the structural features of paramagnetic states using EPR and ENDOR spectroscopies.

While methods for generation of these labeled SAMs are similar to that for SAM, the reaction and purification may need to be reoptimized due to differences in the salts/buffer components of the commercially purchased labeled ATP and methionine. Common alterations may include: (1) extended reaction time, (2) higher or lower concentration of HCl required to elute SAM off the cation exchange column, and (3) need to maintain SAM at a lower temperature due to increased susceptibility to degradation. When synthesizing labeled SAMs, therefore, the following approaches are helpful: (1) using TLC (or HPLC) to carefully monitor the SAM synthesis reaction to ensure optimal reaction time, (2) monitoring UV–vis absorbance at 260 nm during purification to ensure the optimal fractions are collected; (3) minimizing the total time between reaction initiation and final freezing of purified SAM; and (4) always keeping SAM cool. These precautions will ensure successful generation of labeled SAM.

3.4 SAM Analogs

As SAM analogs can exhibit altered reactivity and/or spectroscopic characteristics, these analogs have been employed in spectroscopic and mechanistic studies of radical SAM enzymes. The distinct properties of these SAM analogs are due to minor structural differences, such as the presence of a double bond or selenium in place of sulfur. In some cases, such as the synthesis of seleno-SAM, only minor adjustments (akin to those described in Section 3.3) to the standard SAM synthesis (Section 3.2) are required. In other cases, as with the synthesis of 3′,4′-anhydroadenosyl-L-methionine (anSAM), more elaborate procedures are required. Thus, the following sections provide details for the preparation of anSAM.

The SAM surrogate, anSAM, has a double bond in the sugar moiety between the 3′ and 4′ carbons, such that upon cleavage of the S—C(5′) bond, an allylically stabilized radical is formed (Fig. 15). Owing to the stability of this allylic anAdo radical, it reacts with substrate more slowly than the dAdo radical and thus builds up to spectroscopically observable quantities during turnover in some radical SAM enzymes, while remaining competent as an intermediate in catalysis (Horitani et al., 2015; Magnusson et al., 1999).

The following sections discuss methods for preparing anSAM, including the preparation of anATP (Section 3.4.3) and the altered conditions required for the anSAM synthesis reaction (Section 3.4.4). The preparation of anATP starts with the organic synthesis of anhydroadenosine (anAdo), which is

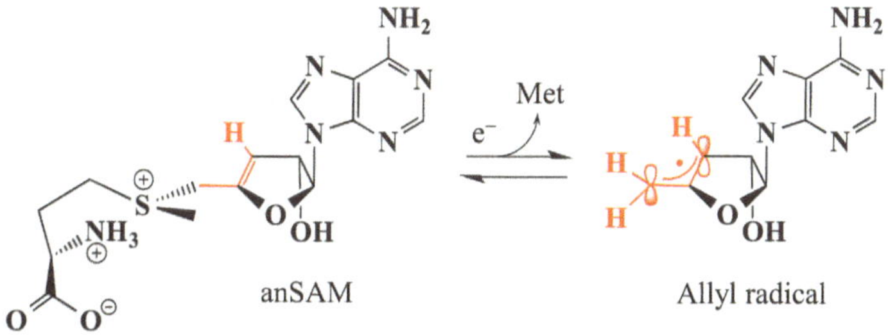

Fig. 15 SAM analog—anSAM. The structure of anSAM is depicted on the left illustrating the double bond in the between the 3′C and 4′C of the ribose; upon a 5′C—S bond cleavage during radical SAM enzyme catalysis, an allylic radical is generated and stabilized across the 3′, 4′, and 5′ carbons (as depicted in *red*) allowing for spectroscopic observation.

described elsewhere (Chandra, Broderick, & Broderick, 2009, 2010; Chandra et al., 2009). Phosphorylation of anAdo to generate anATP (Sections 3.4.1–3.4.3) is performed enzymatically using three enzymes: (1) adenosine kinase (AdoK), to phosphorylate adenosine, (2) adenolate kinase (AK), to phosphorylate AMP, and (3) phosphocreatine kinase (PCK), to phosphorylate ADP. While AK (from rabbit muscle) and the PCK (from rabbit muscle) can be obtained from commercial sources and prepared with minor modifications (Section 3.4.2), as of this writing, AdoK is not easily sourced; thus, AdoK enzyme overexpression and purification are required (Section 3.4.1).

3.4.1 AdoK Preparation

The adenosine kinase gene (*AdoK*), from human T cell cDNA (MOLT-14), is isolated and amplified with the following promoters (*forward*: 5′-CGG GCA GCA TAT GAC GTC AGT CAG AGA AA-3′; *reverse*: 5′-GCG TTC GGA TCC ATC AGT GGA AGT CTG G-3′). Upon amplification, the *AdoK* gene can be ligated into any expression vector employing standard molecular biology techniques. Here, the plasmid construct includes the described gene inserted in a T7-based expression vector pET-14b (Novagen) and is transformed into XL1 Blue supercompetent cells, followed by BL21(DE3)pLysS for overexpression. As AdoK is not air sensitive, all growth and purification steps can be carried out under aerobic conditions.

3.4.1.1 Growth

From glycerol stocks or a new transformation, a culture swab is transferred from the glycerol stock onto an LB/*amp* (50 μg/mL) plate and the plate is grown at 37°C overnight. One colony is removed from this plate with sterile technique and placed in a 250mL Erlenmeyer flask of LB (50mL) with *amp* (50μg/mL) and grown overnight (~15h) with shaking (250rpm). For large-scale growths, media (6L of sterile LB) is prepared in four Fernbach flasks (2.8L). This LB media are warmed at 37°C for an hour with shaking, and then, *amp* (50μg/mL final concentration) along with 10mL of the overnight growth is added to each Fernbach flask. These cultures are grown for 2.5h at 37°C with shaking (240rpm) until the OD_{600nm} reaches ~0.35, at which point the temperature is dropped to 22°C so the culture is near 22°C at induction. Keeping the temperature low at induction minimizes percentage of AdoK protein in inclusion bodies that remain in the pellet after centrifugation; however, the initial growth phase is the best to perform at 37°C to minimize total growth time. Cultures are induced at an OD_{600nm} of 0.5 by adding IPTG (1m*M*, final concentration), and 8h after inoculation (not induction) glucose (4.5g/1500mL) is added to each flask. After induction, the cells are grown for 22h at 22°C with ~240rpm shaking, after which they are harvested by centrifugation (4000×*g*, 10–12min, 4°C). Cell pellets are flash frozen in liquid nitrogen, and stored at −80°C until lysis.

3.4.1.2 Lysis

Lysis buffer (50m*M* Tris–HCl, 7.5 pH, 1% Triton X-100, 0.2mg/mL lysozyme, 1m*M* PMSF, 20μg/mL DNAse, 10μg/mL RNase, 10m*M* magnesium sulfate, and 1m*M* DTT) is added to cell pellet at a ratio of 10mL per gram of cell pellet. Generally, approximately 25g of cell pellet is lysed, corresponding to the amount from 3L of culture. The lysis is performed chilled, in a 4°C fridge or surrounded by ice packs, with a syringe and 18-gauge needle used to help homogenize the solution. In addition to chemical lysis, use of freeze/thaw cycles yields more protein; thus, the lysis suspension undergoes three freeze/thaw cycles in a 500mL Erlenmeyer flask, with freezing by rotation in liquid nitrogen, and thawing under cold water. Then the solution is centrifuged (38,000×*g*, 30min, 4°C) and the AdoK containing supernatant is saved.

3.4.1.3 Purification

AdoK purification is adapted from previous work (Magnusson et al., 1999; Spychala et al., 1996). The initial purification is via an ammonium sulfate precipitation, in which the first cut at 65% ammonium sulfate (430g ammonium sulfate/1L) precipitates most of the unwanted cellular proteins. After centrifugation (38,000×*g*, 30min, 4°C), the final 100%

cut (266 g ammonium sulfate/1 L) precipitates AdoK, as long as care is taken to add the ammonium sulfate slowly with stirring. A second centrifugation isolates the AdoK in the pellet and this AdoK pellet is resuspended in the least amount of buffer (10 m*M* Tris, 1 m*M* DTT, and pH 7.50) necessary; this volume is usually between 5 and 10 mL if the AdoK protein is from about 25 g of cells (from 3 L of a growth). The resuspended AdoK is then dialyzed against the resuspension buffer to remove residual salt from the ammonium sulfate fractionation. The dialysis proceeds overnight (~15 h) at 4°C in 12–14 kDa MWCO tubing in 1–2 L of buffer; however, due to the high concentration of salt, the AdoK solution volume will increase during dialysis, so care must be taken to leave sufficient dialysis tubing to allow for the expansion.

This dialyzed AdoK solution should be affinity purified immediately. After a HisTrap (5 mL CV) column has been equilibrated with Buffer A (20 m*M* Tris, 10 m*M* imidazole, 100 m*M* KCl, and pH 7.5), the protein is loaded via a 50 mL Superloop at a rate of 0.5 mL/min and eluted with Buffer B (20 m*M* Tris, 500 m*M* imidazole, 100 m*M* KCl, and pH 7.5). To elute the AdoK protein, the following method is used: 0% B (40 mL, eight CVs), 5% B (25 mL, five CVs), linear gradient 5%–80% B (40 mL, eight CVs), and 100% B (25 mL, five CVs). As Fig. 16A illustrates, the AdoK elutes in two peaks starting at around 30%–40% B, or ~182 m*M* imidazole, and all fractions (~12–18) with AdoK are saved for further purification; these fractions are mostly pure (Fig. 16B), but do retain an additional band, around 31 kDa, which renders the AdoK protein inactive without further purification. Upon elution from the HisTrap column, the AdoK fractions are dialyzed into 10 m*M* Tris, 1 m*M* DTT, and pH 7.5 to remove the imidazole. This buffer exchange into a no salt buffer is critical as the next step involves ion exchange chromatography and therefore requires low ionic strength for the protein to adsorb. Once in the low salt buffer, the protein can be frozen if needed.

On a HR 16/10 DEAE Sephacel column equilibrated with new Buffer A (10 m*M* Tris, 1 m*M* DTT, and pH 7.5), the AdoK solution is loaded and then eluted with a new Buffer B (10 m*M* Tris, 1 m*M* DTT, 100 m*M* KCl, and pH 7.5) according to the following method: 0% B (20 mL, one CV), 5% B (20 mL, one CV), linear gradient 5%–80% B (145 mL, 120 min, ~7 CVs), and 100% B (40 mL, two CVs). The increasing ionic strength causes the AdoK to elute off the column around 20%–60% Buffer B as seen in the chromatogram (Fig. 17A). Both peaks, from ~75 to 105 mL and from ~105 to 140 mL (Fig. 17B), exhibit AdoK activity, as tested by assays (described later) that confirm phosphorylation of adenosine.

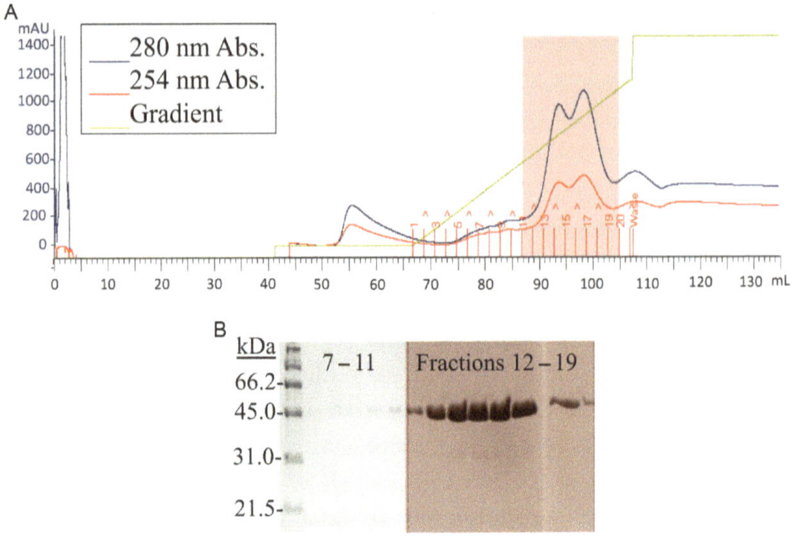

Fig. 16 AdoK affinity purification. (A) AdoK elution profile off the HisTrap™ column has a double peak; all fractions in the *red highlighted section* contain useful AdoK protein and are collected. (B) In the SDS-PAGE of fractions collected, the *red highlighted fractions* show presence of AdoK and appear relatively pure, despite the need for further purification; the BioRad Broad Range Molecular Marker shows weights of 66.2, 45, 31, and 21.5 kDa (*top* to *bottom*) with AdoK protein correlating to ~45 kDa.

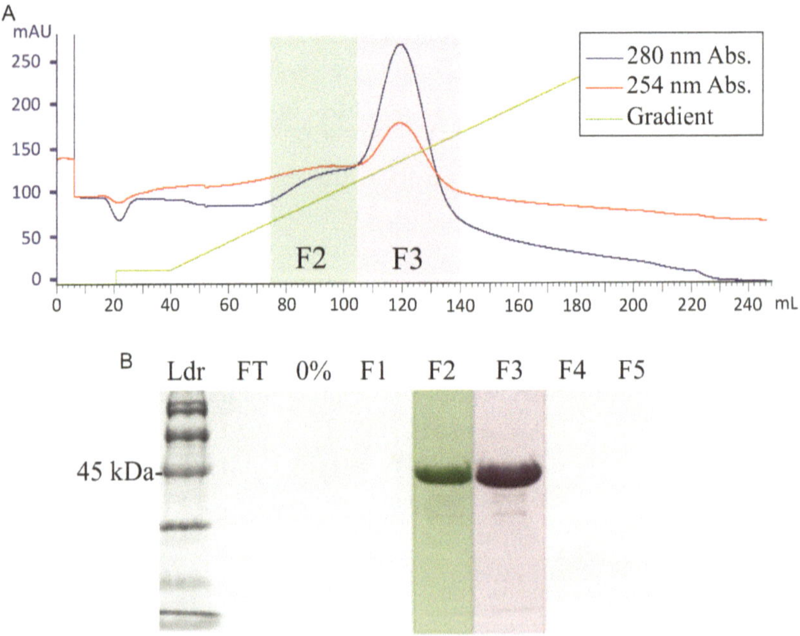

Fig. 17 DEAE sephacel AdoK purification. (A) AdoK elution profile with AdoK fractions highlighted, (B) SDS-PAGE illustration of fraction purity, wherein "Ldr" refers to the same BioRad Broad Range Molecular Marker, "FT" is the flow through during loading of the column, "0%" is the wash with 0% Buffer B, and "F1"–"F5" are fractions. Fractions F2 and F3 both have active AdoK.

3.4.1.4 Activity Assay

The activity of the AdoK is tested to confirm phosphorylation of adenosine to form AMP. The total volume of this is 0.5–1 mL and reaction progress is monitored by HPLC (Fig. 18A and B) or TLC (Fig. 18C) with the same methods described for monitoring the anATP synthesis (Section 3.4.3). The assay components include piperazine-*N*,*N*-bis(2-ethanesulfonic acid) (PIPES) buffer (25 m*M*), $MgCl_2$ (1.25 m*M*), bovine serum albumin (0.01%, w/v), adenosine (1 m*M*), ATP (0.6 m*M*), and AdoK (50 μL in 1 mL total reaction volume). For assays measured by HPLC, aliquots of 100 μL are taken every 3 h. To minimize HPLC column blockages, the assay aliquot is centrifuged (14,000 × *g*, 5 min, 4°C), and the supernatant is applied to a microspin column sufficient to catch AdoK (≤15 kDa MWCO); this flowthrough (15 μL) is then injected onto the HPLC instrument. AdoK activity between 1 and 5 μmol substrate min^{-1} mg^{-1} is sufficient for anSAM preparations.

A

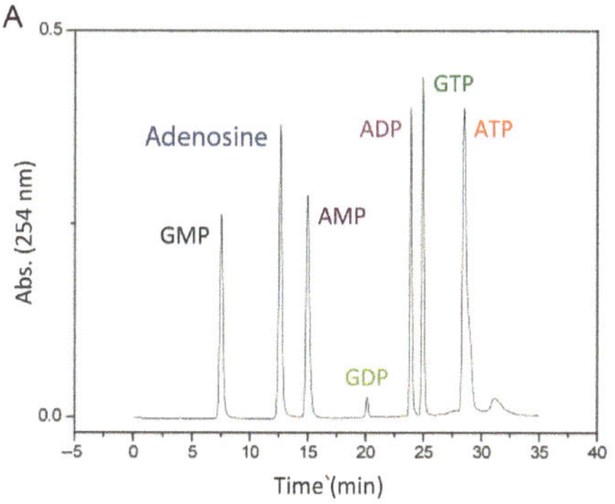

B

Step	Time (min)	Flow rate (mL/min)	%Pump B
1	0	1.2	0
2	1	1.3	0
3	7	1.3	0
4	8	1.2	1
5	9	1.1	2
6	15	1.2	2
7	17	1.2	4
8	19	1.2	7
9	21	1.3	7
10	23	1.3	8
11	25	1.2	8
12	27	1.2	4
13	29	1.1	1
14	31.5	1.0	0

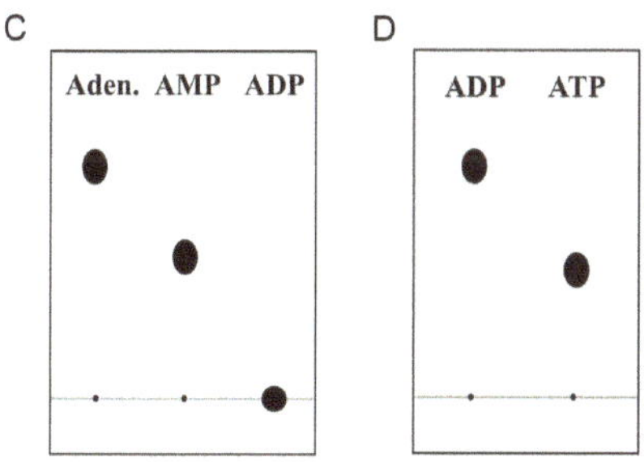

Fig. 18 HPLC and TLC distinguish nucleotides during anATP synthesis. (A) HPLC standards elution profile illustrates clear baseline separation between nucleotides and adenosine; the peak at ~31.5 min is a column impurity. (B) Details of the HPLC method necessary to achieve this separation using the appropriate C_{18} reverse-phase column and buffer system. (C) TLC plate drawing demonstrates distinction between adenosine, AMP, and ADP using a solvent system of formic acid (0.15 M):EtOH (2:3) on a silica solid phase. (D) TLC plate drawing illustrates separation between ADP and ATP using a solvent system of dioxane:isopropanol:NH_4OH:H_2O (4:2:3:4) on a silica solid phase.

3.4.2 Adenylate Kinase (AK) Preparation

AK and PCK phosphorylate AMP and ADP, respectively, and in addition to AdoK, are required to generate anATP; both AK (rabbit muscle) and PCK (rabbit muscle) are available from commercial sources. The PCK solid often can be used, as received, unless high salts are present; however, the AK arrives in concentrated ammonium sulfate and must be dialyzed into conditions close to the assay conditions to minimize conflicts. To do this, AK (200–400 μL) is dialyzed against buffer (25 m*M* PIPES, pH 7.0, 1.25 m*M* $MgCl_2$, 1 mg/mL bovine serum albumin). Owing to the high amount of salt in the AK solution prior to dialysis, the AK concentration must be reassessed after dialysis.

3.4.3 anATP Synthesis and Purification

anAdo is converted to anhydroATP (anATP) through enzymatic synthesis requiring three enzymes: AdoK, AK, and PCK. Usually this reaction is prepared in 20 mL batches at room temperature (19–20°C) and includes: PIPES (25 m*M*), $MgCl_2$ (1.25 m*M*), bovine serum albumin (0.01%, w/v), anAdo (3 m*M*), phosphocreatine (20 m*M*), AdoK (1000 IU), AK (2000 IU), PCK (2000 IU), pH 7.0, and GTP (0.6 m*M*). The anAdo is typically difficult to get into solution, so a little heating (no greater than 42°C) is necessary. As GTP can initiate phosphorylation, it is used here instead of ATP because GTP is not a substrate of SAM synthetase; since ATP and anATP purify together, if ATP were used, then this synthesis reaction would generate a mixture of SAM and anSAM. To maintain reactivity, the pH must remain between 7.0 and 7.5, and be monitored over the full course of the reaction, about 18 h.

Reaction progress is assessed by HPLC (Fig. 18A and B) or TLC (Fig. 18C and D) as delineated here. This HPLC method clearly separates the nucleotides of adenosine and guanosine, in addition to the nucleoside adenosine. This method employs a reverse-phase column (Waters "Symmetry" column [WAT054210], C_{18}, 5 μm, 4.6 mm × 150 mm, reversed phase, 19.1% carbon load, spherical silica with 100 Å pore size, surface area of 335, pH range 2–8, and volume 2.5 mL) with Buffer A as 10 m*M* tetrabutylammonium bromide in 50 m*M* sodium phosphate, pH 6.5, and Buffer B as methanol:isopropanol (1:1, v/v). Though the flow rate varies (Fig. 18B), the column temperature is held at 34°C, and the absorbance is monitored at 254 nm.

An alternate qualitative approach to monitoring the reaction progress is TLC with silica as the solid phase. To distinguish adenosine from AMP and

ADP, a mobile phase of 0.15 *M* formic acid:EtOH (2:3) is used. This set-up will clarify if starting material (adenosine) remains and confirms that AdoK and AK are generating the phosphorylation products, AMP and ADP. To confirm formation of ATP, a mobile phase of dioxane:isopropanol: NH_4OH:H_2O (4:2:3:4) is used; this set-up clearly distinguishes ADP from ATP to determine reaction completion. Both TLC methods use silica gel on aluminum (Silica gel 60/kieselguhr F_{254} EMD 5567-7) as the stationary phase and a handheld UV light is sufficient to see results if the plate is dry (Fig. 18C and D).

Upon completion of the anATP synthesis reaction, the reaction mixture is centrifuged (38,000 × *g*, 10 min, 4°C) and immediately purified using an HR10/30 DEAE Sephadex A_{25} column equilibrated with 10 m*M* NH_4HCO_3, pH 8.0 (50 mL). The reaction is introduced to the column via superloop at 0.5 mL/min. At a flow rate of 3 mL/min, a 20 CV linear gradient from 10 to 500 m*M* NH_4HCO_3 (pH 8.0) is run, eluting anATP around 350 m*M* ammonium bicarbonate. Reaction intermediates can be separated; anAMP and anADP elute around 270 m*M* NH_4HCO_3 and 344 m*M* NH_4HCO_3, respectively, while anATP elutes around 400 m*M* NH_4HCO_3. If GTP initiates the reaction, it elutes around 462 m*M* NH_4HCO_3. The anATP is immediately rotovapped or lyophilized to a solid; to minimize degradation of anATP, the rotovap bath temperature is set to 28°C. The component (anAMP, anADP, and anATP) concentrations can be determined by HPLC (Fig. 18A and B) and/or UV–vis spectroscopic analysis.

3.4.4 anSAM Synthesis and Purification

In a reaction similar to that described for SAM (Section 3.2.1), SAM synthetase can enzymatically transform anATP into anSAM. The optimized anSAM assay components include: EPPS buffer (50 m*M*, pH 8.00), $MgCl_2$ (20 m*M*), L-methionine (5 m*M*), anATP (3.25 m*M*), inorganic pyrophosphatase (~0.25 U), SAM synthetase (1 mL, clarified lysate; see Section 3.1) with a total volume under 20 mL. In contrast to the regular SAM reaction, the anSAM reaction is generally run for only 3–6 h, and should not be run for more than 12 h. This anSAM reaction can be monitored in the same way as a SAM reaction via TLC (Section 3.2.2, Fig. 14A).

Purified following the same procedure as for SAM (Section 3.2), the anSAM is treated slightly differently after purification. To minimize degradation, it is rotovapped to about ~3 mL at a lower temperature (28°C bath) and kept as cold as possible at all times postpurification. Once lyophilized, the anSAM solid should appear white to clearish and be stored at −80°C until it is resuspended in anaerobic buffer in an anaerobic chamber. Resuspension of anSAM in buffer proceeds in a similar manner to that for SAM, including the need for adjusting pH, with care taken to avoid pH extremes. The anSAM concentration is determined by absorbance at 254 nm with the extinction coefficient $\varepsilon_{254} = 15.4\,\text{m}M^{-1}\,\text{cm}^{-1}$. Owing to anSAM's sensitivity to freeze/thaw cycles and the propensity to degrade with mild heating, distributing the anSAM solution into multiple small aliquots (~25 μL) maximizes use.

4. CONCLUSION

Functional radical SAM enzymes require a [4Fe–4S] cluster and SAM be positioned precisely in the active site pocket so that, during catalysis, the dAdo radical is generated in a protected environment in close proximity to the target H-atom of substrate. Here we have outlined our approach to preparing these functional radical SAM enzymes, using as examples two of the most well-characterized radical SAM enzymes, PFL-AE and LAM. While these two canonical radical SAM enzymes use SAM in different ways, as a cosubstrate (PFL-AE) or a cofactor (LAM), the keys to generating productive enzyme are similar and include anaerobic purification, cluster characterization, and appropriate techniques for cluster reduction to the catalytically active state. Methods for synthesizing SAM and SAM analogs are also described herein. These methods should be of use to researchers setting out to investigate one of these fascinating but challenging enzymes.

REFERENCES

Aasa, R., & Vänngård, T. (1975). EPR signal intensity and powder shapes: A reexamination. *Journal of Magnetic Resonance*, *19*, 308–315.

Ashton, W. T., Brown, R. D., & Tolman, R. L. (1978). New routes to 1-deoxy-(3,4-dihydro-7,8-dimethyl-2,4-dioxopyrimido[4,5-b]-quinolin-10(2H)yl)-D-ribitols (5-deazariboflavins). *Journal of Heterocyclic Chemistry*, *15*, 489–491.

Ballinger, M. D., Frey, P. A., & Reed, G. H. (1992). Structure of a substrate radical intermediate in the reaction of lysine 2,3-aminomutase. *Biochemistry*, *31*(44), 10782–10789.

Ballinger, M. D., Reed, G. H., & Frey, P. A. (1992). An organic radical in the lysine 2,3-aminomutase reaction. *Biochemistry*, *31*(4), 949–953.

Baraniak, J., Moss, M. L., & Frey, P. A. (1989). Lysine 2,3-aminomutase. Support for a mechanism of hydrogen transfer involving S-adenosylmethionine. *The Journal of Biological Chemistry*, *264*(3), 1357–1360.

Bradford, M. M. (1976). A rapid and sensitive method for the quantitation of microgram quantities of protein utilizing the principle of protein-dye binding. *Analytical Biochemistry*, *72*, 248–254.

Broderick, J. B. (2003). Iron–sulfur clusters in enzyme catalysis. In J. McCleverty, T. J. Meyer, L. Que, & W. B. Tolman (Eds.), *Comprehensive coordination chemistry II: From biology to nanotechnology: Vol. 8* (pp. 739–757). Oxford: Elsevier.

Broderick, J. B., Duderstadt, R. E., Fernandez, D. C., Wojtuszewski, K., Henshaw, T. F., & Johnson, M. K. (1997). Pyruvate formate-lyase activating enzyme is an iron–sulfur protein. *Journal of the American Chemical Society, 119*, 7396–7397.

Broderick, J. B., Duffus, B. R., Duschene, K. S., & Shepard, E. M. (2014). Radical S-adenosylmethionine enzymes. *Chemical Reviews, 114*, 4229–4317.

Broderick, J. B., Henshaw, T. F., Cheek, J., Wojtuszewski, K., Smith, S. R., Trojan, M. R., et al. (2000). Pyruvate formate-lyase activating enzyme: Strictly anaerobic isolation yields active enzyme containing a $[3Fe–4S]^+$ cluster. *Biochemical and Biophysical Research Communications, 269*, 451–456.

Brush, E. J., Lipsett, K. A., & Kozarich, J. W. (1988). Inactivation of *Escherichia coli* pyruvate formate-lyase by hypophosphate: Evidence for a rate-limiting phosphorus–hydrogen bond cleavage. *Biochemistry, 27*(6), 2217–2222.

Chandra, T., Broderick, W. E., & Broderick, J. B. (2009). Chemoselective deprotection of triethylsilyl ethers. *Nucleosides, Nucleotides & Nucleic Acids, 28*(11 – 12), 1016–1029.

Chandra, T., Broderick, W. E., & Broderick, J. B. (2010). An efficient deprotection of N-trimethylsilylethoxymethyl (SEM) groups from dinucleosides and dinucleotides. *Nucleosides, Nucleotides & Nucleic Acids, 29*, 132–143.

Chandra, T., Silver, S. C., Zilinskas, E., Shepard, E. M., Broderick, W. E., & Broderick, J. B. (2009). Spore photoproduct lyase catalyzes specific repair of the 5*R* but not the 5*S* spore photoproduct. *Journal of the American Chemical Society, 131*(7), 2420–2421.

Chang, C. H., Ballinger, M. D., Reed, G. H., & Frey, P. A. (1996). Lysine 2,3-aminomutase: Rapid mix-freeze-quench electron paramagnetic resonance studies establishing the kinetic competence of a substrate-based radical intermediate. *Biochemistry, 35*, 11081–11084.

Cheek, J., & Broderick, J. B. (2001). Adenosylmethionine-dependent iron–sulfur enzymes: Versatile clusters in a radical new role. *Journal of Biological Inorganic Chemistry, 6*, 209–226.

Chirpich, T. P., Zappia, V., Costilow, R. N., & Barker, H. A. (1970). Lysine 2,3-aminomutase. Purification and properties of a pyridoxal phosphate and S-adenosylmethionine activated enzyme. *The Journal of Biological Chemistry, 245*(7), 1778–1789.

Conradt, H., Hohmann-Berger, M., Hohmann, H.-P., Blaschkowski, H. P., & Knappe, J. (1984). Pyruvate formate-lyase (inactive form) and pyruvate formate-lyase activating enzyme of *Escherichia coli*: Isolation and structural properties. *Archives of Biochemistry and Biophysics, 228*(1), 133–142.

Crain, A. V., & Broderick, J. B. (2013). Flavodoxin cofactor binding induces structural changes that are required for protein–protein interactions with NADP+ oxidoreductase and pyruvate formate-lyase activating enzyme. *Biochimica et Biophysica Acta, 1834*, 2512–2519.

Dong, M., Horitani, M., Dzikovski, B., Pandelia, M.-E., Krebs, C., Freed, J. H., et al. (2016). Organometallic complex formed by an unconventional radical S-adenosylmethionine enzyme. *Journal of the American Chemical Society, 138*, 9755–9758.

Fish, W. W. (1988). Rapid colorimetric micromethod for the quantitation of complexed iron in biological samples. *Methods in Enzymology, 158*, 357–364.

Frey, M., Rothe, M., Wagner, A. F. V., & Knappe, J. (1994). Adenosylmethionine-dependent synthesis of the glycyl radical in pyruvate formate-lyase by abstraction of the glycine C-2 pro-S hydrogen atom. *The Journal of Biological Chemistry, 269*(17), 12432–12437.

Gross, A., Geresh, S., & Whitesides, G. M. (1983). Enzymatic synthesis of S-adenosyl-L-methionine from L-methionine and ATP. *Applied Biochemistry and Biotechnology, 8*(5), 415–422.

Heinrikson, R. L., & Meredith, S. C. (1984). Amino acid analysis by reverse-phase high performance liquid chromatography: Precolumn derivatization with phenylisothiocyanate. *Analytical Biochemistry, 136*(1), 65–74.

Henshaw, T. F., Cheek, J., & Broderick, J. B. (2000). The $[4Fe–4S]^{+}$ cluster of pyruvate formate-lyase activating enzyme generates the glycyl radical on pyruvate formate-lyase: EPR-detected single turnover. *Journal of the American Chemical Society, 122*, 8331–8332.

Hinckley, G. T., & Frey, P. A. (2006). Cofactor dependence of reduction potentials for [4Fe–4S]2+/1+ in lysine 2,3-aminomutase. *Biochemistry, 45*, 3219–3225.

Horitani, M., Byer, A. S., Shisler, K. A., Chandra, T., Broderick, J. B., & Hoffman, B. M. (2015). Why nature uses radical SAM enzymes so widely: Electron nuclear double resonance studies of lysine 2,3-aminomutase show the 5′-dAdo• "free radical" is never free. *Journal of the American Chemical Society, 137*, 7111–7121. https://doi.org/10.1021/jacs.5b00498.

Horitani, M., Shisler, K. A., Broderick, W. E., Hutcheson, R. U., Duschene, K. S., Marts, A. R., et al. (2016). Radical SAM catalysis via an organometallic intermediate with an Fe-[5'-C]-deoxyadenosyl bond. *Science, 352*(6287), 822–825.

Janda, M., & Hemmerich, P. (1976). 5-Deaza- and 5-thiaribovlavins: A simple pathway to antimetabolites of vitamin B2. *Angewandte Chemie (International Ed. in English), 15*(7), 443–444.

Knappe, J., & Blaschkowski, H. P. (1975). Pyruvate formate-lyase from *Escherichia coli* and its activation system. *Methods in Enzymology, 41*, 508–518.

Knappe, J., Bohnert, E., & Brummer, W. (1965). S-adenosyl-L-methionine, a component of the clastic dissimilation of pyruvate in *Escherichia coli*. *Biochimica et Biophysica Acta, 107*(3), 603–605.

Knappe, J., Neugebauer, F. A., Blaschkowski, H. P., & Gänzler, M. (1984). Post-translational activation introduces a free radical into pyruvate formate-lyase. *Proceedings of the National Academy of Sciences of the United States of America, 81*, 1332–1335.

Knappe, J., Schacht, J., Mockel, W., Hopner, T., Vetter, H. J., & Edenharder, R. (1969). Pyruvate formate-lyase reaction in *Escherichia coli*. The enzymatic system converting an inactive form of the lyase into the catalytically active enzyme. *European Journal of Biochemistry, 11*(2), 316–327.

Knappe, J., & Schmitt, T. (1976). A novel reaction of S-adenosyl-L-methionine correlated with the activation of pyruvate formate-lyase. *Biochemical and Biophysical Research Communications, 71*, 1110–1117.

Krebs, C., Broderick, W. E., Henshaw, T. F., Broderick, J. B., & Huynh, B. H. (2002). Coordination of adenosylmethionine to a unique iron site of the [4Fe–4S] of pyruvate formate-lyase activating enzyme: A Mössbauer spectroscopic study. *Journal of the American Chemical Society, 124*(6), 912–913. https://doi.org/10.1021/ja017562i.

Krebs, C., Henshaw, T. F., Cheek, J., Huynh, B.-H., & Broderick, J. B. (2000). Conversion of 3Fe–4S to 4Fe–4S clusters in native pyruvate formate lyase activating enzyme: Mössbauer characterization and implications for mechanism. *Journal of the American Chemical Society, 122*, 12497–12506.

Lepore, B. W., Ruzicka, F. J., Frey, P. A., & Ringe, D. (2005). The x-ray crystal structure of lysine-2,3-aminomutase from Clostridium subterminale. *Proceedings of the National Academy of Sciences of the United States of America, 102*(39), 13819–13824.

Magnusson, O. T., Reed, G. H., & Frey, P. A. (1999). Spectroscopic evidence for the participation of an allylic analogue of the 5′-deoxyadenosyl radical in the reaction of lysine 2,3-aminomutase. *Journal of the American Chemical Society, 121*, 9764–9765.

Mapolelo, D. T., Zhang, B., Naik, S. G., Huynh, B. H., & Johnson, M. K. (2012). Spectroscopic and functional characterization of iron–sulfur cluster-bound forms of Azotobacter vinelandii(Nif)IscA. *Biochemistry*, *51*(41), 8071–8084.

Markham, G. D., DeParasis, J., & Gatmaitan, J. (1984). The sequence of metK, the structural gene for S-adenosylmethionine synthetase in *Escherichia coli*. *The Journal of Biological Chemistry*, *259*(23), 14505–14507.

Markham, G. D., Hafner, E. W., Tabor, C. W., & Tabor, H. (1980). S-adenosylmethionine synthetase from *Escherichia coli*. *The Journal of Biological Chemistry*, *255*, 9082–9092.

Miller, J., Bandarian, V., Reed, G. H., & Frey, P. A. (2001). Inhibition of lysine 2,3-aminomutase by the alternative substrate 4-thialysine and characterization of the 4-thialysyl radical intermediate. *Archives of Biochemistry and Biophysics*, *387*(2), 281–288.

Moss, M., & Frey, P. A. (1987). The role of S-adenosylmethionine in the lysine 2,3-aminomutase reaction. *The Journal of Biological Chemistry*, *262*(31), 14859–14862.

Moss, M. L., & Frey, P. A. (1990). Activation of lysine 2,3-aminomutase by S-adenosylmethionine. *The Journal of Biological Chemistry*, *265*, 18112–18115.

Murib, J. H., & Ritter, D. M. (1952). Decomposition of nitrosyl disulfonate ion. I. Products and mechanism of color fading in acid solution. *Journal of the American Chemical Society*, *74*, 3394–3398.

Petrovich, R. M., Ruzicka, F. J., Reed, G. H., & Frey, P. A. (1991). Metal cofactors of lysine-2,3-aminomutase. *The Journal of Biological Chemistry*, *266*(12), 7656–7660.

Petrovich, R. M., Ruzicka, F. J., Reed, G. H., & Frey, P. A. (1992). Characterization of iron–sulfur clusters in lysine 2,3-aminomutase by electron paramagnetic resonance spectroscopy. *Biochemistry*, *31*, 10774–10781.

Pierrel, F., Hernandez, H. L., Johnson, M. K., Fontecave, M., & Atta, M. (2003). MiaB protein from Thermotoga maritima. Characterization of an extremely thermophilic tRNA-methylthiotransferase. *The Journal of Biological Chemistry*, *278*(32), 29515–29524.

Py, B., & Barras, F. (2010). Building Fe–S proteins: Bacterial strategies. *Nature Reviews Microbiology*, *8*, 436–446.

Ruzicka, F. J., & Frey, P. A. (2010). Kinetic and spectroscopic evidence of negative cooperativity in the action of lysine 2,3-aminomutase. *The Journal of Physical Chemistry. B*, *114*(49), 16118–16124.

Shapiro, S. K., & Ehninger, D. J. (1966). Methods for the analysis and preparation of adenosylmethionine and adenosylhomocysteine. *Analytical Biochemistry*, *15*(2), 323–333.

Shepard, E. M., & Broderick, J. B. (2010). *S*-adenosylmethionine and iron-sulfur clusters in biological radical reactions: The radical SAM superfamily. In L. N. Mander & H. W. Liu (Eds.), *Comprehensive natural products II: Chemistry and biochemistry: Vol. 8* (pp. 625–662). Oxford, UK: Elsevier Press.

Shisler, K. A., & Broderick, J. B. (2014). Glycyl radical activating enzymes: Structure, mechanism, and substrate interactions. *Archives of Biochemistry and Biophysics*, *546*, 64–71.

Shisler, K. A., Hutcheson, R. U., Horitani, M., Duschene, K. S., Crain, A. V., Byer, A. S., et al. (2017). Monovalent cation activation of the radical SAM enzyme pyruvate formate-lyase activating enzyme. *Journal of the American Chemical Society*, *139*(34), 11803–11813.

Smit, P. (1986). Synthesis of 5-deazaflavin adenine dinucleotide (5-dFAD) using a modified triester approach. *Recueil des Travaux Chimiques des Pays-Bas*, *105*, 538–544.

Sofia, H. J., Chen, G., Hetzler, B. G., Reyes-Spindola, J. F., & Miller, N. E. (2001). Radical SAM, a novel protein superfamily linking unresolved steps in familiar biosynthetic pathways with radical mechanisms: Functional characterization using new analysis and information visualization methods. *Nucleic Acids Research*, *29*(5), 1097–1106.

Spychala, J., Datta, N. S., Takabayashi, K., Datta, M., Fox, I. H., Gribbin, T., et al. (1996). Cloning of human adenosine kinase cDNA: Sequence similarity to microbial ribokinases and fructokinases. *Proceedings of the National Academy of Sciences of the United States of America*, *93*, 1232–1237.

Stoll, S., & Schweiger, A. (2006). EasySpin, a comprehensive software package for spectral simulation and analysis in EPR. *Journal of Magnetic Resonance*, *178*(1), 42–55.

Tsuchida, T., Yoshinaga, F., & Okumura, S. (1974). U. S. P. Office.

Ugulava, N. B., Gibney, B. R., & Jarrett, J. T. (2001). Biotin synthase contains two distinct iron–sulfur cluster binding sites: Chemical and spectroelectrochemical analysis of iron–sulfur cluster interconversions. *Biochemistry*, *40*(28), 8343–8351.

Wagner, A. F. V., Frey, M., Neugebauer, F. A., Schäfer, W., & Knappe, J. (1992). The free radical in pyruvate formate-lyase is located on glycine-734. *Proceedings of the National Academy of Sciences of the United States of America*, *89*, 996–1000.

Walsby, C. J., Ortillo, D., Broderick, W. E., Broderick, J. B., & Hoffman, B. M. (2002). An anchoring role for FeS clusters: Chelation of the amino acid moiety of S-adenosylmethionine to the unique iron site of the [4Fe–4S] cluster of pyruvate formate-lyase activating enzyme. *Journal of the American Chemical Society*, *124*(38), 11270–11271.

Walsby, C. J., Ortillo, D., Yang, J., Nnyepi, M., Broderick, W. E., Hoffman, B. M., et al. (2005). Spectroscopic approaches to elucidating novel iron–sulfur chemistry in the "radical SAM" protein superfamily. *Inorganic Chemistry*, *44*(4), 727–741.

Wang, S. C., & Frey, P. A. (2007a). Binding energy in the one-electron reductive cleavage of S-adenosylmethionine in lysine 2,3-aminomutase, a radical SAM enzyme. *Biochemistry*, *46*(45), 12889–12895.

Wang, S. C., & Frey, P. A. (2007b). S-adenosylmethionine as an oxidant: The radical SAM superfamily. *Trends in Biochemical Sciences*, *32*(3), 101–110.

Wollers, S., Layer, G., Garcia-Serres, R., Signor, L., Clemancey, M., Latour, J.-M., et al. (2010). Iron–sulfur (Fe-S) cluster assembly: The SufBCD complex is a new type of Fe–S scaffold with a flavin redox cofactor. *The Journal of Biological Chemistry*, *285*, 23331–23341.

Wright, J. M., & Boyle, S. M. (1984). Intergeneric homology of the speC gene encoding biosynthetic ornithine decarboxylase in Escherichia coli. *Journal of Bacteriology*, *159*(3), 1074–1076.

Wu, W., Booker, S., Lieder, K. W., Bandarian, V., Reed, G. H., & Frey, P. A. (2000). Lysine 2,3-aminomutase and trans-4,5-dehydrolysine: Characterization of an allylic analogue of a substrate-based radical in the catalytic mechanism. *Biochemistry*, *39*, 9561–9570.

Wu, W., Lieder, K. W., Reed, G. H., & Frey, P. A. (1995). Observation of a second substrate radical intermediate in the reaction of lysine 2,3-aminomutase: A radical centered on the beta-carbon of the alternative substrate, 4-thia-L-lysine. *Biochemistry*, *34*, 10532–10537.

Yang, J., Naik, S. G., Ortillo, D. O., Garcia-Serres, R., Li, M., Broderick, W. E., et al. (2009). The iron–sulfur cluster of pyruvate formate-lyase activating enzyme in whole cells: Cluster interconversion and a valence-localized [4Fe-4S]2+ state. *Biochemistry*, *48*, 9234–9241.

CHAPTER ELEVEN

Determining Redox Potentials of the Iron–Sulfur Clusters of the AdoMet Radical Enzyme Superfamily

Stephanie J. Maiocco, Lindsey M. Walker, Sean J. Elliott[1]
Department of Chemistry, Boston University, Boston, MA, United States
[1]Corresponding author: e-mail address: elliott@bu.edu

Contents

Abstract

While protein film electrochemistry (PFE) has proven to be an effective tool in the interrogation of redox cofactors and assessing the electrocatalytic activity of many different enzymes, recently it has been proven to be useful for the study of the redox potentials of the cofactors of AdoMet radical enzymes (AREs). In this chapter, we review the challenges and opportunities of examining the redox cofactors of AREs in a high level of detail, particularly for the deconvolution of redox potentials of multiple cofactors. We comment on how to best assess the electroactive nature of any given ARE, and we see that when applied well, PFE allows for not only determining redox potentials, but also determining proton-coupling and ligand-binding phenomena in the ARE superfamily.

Methods in Enzymology, Volume 606
ISSN 0076-6879
https://doi.org/10.1016/bs.mie.2018.06.002

ABBREVIATIONS

AdoMet	*S*-adenosyl-L-methionine
ARE	AdoMet radical enzyme
CE	counter electrode
CHES	*N*-cyclohexyltaurine
CV	cyclic voltammetry
dAdo	5′-deoxyadenosine
d-Ado•	5′-deoxyadenosyl 5′-radical
DT	dithionite
DTT	dithiothreitol
E. coli	*Escherichia coli*
EPR	electron paramagnetic resonance
ET	electron transfer
FeS	iron–sulfur
HEPES	2-[4-(2-hydroxylethyl)piperazine-1-yl]ethanesulfonic acid
LAM	lysine 2,3-aminomutase
MES	2-morpholin-4-ylethanesulfonic acid
MOPS	3-(*N*-morpholino)propanesulfonic acid
MWCNT	multiwalled carbon nanotubes
NaCl	sodium chloride
PCET	proton-coupled electron transfer
PFE	protein film electrochemistry
PGE	pyrolytic graphite edge
PIPES	piperazine-*N*,*N*′-bis(2-ethanesulfonic acid)
RE	reference electrode
SAH	*S*-adenosylhomocysteine
SAM	self-assembled monolayer
SDS-PAGE	sodium dodecyl sulfate polyacrylamide gel electrophoresis
SWV	square wave voltammetry
TAPS	*N*-tris(hydroxymethyl)methyl-3-aminopropanesulfonic acid
Tris	tris(hydroxymethyl)aminomethane
WE	working electrode

1. INTRODUCTION

1.1 An Electrochemist's View of the AdoMet Radical Enzyme (ARE) Superfamily

As described elsewhere in this volume, the *S*-adenosyl-L-methionine (AdoMet) radical enzyme (ARE) superfamily comprises a diverse set of enzymes that all use AdoMet as a radical initiator for various reactions (Shisler & Broderick, 2012; Wang & Frey, 2007b). While the enzymatic chemistry displayed by

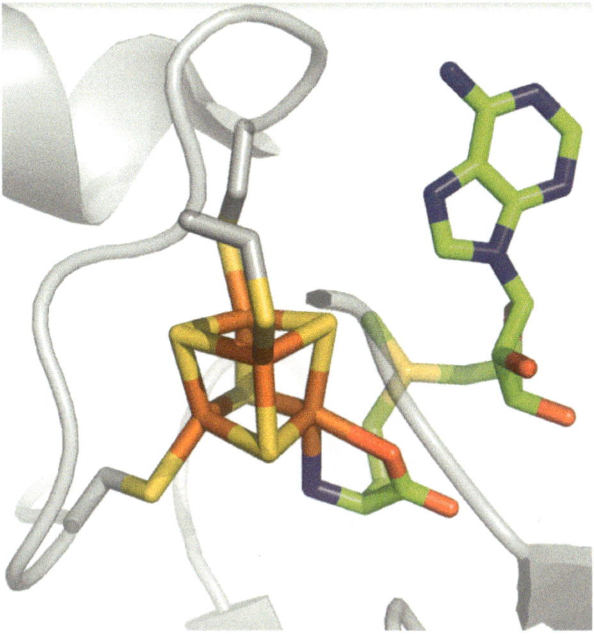

Fig. 1 Depiction of the [4Fe – 4S] AM cluster with AdoMet bound (PDB entry 4M7T).

AREs is diverse, all contain a specific [4Fe–4S] cluster, typically found in a conserved CXXXCXXC motif, with an open coordination site on one of the iron ions that coordinates the α-carboxyl and α-amino groups of AdoMet in a bidentate fashion (Fig. 1). The canonical chemistry of an ARE requires that the AdoMet-binding $[4Fe–4S]^{2+}$ cluster is first reduced to the $[4Fe-4S]^{1+}$ state, allowing for innersphere electron transfer (ET) from the $[4Fe–4S]^{1+}$ cluster to AdoMet, promoting the homolytic cleavage of the S—C5′ bond of AdoMet. This generates methionine and a 5′-deoxyadenosyl 5′-radical (d-Ado•), which then abstracts a hydrogen atom from the substrate to generate a substrate radical and 5′-deoxyadenosine (dAdo) (Landgraf, McCarthy, & Booker, 2016; Shisler & Broderick, 2012; Wang & Frey, 2007b). The fates of the potentially redox-active species are diverse, and there is no single structure–function model for how nature controls the redox reactions of the [4Fe–4S] cluster, or any other ARE-based intermediary. Recently the types of reaction and the potential requirement for redox partners have been reviewed in terms of net oxidative vs redox neutral processes (Ruszczycky, Zhong, & Liu, 2018). In this chapter, we consider the issues of trying to understand the differences in the redox chemistry of the various iron–sulfur-based clusters of the ARE superfamily. While the spectroscopic and structural biology toolkit can be employed to study the function and mechanisms of the ARE superfamily, the ability to measure redox potentials of ARE iron–sulfur clusters has been harder to attain. Protein film electrochemistry (PFE) is one tool that has been recently shown to prove a rapid, robust, and highly sensitive means for measuring the electrochemical properties of AREs. Here we will present some of the latest efforts in using PFE to measure ARE redox potentials, as well as indicating the challenges associated with these cutting-edge experiments.

One of the challenges (and sources of excitement) associated with the study of AREs by any method is that the ARE superfamily is vast in sequence and function space, and only growing. For example, bioinformatics analyses have identified more than many tens of thousands (a number that is hard to keep recent due to the everincreasing sequence data) (Akiva et al., 2014; Grell, Goldman, & Drennan, 2015). Along with these growing number of "representative" AREs, is an appreciation that in addition to the canonical [4Fe–4S] AdoMet-binding cluster (AM cluster) which may be thought of as an active site (Fig. 1), many AREs contain one or more additional redox-active cofactors (Grell et al., 2015), including iron–sulfur clusters ([4Fe–4S] or [2Fe–2S], or other compositions) or possibly cobalamin. Often the additional FeS clusters are found housed in domains called Spasm/Twitch, yet the precise rules of predicting the roles and redox properties of these so-called "Auxiliary" clusters is unclear (Grell et al., 2015; Ruszczycky et al., 2018). And while the tools of electron paramagnetic resonance (EPR and ENDOR) have been employed to sort out the electronic and magnetic structure of the varying states of AREs, one of the primary challenges for the detection of redox potentials is being able to assess multiple cofactors in the same experiment.

PFE is now an established technique that has grown from early days of measuring the redox potential of cytochromes and blue copper proteins that were in diffusional communication with a stationary electrode. PFE has been widely reviewed (Leger & Bertrand, 2008) including reviews in this series (Armstrong, Butt, & Sucheta, 1993; Armstrong, Evans, & Megarity, 2018), the most recent of which highlighted the deployment of PFE as a means to report on FeS clusters, focusing on either simple ferredoxin (Fd) proteins or the catalytic activity of hydrogenase enzymes. As such, here we can avoid a discussion of the theory of PFE aside from when absolutely necessary, for indeed much of it stems from the treatment of an adsorbed species that interacts with an electrode. The adsorbed mode (the "film" in the name PFE) is advantageous when addressing the ARE family. While PFE excels at sample economy, detecting pmol/cm^2 of electroactive coverage, removing the diffusion of analyte molecules (such as proteins) as a factor that contributes to the electrochemical response means that one has the ability to resolve voltammetric line shapes which become sharper than they may appear in a diffusional mode, giving a more precise analysis. With that ability, addressing multiple redox cofactors can be attained for AREs, as it has been attained for many other multicofactor systems, even in cases where the redox potentials are close to one another. Furthermore, PFE is a form of "direct" electrochemistry—where there are no mediators that may influence the redox potentials that are monitored, and the contribution of redox activity from multiple cofactors is observed as current as a function of potential.

In the following case studies, older and newer approaches will be compared in order to highlight the advantages of the PFE method.

1.2 Electron Donors for AREs

The requirement for determining the redox potentials of the [4Fe–4S] cluster(s) of AREs is easy to appreciate, when one considers how their chemistry is dependent on the [4Fe – 4S] cluster being reduced to the +1 oxidation state, yet the biological reductant for AREs has never been directly identified. Previous studies on *Escherichia coli* (*E. coli*) pyruvate formate–lyase activating enzyme found that the FeS cluster could be reduced by *E. coli* flavodoxin (Knappe et al., 1969). This system has also been found to reduce *E. coli* anaerobic ribonucleotide reductase and biotin synthase by supplying reducing equivalents from NADPH via ferredoxin (flavodoxin): NADP+ reductase (Bianchi, Eliasson, et al., 1993; Bianchi, Reichard, et al., 1993; Harder, Eliasson, Pontis, Ballinger, & Reichard, 1992; Ifuku, Koga, Haze, Kishimoto, & Wachi, 1994). The *E. coli* ferredoxin (flavodoxin): NADP+ reductase system has also been used as the reductant in assays with lysine 2,3-aminomutase (LAM) from *Porphyromonas gingivalis* (Brazeau, Gort, Jessen, Andrew, & Liao, 2006). Since the discovery that the *E. coli* system can serve as an electron donor for AREs from different species, it is frequently used as a reductant in ARE activity assays, as it has been capable of supporting turnover. An alternative reductant used in ARE assays is dithionite, as its low reduction potential (−660 mV) (Mayhew, 1978) should be low enough to support turnover. However, assays of AREs with dithionite have often shown unproductive cleavage of AdoMet with dithionite, likely due to the strong reducing power of this reductant. The unproductive cleavage observed is typically much lower when using the *E. coli* flavodoxin, ferredoxin (flavodoxin): NADP+ reductase system, which has a reduction potential of −433 mV (McIver et al., 1998). The potentials of reducing systems capable of supporting turnover provides some indication of the reduction potential of the FeS cluster in AREs, as these reductants would not be able to reduce the cluster if the potentials were not compatible. Given these complications, accurate determination of the reduction potentials of the FeS clusters in AREs is essential.

1.3 LAM and Biotin Synthase as Paradigms

The first extensive study on the redox properties of an ARE was performed by Frey and coworkers on LAM, which catalyzes the conversion of L-α-lysine to L-β-lysine. The reduction potential of the $[4Fe–4S]^{2+/1+}$ couple was measured using the method of EPR-based spectroelectrochemical titrations with various ligands bound to the open coordination site of the cluster (Hinckley & Frey, 2006; Wang & Frey, 2007a). These studies indicated that the reduction potential of the $[4Fe–4S]^{2+/1+}$ couple in LAM is −480 mV and shifts positive in potential to −430 mV with AdoMet bound to the cluster (Fig. 2) (Hinckley & Frey, 2006). This is far removed from the estimated reduction potential of AdoMet in solution (Saeva & Morgan, 1984) representing a 32 kcal/mol barrier. This barrier must be lower in the active site of LAM for catalysis to occur.

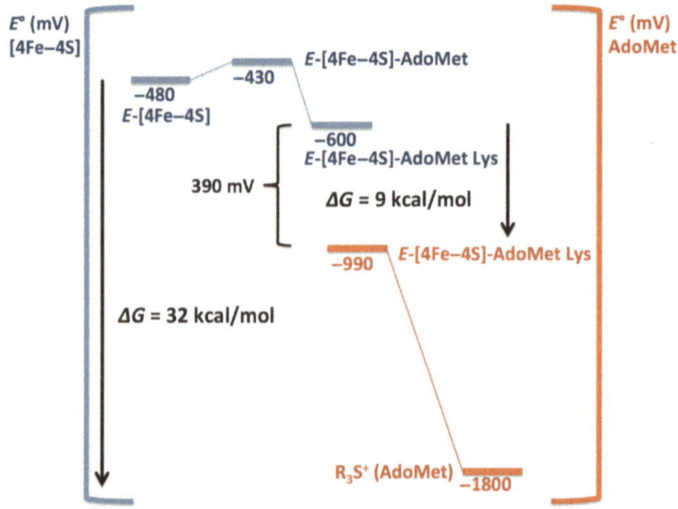

Fig. 2 Energy diagram of lysine 2,3-aminomutase, with the reduction potential of the [4Fe−4S] cluster and AdoMet. *Adapted from Wang, S.C., & Frey, P.A. (2007a). Binding energy in the one-electron reductive cleavage of S-adenosylmethionine in lysine 2,3-aminomutase, a radical SAM enzyme.* Biochemistry, *46(45), 12889–12895.*

The most relevant reduction potential for the AdoMet radical cluster would be for the Michaelis complex poised for reaction; however, the reaction would proceed under these conditions, preventing analysis of the reduction potential. To achieve an estimate for the reduction potential of the Michaelis complex, the reduction potential was determined for structurally similar complexes, where *S*-adenosylhomocysteine (SAH) was used in place of AdoMet and alanine (or alanine and ethylamine) were used in place of lysine. These complexes demonstrated that binding of lysine, or alanine, lowers the midpoint potential of the AdoMet radical cluster. When alanine was used along with AdoMet, a potential of −582 mV was measured, whereas alanine with SAH had a potential of −566 mV. Using lysine with SAH displayed a midpoint potential of −583 mV. The lowest midpoint potential of −594 mV was measured with SAH and alanine with ethylamine. These results taken together indicated that the lysyl side chain is important for a 17–28 mV lowering of the potential. Therefore, the Michaelis complex was expected to have a midpoint potential in the range of −599 to −610 mV, determined by adjusting the midpoint potential of the AdoMet and alanine complex, and −600 mV was used as an estimate (Wang & Frey, 2007a). These Herculean efforts also have caveats, as they were conducted as spectroelectrochemical titrations: external mediators that can influence results, the quantity, and the labor of all of the required LAM means further examination of dependencies upon varying substrate/product or proton concentrations was beyond the scope of the analyses. Also, in using EPR spectroscopy in this way, one can only monitor the EPR-observable $[4Fe-4S]^{1+}$ cluster, and not directly monitoring the concentration of other states of the active site.

The second major historical paradigm for resolving the redox potentials of an ARE comes from biotin synthase (BioB). BioB is a sulfur insertion ARE that contains an obligately sacrificial $[2Fe-2S]$ cluster in addition to the canonical AdoMet radical $[4Fe-4S]$ cluster and catalyzes the conversion of dethiobiotin into biotin through insertion of a sulfur atom between the C6 and C9 carbon atoms of dethiobiotin (Berkovitch, Nicolet, Wan, Jarrett, & Drennan, 2004; Ugulava, Gibney, & Jarrett, 2001). Reconstituted BioB containing two $[4Fe-4S]$ clusters per monomer was reduced with two barely distinguishable potentials of −440 and −505 mV. Here, optical changes were used, where the ability to fit the data to more than one redox couple depended upon the choice of wavelength: changes from 420–480 to 550–625 nm were monitored, and the decrease of absorbance at 550 nm was fit to a single potential of −505 mV, whereas the decrease in absorbance at

420 nm was fit to two potentials of −440 and −505 mV. BioB can also be reconstituted to contain a single [4Fe−4S] cluster, which displayed a single midpoint potential of −440 mV. These results suggested that BioB has a specific binding site for a [4Fe−4S] cluster with a midpoint potential of −440 mV. The native BioB enzyme is believed to contain a [2Fe−2S] cluster and the AdoMet radical [4Fe−4S] cluster. This form of the enzyme reduced with a single potential of −430 mV, but had a Nernstian fit consistent with two-independent one-electron reductions, and the determined experimental error spanned the potential of −440 mV measured for the other two forms of BioB, yielding a value of the AdoMet-binding cluster similar to the value for the cluster in LAM (Ugulava et al., 2001). These results suggest that members of this superfamily may feature common redox properties, but that advances are needed in the precision and robustness of the measurements of redox potentials.

1.3.1 Deploying PFE to Study AREs: Methylthiotransferases

The sulfur insertion AREs also contain the methylthiotransferases, such as MiaB and RimO. These enzymes contain two [4Fe−4S] clusters (Fig. 3A), one of which is the canonical AdoMet radical [4Fe−4S] cluster, and the auxiliary [4Fe−4S] cluster is proposed to be destroyed in catalysis, similar to BioB (Anton et al., 2008; Forouhar et al., 2013; Hernandez et al., 2007; Landgraf, Arcinas, Lee, & Booker, 2013; Lee et al., 2009). MiaB is responsible for the attachment of a methylthio-group on tRNA. The reduction potential of MiaB from *Thermotoga maritima* was originally determined to be −495 mV via differential pulse voltammetry when the protein was believed to contain a single cluster (Pierrel, Hernandez, Johnson, Fontecave, & Atta, 2003). When MiaB was later purified with two [4Fe−4S] clusters, the auxiliary cluster was found in the reduced state as-isolated, indicating a higher reduction potential than the AdoMet radical cluster (Hernandez et al., 2007). Recently we have found that MiaB from *Bacteroides thetaiotaomicron* (Bt) is a amenable to PFE, and we have determined the potentials for both clusters (Maiocco et al., 2016). PFE of Bt MiaB displayed a single broad, but reversible feature that indicates that not one, but two voltammetric components contribute to the overall response that correspond to pure redox reactions. The observed feature is consistent with two one-electron centers with reduction potentials of −450 and −390 mV (shown in Fig. 3B as CV and 3C as square wave voltammetry (SWV) data). Assignment of the potentials was achieved using the triple variant devoid of the AdoMet radical cluster, displaying only the higher potential, confirming the assignment of the −450 mV feature as the AdoMet radical cluster.

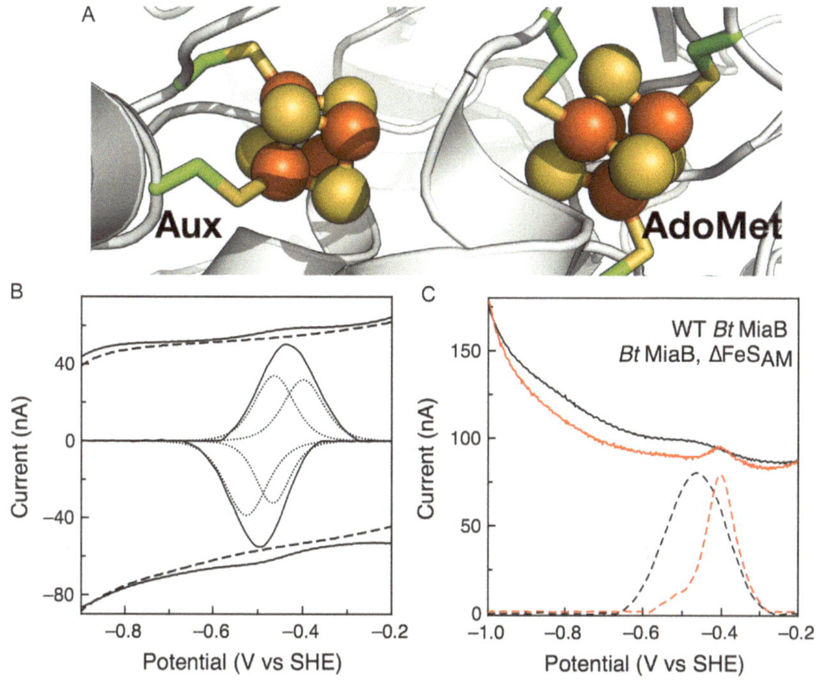

Fig. 3 (A) The two clusters of MiaB and (B) can be resolved with CV PFE. *The inset* shows the baseline (*dash*) corrected data, deconvoluted into two separate one-electron redox couples (*dot*). (C) Square wave voltammograms of wild type (*black*) and the Cys-171,175,178-Ala mutant (*red*) of MiaB from *Bacteroides thetaiotaomicron*. The raw data are shown in the *solid lines* and the baseline-corrected data are shown as the *dashed lines*. Data were collected at 10 Hz, $E_{amp} = 20$ mV, pH 8, 10°C.

As MiaB uses two molecules of AdoMet, one as a methyl donor and one as the canonical ARE radical initiator, the impact of AdoMet binding is complicated. We demonstrated that addition of AdoMet resulted in a time-dependent reaction: narrowing of the initial broad feature occurred upon introduction of AdoMet, and at the same time, a new electrochemical signature grew in at lower potentials, −620 mV (Maiocco et al., 2016). To analyze these features, SAH was used as an analog of AdoMet, and caused a narrowing of the feature observed, along with an increase in current, consistent with the AdoMet radical cluster shifting positive in potential upon binding SAH. This shift moving to more positive in potential was reported for LAM in the presence of both SAH and AdoMet (Hinckley & Frey, 2006). Sodium methanethiolate ($NaSCH_3$) was used as an analog of the methyl group formed from the demethylation of AdoMet. Addition of $NaSCH_3$ resulted in a slow shift to low potential for the feature observed,

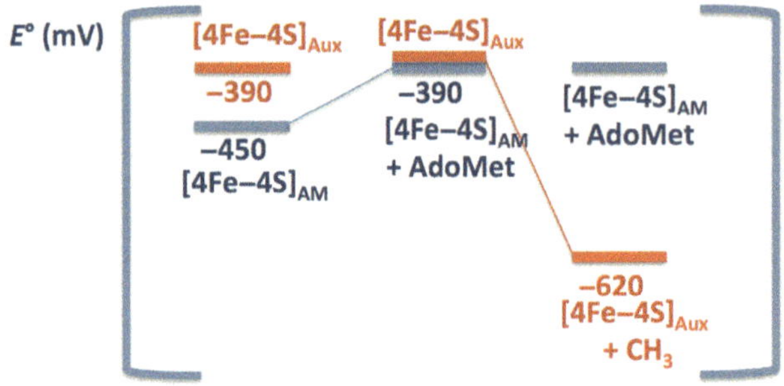

Fig. 4 $[4Fe\text{–}4S]^{2+/1+}$ reduction potentials of the two FeS clusters in MiaB.

suggesting that the methyl group binds to, or near, the auxiliary cluster and lowers the potential (Maiocco et al., 2016). Overall, this has provided for a new view of the modulation of redox potential in AREs such as MiaB, as depicted in Fig. 4.

1.3.2 Deploying PFE to Study AREs: Dehydrogenases

All of the characterized members of the dehydrogenase subclass of AREs to date have been found to contain at least one additional [4Fe – 4S] cluster (Benjdia et al., 2010; Goldman, Grove, Booker, & Drennan, 2013; Goldman et al., 2013; Grove et al., 2013; Grove, Ahlum, Sharma, Krebs, & Booker, 2010). BtrN from *Bacillus circulans* contains an auxiliary [4Fe – 4S] cluster in addition to the AdoMet-binding cluster, and it catalyzes the conversion of deoxyinosamine to aminodeoxyinosose (Fig. 5A) (Grove et al., 2010; Yokoyama, Numakura, Kudo, Ohmori, & Eguchi, 2007; Yokoyama, Ohmori, Kudo, & Eguchi, 2008). In contrast to the features observed for the auxiliary cluster in MiaB, the EPR characterization of BtrN indicated that the auxiliary cluster was too low in potential to be reduced with dithionite or titanium citrate (Grove et al., 2010). This result could be rationalized based upon our recent PFE study of BtrN, which reveals that the two clusters have very different potentials: one displays a potential of −510 mV and the other −765 mV (Fig. 5B) (Maiocco, Grove, Booker, & Elliott, 2015). Use of the Cys-16,20,22-Ala triple variant, which contained only the auxiliary cluster, confirmed the assignment of the AdoMet radical cluster as the −510 mV feature, and the auxiliary cluster as the low potential feature at −765 mV, the lowest reported reduction potential for a $[4Fe\text{–}4S]^{2+/1+}$ cluster to date (Maiocco et al., 2015). The reports for LAM and BioB were performed at pH 8; however, the potential of the AdoMet radical cluster was reported to be independent of pH. Analysis of the reduction potentials of

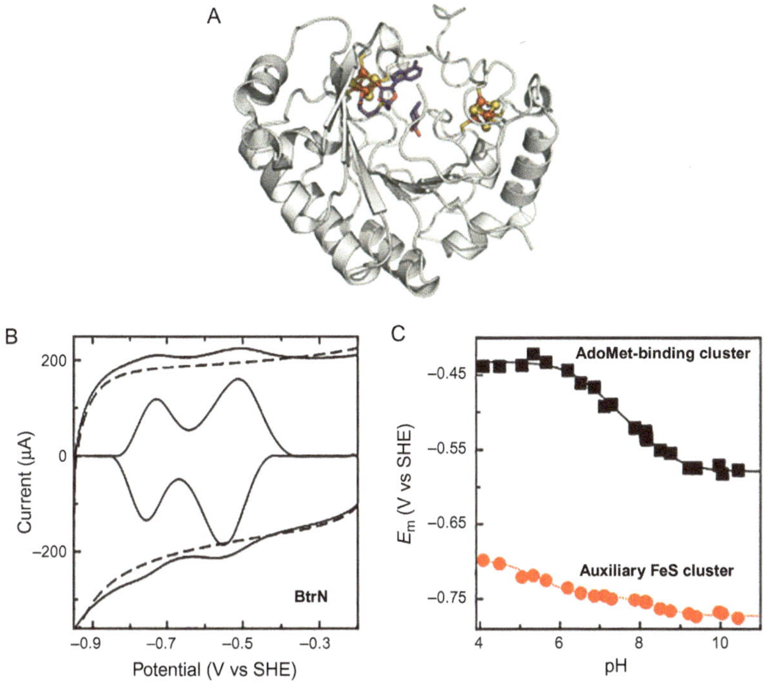

Fig. 5 Structure of BtrN (4M7T.pdb) reveals an AdoMet-bound cluster and an auxiliary. (A) Cyclic voltammograms of wild-type BtrN and (B) reveal two distinct redox couples. (The baseline response of the electrode is shown as the *dashed line* and the baseline-corrected data are shown as the *inset*). Data were collected at 100 mV/s, pH 8, 10°C. (C) PFE's rapid response and sample economy facilitate collecting pH dependencies. Here, the AdoMet-binding cluster engages in a $1e^-$:$1H^+$ reaction.

BtrN as a function of pH indicated proton-coupled electron transfer (PCET) was at work, due to the Nernstian behavior of a clear pH dependence of the redox potential of the AdoMet-binding [4Fe – 4S] cluster (Fig. 5C), the first reported evidence for PCET for an ARE to date. The point of the coupling of proton transfer is unclear currently, and it is has yet to be observed for any other enzyme in the ARE superfamily; for example, the methythiotransferases RimO and MiaB did not display a pH dependence in our hands. Notably, the low potential, auxiliary cluster has a weak pH dependence, if any. This behavior can be interpreted in terms of either a series of pK_a values associated with the ET event (a fit to two subsequent one –H^+ steps is shown here), which may be best thought of as nonspecific, ionic interactions that slightly modulate the redox potential of the auxiliary FeS cluster, due to changes in the protonation state of residues further removed from the cluster environment (Maiocco et al., 2015).

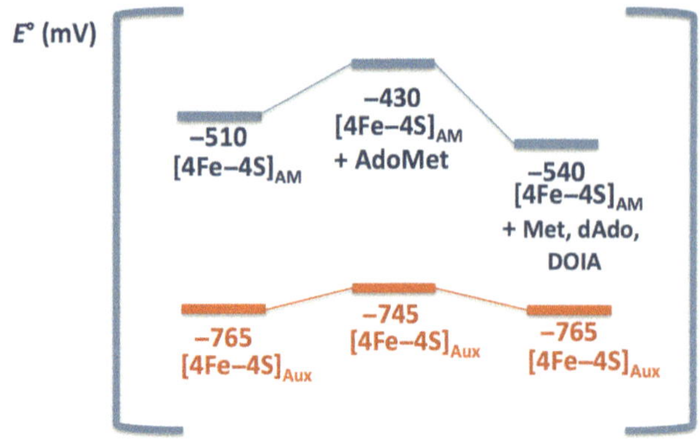

Fig. 6 $[4Fe-4S]^{2+/1+}$ reduction potentials of the AdoMet (AM) and auxiliary (Aux) FeS clusters in BtrN.

To further characterize the redox properties of BtrN, the impact of substrate was assessed (Maiocco et al., 2015). As observed with LAM, in the presence of AdoMet, the potential of the AdoMet radical cluster shifts more positive in potential by +80 mV, ensuring reduction by biological reducing systems (Fig. 6). Using the products of the cleavage reaction, dAdo and methionine, along with the substrate DOIA displayed a −30 mV shift lower in potential for the AdoMet radical cluster, while the auxiliary cluster maintained the −765 mV potential, further indicating control of the redox properties throughout the catalytic cycle. Use of the product, amino-DOI, along with SAH (or dAdo and Met) would provide further insight into the redox properties of BtrN; however, amino-DOI is not commercially available. These results further demonstrate the fine control AREs have over the redox properties of the FeS clusters.

2. TECHNICAL DETAILS

2.1 Instrumentation and Electrodes

The general methods conventionally employed here include the confined mode of dc cyclic voltammetry (CV) and SWV, both in a three-electrode mode, which are well described in textbooks (Bard & Faulkner, 2001; Compton & Banks, 2011). And the general aspects of electrochemical cells employed successfully in PFE have been similarly reviewed (Armstrong et al., 1993; Hill & Hunt, 1993). Such cells are not commercially available, but the traits that will be needed in an electrochemical cell that will be useful for PFE analyses of the FeS clusters of AREs include (1) temperature control at the working electrode (WE), such as with a water-jacketed cell and (2) an ability to control the spacing between the Luggin capillary connection from the reference electrode (RE) and the WE, which can be optimized to minimize electrochemical noise.

An electrochemical analyzer (typically, a potentiostat) is required for any kind of PFE analyses. It can be external to an anaerobic chamber, where the actual experiments must occur due to the potential issues of oxygen-based lability of the AREs FeS clusters. In principle, any analyzer will suffice (instruments are available from many manufacturers now), but it is preferable that the potentiostat has a sensitivity and good signal to noise that extends to the pA (if not fA) range, as the overall current monitored for PFE analyses will be low.

While the RE can be anything that is stable (such as calomel or Ag/AgCl, Accumet), and the CE is typically a Pt wire, there is much more variability in the description of the WE. Typically pyrolytic graphite edge (PGE) electrodes are home fabricated, where high-quality graphite is cut into rods that can be mounted on one edge face, using silver epoxy, and encapsulated in epoxy such that a single face of the graphite edges is exposed to protein solutions, and ultimately, the electrochemical cell. However, other materials or electrodes (even commercially available carbon-based screen-printed electrodes) could work in principle. The benefit of the methods used with PGE-based electrodes are that they can be polished with sandpaper and alumina in order to expose a fresh face of lightly oxidized graphitic edges for multiple uses (Armstrong et al., 1993; Hill & Hunt, 1993). Notably, the redox potentials of all ARE-based FeS clusters are quite low, with respect to the standard hydrogen electrode (less than −400 mV), which indicates why graphite may be a good choice. Gold-based electrodes, with the associated self-assembled monolayer (SAM) technology, have been successfully employed for many heme-based redox systems, which are simply of a higher range of redox potential (Firer-Sherwood, Pulcu, & Elliott, 2008; Levin, Can, Bowman, Bren, & Elliott, 2011; Levin, Walsh, Sullivan, Bren, & Elliott, 2015; Pulcu, Elmore, Arciero, Hooper, & Elliott, 2007; Ye, Kaur, Wen, Bren, & Elliott, 2005). However, in these and many other literature cases, the SAM itself, which passivates the bare gold through a thiolate–Au bond, is not reductively stable. Inevitably, even with long alkane thiol chains which enhance the reductive stability of the SAM upon the electrode surface, there is a limit to where the thiolate bond will be reductively cleaved, exposing the bare gold, and likely denaturing whatever

ARE is being studied. As such, carbon-based electrodes appear to be the most suitable at this time for the study of all AREs. Glassy carbon, at the time of this writing, has yet to be deployed successfully.

2.2 Square Wave vs Cyclic Voltammetry

While dc CV experiments are the norm for measuring redox potentials of metalloproteins, through PFE methods, square wave voltammetry (SWV) has been found to be useful in the reports of ARE electrochemistry (Blaszczyk et al., 2016; Maiocco et al., 2016, 2015). Why might this be so? Simply put, SWV is a difference-based method, where the applied potential is ramped in one direction, and then also jumped in a forward and backward direction (a variable that can be controlled by the experimentalist). From the stand point of absolute sensitivity, this approach can help eliminate noise and enhance the faradaic current associated with a given protein. However, the reversible nature of CV means that it is easy to determine if the electrochemical signal generated corresponds to a simple redox process, or a more complicated coupled reaction (Armstrong et al., 2018; Bard & Faulkner, 2001; Compton & Banks, 2011). The final trace of a SWV experiment (which is the difference of the forward and backward steps of potential), will likely mask this behavior, though one can plot the forward and backward currents independently to verify that the current observed is symmetrical, indicating a simple redox process (Bard & Faulkner, 2001). In addition to manipulating the amplitude or amount of potential used in the square wave, one can also change the timescale (or frequency) used for these jumps, where systems that possess fundamentally fast ET can given enhanced SWV signals with fast (>100 Hz) frequencies (Jeuken, Jones, Chapman, Cecchini, & Armstrong, 2002). In contrast, the buried redox cofactors of AREs tend to yield slower ET rates (which cannot be further optimized) and therefore small frequencies (<50 Hz) are used (Maiocco et al., 2016). While SWV can be a highly useful method, it must be always paired with (and reported alongside) data from CV experiments, where the signals may seem less intense due to the background capacitance of the carbon electrodes, but pure redox reaction can be observed.

2.3 Generating Protein Films

The primary concern of most investigators, and the aspect of PFE analyses that is often not well reported in prior reviews, is the process of generating electrochemically active protein films. Much like the generation of high-quality protein crystals, there are only guidelines that can be offered, and protocols suggested, instead of recipes prescribed. The challenge for PFE is that surface interactions between the electrode and the ARE of interest will determine if the correct, electroactive conformation of the protein can be attained. It is always assumed that only a small fraction of any enzyme sits upon an electrode surface in a mode that gives rise to a short ET tunneling distance, and therefore sufficient ET rates that can support detection of an electrochemical signal.

In all cases, experiments must begin with a clean, protein-free electrode that is used to generate buffer–blank voltammograms over all of the conditions that will be used in the course of subsequent experiments. Whether using CV or SWV methods for data collection, this step is essential due to the possibility for graphite-based surface artifacts to appear in one electrode (or one batch of graphite), but to be absent on another. Such artifacts are typically of a current magnitude approximately that of protein-based signals. These control experiments may reveal that some aspect of the electrochemical cell solution is itself electroactive. And finally, some PFE analyses of AREs have revealed potentials that are very close to the reductive limit of aqueous buffers (~ -0.8 V vs SHE) (Blaszczyk et al., 2016; Maiocco et al., 2015); as such, further analysis of those electrochemical features will always entail an experimental voltammogram which will contain the background response of a highly curved (reductive) shape at low potentials, due to background proton reduction. Thus, collecting "baselines" is essential.

The method of adsorption of the ARE is variable: frequently in the literature a "direct" deposit method is used where 1 μL of concentrated protein solution is applied directly to a clean electrode, adsorbed first onto the surface of a WE, and then an electrochemical scan (or series of scans) is collected in a buffer that supports protein adsorption (Blaszczyk et al., 2016; Dowling et al., 2016; Maiocco et al., 2016, 2015). However, other methods of generating such films may be followed: soaking WEs in concentrated or dilute solutions of proteins for 4 or more hours has proven useful, as has using a rotating electrode that can draw fresh protein molecules up to the electrode surface (Elliott et al., 2004; Jeuken et al., 2002). Temperature is also considered to be a general factor where colder protein solutions may lead to more stable protein: electrode interactions. And has been noted in other reviews, coadsorbates (aminoglycosides, divalent counterions, multiwalled carbon nanotubes (MWCNT), and carbon particles) may all be useful components to include in the immobilization solution (Hill & Hunt, 1993).

Thus, determining which conditions give rise to useful electrochemical signatures are a principle goal, and the following factors may each need to be optimized in terms of determining conditions that work best.

2.3.1 Sample Purity and Concentration

The most significant experimental variables may be the actual ARE preparation and the potential oxygen reactivity. Purity of the sample is paramount, as other (even trace) contaminants may have preferential interactions with the electrode surface, when compared to the ARE of interest. If that is the case, then passivation of the electrode will likely occur as sites on the surface are occupied by nonspecific (and likely nonelectroactive) interactions. The coupled variable of enzyme concentration is also a concern. The recent successes with PFE of AREs have used proteins that are highly concentrated initially (m*M*), where only a very small volume of that concentrated protein need to be deposited, briefly, and never allowed to dry fully. However, an ARE may exist in a quaternary structure that precludes the correct orientation at an electrode, and more dilute protein could be optimal (Bateman, Leger, Goodin, & Armstrong, 2001). Finally, as protein purification and preparation will likely occur inside of one anaerobic chamber, and the electrochemical characterization will typically be set up inside of a different anaerobic chamber, extra care in sample handling and maintaining an air-free environment will be yet-another concern. (If one enzyme prep does not work, it may be worthwhile to try another.)

2.3.2 Buffer Components

Many ARE enzymes are recombinantly produced in *E. coli* and purified and reconstituted to house the correct complement of FeS clusters. Through this, or wild-type purification procedures, reducing agents are often added as components in the purification buffer to stabilize the clusters. Some of these soluble species are more problematic for subsequent PFE. Dithionite (DT) must be removed through dialysis or running protein down a desalting column, prior to PFE, as it is electroactive in its own regard. Dithiothreitol (DTT) is much less problematic. Low m*M* concentrations (2–5 m*M*) can be tolerated, though at concentrations much higher than that, it too can contribute to PFE signals observed at graphite electrodes.

In order to enhance and stabilize the electroactive protein: electrode conformations, it is helpful to consider the other buffer components. Typical Good's buffers are suitable for protein electrochemistry, and a cocktail of 5–10 m*M* of a suite of buffers is useful to keep all buffer components the same, and visit a wide range of pH values. Typically, a cocktail of acetate, MES, HEPES, PIPES, TAPS, CAPS, and CHES (all at 5 m*M*) give good buffering capacity. (Tris is avoided entirely as it is implicated in the binding of metals, and it is known to have a poor stability to changes in temperature.) Other components common to buffers (such as ethylene glycol) are typically nonelectroactive.

For efficient electrochemistry, supporting electrolyte is required (either typically NaCl or KCl). Yet, high concentrations of electrolyte may disrupt ionically driven interactions with a PGE electrode. Thus, like temperature, ionic strength can be a variable that requires optimization.

2.3.3 Scan Range

It can be anticipated that the FeS clusters of any ARE will be low in potential, thus scan ranges visited in either CV or SWV experiments should be kept at −100 mV vs SHE or lower; overoxidation of the FeS clusters can lead to cluster loss and voltammetric changes associated with their destruction (Mitra & Elliott, 2009).

2.4 Data Assessment

Any electrochemical data are typically collected on a proprietary computer-based interface associated with the potentiostat that is used. Upon data collection, it is conventional that analysis may require removal of electronic noise and the non-Faradaic component of the current, to arrive at the type of data that is shown as the insets in the figures above. These manipulations can be routinely achieved with QSoas (Fourmond, 2016). Important aspects of data assessment for pure electrochemical signals of an adsorbed protein include the separation between the oxidative and the reductive peaks, which should be zero at optimally fast ET rates, but for all AREs (and most metalloproteins that have been examined in our hands) are typically 10–20 mV (which is still smaller than what would be observed for an ideal, diffusional voltammogram; Bard & Faulkner, 2001; Compton & Banks, 2011; Leger & Bertrand, 2008). In CV, peaks should be symmetrical, and if they are not symmetrical, then coupled processes may be at work. Similarly, if SWV is being used, the forward and backward currents should be examined independently. If asymmetry persists, the timescale of experiments can be altered (e.g., changing the scan rate in CV from 10^2 to 3–5 mV/s or upward of 1 V/s) to assess if there are changes in the shape that can be attributed to other processes. Finally, new investigators should understand that thus far, PGE electrodes have yet to realize an ARE-based voltammogram that has current magnitudes that are greater than a few hundred nA in magnitude; and that is not labile over the timescale of hours, due to desorption from an electrode. Films may need to be reapplied over time to collect many data traces to construct a pH dependence, and the behavior of an ARE (or any metalloprotein) should be reproducible. Thus, if any kind of electrochemical response is not reproducible from experiment to experiment, or from preparation to preparation of a given ARE, it is to be viewed with extreme caution.

3. CONCLUSION

Here we have presented guidelines and details that will hopefully be of use to any investigator tackling the study of the direct electrochemistry of an ARE, or a similar O_2-sensitive metalloprotein of low redox potential. As we have found in our own recent investigations, the ability of PFE to distinguish individual redox couples with a high degree of sensitivity will make it a useful component of any ARE-centered tool kit for years to come. PFE continues to provide surprises about the reactivity, PCET, time dependence, and redox properties of AREs. As our knowledge about the superfamily grows, the electrochemical perspective will only become more important.

ACKNOWLEDGMENT

This work has been supported by the National Institutes of Health (NIGMS GM101957 and NIGMS GM120283).

REFERENCES

Akiva, E., Brown, S., Almonacid, D. E., Barber, A. E., Custer, A. F., Hicks, M. A., et al. (2014). The structure–function linkage database. *Nucleic Acids Research*, *42*(D1), D521–D530. https://doi.org/10.1093/nar/gkt1130.

Anton, B. P., Saleh, L., Benner, J. S., Raleigh, E. A., Kasif, S., & Roberts, R. J. (2008). RimO, a MiaB-like enzyme, methylthiolates the universally conserved Asp88 residue of ribosomal protein S12 in Escherichia coli. *Proceedings of the National Academy of Sciences*, *105*(6), 1826–1831.

Armstrong, F. A., Butt, J. N., & Sucheta, A. (1993). Voltammetric studies of redox-active centers in metalloproteins adsorbed on electrodes. *Methods in Enzymology*, *227*, 479–500.

Armstrong, F. A., Evans, R. M., & Megarity, C. F. (2018). Protein film electrochemistry of iron–sulfur enzymes. *Methods in Enzymology*, *599*, 387–407. https://doi.org/10.1016/bs.mie.2017.11.001.

Bard, A. J., & Faulkner, L. R. (2001). *Electrochemical methods: Fundamentals and applications*. Wiley.

Bateman, L., Leger, C., Goodin, D. B., & Armstrong, F. A. (2001). A distal histidine mutant (H52Q) of yeast cytochrome c peroxidase catalyzes the oxidation of H(2)O(2) instead of its reduction. *Journal of the American Chemical Society*, *123*(38), 9260–9263.

Benjdia, A., Subramanian, S., Leprince, J., Vaudry, H., Johnson, M. K., & Berteau, O. (2010). Anaerobic sulfatase-maturating enzyme—A mechanistic link with glycyl radical-activating enzymes? *FEBS Journal*, *277*(8), 1906–1920.

Berkovitch, F., Nicolet, Y., Wan, J. T., Jarrett, J. T., & Drennan, C. L. (2004). Crystal structure of biotin synthase, an S-adenosylmethionine-dependent radical enzyme. *Science*, *303*(5654), 76–79. https://doi.org/10.1126/science.1088493.

Bianchi, V., Eliasson, R., Fontecave, M., Mulliez, E., Hoover, D. M., Matthews, R. G., et al. (1993). Flavodoxin is required for the activation of the anaerobic ribonucleotide reductase. *Biochemical and Biophysical Research Communications*, *197*(2), 792–797. https://doi.org/10.1006/bbrc.1993.2548.

Bianchi, V., Reichard, P., Eliasson, R., Pontis, E., Krook, M., Jornvall, H., et al. (1993). Escherichia coli ferredoxin NADP+ reductase: Activation of E. coli anaerobic ribonucleotide reduction, cloning of the gene (fpr), and overexpression of the protein. *Journal of Bacteriology*, *175*(6), 1590–1595.

Blaszczyk, A. J., Silakov, A., Zhang, B., Maiocco, S. J., Lanz, N. D., Kelly, W. L., et al. (2016). Spectroscopic and electrochemical characterization of the iron–sulfur and cobalamin cofactors of TsrM, an unusual radical S-adenosylmethionine methylase. *Journal of the American Chemical Society*, *138*(10), 3416–3426. https://doi.org/10.1021/jacs.5b12592.

Brazeau, B. J., Gort, S. J., Jessen, H. J., Andrew, A. J., & Liao, H. H. (2006). Enzymatic activation of lysine 2,3-aminomutase from Porphyromonas gingivalis. *Applied and Environmental Microbiology*, *72*(9), 6402–6404.

Compton, R. G., & Banks, C. E. (2011). *Understanding voltammetry*. Imperial College Press.

Dowling, D. P., Miles, Z. D., Kohrer, C., Maiocco, S. J., Elliott, S. J., Bandarian, V., et al. (2016). Molecular basis of cobalamin-dependent RNA modification. *Nucleic Acids Research*, *44*(20), 9965–9976. https://doi.org/10.1093/nar/gkw806.

Elliott, S. J., Hoke, K. R., Heffron, K., Palak, M., Rothery, R. A., Weiner, J. H., et al. (2004). Voltammetric studies of the catalytic mechanism of the respiratory nitrate reductase from Escherichia coli: How nitrate reduction and inhibition depend on the oxidation state of the active site. *Biochemistry*, *43*(3), 799–807. https://doi.org/10.1021/bi035869j.

Firer-Sherwood, M., Pulcu, G. S., & Elliott, S. J. (2008). Electrochemical interrogations of the Mtr cytochromes from Shewanella: Opening a potential window. *Journal of Biological Inorganic Chemistry*, *13*(6), 849–854. https://doi.org/10.1007/s00775-008-0398-z.

Forouhar, F., Arragain, S., Atta, M., Gambarelli, S., Mouesca, J. M., Hussain, M., et al. (2013). Two Fe–S clusters catalyze sulfur insertion by radical-SAM methylthiotransferases. *Nature Chemical Biology*, *9*(5), 333–338. https://doi.org/10.1038/nchembio.1229.

Fourmond, V. (2016). QSoas: A versatile software for data analysis. *Analytical Chemistry*, *88*(10), 5050–5052.

Goldman, P. J., Grove, T. L., Booker, S. J., & Drennan, C. L. (2013). X-ray analysis of butirosin biosynthetic enzyme BtrN redefines structural motifs for AdoMet radical chemistry. *Proceedings of the National Academy of Sciences of the United States of America*, *110*(40), 15949–15954. https://doi.org/10.1073/pnas.1312228110.

Goldman, P. J., Grove, T. L., Sites, L. A., McLaughlin, M. I., Booker, S. J., & Drennan, C. L. (2013). X-ray structure of an AdoMet radical activase reveals an anaerobic solution for formylglycine posttranslational modification. *Proceedings of the National Academy of Sciences*, *110*(21), 8519–8524.

Grell, T. A. J., Goldman, P. J., & Drennan, C. L. (2015). SPASM and twitch domains in S-adenosylmethionine (SAM) radical enzymes. *Journal of Biological Chemistry, 290*(7), 3964–3971.

Grove, T. L., Ahlum, J. H., Qin, R. M., Lanz, N. D., Radle, M. I., Krebs, C., et al. (2013). Further characterization of Cys-type and Ser-type anaerobic sulfatase maturating enzymes suggests a commonality in the mechanism of catalysis. *Biochemistry, 52*(17), 2874–2887.

Grove, T. L., Ahlum, J. H., Sharma, P., Krebs, C., & Booker, S. J. (2010). A consensus mechanism for radical SAM-dependent dehydrogenation? BtrN contains two [4Fe–4S] clusters. *Biochemistry, 49*(18), 3783–3785.

Harder, J., Eliasson, R., Pontis, E., Ballinger, M. D., & Reichard, P. (1992). Activation of the anaerobic ribonucleotide reductase from Escherichia coli by S-adenosylmethionine. *The Journal of Biological Chemistry, 267*(35), 25548–25552.

Hernandez, H. L., Pierrel, F., Elleingand, E., Garcia-Serres, R., Huynh, B. H., Johnson, M. K., et al. (2007). MiaB, a bifunctional radical-S-adenosylmethionine enzyme involved in the thiolation and methylation of tRNA, contains two essential [4Fe–4S] clusters. *Biochemistry, 46*(17), 5140–5147. https://doi.org/10.1021/bi7000449.

Hill, H. A. O., & Hunt, N. I. (1993). Direct and indirect electrochemical investigations of metalloenzymes. *Methods in Enzymology, 227*, 501–522.

Hinckley, G. T., & Frey, P. A. (2006). Cofactor dependence of reduction potentials for [4Fe–4S] 2+/1+ in lysine 2,3-aminomutase. *Biochemistry, 45*(10), 3219–3225.

Ifuku, O., Koga, N., Haze, S.-i., Kishimoto, J., & Wachi, Y. (1994). Flavodoxin is required for conversion of dethiobiotin to biotin in Escherichia coli. *European Journal of Biochemistry, 224*(1), 173–178. https://doi.org/10.1111/j.1432-1033.1994.tb20009.x/pdf.

Jeuken, L. J., Jones, A. K., Chapman, S. K., Cecchini, G., & Armstrong, F. A. (2002). Electron-transfer mechanisms through biological redox chains in multicenter enzymes. *Journal of the American Chemical Society, 124*(20), 5702–5713.

Knappe, J., Schacht, J., Mockel, W., Hopner, T., Vetter, H., & Edenharder, R. (1969). Pyruvate formate–lyase reaction in Escherichia coli—Enzymatic system converting an inactive form of lyase into catalytically active enzyme. *European Journal of Biochemistry, 11*(2), 316–327. https://doi.org/10.1111/J.1432-1033.1969.Tb00775.X.

Landgraf, B. J., Arcinas, A. J., Lee, K. H., & Booker, S. J. (2013). Identification of an intermediate methyl carrier in the radical S-adenosylmethionine methylthiotransferases RimO and MiaB. *Journal of the American Chemical Society, 135*(41), 15404–15416. https://doi.org/10.1021/ja4048448.

Landgraf, B. J., McCarthy, E. L., & Booker, S. J. (2016). Radical S-adenosylmethionine enzymes in human health and disease. *Annual Review of Biochemistry, 85*, 485–514.

Lee, K. H., Saleh, L., Anton, B. P., Madinger, C. L., Benner, J. S., Iwig, D. F., et al. (2009). Characterization of RimO, a new member of the methylthiotransferase subclass of the radical SAM superfamily. *Biochemistry, 48*(42), 10162–10174. https://doi.org/10.1021/bi900939w.

Leger, C., & Bertrand, P. (2008). Direct electrochemistry of redox enzymes as a tool for mechanistic studies. *Chemical Reviews, 108*(7), 2379–2438. https://doi.org/10.1021/cr0680742.

Levin, B. D., Can, M., Bowman, S. E., Bren, K. L., & Elliott, S. J. (2011). Methionine ligand lability in bacterial monoheme cytochromes c: An electrochemical study. *The Journal of Physical Chemistry B, 115*(40), 11718–11726. https://doi.org/10.1021/jp203292h.

Levin, B. D., Walsh, K. A., Sullivan, K. K., Bren, K. L., & Elliott, S. J. (2015). Methionine ligand lability of homologous monoheme cytochromes c. *Inorganic Chemistry, 54*(1), 38–46. https://doi.org/10.1021/ic501186h.

Maiocco, S. J., Arcinas, A. J., Landgraf, B. J., Lee, K. H., Booker, S. J., & Elliott, S. J. (2016). Transformations of the FeS clusters of the methylthiotransferases MiaB and RimO, detected by direct electrochemistry. *Biochemistry*, *55*(39), 5531–5536. https://doi.org/10.1021/acs.biochem.6b00670.

Maiocco, S. J., Grove, T. L., Booker, S. J., & Elliott, S. J. (2015). Electrochemical resolution of the [4Fe-4S] centers of the AdoMet radical enzyme BtrN: Evidence of proton coupling and an unusual, low-potential auxiliary cluster. *Journal of the American Chemical Society*, *137*(27), 8664–8667. https://doi.org/10.1021/jacs.5b03384.

Mayhew, S. G. (1978). The redox potential of dithionite and $SO'_2{}^-$ from equilibrium reactions with flavodoxins, methyl viologen and hydrogen plus hydrogenase. *European Journal of Biochemistry*, *85*(2), 535–547. https://doi.org/10.1111/j.1432-1033.1978.tb12269.x/pdf.

McIver, L., Leadbeater, C., Campopiano, D. J., Baxter, R. L., Daff, S. N., Chapman, S. K., et al. (1998). Characterisation of flavodoxin NADP+ oxidoreductase and flavodoxin: Key components of electron transfer in Escherichia coli. *European Journal of Biochemistry*, *257*(3), 577–585. https://doi.org/10.1046/j.1432-1327.1998.2570577.x/pdf.

Mitra, S., & Elliott, S. J. (2009). Oxidative disassembly of the [2Fe–2S] cluster of human Grx2 and redox regulation in the mitochondria. *Biochemistry*, *48*(18), 3813–3815. https://doi.org/10.1021/bi900112m.

Pierrel, F., Hernandez, H. L., Johnson, M. K., Fontecave, M., & Atta, M. (2003). MiaB protein from Thermotoga maritima. Characterization of an extremely thermophilic tRNA-methylthiotransferase. *The Journal of Biological Chemistry*, *278*(32), 29515–29524. https://doi.org/10.1074/jbc.M301518200.

Pulcu, G. S., Elmore, B. L., Arciero, D. M., Hooper, A. B., & Elliott, S. J. (2007). Direct electrochemistry of tetraheme cytochrome c(554) from Nitrosomonas europaea: Redox cooperativity and gating. *Journal of the American Chemical Society*, *129*(7), 1838–1839. https://doi.org/10.1021/ja065657k.

Ruszczycky, M. W., Zhong, A., & Liu, H. W. (2018). Following the electrons: Peculiarities in the catalytic cycles of radical SAM enzymes. *Natural Product Reports*. https://doi.org/10.1039/c7np00058h.

Saeva, F. D., & Morgan, B. P. (1984). Mechanism of one-electron electrochemical reductive cleavage reactions of sulfonium salts. *Journal of the American Chemical Society*, *106*(15), 4121–4125. https://doi.org/10.1021/Ja00327a010.

Shisler, K. A., & Broderick, J. B. (2012). Emerging themes in radical SAM chemistry. *Current Opinion in Structural Biology*, *22*(6), 701–710. https://doi.org/10.1016/j.sbi.2012.10.005.

Ugulava, N. B., Gibney, B. R., & Jarrett, J. T. (2001). Biotin synthase contains two distinct iron–sulfur cluster binding sites: Chemical and spectroelectrochemical analysis of iron–sulfur cluster interconversions. *Biochemistry*, *40*(28), 8343–8351.

Wang, S. C., & Frey, P. A. (2007a). Binding energy in the one-electron reductive cleavage of S-adenosylmethionine in lysine 2,3-aminomutase, a radical SAM enzyme. *Biochemistry*, *46*(45), 12889–12895.

Wang, S. C., & Frey, P. A. (2007b). S-adenosylmethionine as an oxidant: The radical SAM superfamily. *Trends in Biochemical Sciences*, *32*(3), 101–110.

Ye, T., Kaur, R., Wen, X., Bren, K. L., & Elliott, S. J. (2005). Redox properties of wild-type and heme-binding loop mutants of bacterial cytochromes c measured by direct electrochemistry. *Inorganic Chemistry*, *44*(24), 8999–9006. https://doi.org/10.1021/ic051003l.

Yokoyama, K., Numakura, M., Kudo, F., Ohmori, D., & Eguchi, T. (2007). Characterization and mechanistic study of a radical SAM dehydrogenase in the biosynthesis of butirosin. *Journal of the American Chemical Society*, *129*(49), 15147–15155.

Yokoyama, K., Ohmori, D., Kudo, F., & Eguchi, T. (2008). Mechanistic study on the reaction of a radical SAM dehydrogenase BtrN by electron paramagnetic resonance spectroscopy. *Biochemistry*, *47*(34), 8950–8960.

CHAPTER TWELVE

Radical *S*-Adenosyl-L-Methionine (SAM) Enzyme Involved in the Maturation of the Nitrogenase Cluster

Lee Rettberg*, Kazuki Tanifuji*, Andrew Jasniewski*, Markus Walter Ribbe*,†,1, Yilin Hu*

*Department of Molecular Biology and Biochemistry, University of California, Irvine, CA, United States
†Department of Chemistry, University of California, Irvine, CA, United States
[1]Corresponding author: e-mail address: mribbe@uci.edu

Contents

Abstract

Nitrogenase is the only known enzymatic system that converts atmospheric dinitrogen (N_2) into bioavailable ammonia (NH_3). The active-site cofactor responsible for this reactivity is a [(*R*-homocitrate)$MoFe_7S_9C$] cluster that is designated as the M-cluster. This important cofactor is assembled stepwise from a pair of [Fe_4S_4] clusters that become fused into a [Fe_8S_9C] core before additional refinements take place to complete the biosynthesis. NifB, a member of the radical *S*-adenosyl-L-methionine (SAM) superfamily, facilitates the conversion of the [Fe_4S_4] clusters (called the K-cluster) to the [Fe_8S_9C] core (called the L-cluster). This transformation includes a SAM-dependent carbide insertion with concomitant incorporation of an additional sulfur. While difficulties with the purification of NifB have historically prevented detailed biochemical analyses, we have developed a heterologous expression system in *Escherichia coli* that yields stable NifB proteins from various N_2-fixing methanogenic organisms that can be used for studies.

Methods in Enzymology, Volume 606
ISSN 0076-6879
https://doi.org/10.1016/bs.mie.2018.04.009

This chapter details the procedures necessary to prepare an active NifB protein. The methods used for the biochemical characterization of the SAM-dependent carbide insertion reactions are also described.

ABBREVIATIONS

5′-dA•	5′-deoxyadenosyl radical
5′-dAD	deuterated 5′-deoxyadenosine
5′-dAH	5′-deoxyadenosine
A. vinelandii	*Azotobacter vinelandii*
ADP	adenosine diphosphate
ATP	adenosine triphosphate
CH_3SeH	methaneselenol
CH_3SH	methanethiol
CD_3SH	methane-d_3-thiol
DT	dithionite
DTT	dithiothreitol
E. coli	*Escherichia coli*
EPR	electron paramagnetic resonance
GC–MS	gas chromatography-mass spectrometry
H•	hydrogen atom
H_2	hydrogen gas
IPTG	isopropyl-β-D-1-thiogalactopyranoside
K-cluster	pair of precursor [Fe_4S_4] clusters to the L-cluster
LB	Lysogeny broth
L-cluster	[Fe_8S_9C] precursor cluster to the M-cluster
M-cluster	[(*R*-homocitrate)$MoFe_7S_9C$] cofactor of nitrogenase
MoO_4^{2-}	molybdate
N_2	dinitrogen
NH_3	ammonia
NifB	SAM-dependent scaffold protein for the nitrogenase system that is responsible for the maturation of the K-cluster to the L-cluster
$NifB^{Ma}$	NifB protein from *Methanosarcina acetivorans*
NifEN	scaffold protein for nitrogenase that facilitates the conversion of the L-cluster to the M-cluster
NifEN-B	covalent fusion of NifEN and NifB proteins
NifEN-B′	covalent fusion of NifEN and a truncated version of NifB
NifDK (also MoFe protein)	catalytic component of nitrogenase, houses the M- and P-clusters
NifH (also Fe protein)	[Fe_4S_4]-containing reductase component of nitrogenase
P-cluster	an [Fe_8S_7] cluster that behaves as an internal electron mediator in nitrogenase
PMSF	phenylmethane sulfonylfluoride
SAH	*S*-adenosyl-L-homocysteine
SAM	*S*-adenosyl-L-methionine
TFA	trifluoroacetic acid
UHPLC	ultrahigh-performance liquid chromatography

1. INTRODUCTION

Diazotrophic bacteria and archaea play a key role in the global nitrogen cycle, as these organisms utilize an enzyme called nitrogenase to convert the inert dinitrogen (N_2) molecule into the more biologically available ammonia (NH_3). This process, called nitrogen fixation, is critical for the propagation of plants and other organisms that are unable to use N_2 directly as a source of nitrogen. The best-studied nitrogenase system to date comes from the organism *Azotobacter vinelandii* and is composed of the catalytic substrate reduction component, NifDK (Fig. 1, blue), as well as a reductase component, NifH (Fig. 1, orange), and these proteins are produced from the *nifD*, *nifK*, and *nifH* genes, respectively. NifDK consists of a heterotetrameric $\alpha_2\beta_2$ structure that houses two metallocofactors: a [(*R*-homocitrate)$MoFe_7S_9C$] cluster, called the M-cluster, that binds in the active site (Lancaster et al., 2011;

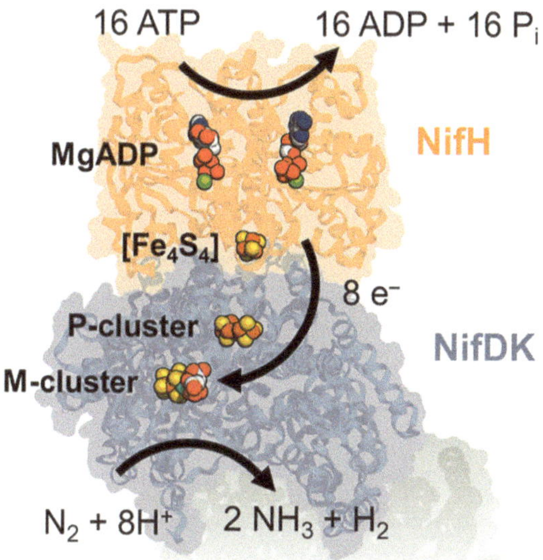

Fig. 1 Crystal structure of NifH with MgADP bound in complex with NifDK. Two subunits of NifH homodimer are shown as *red ribbons* and surface outline. One α-subunit and one β-subunit of NifDK, depicting one-half of the biologically functional heterotetramer, are shown as *blue ribbons* and surface outline. MgADP and the clusters are shown with atoms represented by *space-filling spheres*. Atoms are colored as follows: C, *light gray*; Fe, *orange*; Mg, *green*; Mo, *cyan*; N, *blue*; O, *red*; S, *yellow*. PyMOL was used to create this figure from PDB ID: 2AFI (Tezcan et al., 2005). *MgADP*, magnesium adenosine diphosphate.

Spatzal et al., 2011) as well as an $[Fe_8S_7]$ cluster, called the P-cluster, that serves as internal electron mediator (Peters et al., 1997). NifH has a γ_2 homodimeric protein structure that binds an $[Fe_4S_4]$ cubane cluster at the subunit interface and is also capable of the binding and hydrolysis of adenosine triphosphate (ATP) (Schindelin, Kisker, Schlessman, Howard, & Rees, 1997). To facilitate substrate reduction, NifDK and NifH bind to form a protein complex. The N_2 substrate binds to the M-cluster on NifDK, and NifH transfers electrons to the M-cluster, via the P-cluster, with concomitant hydrolysis of ATP (Fig. 1). The nitrogenase reaction also produces one hydrogen (H_2) molecule, and this is widely accepted as the catalytic stoichiometry for the reaction (for example, see Hoffman, Lukoyanov, Yang, Dean, & Seefeldt, 2014); however, the exact mechanism of H_2 formation remains unclear. Based on this information, the most plausible overall reaction for nitrogenase is as follows:

$$N_2 + 8H^+ + 8e^- + 16ATP \rightarrow 2NH_3 + H_2 + 16ADP + 16Pi$$

where ADP is adenosine diphosphate and P_i is inorganic phosphate.

The structure of the M-cluster is complex, and the biosynthesis of the cluster inherently involves many steps featuring several different proteins (Ribbe, Hu, Hodgson, & Hedman, 2014) (Fig. 2). The *S*-adenosyl-L-methionine (SAM)-dependent protein, NifB (Fay, Wiig, Lee, & Hu, 2015; Wiig, Hu, Lee, & Ribbe, 2012; Wiig, Hu, & Ribbe, 2011; Wilcoxen et al., 2016) is a scaffold protein that houses the first observable cofactor intermediate, the K-cluster. The K-cluster is a pair of $[Fe_4S_4]$ clusters that are fused together in the presence of SAM to generate an $[Fe_8S_9C]$ iron–sulfur precursor of the M-cluster, designated the L-cluster. The L-cluster is then transferred from NifB to another scaffold protein called NifEN, which is a NifDK homolog (Fay et al., 2011; Hu, Fay, & Ribbe, 2005; Yoshizawa et al., 2009). The L-cluster on NifEN is then converted into the M-cluster in the presence of NifH, molybdate (MoO_4^{2-}), *R*-homocitrate, and ATP. After the maturation of the M-cluster on NifEN, the cluster is delivered to the active site in NifDK making the enzyme competent for substrate reduction. As the primary focus of this chapter is providing experimental perspective for the SAM-related biosynthesis of the nitrogenase cofactor, greater emphasis will be placed on the NifB-facilitated cluster conversions in the sections below. A more detailed and comprehensive review of M-cluster biosynthesis, including the reactions that are facilitated by NifEN and NifH, can be found elsewhere (Jasniewski, Sickerman, Hu, & Ribbe, 2018; Ribbe et al., 2014; Sickerman, Rettberg, Lee, Hu, & Ribbe, 2017; Sickerman, Ribbe, & Hu, 2017).

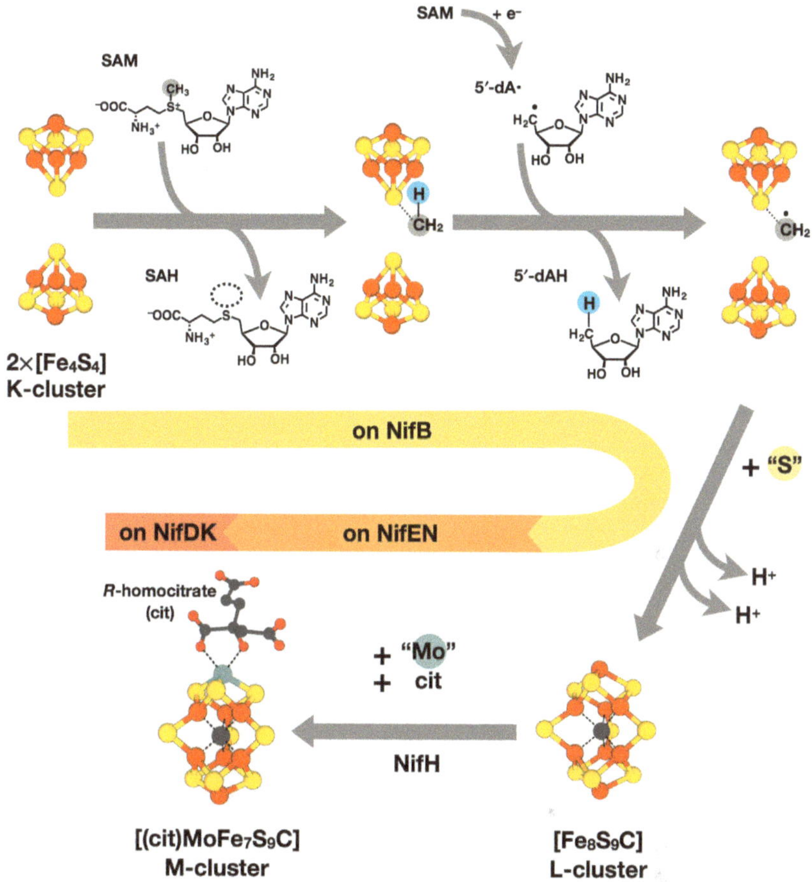

Fig. 2 A model of the M-cluster biosynthesis from the K-cluster. The SAM-dependent conversion of the K-cluster occurs on NifB, then the resultant L-cluster is transferred to NifEN, on which insertion of Mo and *R*-homocitrate takes place. The complete M-cluster is finally delivered to its target location on NifDK. Color legend: C, *gray*; Fe, *orange*; Mo, *cyan*; O, *red*; S, *yellow*. *5′-dA•*, 5′-deoxyadenosyl radical; *5′-dAH*, 5′-deoxyadenosyine; *SAH*, *S*-adenosyl-L-homocysteine; *SAM*, *S*-adenosyl-L-methionine.

NifB has long been recognized as a member of radical SAM superfamily, based on the overall sequence homology and the characteristic CXXXCXXC motif (SAM-motif) at the N-terminus (Sofia, Chen, Hetzler, Reyes-Spindola, & Miller, 2001). However, biochemical characterization of this essential protein was hindered by several factors, including the instability of the isolated protein from well-studied diazotrophs (e.g., *A. vinelandii*) and the lack of understanding of the exact roles of the protein and SAM in the cluster maturation. In an effort to address the protein instability, one successful approach employed was to fuse NifB to the C-terminus of NifN and subsequently isolate NifB as a NifEN-B association complex in *A. vinelandii* (Wiig et al., 2011); such a fusion is naturally found in the *nif* genes of the organism *Clostridium pasteurianum* (Dean & Jacobson, 1992). Purified, soluble NifEN-B is competent at restoring the activity of M-cluster-free NifDK in cluster maturation assays, proposed to be the result of sequential cluster conversions from K- to M-cluster in the same protein complex. The detailed biochemical analyses of this engineered species have provided crucial insights into the K- to L-cluster conversion, which are summarized below.

In addition to the SAM-motif that houses a $[Fe_4S_4]$ cluster (for representative reviews, see Broderick, Duffus, Duschene, & Shepard, 2014; Dowling, Vey, Croft, & Drennan, 2012; Vey & Drennan, 2011), NifB bears conserved cysteine and histidine residues which bind two additional $[Fe_4S_4]$ clusters (the K-cluster) and these clusters are most likely supplied by the NifS/NifU proteins (Johnson, Dean, Smith, & Johnson, 2005; Ribbe et al., 2014). Once on NifB, the two $[Fe_4S_4]$ clusters of the K-cluster are coupled and dynamically rearranged into the $[Fe_8S_9C]$ inorganic core of the L-cluster via concurrent incorporation of the interstitial carbide and an additional sulfur atom. These modifications were proposed using electron paramagnetic resonance (EPR) spectroscopy, by observing the SAM-dependent loss of the EPR signals assigned to $[Fe_4S_4]$ clusters, as well as the appearance of the characteristic signal for the L-cluster (Ribbe et al., 2014; Wiig et al., 2011). While the conversion of the K- to L-cluster likely consists of many steps, it is difficult to observe intermediates associated with the reaction because the only EPR-detectable species that have been reported upon treating the protein with SAM are the K- and L-clusters.

Despite the technical challenges mentioned above, efforts to characterize the reactions facilitated by NifB have involved the analysis of the products of the reaction in addition to isotopic labeling experiments, and these studies have revealed key details about the composition and mechanism of the NifB reactions (Fig. 2). The first crucial details gleaned from these experiments were that SAM functions in a dual role as the carbon-atom supplier for the L-cluster and serves as a precursor to an organic radical important for the conversion of the K- to L-cluster. Treatment of NifB with SAM gives rise to the formation of *S*-adenosyl-L-homocysteine (SAH), by delivering a methyl group from one SAM molecule to the K-cluster (Fig. 2, top left). A second molecule of SAM undergoes homolytic cleavage to generate a 5′-deoxyadenosyl radical (5′-dA$^{\bullet}$) capable of performing hydrogen atom abstraction, producing the

5′-deoxyadenosine (5′-dAH) product (Fig. 2, top right). Both of these steps are similar to the methylation and methylthiolation reactions mediated by other radical SAM enzymes (Atta et al., 2012; Bauerle, Schwalm, & Booker, 2015; Zhang, van der Donk, & Liu, 2012). A series of labeling experiments further demonstrated that 5′-dA$^{\bullet}$ abstracts a hydrogen atom (H$^{\bullet}$) from the methyl group that is transferred to the K-cluster, and the carbon atom of this methyl group is incorporated into the resultant L- and, eventually, M-cluster (Wiig et al., 2012; Wiig, Lee, Hu, & Ribbe, 2013). Another important revelation was that the methyl group of SAM is transferred to a bridging sulfide ligand of the K-cluster (Wiig, Hu, & Ribbe, 2015). When SAM-treated NifEN-B was quenched with acid to denature the protein, decomposition of the iron–sulfur species produced methanethiol and labeling determined that the carbon originated from SAM. In addition, treatment of NifEN-B with the SAM analog allyl-SAM (allyl = $-CH_2-CH=CH_2$) produced the corresponding allyl thiol following the same acid quench procedure. These results strongly support the proposal that the methyl group is delivered from SAM to a μ_3-sulfide ligand of the K-cluster.

While the NifEN-B fusion approach shed light on significant aspects of M-cluster biosynthesis, the presence of the NifEN moiety prevents the isolation of the specific contributions from NifB. As an alternative strategy, heterologous expression of NifB proteins from methanogenic, diazotrophic archaea in *Escherichia coli* has enabled an even more detailed view of cluster maturation and function (Fay et al., 2015; Wilcoxen et al., 2016). The advantage of this approach is that it provides the opportunity to carry out spectroscopic and biochemical characterizations of NifB without the presence of a NifEN component. Additionally, there is potential utility for heterologous expression of fully active nitrogenase, including its biosynthetic machinery, in nondiazotrophic bacteria; a feat which has not yet been demonstrated in a heterologous expression system. With currently available technology, the NifB protein can be expressed and isolated in a soluble form, but the $[Fe_4S_4]$ clusters still need to be reconstituted to make the protein functional, and this requires careful handling of the proteins under strictly anaerobic conditions.

This chapter will highlight the methods used in the biochemical characterization of a NifB from *Methanosarcina acetivorans* (NifB^{Ma}). These techniques include the heterologous expression of NifB^{Ma} in *E. coli*, an ultrahigh-performance liquid chromatography (UHPLC) assay for analyzing SAM cleavage products, and the detection of acid labile methanethiol by gas chromatography-mass spectrometry (GC–MS) measurements (Fay et al., 2015; Wiig et al., 2012, 2015).

2. PREPARATION OF FUNCTIONAL, RECOMBINANT NifB^{Ma} EXPRESSED IN *E. coli*

A. vinelandii and many other bacterial diazotrophs contain a NifB protein with two identified structural domains, one being the core radical SAM domain (see Section 1) and the other is designated as a NifX-like domain from sequence similarity to the *nifX* gene product, NifX. However, truncation of the NifX-like domain from NifB proteins has been identified in several species, including *Euryarchaeotes*, *Firmicutes*, and green nonsulfur bacteria (Hu & Ribbe, 2016), suggesting that this domain is not necessarily required for NifB activity. Indeed, this domain was found not to be essential for the in vitro maturation of the M-cluster, as when a NifX-truncated version of NifEN-B was used (called NifEN-B′), the activity observed was not different than the untruncated version (see supporting information of Fay et al., 2015). The NifB proteins from *M. acetivorans* (NifB^{Ma}), *Methanococcus thermolithotrophicus* (NifB^{Mt}), and *Methanocaldococcus infernus* (NifB^{Mi}) have been expressed in *E. coli* (Fay et al., 2015; Wilcoxen et al., 2016) and do not have the NifX-like domain in contrast to NifB from *A. vinelandii*. These proteins have been purified and have also been shown to be reactive, demonstrating the same SAM-dependent K- to L-cluster conversion as the protein from *A. vinelandii* but with fewer complications.

Herein, we describe experimental protocols for the expression, purification, and reconstitution of NifB^{Ma}. Using a 6xHis tag inserted at the N-terminus of NifB^{Ma}, the protein can be purified by using Ni Sepharose affinity chromatography. An $FeCl_3/Na_2S$ reconstitution of the as-purified NifB^{Ma} provides access to the active form of the protein for further biochemical analyses. The strain which expresses His-tagged NifB^{Ma} can be constructed using the methods described elsewhere (Fay et al., 2015).

2.1 Heterologous Expression of NifB^{Ma} in *E. coli*

E. coli can be grown in shaker flasks, but bioreactors can provide higher yields of cells over the same time. Generally, 50 g of wet cell paste can be recovered from a growth in a 10-L bioreactor. Cell expression at this scale gives sufficient quantities of protein for numerous experiments and this facilitates reproducible and comparable results, as protein activity levels can vary batch-to-batch.

2.1.1 Equipment

- New Brunswick Innova 40 shaker (Eppendorf)
- New Brunswick Bioflo 415 fermenter (Eppendorf; 10-L volume)
- Genesys 20 monochrometer (Thermo Spectronic)
- Sorvall Legend XTR centrifuge equipped with Fiberlite F14-6x250 LE rotor (Thermo Scientific)
- Centrifuge bottles (250-mL volume)

2.1.2 Medium and Reagent

- Lysogeny broth medium (LB medium; 10 g/L tryptone, 5 g/L yeast extract, 10 g/L NaCl)
- 100 mg/mL ampicillin
- M isopropyl-β-D-1-thiogalactopyranoside (IPTG)

2.1.3 Procedure

1. Inoculate from a frozen stock of an *E. coli* strain expressing His-tagged $NifB^{Ma}$ (YM114EE; Fay et al., 2015) into 250 mL of LB medium supplemented with 250 μL of 100 mg/mL ampicillin, and grow this preculture in a shaker at 37°C with 225 rpm agitation.
2. Prepare a 10-L batch of LB medium containing 100 μg/mL ampicillin in the fermenter, inoculate with 100 mL of the preculture, and grow the cells at 37°C, with 200 rpm agitation and 10 L/min airflow. To check the cell growth, measure optical density at 600 nm (OD_{600}) occasionally by using a spectrophotometer.
3. When the OD_{600} reaches 0.6, cool down the culture to 25°C. After the temperature is stabilized, start expressing the protein by adding 300 μL of 1 *M* IPTG and keep growing the cells for 16 h at the same temperature (25°C).
4. Harvest the grown cells by repeated centrifugation. Store the recovered cell paste as a frozen stock until used.

2.1.4 Notes

1. Typically, two 10-L fermenters are used to grow cells at one time, and the recovered cells are combined into one batch for further purification described below.
2. After removing the cell culture from the fermenter, store it at 4°C in a cold room until it is pelleted by centrifugation.

2.2 Purification of NifB^{Ma} Using Ni Sepharose Affinity Chromatography

The purification of NifB^{Ma} is similar to that for other His-tagged recombinant proteins. A single-step purification by Ni Sepharose affinity chromatography can provide a sufficient amount of NifB^{Ma} in good purity. A protease inhibitor, phenylmethane sulfonylfluoride (PMSF), is added to the cell suspension to avoid the breakdown of the target protein.

To protect the cysteine residues that bind [Fe_4S_4] clusters from undesirable oxidation, it is preferable that NifB^{Ma} is handled under anaerobic conditions. Thus, we note that the following procedure is performed using an Ar-purged glove box (typical O_2 level $<$ 3 ppm) and Schlenk techniques, unless otherwise stated. All buffers used are degassed under vacuum for 1 h at ambient temperature, and if applicable, subsequently treated with dithionite (DT).

2.2.1 Equipment

- OMNI-LAB glove box (VAC) purged with a high-purity Ar
- Schlenk line equipped with a high-purity Ar flow and a vacuum pump (Edward)
- M110L microfluidizer (Microfluidics)
- Sorvall Legend XTR centrifuge equipped with Fiberlite F14-6x250 LE rotor (Thermo Scientific)
- Spectra/Chrom chromatography column (Spectrum Chromatography; 2.5-cm diameter, 30-cm length)
- MINIPLUS2 peristaltic pump (Gilson) equipped with plastic tubes
- VHC 35 cryogenic storage container (Taylor-Wharton) filled with liquid N_2

2.2.2 Buffers and Reagent

- Ni Sepharose high performance (GE Healthcare)
- Ni-equilibration buffer: 25 m*M* Tris(hydroxymethyl)aminomethane hydrochloride (Tris–HCl; pH 8.0), 500 m*M* NaCl, 10% volume per volume (v/v) glycerol, and 2 m*M* DT
- Ni-wash buffer: Ni-equilibration buffer containing 40 m*M* imidazole
- Ni-elution buffer: Ni-equilibration buffer containing 250 m*M* imidazole
- DNase
- PMSF

2.2.3 Procedure

1. Resuspend the frozen cell paste (see Section 2.1.3) with a suitable volume of Ni-equilibration buffer (roughly 2 mL of the buffer per 1 g of the cells) supplemented with a small amount (2–3 mg) of DNase. While thawing the frozen paste, degas the cell suspension mixture for 1 h under vacuum.
2. Place the cell suspension under an argon atmosphere, then add a suitable amount of PMSF (final concentration of 1 m*M*).
3. Lyse the cells by using a microfluidizer at ~800 atm.
4. Degas the cell lysate for 1 h under vacuum, then centrifuge at 14,000 × *g* for 45 min at 4°C.
5. In the meantime, pack Ni Sepharose resin (~12.5-mL packed volume) into the chromatography column and equilibrate with 5 column volumes of Ni-equilibration buffer.
6. Collect the resultant supernatant from centrifugation into a flask by decantation. This recovered crude extract is immediately used for further purification.
7. Load the crude extract onto the preequilibrated Ni Sepharose column, then wash the loaded column with 5–10 column volumes of Ni-wash buffer to eliminate nonspecifically binding species.
8. Elute the bound protein from the column with Ni-elution buffer. The eluted protein is monitored with a flow-through spectrophotometer by following the growth of an absorbance peak at 280 nm.
9. Freeze the pure $NifB^{Ma}$ solution as pellets in liquid N_2. Store the frozen pellets in a cryogenic storage container filled with liquid N_2.

2.2.4 Notes

1. To assess if the buffers and columns used are under anaerobic conditions, a solution of methyl viologen can be utilized as an indicator, as the solution will turn blue when reduced by DT.
2. A large-scale preparation scale was described earlier mainly using Schlenk techniques, however, the entire purification procedure may also be carried out in a glove box on a smaller scale to prevent contact with oxygen.
3. A concentration of the recovered protein is determined by the Bradford protein assay. The dye and the bovine serum albumin standard can be purchased from Bio-Rad.
4. The liquid N_2 helps to prevent the protein from coming in direct contact with O_2 from the air. Caution needs to be taken when removing protein for use in assays, as water condensing on the liquid N_2 cooled protein pellets can introduce dissolved O_2 into the system.

2.3 Reconstitution of NifB^{Ma} With $FeCl_3$ and Na_2S

There are several methods that can be used to obtain active heterologously expressed [Fe_4S_4]-containing proteins. Proteins overexpressed from a plasmid can be coexpressed with *E. coli* IscSUA proteins from another plasmid (Zheng, Cash, Flint, & Dean, 1998). The presence of this IscSUA plasmid enhances the in vivo synthesis of the [Fe_4S_4] clusters that are inserted into overexpressed target [Fe_4S_4] proteins. In some cases, this results in proteins that are active as-purified (for example, see Hiller, Stiebritz, Lee, Liedtke, & Hu, 2017). However, some proteins, especially ones that require multiple [Fe_4S_4] units to function (e.g., NifB), are difficult to recover in an active form due to incomplete cluster incorporation.

Overexpressed proteins that are isolated from cells in an inactive form can be reconstituted with Fe—S clusters in vitro (Lanz et al., 2012). Combined biochemical and spectroscopic data of NifB^{Ma} reconstituted in this way indicate that NifB^{Ma} contains multiple Fe—S clusters that are competent for the generation of the L-cluster in the presence of SAM. The resultant reconstituted NifB^{Ma} is vulnerable to oxidation by oxygen, so note that the protein and buffers are handled under strictly anaerobic conditions as described in Section 2.2.

2.3.1 Equipment

- Glove box (see Section 2.2.1)
- Schlenk line (see Section 2.2.1)
- Gravity column (Bio-Rad; 0.7- to 1.5-cm diameter, 30-cm length)
- Chromatography column (see Section 2.2.1)
- Cryogenic storage container (see Section 2.2.1)

2.3.2 Buffers and Reagent

- Ni Sepharose high performance (see Section 2.2.1)
- Ni-equilibration/wash/elution buffers (see Section 2.2.1)
- Desalting buffer: 25 m*M* Tris–HCl (pH 8.0) and 5% v/v glycerol
- Sephadex G-25 gel filtration media (GE Healthcare) preswollen and washed with desalting buffer
- 500 m*M* dithiothreitol (DTT)
- 10 m*M* $FeCl_3$
- 10 m*M* Na_2S

2.3.3 Procedure

1. Pack the G-25 resin into the Bio-Rad gravity column, then equilibrate with 5 column volumes of desalting buffer.
2. Desalt the as-purified $NifB^{Ma}$ (97 mg; see Section 2.2.3) by using the G-25 column.
3. Dilute the desalted $NifB^{Ma}$ solution to ~50 mL with desalting buffer, then add 1.28 mL of 500 m*M* DTT.
4. On an ice bath, add 6.4 mL of 10 m*M* $FeCl_3$ and the same volume of 10 m*M* Na_2S slowly to the diluted $NifB^{Ma}$ followed by an incubation with gentle stirring for 1 h.
5. Meanwhile, pack Ni Sepharose resin (~12.5-mL packed volume) into the chromatography column and equilibrate the column with 5 column volumes of Ni-equilibration buffer.
6. Dilute the reconstitution mixture of $NifB^{Ma}$ with at least 130 mL of Ni-equilibration buffer, then load the diluted mixture onto the preequilibrated Ni Sepharose column.
7. Wash the loaded column with 5–10 column volumes of Ni-wash buffer.
8. Elute the bound $NifB^{Ma}$ protein with Ni-elution buffer and collect the darkly colored fraction into a flask.
9. Freeze the reconstituted $NifB^{Ma}$ solution as pellets in liquid N_2, then store the frozen pellets in the cryogenic container.

2.3.4 Notes

1. The reconstituted $NifB^{Ma}$ is typically dark-colored; hence, the elution of the protein can be observed by the naked eye.
2. The reconstituted $NifB^{Ma}$ is quite sensitive to oxygen, precipitously losing activity in the presence of any air contamination. If buffers are used without added DT (i.e., desalting buffer, see Section 2.3.2), they should be degassed under vacuum for 1 h and, prior to use, equilibrated under argon in the glove box at least overnight.

3. SAM-RELATED BIOCHEMICAL ANALYSIS OF THE RECONSTITUTED $NifB^{Ma}$

As described earlier, SAM plays two main roles in the reactions facilitated by NifB. In the first role, SAM delivers a methyl group, likely through an S_N2 displacement (Boal et al., 2011; Grove et al., 2011) to a bridging sulfide ligand of the K-cluster, and this cleaves SAM heterolytically and releases SAH as a product. In the second role, another SAM molecule receives an electron from the [Fe_4S_4] cluster of NifB bound to the SAM-motif, and this cleaves SAM homolytically to generate 5′-dA$^{\bullet}$. This radical species abstracts a hydrogen atom from the methyl group bound to the K-cluster to generate 5′-dAH as a product (Fig. 3). The hydrogen atom abstraction generates a radical intermediate on the K-cluster (i.e., methylene radical: —$CH_2^{\bullet}$—) which undergoes further yet uncharacterized reactions, and eventually serves as the source of the interstitial carbide. When NifB is incubated with [methyl-d_3] SAM, deuterated 5′-deoxyadenosine (5′-dAD) is produced, providing evidence of hydrogen abstraction from the methyl group bound to the K-cluster. Further, ^{14}C labeling of this methyl group demonstrates that the methyl carbon ends up in the L- and M-clusters, as the radioactivity of the proteins and extracted clusters can be clearly measured (Wiig et al., 2012, 2013).

Further insights into the L-cluster formation came from efforts to characterize the intermediate species by acid quenching the SAM-treated NifB (Wiig et al., 2015). Incubation of NifB with nonlabeled and [methyl-d_3] SAM afforded methanethiol (CH_3SH) (Fig. 4A) and methane-d_3-thiol (CD_3SH) (Fig. 4B), respectively, upon acid quench. Combined with the formation of methaneselenol (CH_3SeH) from Fe/Se-reconstituted NifB, these results further support the transfer of the SAM-derived methyl group to the acid labile sulfur atom of the [Fe_4S_4] component. Interestingly, methanethiol formation was not observed when NifB was oxidized, which may

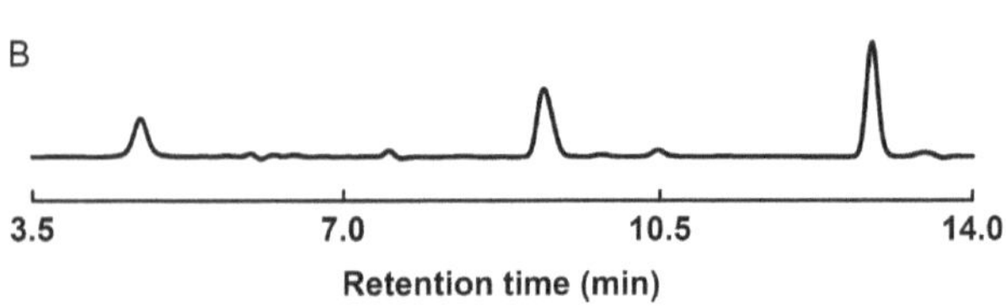

Fig. 3 HPLC elution profile from SAM cleavage assays. (A) Trace (i) is a standard consisting of SAM, SAH, and 5′-dAH. (B) Trace (ii) shows products of an assay containing reduced NifB^{Ma} incubated with SAM, resulting in SAM, SAH, and 5′-dAH. *5′-dAH*, 5′-deoxyadenosyine; *SAH*, *S*-adenosyl-L-homocysteine; *SAM*, *S*-adenosyl-L-methionine.

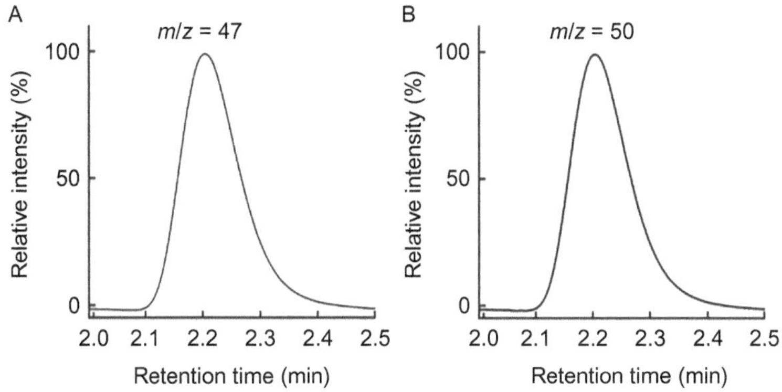

Fig. 4 GC–MS analysis of acid-quenched NifB^{Ma} after incubation with SAM (A) or [methyl-d_3] SAM (B). The products were identified as methanethiol and methane-d_3-thiol by their GC retention times and GC–MS *m/z* ratios of 47 and 50, respectively. *SAM*, *S*-adenosyl-L-methionine.

be due to either a low affinity of SAM to the oxidized clusters or to the lower nucleophilicity of the bridging sulfur atoms in the oxidized cluster.

In this section, a SAM cleavage assay and acid labile methanethiol detection will be described for the analysis of the SAM-related reactivities of NifB. The protocols shown here can be applied to labeling experiments as needed. Again, note that the protein and buffers are handled under strictly anaerobic conditions as described in Section 2.2.

3.1 SAM Cleavage Assay

UHPLC-based SAM cleavage assays are used to monitor the methyl group transfer and hydrogen atom abstraction in which SAH and 5′-dAH are produced, respectively. Reverse-phase chromatography using a C18 column gives good separation between SAM, SAH, and 5′-dAH. All of these components can be monitored by following the absorbance spectrum at 254 nm, a transition that is due to the aromatic rings of the SAM-derived products. Commercially purchased standards of SAM, SAH, and 5′-dAH are required to determine the retention time of each compound. If required, the separated flow-through sample is applied to a mass spectrometer for further analysis.

3.1.1 Equipment

- Glove box (see Section 2.2.1)
- Poly-prep chromatography column (Bio-Rad; 2-mL bed volume)

- AmiconUltra centrifugal filter (MilliporeSigma; 500-μL volume, 30,000 molecular weight cut-off)
- 300-μL polypropylene vial with septum screw cap (Thermo Scientific)
- Dionex Ultimate 3000 UHPLC system (Thermo Scientific) equipped with an Acclaim120 C18 column (Thermo Scientific; 4.6 × 100 mm, 5-μm particle size)

3.1.2 Buffers and Reagent

- Sephadex G-25 gel filtration media (see Section 2.3.2)
- Desalting buffer (see Section 2.3.2)
- 1 m*M* SAM
- 20% weight per volume (w/v) trifluoroacetic acid (TFA)
- Mobile phase A: 50 m*M* KH_2PO_4 (pH 6.6)
- Mobile phase B: methanol

3.1.3 Procedure

1. Pack G-25 resin into the poly-prep chromatography column followed by equilibration with 5 column volumes of desalting buffer.
2. Desalt the reconstituted $NifB^{Ma}$ (0.32 mg; see Section 2.3.3) by a single passage through the preequilibrated G-25 column, then aliquot a part of the desalted $NifB^{Ma}$ (0.29 mg) for an assay.
3. Increase the volume of the aliquoted $NifB^{Ma}$ solution to 495 μL with desalting buffer.
4. Start the reaction by addition of 5 μL of 1 m*M* SAM to the $NifB^{Ma}$ portion, then incubated at 25°C for 30 min.
5. Under air, transfer the reaction mixture into an AmiconUltra centrifugal filter and spin at 13,000 rpm, at 4°C, for 10 min to remove the protein from the measurement sample.
6. Transfer 200 μL of the flow-through sample into a 300-μL microvial followed by treatment with 1.4 μL of 20% w/v TFA to stabilize the analytes. Then, cap the vial to avoid sample evaporation.
7. Inject 100 μL of the prepared aqueous sample into the UHPLC system (see Section 3.1.1), then analyze under the following conditions: flow rate, 0.5 mL/min; column temperature, 30°C; detection, absorbance at 254 nm. The mobile phase composition is programmed as follows: preequilibration, 98% mobile phase A and 2% mobile phase B; 0–20 min after sample injection (SI), a linear gradient of 2%–60% mobile phase B; 20–28 min after SI, 40% mobile phase A and 60% mobile phase B; 28–32 min after SI, a linear gradient of 60%–2% mobile phase B.

3.1.4 Notes

1. For background subtraction, run and analyze a sample containing all components except SAM.
2. For experimental controls, the reconstituted NifB^{Ma} can be substituted by, for example, no protein and the as-purified NifB^{Ma} (see Section 2.2.3).
3. Standards of SAM, SAH, and 5′-dAH should be prepared from commercially available reagents.
4. The SAM used for the cleavage assays and acid labile methanethiol detection is unstable in solution particularly at high pH values (Parks & Schlenk, 1958). Thus, the stock solutions of SAM (see Sections 3.1.2 and 3.2.2) should be prepared on an ice bath and stored at −20°C until use.
5. The liquid sample used for the SAM cleavage product analysis should be treated with 20% w/v TFA immediately after the completion of the assay to stabilize SAM under acidic conditions. If caution is not taken, the SAM remaining in the reaction mixture can be hydrolyzed prior to injecting the sample.

3.2 Acid Labile Methanethiol Detection

The acid labile methanethiol (boiling point, 5.95°C) can be detected through a gas phase analysis using a GC–MS. Since methanethiol is volatile and can be potentially oxidized, the experimental samples should be immediately analyzed once the reactions are completed. The analytical method described here can be directly applied to labeling experiments because isotopic mass shifts can be detected by a GC–MS instrument.

3.2.1 Equipment

- Glove box (see Section 2.2.1)
- Poly-prep chromatography column (see Section 3.1.1)
- 400-μL glass vial with aluminum crimp seal (Thermo Scientific)
- Trace 1300 gas chromatography connected to ISQ QD single quadrupole mass spectrometry (Thermo Fisher Scientific) with a Rxi-1ms column (Restek; 30-m length, 0.32-mm internal diameter, 4.0-μm film thickness)
- MGW Lauda RM6 isothermal bath (Brinkmann)

3.2.2 Buffers and Reagent

- Sephadex G-25 gel filtration media (see Section 2.3.2)
- Desalting buffer (see Section 2.3.2)
- 100 m*M* SAM
- 1 *M* hydrochloric acid (HCl)

3.2.3 Procedure

1. Prepare a G-25 column and desalt the reconstituted NifB^{Ma} (0.4 mg; see Section 2.3.3) as described in Section 3.1.3. Then, aliquot a part of the desalted NifB^{Ma} (0.36 mg) into a 400-μL glass vial for an assay.
2. Dilute the aliquoted NifB^{Ma} to 98 μL with desalting buffer.
3. Start the reaction by addition of 2 μL of 100 m*M* SAM, mix immediately by pipetting, and crimp the vial with an aluminum seal.
4. Incubate the reaction mixture at 25°C for 30 min, and then quench by the addition of 25 μL of 1 *M* HCl.
5. Incubate the quenched sample at 60°C in an isothermal bath for 15 min followed by equilibration at 25°C for 10 min.
6. Inject 100 μL of the vial's headspace gas into a GC–MS (see Section 3.2.1). The temperature program for the analysis is as follows: preequilibration, 30°C; 0–8 min after SI, 30°C; 8–12.25 min after SI, a linear heating up to 200°C; 12.25–13.25 min after SI, 200°C; 13.25 min after SI, cooling down to 30°C.

3.2.4 Notes

1. Samples without SAM or without NifB^{Ma} can be run and analyzed as experimental controls.
2. Standards for calibration are prepared from aqueous solutions of sodium methanethiolate. The samples are quenched, treated, and analyzed in the same manner as the experimental samples.

4. SUMMARY AND CONCLUSION

The NifB protein is a critical component of the nitrogenase operon and is the first scaffold onto which the complex metallocofactor, the M-cluster, is synthesized. While isolation of NifB from *A. vinelandii* came with a host of challenges, the expression and purification of NifB^{Ma} was less complicated, employing a standard protocol used for recombinant proteins with the inclusion of air-free techniques. Maintaining a high degree of control over the anaerobic components significantly affects reproducibility and reliability of the reconstitution and biochemical assays associated with NifB^{Ma}.

The advances made with the heterologous expression, stable preparation, and functional reconstitution of NifB^{Ma} provided opportunities to specifically investigate the biochemical and spectroscopic properties of NifB. The methods and techniques described for reactivity assays of NifB^{Ma} in the presence of SAM have been used to shed light on the function of this versatile cofactor, determining the origin and initial binding location of the eventual carbide atom. Recently, this single-component enzyme was also used to elucidate the enigmatic sulfur insertion process during the conversion of the K-cluster to the L-cluster (Tanifuji et al., 2018). Even though NifB^{Ma} is not expressed in an active form, the expression of NifB in *E. coli* shows the potential utility of this approach and provides the first steps toward transplanting fully active biosynthetic machinery for nitrogen fixation into desirable host cells. Combined with the successful expression of functional NifH from *M. acetivorans* in *E. coli* (Hiller et al., 2017), future research efforts will focus on developing in vivo systems using heterologous expression techniques.

ACKNOWLEDGMENTS

This work is supported by supported by NIH-NIGMS Grant GM67626 (to M.W.R. and Y.H.).

REFERENCES

Atta, M., Arragain, S., Fontecave, M., Mulliez, E., Hunt, J. F., Luff, J. D., et al. (2012). The methylthiolation reaction mediated by the radical-SAM enzymes. *Biochimica et Biophysica Acta—Proteins and Proteomics*, *1824*(11), 1223–1230. https://doi.org/10.1016/j.bbapap.2011.11.007.

Bauerle, M. R., Schwalm, E. L., & Booker, S. J. (2015). Mechanistic diversity of radical S-adenosylmethionine (SAM)-dependent methylation. *Journal of Biological Chemistry*, *290*(7), 3995–4002. https://doi.org/10.1074/jbc.R114.607044.

Boal, A. K., Grove, T. L., McLaughlin, M. I., Yennawar, N. H., Boomsma, D. I., & Rosenzweig, A. C. (2011). Structural basis for methyl transfer by a radical SAM enzyme. *Science*, *332*(6033), 1089–1092. https://doi.org/10.1126/science.1205358.

Broderick, J. B., Duffus, B. R., Duschene, K. S., & Shepard, E. M. (2014). Radical S-adenosylmethionine enzymes. *Chemical Reviews*, *114*(8), 4229–4317. https://doi.org/10.1021/cr4004709.

Dean, D. R., & Jacobson, M. R. (1992). Biochemical genetics of nitrogenase. In G. Stacey, R. H. Burris, & H. J. Evans (Eds.), *Biological nitrogen fixation* (pp. 763–834). London: Chapmann and Hall.

Dowling, D. P., Vey, J. L., Croft, A. K., & Drennan, C. L. (2012). Structural diversity in the AdoMet radical enzyme superfamily. *Biochimica et Biophysica Acta—Proteins and Proteomics*, *1824*(11), 1178–1195. https://doi.org/10.1016/j.bbapap.2012.04.006.

Fay, A. W., Blank, M. A., Lee, C. C., Hu, Y., Hodgson, K. O., Hedman, B., et al. (2011). Spectroscopic characterization of the isolated iron-molybdenum cofactor (FeMoco) precursor from the protein NifEN. *Angewandte Chemie International Edition*, *50*(34), 7787–7790. https://doi.org/10.1002/anie.201102724.

Fay, A. W., Wiig, J. A., Lee, C. C., & Hu, Y. (2015). Identification and characterization of functional homologs of nitrogenase cofactor biosynthesis protein NifB from methanogens. *Proceedings of the National Academy of Sciences of the United States of America*, *112*(48), 14829–14833. https://doi.org/10.1073/pnas.1510409112.

Grove, T. L., Benner, J. S., Radle, M. I., Ahlum, J. H., Landgraf, B. J., Krebs, C., et al. (2011). A radically different mechanism for S-adenosylmethionine—Dependent methyltransferases. *Science*, *332*(6029), 604–607. https://doi.org/10.1126/science.1200877.

Hiller, C. J., Stiebritz, M. T., Lee, C. C., Liedtke, J., & Hu, Y. (2017). Tuning electron flux through nitrogenase with methanogen iron protein homologues. *Chemistry—A European Journal*, *23*(64), 16152–16156. https://doi.org/10.1002/chem.201704378.

Hoffman, B. M., Lukoyanov, D., Yang, Z., Dean, D. R., & Seefeldt, L. C. (2014). Mechanism of nitrogen fixation by Nitrogenase: The next stage. *Chemical Reviews*, *114*(8), 4041–4062. https://doi.org/10.1021/cr400641x.

Hu, Y., Fay, A. W., & Ribbe, M. W. (2005). Identification of a nitrogenase FeMo cofactor precursor on NifEN complex. *Proceedings of the National Academy of Sciences of the United States of America*, *102*(9), 3236–3241. https://doi.org/10.1073/pnas.0409201102.

Hu, Y., & Ribbe, M. W. (2016). Maturation of nitrogenase cofactor—The role of a class E radical SAM methyltransferase NifB. *Current Opinion in Chemical Biology*, *31*, 188–194. https://doi.org/10.1016/j.cbpa.2016.02.016.

Jasniewski, A. J., Sickerman, N. S., Hu, Y., & Ribbe, M. W. (2018). The Fe protein: An unsung hero of nitrogenase. *Inorganics*, *6*(1), 25. https://doi.org/10.3390/inorganics6010025.

Johnson, D. C., Dean, D. R., Smith, A. D., & Johnson, M. K. (2005). Structure, function, and formation of biological iron-sulfur clusters. *Annual Review of Biochemistry*, *74*(1), 247–281. https://doi.org/10.1146/annurev.biochem.74.082803.133518.

Lancaster, K. M., Roemelt, M., Ettenhuber, P., Hu, Y., Ribbe, M. W., Neese, F., et al. (2011). X-ray emission spectroscopy evidences a central carbon in the nitrogenase iron-molybdenum cofactor. *Science*, *334*(6058), 974–977. https://doi.org/10.5061/dryad.6m0f6870.

Lanz, N. D., Grove, T. L., Gogonea, C. B., Lee, K. H., Krebs, C., & Booker, S. J. (2012). RlmN and AtsB as models for the overproduction and characterization of radical SAM proteins. In D. A. Hopwood (Ed.), *Methods in enzymology: Vol. 516. Natural product biosynthesis by microorganisms and plants, part B* (pp. 125–152). Amsterdam: Elsevier. https://doi.org/10.1016/B978-0-12-394291-3.00030-7.

Parks, L. W., & Schlenk, F. (1958). The stability and hydrolysis of S-adenosylmethionine; isolation of S-ribosylmethionine. *The Journal of Biological Chemistry*, *230*(1), 295–305. Retrieved from https://www.jbc.org/content/230/1/295.citation.

Peters, J. W., Stowell, M. H. B., Soltis, S. M., Finnegan, M. G., Johnson, M. K., & Rees, D. C. (1997). Redox-dependent structural changes in the nitrogenase P-cluster. *Biochemistry*, *36*(6), 1181–1187. https://doi.org/10.1021/bi9626665.

Ribbe, M. W., Hu, Y., Hodgson, K. O., & Hedman, B. (2014). Biosynthesis of nitrogenase metalloclusters. *Chemical Reviews*, *114*(8), 4063–4080. https://doi.org/10.1021/cr400463x.

Schindelin, H., Kisker, C., Schlessman, J. L., Howard, J. B., & Rees, D. C. (1997). Structure of ADP·AlF4—Stabilized nitrogenase complex and its implications for signal transduction. *Nature*, *387*(6631), 370–376. https://doi.org/10.1038/387370a0.

Sickerman, N. S., Rettberg, L. A., Lee, C. C., Hu, Y., & Ribbe, M. W. (2017). Cluster assembly in nitrogeanse. *Essays in Biochemistry*, *61*(2), 271–279. https://doi.org/10.1042/EBC20160071.

Sickerman, N. S., Ribbe, M. W., & Hu, Y. (2017). Nitrogenase cofactor assembly: An elemental inventory. *Accounts of Chemical Research*, *50*(11), 2834–2841. https://doi.org/10.1021/acs.accounts.7b00417.

Sofia, H. J., Chen, G., Hetzler, B. G., Reyes-Spindola, J. F., & Miller, N. E. (2001). Radical SAM, a novel protein superfamily linking unresolved steps in familiar biosynthetic pathways with radical mechanisms: Functional characterization using new analysis and information visualization methods. *Nucleic Acids Research*, *29*(5), 1097–1106. https://doi.org/10.1093/nar/29.5.1097.

Spatzal, T., Aksoyoglu, M., Zhang, L., Andrade, S. L. A., Schleicher, E., Weber, S., et al. (2011). Evidence for interstitial carbon in nitrogenase FeMo cofactor. *Science*, *334*(6058), 940. https://doi.org/10.1126/science.1214025.

Tanifuji, K., Lee, C. C., Sickerman, N. S., Tatsumi, K., Ohki, Y., Hu, Y., et al. (2018). Tracing the '9th sulfur' of the nitrogenase cofactor via a semi-synthetic approach. *Nature Chemistry*, *10*(5), 568–572. https://doi.org/10.1038/s41557-018-0029-4.

Tezcan, F. A., Kaiser, J. T., Mustafi, D., Walton, M. Y., Howard, J. B., & Rees, D. C. (2005). Nitrogenase complexes: Multiple docking sites for a nucleotide switch protein. *Science*, *309*(5739), 1377–1380. https://doi.org/10.1126/science.1115653.

Vey, J. L., & Drennan, C. L. (2011). Structural insights into radical generation by the radical SAM superfamily. *Chemical Reviews*, *111*(4), 2487–2506. https://doi.org/10.1021/cr9002616.

Wiig, J. A., Hu, Y., Lee, C. C., & Ribbe, M. W. (2012). Radical SAM-dependent carbon insertion into the nitrogenase M-cluster. *Science*, *337*(6102), 1672–1675. https://doi.org/10.1126/science.1224603.

Wiig, J. A., Hu, Y., & Ribbe, M. W. (2011). NifEN-B complex of *Azotobacter vinelandii* is fully functional in nitrogenase FeMo cofactor assembly. *Proceedings of the National Academy of Sciences of the United States of America*, *108*(21), 8623–8627. https://doi.org/10.1073/pnas.1102773108.

Wiig, J. A., Hu, Y., & Ribbe, M. W. (2015). Refining the pathway of carbide insertion into the nitrogenase M-cluster. *Nature Communications*, *6*, 8034. https://doi.org/10.1038/ncomms9034.

Wiig, J. A., Lee, C. C., Hu, Y., & Ribbe, M. W. (2013). Tracing the interstitial carbide of the nitrogenase cofactor during substrate turnover. *Journal of the American Chemical Society*, *135*(13), 4982–4983. https://doi.org/10.1021/ja401698d.

Wilcoxen, J., Arragain, S., Scandurra, A. A., Jimenez-Vicente, E., Echavarri-Erasun, C., Pollmann, S., et al. (2016). Electron paramagnetic resonance characterization of three iron-sulfur clusters present in the nitrogenase cofactor maturase NifB from *Methanocaldococcus infernus*. *Journal of the American Chemical Society*, *138*(24), 7468–7471. https://doi.org/10.1021/jacs.6b03329.

Yoshizawa, J. M., Blank, M. A., Fay, A. W., Chi, C. L., Wiig, J. A., Hu, Y., et al. (2009). Optimization of FeMoco maturation on NifEN. *Journal of the American Chemical Society*, *131*(26), 9321–9325. https://doi.org/10.1021/ja9035225.

Zhang, Q., van der Donk, W. A., & Liu, W. (2012). Radical-mediated enzymatic methylation: A tale of two SAMS. *Accounts of Chemical Research*, *45*(4), 555–564. https://doi.org/10.1021/ar200202c.

Zheng, L., Cash, V. L., Flint, D. H., & Dean, D. R. (1998). Assembly of iron-sulfur clusters. *The Journal of Biological Chemistry*, *273*(21), 13264–13272. https://doi.org/10.1074/jbc.273.21.13264.

CHAPTER THIRTEEN

Purification, Characterization, and Biochemical Assays of Biotin Synthase From *Escherichia coli*

Julia D. Cramer, Joseph T. Jarrett[1]
Department of Chemistry, University of Hawaii at Manoa, Honolulu, HI, United States
[1]Corresponding author: e-mail address: jtj@hawaii.edu

Contents

Abstract

Biotin synthase (BioB) catalyzes the oxidative insertion of a sulfur atom between the C6 methylene and the C9 methyl positions in dethiobiotin. The enzyme couples oxidation of each carbon position to reduction of the *S*-adenosyl-L-methionine (SAM) sulfonium center, generating 5′-deoxyadenosine and L-methionine, products that are characteristic of enzymes from the radical SAM superfamily. In bacteria, biotin biosynthesis is tightly regulated by the dual-function BirA repressor/holocarboxylase synthetase, resulting in very low levels of all biotin biosynthetic enzymes such that activity-based purification of BioB from the native organism is virtually impossible. However, overexpression and purification of recombinant BioB from *E. coli* are straight forward and, in contrast with many radical SAM enzymes, can be carried out under aerobic conditions. The active enzyme contains two iron–sulfur clusters, and the characterization and manipulation of these clusters are essential for a thorough understanding of enzyme catalysis and stability.

Methods in Enzymology, Volume 606
ISSN 0076-6879
https://doi.org/10.1016/bs.mie.2018.06.003

An optimized in vitro assay for BioB is described herein that requires use of an auxiliary protein reducing system and must be carried out under anaerobic conditions to prevent oxidative damage to the reduced iron–sulfur clusters. Three methods for detection of biotin are described, with discussion of the advantages and limitations of each method. Challenges that may be encountered in adapting these assays to other organisms are also discussed.

1. INTRODUCTION

Biotin (vitamin H or B7) was one of the last water-soluble vitamins to be discovered (du Vigneaud, Melville, Gyorgy, & Rose, 1940) and groundbreaking work by Rolfe and Eisenberg (1968), and separately by Campbell and coworkers (Del Campillo-Campbell, Kayajanian, Campbell, & Adhya, 1967) used *Escherichia coli* mutants to establish a pathway from pimeloyl CoA and L-alanine that required only four enzymes: BioF, BioA, BioD, and BioB (Marquet, Tse Sum Bui, & Florentin, 2001). BioF and BioA are pyridoxal phosphate-dependent enzymes that couple pimeloyl CoA and L-alanine together and transfer an amino group from SAM, respectively (Marquet et al., 2001; Ploux & Marquet, 1992). The resulting pathway intermediate, 7,8-diaminopelargonic acid, is converted to dethiobiotin (DTB) through an ATP-dependent condensation with CO_2 as catalyzed by BioD (Gibson, 1997; Marquet et al., 2001). Each of these reactions has been characterized using recombinant enzymes and follows well-precedented chemical mechanisms. The final step, conversion of DTB to biotin, was characterized in *E. coli* using feeding studies with radiolabeled precursors (DeMoll & Shive, 1983; Parry & Kunitani, 1976), but the conversion could not be replicated in vitro. Sequencing of the *bio* operon led to the recognition that BioB shared partial sequence homology with the known *S*-adenosyl-L-methionine (SAM)-dependent radical enzymes pyruvate formate-lyase activating enzyme (PflB) and type III ribonucleotide reductase activating subunit (NrdG) (Otsuka et al., 1988). Several laboratories were then able to produce limited substoichiometric activity using recombinant enzyme in cell-free bacterial extracts by addition of SAM, a reductant, and a number of other factors that were not essential but improved activity (Birch, Fuhrmann, & Shaw, 1995; Ifuku, Kishimoto, Haze, Yanagi, & Fukushima, 1992; Sanyal, Gibson, & Flint, 1996; Tse Sum Bui et al., 1998).

Recombinant BioB from *E. coli*, purified under aerobic conditions, is a dimer that contains one air-stable $[2Fe–2S]^{2+}$ cluster in each monomer

(Cosper et al., 2004; Sanyal, Cohen, & Flint, 1994; Ugulava, Gibney, & Jarrett, 2000). In contrast, other known radical SAM enzymes utilize an air-sensitive $[4Fe–4S]^{2+/+}$ cluster to catalyze radical generation through reductive cleavage of SAM (Broderick, Duffus, Duschene, & Shepard, 2014; Frey & Magnusson, 2003). Further progress in establishing a defined in vitro assay for BioB required thorough characterization and manipulation of the iron–sulfur content of the enzyme; these methods are described later. Under mild reducing conditions, the $[4Fe–4S]^{2+}$ cluster can be reconstituted while the $[2Fe–2S]^{2+}$ cluster remains intact (Ugulava, Gibney, & Jarrett, 2001; Ugulava, Surerus, & Jarrett, 2002), and the resulting holoenzyme is active for one turnover (Ugulava, Sacanell, & Jarrett, 2001). In addition to DTB and SAM, in vitro activity required a reducing system to provide electrons for reductive cleavage of SAM, and with the *E. coli* enzyme, improved activity was obtained using a native protein-based reducing system consisting of flavodoxin, ferredoxin(flavodoxin):$NADP^{+}$ oxidoreductase, and NADPH (Ifuku, Koga, Haze, Kishimoto, & Wachi, 1994; Sanyal et al., 1996; Ugulava, Sacanell, et al., 2001).

With a thorough knowledge of the components needed for a defined in vitro assay, and following the elucidation of a structure of the holoenzyme with substrates bound (Berkovitch, Nicolet, Wan, Jarrett, & Drennan, 2004), the mechanism of thiophane ring formation could be further elucidated (Fig. 1). Deuterium atoms at the C6 and C9 positions were shown to be transferred to the product 5′-deoxyadenosine (dAH), suggesting that reduction of the SAM sulfonium is coupled with direct abstraction of a hydrogen atom from these positions (Escalettes, Florentin, Tse Sum Bui, Lesage, & Marquet, 1999). Sulfur incorporated at these positions to generate the thiophane ring is derived from the $[2Fe–2S]^{2+}$ cluster (Farrar, Siu, Howell, & Jarrett, 2010; Ugulava, Gibney, et al., 2001). However, the enzyme active site contains only one equivalent of SAM, indicating that the reaction must proceed in a stepwise manner, with a discrete intermediate that persists on the enzyme while 5′-dAH and methionine generated during the first C—S bond formation step dissociate and a second equivalent of SAM binds to catalyze the subsequent ring closure step (Taylor, Farrar, & Jarrett, 2008). This intermediate was shown to be 9-mercaptodethiobiotin (9-MDTB) (Taylor et al., 2008), bound as a ligand to a $[2Fe–2S]^{+}$ cluster (Fugate et al., 2012), with formation of this intermediate rate-limiting and ring closure to generate biotin approximately fivefold faster. A single enzyme turnover of biotin formation results in the destruction of the $[2Fe–2S]^{2+}$ cluster and subsequent turnover presumably requires restoring the enzyme to the original

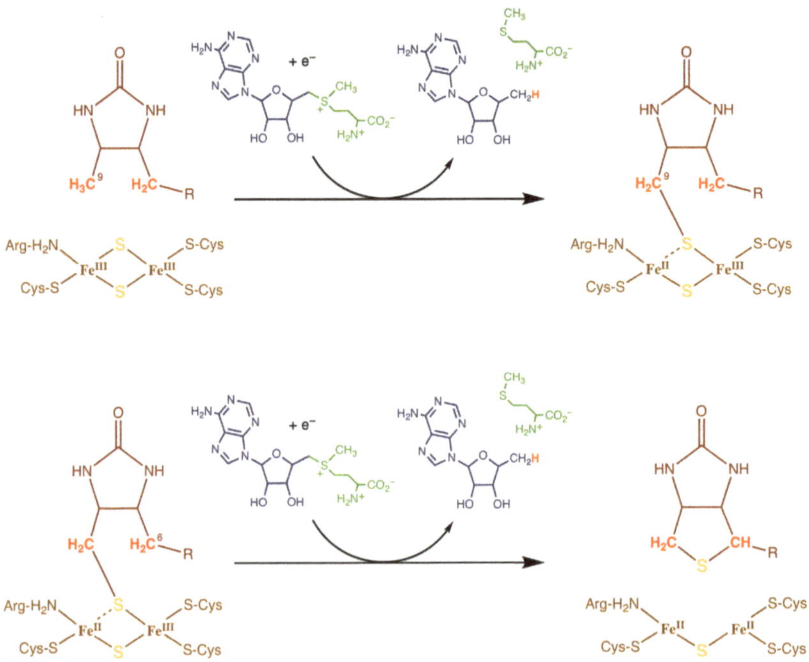

Fig. 1 The proposed BioB-catalyzed reaction. Reduction of the SAM sulfonium is coupled with abstraction of a hydrogen atom from the C9 position of DTB, resulting in formation of 5′-dAH, Met, and a putative C9 DTB radical. This radical reacts with the $[2Fe–2S]^{2+}$ cluster to generate 9-MDTB, which stays bound as 5′-dAH dissociates and a second equivalent of SAM binds. A similar reaction directed at C6 leads to ring closure and formation of biotin.

holoenzyme form. Multiple in vitro turnovers have been detected in the presence of excess sulfide but proceed at a rate ($k_{\text{steady state}} = 0.009\,\text{min}^{-1}$) that is at least 10-fold slower than the first turnover ($k_{\text{burst}} = 0.1\,\text{min}^{-1}$) (Farrar et al., 2010). In vivo, the iron–sulfur cluster assembly system likely plays an important role in regenerating active enzyme after each turnover (Reyda, Fugate, & Jarrett, 2009).

This chapter will cover methods for purifying recombinant *E. coli* BioB and characterizing and manipulating the iron–sulfur clusters present in the recombinant enzyme. Detailed methods for in vitro assays based on the quantification of biotin will be described. Finally, challenges that may be encountered in trying to adapt these protocols to BioB from other species will be discussed.

2. EXPRESSION AND PURIFICATION OF RECOMBINANT BioB

BioB has been expressed and purified, both from the native sequence, and using an N-terminal hexahistidine tag (Ugulava et al., 2000). IPTG-inducible expression using a T7 promoter results in BioB being present at ca. 5%–10% of total cytosolic protein. The native (nontagged) enzyme can be purified to ~80%–90% purity by anion exchange chromatography followed by gel filtration chromatography, but further purification to homogeneity by hydrophobic interaction chromatography results in a significant decrease in activity. Purification of His_6–BioB by standard methods results in protein with >98% purity. The activity of the partially purified untagged enzyme and His_6–BioB are similar and there appears to be no need to remove the fusion tag for activity assays (Ugulava, Sacanell, et al., 2001).

2.1 Construction of Expression Strains

2.1.1 Cloning of the Native BioB Enzyme

The *bioB* gene was cloned by PCR from Kohara phage [202]3D4 containing the *bio* operon and incorporated between the NdeI and HindIII sites in pET23b. The resulting plasmid pJJ07 was transformed into *E. coli* BL21 (DE3)pLysS (Ugulava et al., 2000).

2.1.2 Cloning as a Hexahistidine Fusion Protein

The *bioB* gene was subcloned by PCR using pJJ07 as a template and inserted between the BamHI and XhoI sites in a custom plasmid derived from pET21d (a gift from G. Van Duyne, University of Pennsylvania). The final construct has an N-terminal hexahistidine tag followed by a TEV protease site. Following cleavage with TEV protease, the N-terminal methionine is replaced by glycine–serine specified by the BamHI restriction site. The resulting plasmid pJJ14 was transformed into *E. coli* BL21(DE3)pLysS (Ugulava et al., 2000).

2.2 Recombinant BioB Expression

2.2.1 Equipments

Erlenmyer-baffled culture flasks (200 mL and 2 L)
Centrifuge bottles (500 mL)
Temperature-controlled shaker
Temperature-controlled centrifuge

2.2.2 Materials

E. coli expression strains pJJ07/BL21(DE3)pLysS or pJJ14/BL21(DE3) pLysS
Luria Broth, Miller (LB)
LB agar plates containing 100 μg/mL ampicillin
Ampicillin (AMP), 100 mg/mL in H_2O, filter sterilized
Iron (III) citrate, 0.1 *M* in 1 m*M* HCl
Isopropyl-β-D-thiogalactopyranoside (IPTG), 1 *M* solution in H_2O, filter sterilized
50 mM Tris, adjusted to pH 8.0 with 12 *M* HCl

2.2.3 Procedure

1. A freezer stock of pJJ07/BL21(DE3)pLysS (for nontagged BioB) or pJJ14/BL21(DE3)pLysS (for His_6–BioB) is plated onto a LB agar plate containing 100 μg/mL ampicillin. The plate is incubated overnight at 37°C.
2. A single colony is selected and used to inoculate 50 mL of LB containing 100 μg/mL ampicillin in a 200 mL culture flask. The flask is incubated overnight at 37°C with shaking at ~250 rpm.
3. Culture flasks (typically 4–8 × 2 L baffled flasks) are prepared containing 1 L of LB, 50 μ*M* iron (III) citrate, and 100 μg/mL ampicillin. The flasks are inoculated with 5 mL of the overnight culture and incubated at 37°C with shaking at 250 rpm until the cultures have reached an OD_{600} = 1.0–1.5, approximately 4–5 h.
4. The incubator is cooled to 25°C with shaking at 250 rpm for 30 min.
5. IPTG is added to a final concentration of 0.1 m*M*, and the flasks are incubated for 3–5 h at 25°C with slower shaking at 100 rpm.
6. Flasks are cooled on ice and cultures are distributed between 500 mL centrifuge bottles.
7. Cells are pelleted by centrifugation (5000 × *g*, 10 min, 4°C) and the media decanted. This step is repeated with the same centrifuge bottles until all cells are collected.
8. Cells are resuspended in cold 50 m*M* Tris, pH 8.0, and transferred into a single centrifuge bottle. The cells are again pelleted by centrifugation (5000 × *g*, 10 min, 4°C) and the buffer decanted.
9. Cell pellet is weighed and frozen at −80°C for later protein purification.

2.2.4 Notes

1. Freezer stocks of expression strains should be replaced approximately every 6 months, using fresh competent BL21(DE3)pLysS, and purified plasmid to regenerate the strain.
2. Protein expression at 37°C or for longer than 5 h at 25°C leads to inclusion body formation and severely decreased recovery of soluble protein.
3. Slower shaking during protein expression minimizes oxidative damage to iron–sulfur clusters in the iron–sulfur assembly system and is essential for optimal yield of enzyme containing $[2Fe\text{–}2S]^{2+}$ clusters.
4. Cells containing the pLysS plasmid must be frozen to release lysozyme. Alternatively, solid lysozyme can be added prior to the cell lysis step (below).

2.3 Purification of nontagged BioB

2.3.1 Equipments

Glass or metal beaker, 100 mL
Sonicator
Centrifuge tubes (50 mL)
Temperature-controlled centrifuge
FPLC system
DEAE fast-flow sepharose anion exchange column, 2.5 × 20 cm (GE Life Sciences)
Mono-Q anion exchange column, 1 × 10 cm (GE Life Sciences)
Sephacryl S-300 HR gel filtration column, 1.6 × 60 cm (GE Life Sciences)
BioGel P2 desalting column, 2.5 × 20 cm (BioRad)
Amicon Ultra-15 30 kDa centrifugal filters

2.3.2 Materials

Lysis buffer: 10 m*M* Tris, pH 8.0
Buffer A: 50 mM Tris, pH 8.0
Buffer B: 50 m*M* Tris, pH 8.0, 600 m*M* NaCl
Storage buffer: 25 m*M* Tris, pH 8.0, 100 m*M* NaCl

2.3.3 Procedure

1. The cell pellet is thawed and resuspended in lysis buffer (2–4 mL per g of cells) and incubated on ice for 30 min to allow lysozyme to degrade the bacterial cell walls.

2. Cells are transferred to a 100 mL glass or metal beaker, cooled on ice, and lysed by sonication using a Fisher model 505 sonic dismembrator with a 0.5 in. diameter probe at 70% power with cycles of 30 s on/120 s off for a total on time of 3 min.
3. Cell lysates are transferred to centrifuge bottles and insoluble debris pelleted by centrifugation (20,000 × *g*, 30 min, 4°C).
4. A DEAE fast-flow sepharose column is washed with 2 *M* NaCl and equilibrated in buffer A. The clarified lysate is loaded onto the column and washed with 200 mL of buffer A or until the UV detector (280 nm) shows a flat baseline. The column should show a dark brown band in the upper 3–4 cm of the resin bed.
5. Protein is eluted with a linear gradient to 100% buffer B over 500 mL, collecting 10 mL fractions.
6. Dark brown fractions eluting at ~0.4 M NaCl are pooled and concentrated to <20 mL by centrifugation in Amicon Ultra-15 30 kDa centrifugal filters.
7. A BioGel P2 desalting column is equilibrated with buffer A. The concentrated protein is loaded and eluted with buffer A. The dark brown fractions are pooled and the concentration estimated using $\varepsilon_{452} = 8400\ M^{-1}\ cm^{-1}$.
8. A Mono-Q anion exchange column is washed with 2 *M* NaCl and equilibrated in buffer A. The protein (~100 mg) is loaded and washed with buffer A. Protein is eluted with a linear gradient to 100% buffer B over 250 mL, collecting 5 mL fractions.
9. Dark brown fractions are pooled and concentrated to ~5 mL by centrifugation in Amicon Ultra-15 30 kDa centrifugal filters.
10. A Sephacryl S-300 HR 16/60 gel filtration column is equilibrated with storage buffer. The concentrated protein (~2 mL containing 20 mg protein) is loaded and eluted with 200 mL of the same buffer, collecting 2 mL fractions. Purity of brown fractions is evaluated by SDS polyacrylamide electrophoresis, and the highest purity fractions are pooled.
11. The pooled fractions are concentrated to <5 mL by centrifugation in Amicon Ultra-15 30 kDa centrifugal filters, and the concentration estimated using $\varepsilon_{452} = 8400\ M^{-1}\ cm^{-1}$. The protein is diluted with storage buffer to 500 μ*M*, divided into 2 mL screw-top vials, flash frozen, and stored at −80°C.

2.3.4 Notes

1. The Mono-Q column will separate three overlapping fractions that contain no FeS clusters, one $[2Fe–2S]^{2+}$ cluster, and two $[2Fe–2S]^{2+}$

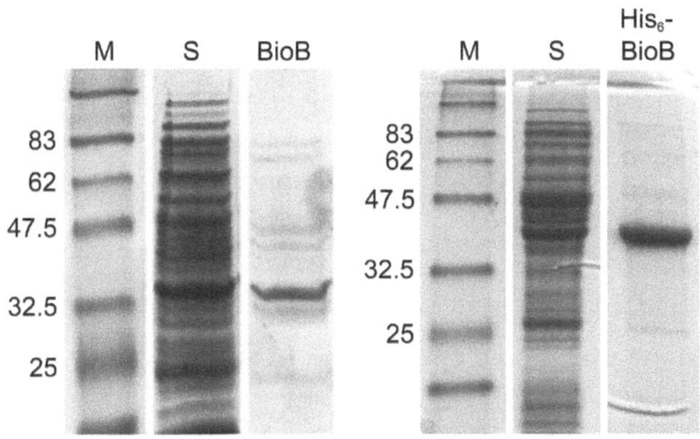

Fig. 2 SDS polyacrylamide gel electrophoresis (15%) of BioB and His_6–BioB purification. (*Left*) Nontagged BioB (~1 μg) as-purified by two rounds of anion exchange chromatography and gel filtration chromatography to ~80%–90% purity. The minor band below BioB corresponds to proteolytic cleavage of ~4 kDa from the C-terminus of the polypeptide during purification. (*Right*) His_6–BioB (~3 μg) is purified by nickel NTA Superflow affinity chromatography to >95% purity. For both purification examples, lanes derive from nonadjacent positions on the same gel are separated by a white bar. M, molecular weight markers; S, sonicate (~25 μg).

clusters, in the order of elution. The more dilute earlier fractions should be discarded to increase the FeS cluster content of the purified protein.

2. The Sephacryl S-300 HR 16/60 gel filtration column will separate nonnative crosslinked tetramer from the native dimer (~1:10 ratio). The later fractions in the major peak have the highest purity and activity.
3. At each stage, only a portion of the protein is carried into the next step. Overall yield is 10–15 mg of protein with 80%–90% purity as judged by SDS-PAGE (Fig. 2).

2.4 Purification of His_6–BioB

2.4.1 Equipments

Glass or metal beaker, 100 mL
Sonicator
Centrifuge tubes (50 mL)
Temperature-controlled centrifuge
FPLC system
Nickel NTA Superflow affinity column, $2.5 \times 10\,cm^2$ (Qiagen)
BioGel P2 desalting column, $2.5 \times 20\,cm^2$ (BioRad)
Amicon Ultra-15 30 kDa centrifugal filters

2.4.2 Materials

Lysis buffer: 10 m*M* Tris, pH 8.0
Buffer A: 50 m*M* Tris, pH 8.0, 500 m*M* NaCl, 20 m*M* imidazole
Buffer B: 50 m*M* Tris, pH 8.0, 500 m*M* NaCl, 200 m*M* imidazole
Storage buffer: 25 m*M* Tris, pH 8.0, 100 m*M* NaCl

2.4.3 Procedure

1. The cell pellet is thawed and resuspended in lysis buffer (2–4 mL per g of cells) and incubated on ice for 30 min to allow lysozyme to degrade the bacterial cell walls.
2. Cells are transferred to a 100 mL glass or metal beaker, cooled on ice, and lysed by sonication using a Fisher model 505 sonic dismembrator with a 0.5 in. diameter probe at 70% power with cycles of 30 s on/120 s off for a total on time of 3 min.
3. Cell lysates are transferred to centrifuge bottles and insoluble debris pelleted by centrifugation (20,000 × *g*, 30 min, 4°C).
4. A Nickel NTA Superflow column (Qiagen) is washed with 0.1 *M* tetrasodium EDTA, recharged with 0.1 *M* $NiCl_2$, and equilibrated in buffer A. The clarified lysate is loaded onto the column and washed with 200 mL of buffer A or until the UV detector (280 nm) shows a flat baseline. The column is typically completely brown throughout the resin bed.
5. Protein is eluted with 200 mL of buffer B, collecting 10 mL fractions.
6. Dark brown fractions are pooled and concentrated to <20 mL by centrifugation in Amicon Ultra-15 30 kDa centrifugal filters.
7. A BioGel P2 desalting column is equilibrated with storage buffer. The concentrated protein is loaded and eluted with the same buffer.
8. Dark brown fractions are pooled, concentrated to ~10 mL by centrifugation in Amicon Ultra-15 30 kDa centrifugal filters, and the concentration determined using $\varepsilon_{452} = 8400\ \mathrm{M}^{-1}\ \mathrm{cm}^{-1}$. The protein is diluted with storage buffer to 500 μ*M*, divided into 2 mL screw-top vials, flash frozen, and stored at −80°C.

2.4.4 Notes

1. Once the protein has been eluted in buffer B, the protein must be desalted as soon as possible. Prolonged incubation in high imidazole buffer at 4°C will result in gradual depletion of the FeS cluster content of the protein. Freezing in high imidazole buffer will result in precipitation of the protein upon thawing.

2. For use in crystallography, passage through the Sephacryl S-300 HR 16/60 gel filtration column (as described in Section 2.3.3) will remove trace amounts (<5%) of higher molecular weight complexes.
3. Overall yield is 100–200 mg of protein with >95% purity as judged by SDS-PAGE (Fig. 2).

3. CHARACTERIZATION AND MANIPULATION OF BioB COFACTOR COMPOSITION

BioB purified under aerobic conditions following the protocols in Sections 2.3 and 2.4 contains predominantly $[2Fe–2S]^{2+}$ clusters, although the stoichiometry can vary depending on expression conditions (Ugulava et al., 2000). Since the $[2Fe–2S]^{2+}$ cluster is the sulfur donor for the catalytic reaction (Farrar et al., 2010; Ugulava, Sacanell, et al., 2001), enzyme activity depends directly on cluster content, with small but significant amounts of inactive protein present with 0 or 1 cluster per dimer. Routine assays can be carried out while ignoring this issue, but any attempt to conduct quantitative kinetic or spectroscopic studies must be based on enzyme for which the cofactor content has been thoroughly characterized. While the only completely thorough method for simultaneously detecting all iron–sulfur clusters present in a sample is a combination of electron paramagnetic resonance and Mössbauer spectroscopies (Cosper et al., 2004; Ugulava et al., 2002), a great deal can be learned from careful analytical analysis and UV/visible spectroscopy.

The active form of BioB should contain one $[4Fe–4S]^{2+}$ and one $[2Fe–2S]^{2+}$ cluster per monomer (Ugulava, Sacanell, et al., 2001), a state that is only stable under anaerobic conditions (Ugulava, Frederick, & Jarrett, 2003). A number of prior studies have focused on studying the properties of each cluster in both the oxidized and the reduced states, and the procedures described below can be used to obtain enzyme containing all possible relevant FeS cluster states.

3.1 Characterization of the Iron–Sulfur Clusters in BioB

3.1.1 *Polypeptide Analysis*

Protein polypeptide concentration of the samples can be determined in two ways. First, the Bradford assay using a commercially available dye (BioRad) and BSA standards provides a reliable estimate of the protein concentration; however, the result should be multiplied by the correction factor 0.90, previously determined by quantitative amino acid analysis. Alternatively, the protein can be diluted 10-fold into 0.1 *M* EDTA, 7.5 *M* guanidinium hydrochloride, 100 m*M* Tris, and pH 8 to remove all iron–sulfur clusters and denature the protein. The polypeptide absorbance at 280 nm is then measured relative to a buffer control and the polypeptide concentration determined from $\varepsilon_{280} = 30{,}370\,M^{-1}\,cm^{-1}$ (Gill & von Hippel, 1989).

3.1.2 Iron Analysis

The iron content of the protein samples is determined in a complex with bathophenanthroline according to the method described by Fish (1988), except that protein samples (10 nmol polypeptide) do not need to be digested with $KMnO_4$ but can simply be acidified with 0.2 N HCl to release bound iron. Iron standards (5–25 nmol) are prepared from a commercially available atomic absorption standard (Aldrich). Samples of *E. coli* ferredoxin (2Fe per polypeptide) are occasionally used to validate the accuracy of the assay. Typical values obtained for BioB range from 1.2 to 2.0 Fe per polypeptide.

3.1.3 Sulfide Analysis

The sulfide content of the protein samples (~1 nmol per sample) is determined following extraction with zinc acetate/NaOH and subsequent derivatization with *N*,*N*-dimethyl-*p*-phenylenediamine (DMPD) as described by Broderick et al. (2000). A sulfide stock solution (0.1 *M*) is prepared gravimetrically from $Na_2S{\cdot}9H_2O$ dissolved in H_2O and standards for analysis (0.5–5 nmol) are prepared by dilution into 10 m*M* NaOH. Samples of *E. coli* ferredoxin (2Fe per polypeptide) were used to validate the assay. Typical values obtained for BioB range from 1.5 to 2.5 S^{2-} per polypeptide.

3.1.4 UV/Visible Spectra Analysis

The UV/visible spectrum of as-isolated BioB exhibits four major overlapping absorption bands at 280, 320, 420, and 452 nm. The peak at 280 nm is predominantly due to Tyr and Trp residues in the polypeptide, while the peaks at 320 and 452 nm are characteristic ligand-to-metal charge transfer bands associated with a cysteinyl-ligated $[2Fe–2S]^{2+}$ cluster. The peak or shoulder at 420 nm can be due to a combination of a weak band associated with the $[2Fe–2S]^{2+}$ cluster and stronger bands associated with residual $[4Fe–4S]^{2+}$ or $[3Fe–4S]^{+}$ clusters bound in the radical SAM cluster-binding site.

BioB is diluted into 25 m*M* Tris, pH 8.0, 100 m*M* NaCl to a polypeptide concentration of 20 μ*M*, the sample clarified by centrifugation

(5 min, 15,000 × *g*, 20°C), and the UV/visible spectrum recorded from 250 to 800 nm. Information concerning the cofactor content can be gleaned by examining the following absorption ratios. The stoichiometry of the $[2Fe–2S]^{2+}$ cluster to polypeptide can be determined by the ratio A_{452}/A_{280}, where a value of 0.277 is obtained for a 1:1 stoichiometry and lower values indicate the presence of some apoprotein. The presence of residual $[4Fe–4S]^{2+}$ or $[3Fe–4S]^{+}$ clusters can be detected by the ratio A_{420}/A_{452}, where the $[2Fe–2S]^{2+}$ cluster alone gives a value of 0.97, while higher values may indicate the presence of other clusters (or possibly a contaminating cofactor-containing protein). For reference purposes, the presence of one $[4Fe–4S]^{2+}$ cluster and one $[2Fe–2S]^{2+}$ cluster per polypeptide will give a ratio $A_{420}/A_{452} = 1.18$.

3.2 Manipulation of the Iron–Sulfur Clusters in BioB

3.2.1 Materials

Ferrous ammonium sulfate, 0.05 *M*, ferric ammonium sulfate, 0.05 *M*
Sodium sulfide, 0.1 *M* in H_2O under N_2
Dithiothreitol (DTT), 0.1 *M* in H_2O
Sodium dithionite, 1.0 and 0.05 *M* in H_2O, prepared and stored under N_2 or Ar
Methyl viologen, 1 m*M* in H_2O
Ethylene glycol
1 *M* Tris–HCl, pH 8.5
Buffer A: 25 m*M* Tris–HCl, pH 8.0, 0.1 *M* NaCl
Note: All materials should be stored inside the N_2-filled glovebox.

3.2.2 Equipments

Quartz cuvette with ground glass joint and tightly sealed rubber septum
Sephadex G-25 column, 1.0 × 25 cm^2, washed with 200 mL buffer A
Peristaltic pump (or FPLC) in the glovebox
UV/visible spectrophotometer
Nitrogen-filled glovebox or Coy chamber with <5 ppm O_2

3.2.3 Procedure to Obtain Apoprotein

1. BioB (100–500 μ*M* in 1–2 mL total volume) is dissolved in buffer A containing 10 m*M* EDTA under anaerobic conditions. Sodium dithionite (2 m*M*) and methyl viologen (5 m*M*) are added and the sample mixed gently. The sample is incubated at room temperature for 30 min, after which it should appear clear to slightly tan.

2. The Sephadex G-25 column is equilibrated with buffer A in the glovebox. The protein sample is loaded onto the column and eluted with 30 mL of buffer A while collecting 1 mL fractions.
3. Fractions that absorb at 280 nm are pooled and concentrated to <1 mL. The polypeptide concentration is determined as described in Section 3.1.1.

3.2.4 Procedure to Obtain 1:1 $[2Fe\text{–}2S]^{2+}$ and $[4Fe\text{–}4S]^{2+}$ Clusters

1. BioB (200 μ*M* in 2 mL total volume) is dissolved in 200 m*M* Tris, pH 8.5 in the glovebox. DTT (5 m*M*) and Na_2S (800 μ*M*) are added and the sample mixed gently.
2. Fe (II/III) ammonium sulfate (800 μ*M* final) is added in eight equal portions. An initial black precipitate will redissolve as the sample is gently swirled after each addition.
3. The Sephadex G-25 column is equilibrated with buffer A in the glovebox. The protein sample is loaded onto the column and eluted with 30 mL of buffer A while collecting 1 mL fractions.
4. An initial black–gray fraction may elute, but this contains ferric sulfide aggregates and should be discarded. Brown–orange fractions are pooled, a sample transferred to a cuvette and sealed with a septum, and the UV/visible spectrum evaluated. Full reconstitution should result in $A_{420}/A_{452} = 1.18$ (Fig. 3A).
5. The reconstituted sample is concentrated to <2 mL. The concentration can be estimated from the UV/visible spectrum ($\varepsilon_{420} = 20{,}500\ M^{-1}\ cm^{-1}$).

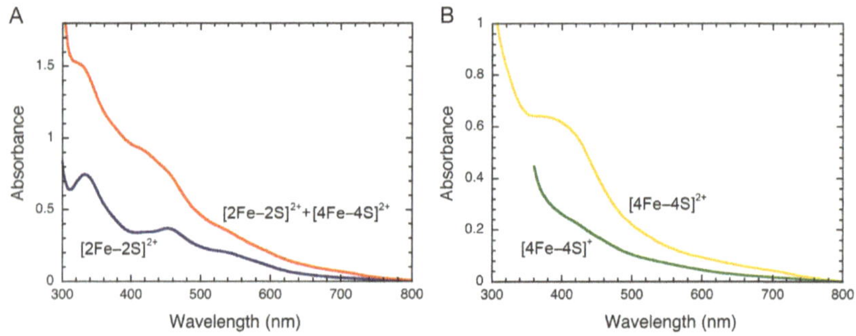

Fig. 3 UV/visible spectra of BioB containing different iron–sulfur clusters (all samples at 45 μM). (A) As-purified BS containing one $[2Fe\text{–}2S]^{2+}$ cluster and reconstituted BS containing one $[2Fe\text{–}2S]^{2+}$ cluster and one $[4Fe\text{–}4S]^{2+}$ cluster per monomer. (B) BS containing one $[4Fe\text{–}4S]^{2+}$ cluster, prepared as described in buffer containing 60% ethylene glycol and reduced BS containing one $[4Fe\text{–}4S]^{+}$ cluster per monomer (spectrum truncated at 360 nm due to high absorbance of dithionite).

3.2.5 Procedure to Obtain [4Fe–4S]$^{2+}$ Cluster Only

1. BioB (200 μ*M* in 2 mL total volume) is dissolved in 200 m*M* Tris, pH 8.5 containing 60% ethylene glycol in the glovebox. DTT (5 m*M*) and Na_2S (400 μ*M*) are added and the sample mixed gently.
2. Fe (II/III) ammonium sulfate (400 μ*M* final) is added in four equal portions. An initial black precipitate will redissolve as the sample is gently swirled after each addition.
3. Sodium dithionite (5 m*M*) and methyl viologen (5 μ*M*) are added and the sample incubated for 4 h at room temperature. The sample should gradually turn bright yellow.
4. The Sephadex G-25 column is equilibrated with buffer A in the glovebox. The protein sample is loaded onto the column and eluted with 30 mL of buffer A while collecting 1 mL fractions.
5. Yellow fractions are pooled, a sample transferred to a cuvette and sealed with a septum, and the UV/visible spectrum evaluated. Full conversion should result in the disappearance of the peaks at 332 and 452 nm, and the appearance of a broad peak at 420 nm ($\varepsilon_{420} = 13{,}500\ M^{-1}\ cm^{-1}$) (Fig. 3B).

3.2.6 Procedure to Obtain [4Fe–4S]$^{+}$ Cluster Only

1. BioB (200 μ*M* in 2 mL total volume) is dissolved in 200 m*M* Tris, pH 8.5 in the glovebox. DTT (5 m*M*) and Na_2S (400 μ*M*) are added and the sample mixed gently.
2. Fe (II/III) ammonium sulfate (400 μ*M* final) is added in four equal portions. An initial black precipitate will redissolve as the sample is gently swirled after each addition.
3. Sodium dithionite (5 m*M*) and methyl viologen (5 μ*M*) are added and the sample incubated for 1 h at room temperature. The sample should gradually turn blue–gray.
4. The Sephadex G-25 column is equilibrated with buffer A containing 0.5 m*M* dithionite in the glovebox. The protein sample is loaded onto the column and eluted with 30 mL of buffer A containing 0.5 m*M* dithionite while collecting 1 mL fractions.
5. Blue–gray fractions are pooled, a sample transferred to a cuvette and sealed with a septum, and the UV/visible spectrum evaluated. Full conversion should result in the disappearance of all distinct peaks, with a broad increasing absorbance from 650 to 370 nm (Fig. 3B).

3.2.7 Notes

1. Apoprotein should be prepared and desalted under N_2 to prevent oxidative damage to the protein. Once the protein is desalted, it should be used immediately or mixed with 1 m*M* DTT to minimize disulfide formation. Apoprotein is also highly susceptible to proteolysis by trace contaminants.
2. Reconstitution and conversion reactions are carried out in high concentration, high pH, Tris buffer to minimize the chance of protein precipitation during addition of $FeCl_3$.
3. Protein containing 1:1 $[2Fe–2S]^{2+}$ and $[4Fe–4S]^{2+}$ clusters can be stabilized by mixing with a slight excess of DTB and SAM. These will not participate in a catalytic reaction unless dithionite or the protein reducing system is added.
4. Protein containing a $[4Fe–4S]^{+}$ cluster is highly reactive toward even trace residual O_2. It is recommended that all buffers contain 0.5 m*M* dithionite and that the protein sample is used immediately and exposure to the glove box atmosphere is kept to a minimum.

4. DETERMINATION OF BioB ENZYME ACTIVITY

4.1 In Vitro Enzyme Assay Based on Biotin Detection

E. coli BioB purified under aerobic conditions is inactive, will not bind DTB or SAM, and will not accept electrons from the native reductant flavodoxin. While many radical SAM enzymes become active in the presence of a strong reductant such as sodium dithionite, BS is an exception, as the essential $[2Fe–2S]^{2+}$ cluster is destroyed by strong chemical reductants. Maximal activity requires reconstitution of the $[4Fe–4S]^{2+}$ cluster under conditions that leave the $[2Fe–2S]^{2+}$ cluster intact (Ugulava, Gibney, et al., 2001), as described in Section 3.2.4. For convenience, this reconstitution can be accomplished in situ, with no need to desalt the protein prior to continuing the assay.

The destructive effect of chemical reductants on the $[2Fe–2S]^{2+}$ cluster is an additional challenge in assay design. Early research in *E. coli* cell extracts identified flavodoxin (FldA) and ferredoxin:NADP^{+} oxidoreductase (Fpr) as the native reducing system (Ifuku et al., 1994). We have found that *E. coli* BioB is highly specific for *E. coli* FldA and has much lower activity when other flavodoxins or ferredoxins are substituted (Ugulava, Sacanell, et al., 2001; Wan & Jarrett, 2002). Ferredoxin:NADP^{+} oxidoreductase is relatively promiscuous toward flavodoxins and ferredoxins

(Wan & Jarrett, 2002), including those from other species, and we have found that the homologous enzyme from *Thermosynechococcus elongatus* (GenBank accession WP_011057053) is much more active and stable under BioB assay conditions, and so we have substituted the use of that enzyme for that native *E. coli* enzyme.

A final factor that must be overcome to achieve maximum activity is inhibition, both by *S*-adenosyl-L-homocysteine (SAH), a contaminant in commercial SAM samples, as well as by 5′-dAH, the product of the catalytic reaction (Farrar et al., 2010). The former issue is solved by using SAM produced in-house using *E. coli* SAM synthetase. Producing SAM also allows minimal contamination by the nonbiological (*R*,*S*)-SAM diastereomer, a product of thermal racemization of the chiral sulfonium center, and 5′-methylthioadenosine (MTA), a product of spontaneous cycloserine formation from SAM at neutral pH. Product inhibition by 5′-dAH can be prevented by addition of *E. coli* MTA/SAH/5′-dAH nucleosidase (MtnN), which cleaves all three potential inhibitors to adenine and the respective sugar analogs. The action of MtnN is essential in order to observe multiple catalytic turnovers.

4.1.1 Materials

0.5 *M* Tris–HCl buffer, pH 8.0, 100 m*M* sodium chloride
500 μ*M* BioB in storage buffer (BioB or His_6–BioB)
250 m*M* dithiothreitol (DTT)
Ferrous ammonium sulfate, 0.005 *M*, ferric ammonium sulfate, 0.005 *M*
10 m*M* sodium sulfide (Na_2S)
500 μ*M* flavodoxin solution (FldA)
100 μ*M* ferredoxin:$NADP^+$ oxidoreductase (FNR)
20 m*M* NADPH
10 m*M* dethiobiotin in 10 m*M* NaOH (DTB)
20 m*M* SAM in 1 m*M* HCl
70 μ*M* nucleosidase (MtnN)
4 *M* acetic acid/sodium acetate buffer, pH 4.5
H_2O with 1% H_3PO_4 or 0.1% (v/v) formic acid
Acetonitrile with 1% H_3PO_4 or 0.1% (v/v) formic acid

4.1.2 Equipments

Nitrogen-filled glovebox or Coy chamber with <5 ppm O_2
Heating block in glovebox, set to 37°C
High-speed refrigerated microcentrifuge

HPLC system with UV detector
Waters Atlantis T3 reversed-phase column ($2.1 \times 150\,mm^2$, 3 μm)
HPLC system coupled to ES-TOF or ES-QQQ mass spectrometer
Eclipse XDB-C_{18} column (4.6 mm × 50 mm, 5 μm, Agilent)

4.1.3 Assay Procedure

1. Bring all assay components into N_2-filled glovebox and equilibrate open vials/tubes under N_2 for >1 h to displace any residual contaminating O_2.
2. Add in order: H_2O to bring final volume to 200 μL, Tris–NaCl buffer (20 μL), BioB (10–25 μ*M* monomer final concentration), DTT (5 m*M*), Na_2S (4 × [BioB] μ*M*), Fe (II/III) ammonium sulfate (4 × [BioB] μ*M*), and incubate for 10 min at room temperature. Sample will be brown–orange.
3. Add FldA (25 μ*M*), FNR (5 μ*M*), NADPH (1 m*M*), DTB (100 μ*M*), and MtnN (70 n*M*) to the reconstituted enzyme. Incubate assay mixture at 37°C for 10 min. FldA will turn from orange to blue indicating successful reduction.
4. Initiate reaction with SAM (100 μ*M*) and incubate at 37°C for 5–120 min. Sample will become noticeably bleached as $[2Fe–2S]^{2+}$ cluster is destroyed.
5. Quench assay with 20 μL acetic acid/sodium acetate, pH 4.5, and vortex to mix thoroughly. BioB and FNR will precipitate.
6. Remove samples from anaerobic chamber and incubate on ice for 20 min to ensure full precipitation of protein.
7. Centrifuge samples (15 min, 18,000 × *g*, 4°C) to pellet precipitated protein and transfer supernatant to autosampler vials for analysis.

4.2 Biotin Quantitation in Assay Samples

Several methods are available that can determine the concentration of biotin, but most of these methods suffer from a high background from DTB. We have found that direct detection of biotin by UV absorbance (Farrar & Jarrett, 2009) or LCMS (Taylor et al., 2008) is the most reliable and unambiguous analytical technique. A standard microbiological method has been used for decades to measure biotin (DeMoll & Shive, 1986), but with the usual organism *Lactobacillus plantarum*, DTB also supports growth at roughly 100-fold higher concentrations. Substitution of a true Δ *bioB* strain allows this assay to be used for measuring biotin in the presence of 100 μ*M* DTB. The detection range for each method is approximately: UV detection, 100–2000 pmol; LCMS detection, 30–1000 pmol; and microbiological detection, 5–500 pmol.

4.2.1 Biotin Quantitation With UV Detection

1. A standard curve is constructed using commercially available biotin and DTB (Sigma-Aldrich) diluted to a range of concentrations (5–100 μ*M*) in 200 μL H_2O with 20 μL acetic acid/sodium acetate, pH 4.5.
2. A Waters Atlantis T3 reversed-phase column ($2.1 \times 150\,mm^2$, 3 μm) is equilibrated in 97:3 H_2O:acetonitrile (1% H_3PO_4). Samples (25 μL) are loaded and eluted with a linear gradient to 70:30 H_2O:acetonitrile (0.1% H_3PO_4) over 25 min at a flow rate of 0.7 mL/min.
3. The Waters Breeze HPLC system includes a model 2489 dual-wavelength UV/visible detector set to 210 and 260 nm. Biotin ($t_R = 17.6$ min) and DTB ($t_R = 21.8$ min) are quantified by integration of peak areas based on absorbance at 210 nm (Fig. 4). Possible mis-identification of peaks from FAD, FMN, or riboflavin is avoided by eliminating any peaks with significant absorbance at 260 nm. 9-Mercaptodethiobiotin (9-MDTB, $t_R = 23.6$ min) is observed in some samples and is quantified based on the biotin standards, with the result multiplied by:

$$\frac{\varepsilon_{210}^{\text{biotin}}}{\varepsilon_{210}^{\text{MDTB}}} = 1.69$$

4.2.2 Biotin Quantitation With MS Detection

1. Caffeine (10 μL, 0.1 n*M* final concentration) is added to each sample as an internal standard.
2. A standard curve was constructed using commercially available biotin and DTB (Sigma-Aldrich) diluted to a range of concentrations (1–50 μ*M*) in 200 μL H_2O with 20 μL acetic acid/sodium acetate, pH 4.5, also with 10 μL caffeine as an internal standard.
3. An Agilent XDB-C_{18} column is equilibrated in 95:5 H_2O:acetonitrile (0.1% formic acid). Samples (10 μL) are loaded and eluted with a linear gradient to 70:30 H_2O:acetonitrile (0.1% formic acid) over 9 min at a flow rate of 0.7 mL/min.

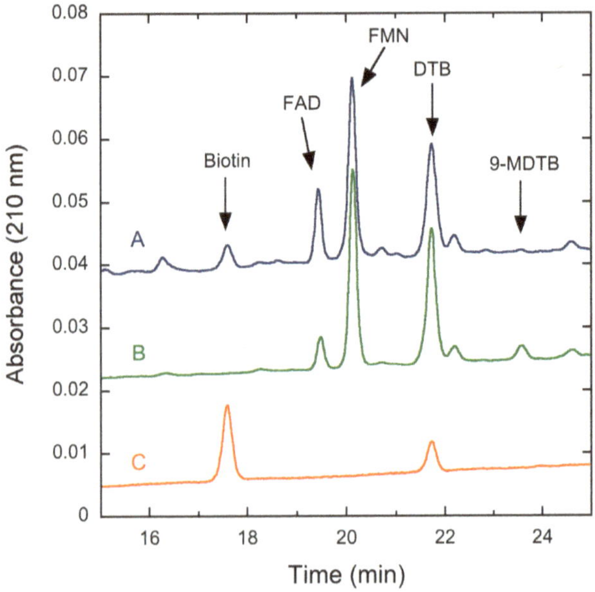

Fig. 4 HPLC chromatograms for BS assays detected by absorbance at 210 nm. (*Trace A*) A typical assay as described in Section 4.1.3, with 25 μ*M* His_6–BioB, 100 μ*M* DTB, and 500 μ*M* SAM incubated at 37°C for 120 min. (*Trace B*) A parallel assay designed to produce the monothiolated 9-MDTB intermediate, conducted with 25 μ*M* His_6-BioB, 100 μ*M* DTB, and 20 μ*M* SAM, incubated at 37°C for 30 min. (*Trace C*) Biotin and DTB standard containing 25 μ*M* (625 pmol) of each component. FAD and FMN observed in BioB assay traces are cofactors in FNR and FldA, respectively, and are released from those proteins upon acidification of the samples.

4. The LC system is coupled to an Agilent 6400 triple–quadrupole mass spectrometer equipped with an electrospray–ionization source operating in positive ion mode. Dry nitrogen is used as the nebulizer drying gas with pressure = 30 psi, flow rate = 7 L/min, and T = 350°C. Additional ESI–MS parameters are set as follows: spray capillary voltage, 3500 V; skimmer voltage, 65 V; and fragmentor voltage, 100 V. Mass spectra were acquired over the range m/z = 100–400 at 0.99 scan/s. Biotin (t_R = 6.9 min, $[M+H]^+$ = 245 m/z, $[M-OH]^+$ = 227 m/z, $[M+Na]^+$ = 267 m/z) and DTB (t_R = 7.5 min, $[M+H]^+$ = 215 m/z, $[M-OH]^+$ = 197 m/z, $[M+Na]^+$ = 237 m/z) were quantified by integration of peak areas in the extracted ion chromatogram (EIC) for each analyte and summation of the areas for the three major ions listed (Figs. 5 and 6). Caffeine was used as an internal standard (t_R = 4.5 min, $[M+H]^+$ = 195 m/z) to correct for variability in sample injection and recovery.

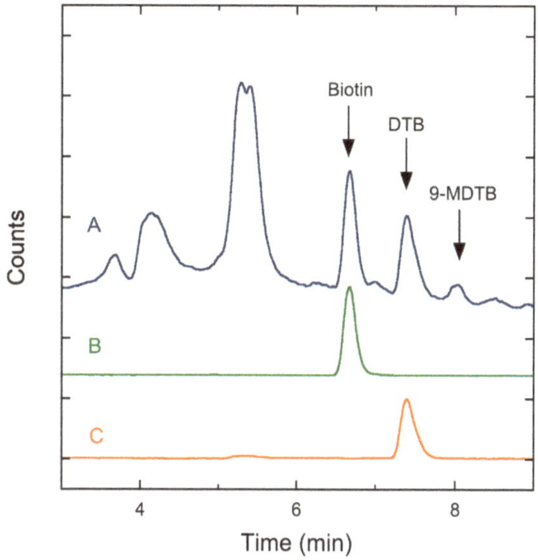

Fig. 5 LCMS detected HPLC chromatograms for a BioB assay as described in Section 4.1.3, with 10 μ*M* His_6–BioB, 25 μ*M* DTB, and 100 μ*M* SAM incubated at 37°C for 30 min. (*Trace A*) Total ion chromatogram (*m/z* = 100–400). (*Trace B*) Extracted ion chromatogram for biotin showing the sum of counts at *m/z* = 227, 245, and 267. (*Trace C*) Extracted ion chromatogram for DTB showing the sum of counts at *m/z* = 197, 215, and 237. Sample contains ca. 8 μ*M* biotin, 15 μ*M* DTB, and 1 μ*M* 9-MDTB.

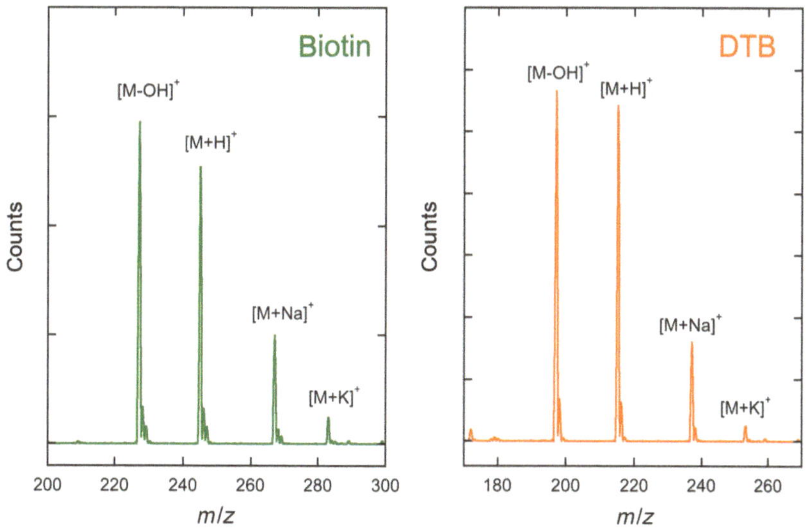

Fig. 6 Mass spectra for biotin and DTB standards (250 pmol each). Biotin and DTB both exhibit strong ions generated by loss of OH^- from the carboxylic acid, as well as the expected ions generated by binding of H^+, Na^+, and K^+.

4.2.3 Biotin Quantitation Using a Microbiological Assay

This method is adapted from DeMoll and Shive, except *L. plantarum* is substituted by an *E. coli* Δ *bioB* strain (Del Campillo-Campbell et al., 1967) (R875, CGSC#9954, Yale Genetic Stock Center). *L. plantarum* will grow on high concentrations of DTB (DeMoll & Shive, 1986), causing some background in the typical microbiological biotin assay, particularly for analysis of BS enzyme assay samples containing unreacted DTB substrate.

1. Growth medium is prepared from BD Difco Biotin Assay Medium (ThermoFisher), and 2 mL medium is dispensed into new 10 mL loosely capped disposable culture tubes.
2. An LB agar plate is used to grown colonies of *E. coli* strain R875. A 2 mL culture in biotin assay medium containing 25 n*M* biotin is inoculated with a single colony and grown overnight at 37°C with shaking at 200 rpm.
3. Standards are prepared containing 0.5–5 μ*M* biotin together with 100 μ*M* DTB. A negative control contains only 100 μ*M* DTB. Ten microliter of each standard is added to a culture tube containing 2 mL of medium.
4. Assay samples from Section 4.1.3 are diluted fivefold in sterile H_2O, and 10 μL of each is added to a culture tube containing 2 mL of medium. Each sample should be measured in triplicate.
5. Culture tubes are inoculated with 10 μL of the overnight culture and incubated for >15 h at 37°C with shaking at 200 rpm.
6. Cell growth is assessed by measuring OD_{600}. The OD_{600} for the DTB-only sample should be subtracted from the other results, as some growth will occur due to residual biotin in the cells derived from the overnight culture. A standard curve is constructed based upon the standards and the concentration of biotin in assay samples is calculated by comparison to this curve and correcting for dilution.

4.2.4 Notes

1. The concentration of assay components that are required as molar equivalents should be based on the BioB monomer concentration, determined from $\varepsilon_{452} = 8400\ M^{-1}\ cm^{-1}$ (e.g., for 10 μ*M* BioB monomer use 40 μ*M* $FeCl_3$ and Na_2S).
2. Commercially available biotin and DTB (Sigma-Aldrich) are freeze dried to remove bound water for >24 h immediately prior to preparing stock solutions and standards.

3. SAM is prepared using methionine, ATP, and SAM synthetase, and purified by cation exchange chromatography as previously described (Farrar et al., 2010). The concentration of SAM stock solutions is determined by UV absorption at 260 nm ($\varepsilon_{260} = 15{,}400\, M^{-1}\, cm^{-1}$).
4. The nucleosidase (MtnN) results in ~20% higher activity during the first turnover, and roughly twofold higher activity after 120 min.
5. Precipitation of the assay components with 10% trichloroacetic acid is more effective but gives very strong interfering peaks in both UV and MS detection.
6. Phosphoric acid is used in HPLC solvents for UV detection to avoid background absorbance at 210 nm observed with formic acid. Formic acid is preferred for ESI–MS detection due to its volatility in the nebulizer.
7. Column selection is critical for separation of biotin from FAD. The Waters Atlantis T3 column is a C18 column with polar-end capping that is optimal for vitamin and nucleotide analysis.
8. We have not examined whether the microbiological assay can distinguish 9-MDTB from biotin. 9-MDTB has been reported to be effective at promoting the growth of a biotin auxotroph, through an unknown mechanism (Tse Sum Bui et al., 1998).

5. CONCLUSIONS AND FUTURE DIRECTIONS

In vitro, *E. coli* BioB catalyzes a single turnover of biotin production per homodimer in ca. 10 min (Farrar et al., 2010), a rate that appears to be physiologically relevant based upon an estimate of the rate of in vivo biotin production in *E. coli* during steady-state growth in minimal media (Choi-Rhee & Cronan, 2005). However, subsequent in vitro turnovers of biotin synthase occur at a much slower rate, approximately one turnover every hour (Farrar et al., 2010), a rate that is clearly not physiologically relevant for bacteria. Our studies suggest the $[2Fe–2S]^{2+}$ cluster plays a critical role in supplying sulfur for the reaction and reassembly of this cluster could be limiting multiple turnovers in vitro. Future progress on understanding the physiologically relevant reactions must include understanding how BioB interacts with the iron–sulfur cluster assembly systems, either the Isc or the Suf systems (Reyda et al., 2009). *E. coli* has both systems and may not be the best organism in which to study these interactions due to the promiscuity and leakiness of these iron–sulfur cluster assembly reactions.

An optimized assay for the activity of BioB obtained from organisms other than *E. coli* has not been reported. The primary obstacle to studying BioB in most organisms is a poor understanding of the redox partners for radical SAM enzymes. Flavodoxins homologous to *E. coli* FldA are only found in roughly 1/3 of proteobacteria, and essentially not found in actinobacteria or firmicutes, suggesting that this protein is replaced by another redox protein, possibly a ferredoxin. However, most species have 10–20 ferredoxins, and a general method for identifying the specific redox partner that interacts with BioB in each organism has not been identified. Similar problems would likely exist for the interaction of BioB with proteins from the iron–sulfur cluster assembly systems. Future progress in understanding biotin biosynthesis in organisms other than *E. coli* will depend critically on being able to identify the complex web of weak protein:protein interactions that support the activity of this complex enzyme.

ACKNOWLEDGMENT

Work on this chapter was supported by a grant from the NSF (MCB 09-23829).

REFERENCES

Berkovitch, F., Nicolet, Y., Wan, J. T., Jarrett, J. T., & Drennan, C. L. (2004). Crystal structure of biotin synthase, an *S*-adenosylmethionine-dependent radical enzyme. *Science*, *303*, 76–79.

Birch, O. M., Fuhrmann, M., & Shaw, N. M. (1995). Biotin synthase from *Escherichia coli*, an investigation of the low molecular weight and protein components required for activity *in vitro*. *Journal of Biological Chemistry*, *270*, 19158–19165.

Broderick, J. B., Duffus, B. R., Duschene, K. S., & Shepard, E. M. (2014). Radical S-adenosylmethionine enzymes. *Chemical Reviews*, *114*, 4229–4317.

Broderick, J. B., Henshaw, T. F., Cheek, J., Wojtuszewski, K., Smith, S. R., Trojan, M. R., et al. (2000). Pyruvate formate-lyase-activating enzyme: Strictly anaerobic isolation yields active enzyme containing a $[3Fe–4S]^+$ cluster. *Biochemical and Biophysical Research Communications*, *269*, 451–456.

Choi-Rhee, E., & Cronan, J. E. (2005). A nucleosidase required for in vivo function of the *S*-adenosyl-L-methionine radical enzyme, biotin synthase. *Chemistry and Biology*, *12*, 589–593.

Cosper, M., Jameson, G., Hernández, H., Krebs, C., Huynh, B., & Johnson, M. (2004). Characterization of the cofactor composition of *Escherichia coli* biotin synthase. *Biochemistry*, *43*, 2007–2021.

Del Campillo-Campbell, A., Kayajanian, G., Campbell, A., & Adhya, S. (1967). Biotin-requiring mutants of *Escherichia coli* K-12. *Journal of Bacteriology*, *94*, 20652066.

DeMoll, E., & Shive, W. (1983). The origin of sulfur in biotin. *Biochemical and Biophysical Research Communications*, *110*, 243–249.

DeMoll, E., & Shive, W. (1986). Assay for biotin in the presence of dethiobiotin with *Lactobacillus plantarum*. *Analytical Biochemistry*, *158*, 55–58.

du Vigneaud, V., Melville, D. B., Gyorgy, P., & Rose, C. (1940). On the identity of vitamin H with biotin. *Science*, *92*, 62–63.

Escalettes, F., Florentin, D., Tse Sum Bui, B., Lesage, D., & Marquet, A. (1999). Biotin synthase mechanism: Evidence for hydrogen transfer from the substrate into deoxyadenosine. *Journal of the American Chemical Society, 121*, 3571–3578.

Farrar, C. E., & Jarrett, J. T. (2009). Protein residues that control the reaction trajectory in *S*-adenosylmethionine radical enzymes: Mutagenesis of asparagine 153 and aspartate 155 in *Escherichia coli* biotin synthase. *Biochemistry, 48*, 2448–2458.

Farrar, C. E., Siu, K. K. W., Howell, P. L., & Jarrett, J. T. (2010). Biotin synthase exhibits burst kinetics and multiple turnovers in the absence of inhibition by products and product-related biomolecules. *Biochemistry, 49*, 9985–9996.

Fish, W. W. (1988). Rapid colorimetric micromethod for the quantitation of complexed iron in biological samples. *Methods in Enzymology, 158*, 357–364.

Frey, P. A., & Magnusson, O. T. (2003). *S*-Adenosylmethionine: A wolf in sheep's clothing, or a rich man's adenosylcobalamin? *Chemical Reviews, 103*, 2129–2148.

Fugate, C. J., Stich, T. A., Kim, E. G., Myers, W. K., Britt, R. D., & Jarrett, J. T. (2012). 9-Mercaptodethiobiotin is generated as a ligand to the [2Fe–2S]+ cluster during the reaction catalyzed by biotin synthase from Escherichia coli. *Journal of the American Chemical Society, 134*, 9042–9045.

Gibson, K. J. (1997). Isolation and chemistry of the mixed anhydride intermediate in the reaction catalyzed by dethiobiotin synthetase. *Biochemistry, 36*, 8474–8478.

Gill, S. C., & von Hippel, P. H. (1989). Calculation of protein extinction coefficients from amino acid sequence data. *Analytical Biochemistry, 182*, 319–326.

Ifuku, O., Kishimoto, J., Haze, S., Yanagi, M., & Fukushima, S. (1992). Conversion of dethiobiotin to biotin in cell-free extracts of *Escherichia coli*. *Bioscience, Biotechnology, and Biochemistry, 56*, 1780–1785.

Ifuku, O., Koga, N., Haze, S., Kishimoto, J., & Wachi, Y. (1994). Flavodoxin is required for conversion of dethiobiotin to biotin in *Escherichia coli*. *European Journal of Biochemistry, 224*, 173–178.

Marquet, A., Tse Sum Bui, B., & Florentin, D. (2001). Biosynthesis of biotin and lipoic acid. *Vitamins and Hormones, 61*, 51–101.

Otsuka, A. J., Buoncristiani, M. R., Howard, P. K., Flamm, J., Johnson, C., Yamamoto, R., et al. (1988). The *Escherichia coli* biotin biosynthetic enzyme sequences predicted from the nucleotide sequence of the *bio* operon. *Journal of Biological Chemistry, 263*, 19577–19585.

Parry, R. J., & Kunitani, M. G. (1976). Biotin biosynthesis. 1. The incorporation of specifically tritiated dethiobiotin into biotin. *Journal of the American Chemical Society, 98*, 4024–4026.

Ploux, O., & Marquet, A. (1992). The 8-amino-7-oxopelargonate synthase from *Bacillus sphaericus*. Purification and preliminary characterization of the cloned enzyme overproduced in *Escherichia coli*. *Biochemical Journal, 283*, 327–331.

Reyda, M. R., Fugate, C. J., & Jarrett, J. T. (2009). A complex between biotin synthase and the iron–sulfur cluster assembly chaperone HscA that enhances in vivo cluster assembly. *Biochemistry, 48*, 10782–10792.

Rolfe, B., & Eisenberg, M. (1968). Genetic and biochemical analysis of the biotin loci of *Escherichia coli* K-12. *Journal of Bacteriology, 96*, 515–524.

Sanyal, I., Cohen, G., & Flint, D. H. (1994). Biotin synthase: Purification, characterization as a [2Fe–2S] cluster protein, and in vitro activity of the Escherichia coli *bioB* gene product. *Biochemistry, 33*, 3625–3631.

Sanyal, I., Gibson, K. J., & Flint, D. H. (1996). *Escherichia coli* biotin synthase: An investigation into the factors required for its activity and its sulfur donor. *Archives of Biochemistry and Biophysics, 326*, 48–56.

Taylor, A. M., Farrar, C. E., & Jarrett, J. T. (2008). 9-Mercaptodethiobiotin is formed as a competent catalytic intermediate by *Escherichia coli* biotin synthase. *Biochemistry, 47*, 9309–9317.

Tse Sum Bui, B., Florentin, D., Fournier, F., Ploux, O., Mejean, A., & Marquet, A. (1998). Biotin synthase mechanism: On the origin of sulphur. *FEBS Letters*, *440*, 226–230.

Ugulava, N. B., Frederick, K. K., & Jarrett, J. T. (2003). Control of adenosylmethionine-dependent radical generation in biotin synthase: A kinetic and thermodynamic analysis of substrate binding to active and inactive forms of BioB. *Biochemistry*, *42*, 2708–2719.

Ugulava, N. B., Gibney, B. R., & Jarrett, J. T. (2000). Iron-sulfur cluster interconversions in biotin synthase: Dissociation and reassociation of iron is required for conversion of [2Fe–2S] to [4Fe–4S] clusters. *Biochemistry*, *39*, 5206–5214.

Ugulava, N. B., Gibney, B. R., & Jarrett, J. T. (2001). Biotin synthase contains two distinct iron–sulfur cluster binding sites: Chemical and spectroelectrochemical analysis of iron–sulfur cluster interconversions. *Biochemistry*, *40*, 8343–8351.

Ugulava, N. B., Sacanell, C. J., & Jarrett, J. T. (2001). Spectroscopic changes during a single turnover of biotin synthase: Destruction of a [2Fe–2S] cluster accompanies sulfur insertion. *Biochemistry*, *40*, 8352–8358.

Ugulava, N. B., Surerus, K. K., & Jarrett, J. T. (2002). Evidence from Mössbauer spectroscopy for distinct $[2Fe–2S]^{2+}$ and $[4Fe–4S]^{2+}$ cluster binding sites in biotin synthase from *Escherichia coli*. *Journal of the American Chemical Society*, *124*, 9050–9051.

Wan, J. T., & Jarrett, J. T. (2002). Electron acceptor specificity of ferredoxin (flavodoxin): $NADP^+$ oxidoreductase from *Escherichia coli*. *Archives of Biochemistry and Biophysics*, *406*, 116–126.

CHAPTER FOURTEEN

Methods for Expression, Purification, and Characterization of PqqE, a Radical SAM Enzyme in the PQQ Biosynthetic Pathway

Wen Zhu*,†, Ana M. Martins*, Judith P. Klinman*,†,‡,1

*California Institute for Quantitative Biosciences, University of California, Berkeley, CA, United States
†Department of Chemistry, University of California, Berkeley, CA, United States
‡Department of Molecular and Cell Biology, University of California, Berkeley, CA, United States
[1]Corresponding author: e-mail address: klinman@berkeley.edu

Contents

Abstract

PqqE is the first enzyme in the biosynthetic pathway of the redox cofactor pyrroloquinoline quinone (PQQ), catalyzing the formation of a carbon–carbon bond in the precursor peptide PqqA. PqqE is a radical *S*-adenosyl-L-methionine (SAM) (RS) enzyme, a family of enzymes that use the reductive cleavage of a [4Fe–4S] cluster-bound SAM molecule to generate a 5′-deoxyadenosyl radical. This radical is then used to initiate an array of reactions that otherwise would be unlikely to occur. PqqE is a founding member of a subset family of RS enzymes that, additionally to the SAM [4Fe–4S] cluster, have a SPASM domain containing additional, auxiliary Fe–S clusters.

Methods in Enzymology, Volume 606
ISSN 0076-6879
https://doi.org/10.1016/bs.mie.2018.04.002

Most radical SAM enzymes are highly sensitive to oxygen, which destroys their Fe–S clusters. This can pose several limitations when working with these enzymes, since most of the work has to be done under anaerobic conditions. Here, we summarize the methods developed in our lab for the expression and purification of PqqE. We also highlight the several methods we have used for the characterization of the enzyme.

ABBREVIATIONS

5′-dA	5′-deoxyadenosine
AHQQ	3*a*-(2-amino-2-carboxyethyl)-4,5-dioxo-4,5,6,7,8,9-hexahydroquinoline-7,9-dicarboxylic acid
Amp	ampicillin
Cam	chloramphenicol
DTH	sodium dithionite
DTT	dithiothreitol
FldA	flavodoxin A
FNR	flavodoxin reductase
GSH	glutathione
IPTG	isopropyl β-D-1-thiogalactopyranoside
ITC	isothermal titration calorimetry
Kan	kanamycin
Klbp	*Klebsiella pneumoniae*
LB	Luria–Bertani medium
LBAmp100	Luria–Bertani medium with 100 μg/mL ampicillin
LBKan50Cam50	Luria–Bertani medium with 50 μg/mL kanamycin and 50 μg/mL chloramphenicol
Mex	*Methylobacterium extorquens*
Nano-ESI-MS	nanoelectrospray ionization mass spectrometry
PQQ	pyrroloquinoline quinone (4,5-dihydro-4,5-dioxo-1*H*-pyrrolo[2,3-*f*] quinolone-2,7,9-tricarboxilic acid)
RiPPs	ribosomally synthesized and posttranslationally modified peptides
RS	radical SAM
SAM	*S*-adenosyl-L-methionine
SPR	surface plasmon resonance
TB	terrific broth
TBKan50Cam50	terrific broth with 50 μg/mL kanamycin 50 μg/mL chloramphenicol
TCEP	Tris(2-carboxyethyl)phosphine

1. INTRODUCTION

Pyrroloquinoline quinone (4,5-dihydro-4,5-dioxo-1*H*-pyrrolo[2,3-*f*] quinolone-2,7,9-tricarboxilic acid, PQQ) is a quinone-containing non-covalently bound redox cofactor first described in 1964 as the "new prosthetic group" of a *Bacterium anitratum* glucose dehydrogenase (Hauge, 1964).

It is used by a wide variety of bacterial enzymes, mainly alcohol and sugar dehydrogenases in the nonglycolytic substrate oxidation for the production of ATP (Duine, 1999).

Quinocofactors are derived from canonical amino acid side chains in either a protein or a peptide, by posttranslational modifications and, besides the peptide-derived PQQ, include the protein-derived trihydroxyphenylalanine quinone (TPQ), tryptophan tryptophylquinone (TTQ), lysine tyrosyl quinone (LTQ), and cysteine tryptophylquinone (CTQ) (Klinman & Bonnot, 2014).

PQQ is a ribosomally synthesized and posttranslationally modified peptide (RiPP), being derived from the peptide precursor PqqA. The enzymes involved in the biosynthesis of PQQ are encoded by genes arranged in the *pqq* operon, with five of the genes—*pqqA*, *pqqB*, *pqqC*, *pqqD*, and *pqqE*—being obligatory for PQQ production (Shen et al., 2012). *pqqF*, encoding a protease, may also be present—in the operon or elsewhere in the genome—but it is not consistently conserved (Fig. 1). The PQQ molecule derives from the evolutionarily conserved glutamate and tyrosine side chains within the ribosomally produced peptide substrate PqqA, encoded by *pqqA* (Velterop et al., 1995).

The biosynthesis of PQQ from PqqA starts with the formation of a new carbon–carbon bond between the γ-carbon of the glutamate and the C5 of the tyrosine ring, catalyzed by the radical *S*-adenosyl-L-methionine (SAM) (RS) enzyme PqqE (Barr et al., 2016) (Fig. 2).

PqqE uses a [4Fe–4S] cluster located in its N-terminal domain to reductively cleave SAM, initiating a free radical cascade that ultimately leads to the formation of the new carbon–carbon bond, creating a cross-linked product. PqqE belongs to a subset of RS enzymes that additionally contain a SPASM domain in their C-terminal (RS-SPASM) comprised of most generally two auxiliary Fe–S clusters that are also essential for the net reaction (Grell, Goldman, & Drennan, 2015). PqqE has been proposed to contain the typical

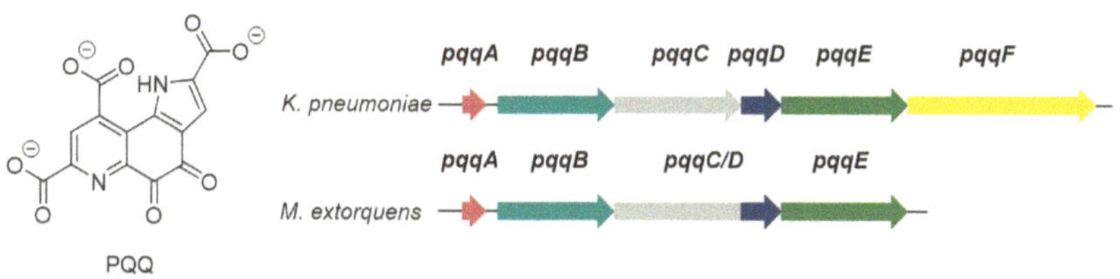

Fig. 1 The structure of PQQ and the *pqq* operons in *K. pneumoniae* and *M. extorquens* (*Mex*). Note that in *Mex*, *pqqD* is fused to the C-terminal of *pqqC*, while in *K. pneumoniae* PqqC and PqqD are encoded by separate genes. In *K. pneumoniae* the gene encoding the PqqF protease is located in the operon, unlike in *Mex*.

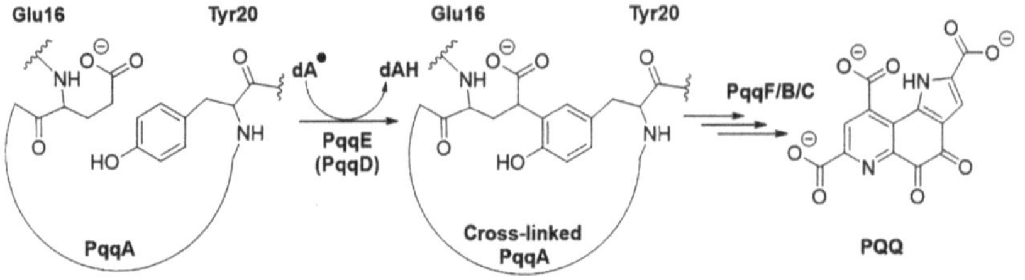

Fig. 2 Simplified scheme of the PQQ biosynthetic pathway. PqqE catalyzes the first step in PQQ biosynthesis. The specific involvement of PqqF and PqqB within the pathway is under investigation. The roles of PqqA, PqqC, PqqE, and PqqD are now well established (see text).

RS site (Wecksler et al., 2009) as well as two auxiliary Fe–S cluster-binding sites, designated AuxI and AuxII (Barr et al., 2016; Saichana et al., 2017). What distinguishes PqqE from other SPASM proteins is the fact that it contains a unique $Cx_2Cx_5Dx_{21}C$ motif with the ligand aspartate (D) in AuxII, while other SPASM proteins contain a conserved $Cx_{(2-4)}Cx_5Cx_{(21-22)}C$, with four cysteines in their AuxII site. Additionally there is a CxC motif present in AuxI of PqqE, a feature observed in both [4Fe–4S] and [2Fe–2S] containing proteins (Banci et al., 2013; Wagner, Koch, Ermler, & Shima, 2017; Fig. 3).

PqqD, a small protein, acts as a chaperone to deliver PqqA to PqqE, as shown by the formation of a PqqA–PqqD complex and its interaction with PqqE (Latham, Iavarone, Barr, Juthani, & Klinman, 2015). The hypothesis is that PqqD shields PqqA from premature proteolytic cleavage catalyzed by cellular proteases.

The cross-linked precursor to PQQ is hypothesized to be cleaved by a protease, which in some cases is encoded by the gene *pqqF*, and the resulting molecule is later acted on by PqqB, a metallo-β-lactamase enzyme, ultimately yielding 3*a*-(2-amino-2-carboxyethyl)-4,5-dioxo-4,5,6,7,8,9-hexahydroquinoline-7,9-dicarboxylic acid, AHQQ (Bonnot, Iavarone, & Klinman, 2013; Magnusson et al., 2004). Finally, the well-studied PqqC, a cofactorless synthase, catalyzes the eight-electron oxidation and ring cyclization of AHQQ to PQQ (Magnusson et al., 2004).

This chapter focuses on the proteins acting in the initial step of PqqA cross-linking, with an emphasis on enzymes from *Methylobacterium extorquens* (*Mex*). PqqE from different sources has different tolerances to oxygen, thus the expression and purification methods differ based on the strain from which the enzyme is being purified (Barr et al., 2016; Latham et al., 2015;

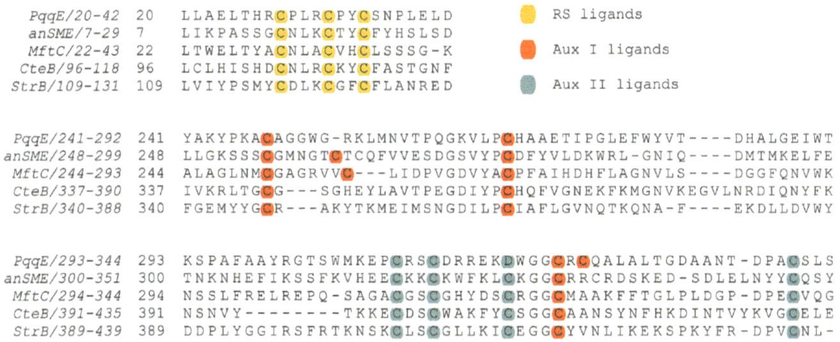

Fig. 3 Partial sequence alignment of SPASM proteins. The protein sequences near the Fe–S clusters in PqqE in *Methylobacterium extorquens*, anaerobic sulfatase-maturating enzyme (anSME) in *Clostridium perfringens*, mycofactocin radical SAM maturase (MftC) in *Mycobacterium tuberculosis*, sactisynthase (CteB) in *Clostridium thermocellum*, and a radical SAM enzyme in biosynthesis of streptide (StrB) *Streptococcus thermophiles*. Note that the assignment of Aux I and Aux II ligands for anSME and CteB is based on crystal structures in the database (PDB: 4K37 and PDB: 5WHY) (Goldman et al., 2013; Grove et al., 2017). The assignment of PqqE, MftC, and StrB is predicted based on the structure model created by SWISSMODEL (http://swissmodel.expasy.org/).

Saichana et al., 2016; Wecksler et al., 2009). Most radical SAM enzymes are highly sensitive to oxygen due to the oxidation of their Fe–S clusters, but the *Mex* enzyme is quite tolerant, making it easier to express and purify under normal lab conditions (Saichana et al., 2016).

2. PROTEIN EXPRESSION, PURIFICATION, AND RECONSTITUTION

The initial attempts to express and purify recombinant PqqE, from *Klebsiella pneumoniae*, were done under aerobic conditions (Wecksler et al., 2009). In these conditions the protein was restricted to inclusion bodies and the purification procedure included a protein refolding step prior to chromatographic purification. The protein obtained was >95% pure and was reconstituted with Fe^{2+}/Fe^{3+} and S^{2-} ions, under anaerobic conditions and in the presence of dithiothreitol (DTT). Although the reconstituted protein exhibited the reddish-brown color characteristic of the Fe–S presence, the EPR spectral characteristics were not in accordance with a radical SAM enzyme and a different approach for enzyme purification had to be pursued (Wecksler et al., 2009). Growth and induction under anaerobic conditions yielded soluble protein but subsequent aerobic purification resulted in an enzyme with EPR spectra similar to the one grown and purified aerobically. Thus, growth and purification of *K. pneumoniae* PqqE must be done in strict anaerobic conditions (see protocol in Sections 2.2.1 and 2.3).

Mex PqqE, on the other hand, is much more tolerant to oxygen, and it can be expressed under aerobic conditions (Barr et al., 2016; Saichana et al., 2016), although it was initially expressed under anaerobic conditions (Latham et al., 2015).

All anaerobic work must be done in a controlled atmosphere chamber (glovebox) with <5 ppm oxygen (Barr et al., 2016). The buffers to be used in anaerobic work are prepared under normal lab conditions and then purged with argon. The reagent solutions are either prepared like the buffers or the powder reagent is brought to the glovebox, weighed, and reconstituted with the anaerobic buffers.

All steps involving work with bacterial cell hosts are done in sterile conditions, using autoclaved media and autoclaved or filter (0.2 μm)-sterilized solutions.

Fig. 4 shows a typical SDS-PAGE result for *Mex* proteins expressed and purified as described in this section.

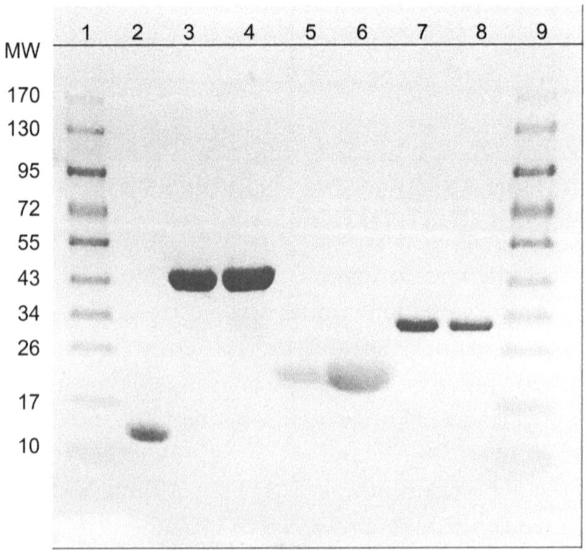

Fig. 4 SDS-PAGE gel of proteins expressed and purified in this work. The lanes are as follows: (1),(9) protein ladder; (2) *Mex* PqqD; (3),(4) *Mex* PqqE; (5),(6) *A. vinelandii* FldA; (7),(8) *Mex* FNR.

2.1 *pqqE* Cloning Strategies

The gene encoding PqqE is amplified from genomic DNA by regular molecular biology techniques, using a set of primers designed in order to insert the required restrictions sites. PqqE has been expressed and purified as a His_6-tag construct (either N-terminal or C-terminal), thus the vectors pET28 (Novagen) have been used for cloning purposes (Barr et al., 2016; Latham et al., 2015; Wecksler et al., 2009). The pET cloning system uses a strong T7 promoter, facilitating a robust overexpression of the protein encoded by the cloned gene. The plasmids are then transformed into host competent cells—*E. coli* XL1 Blue or *E. coli* TOP10 (Invitrogen)—isolated and sequenced. Once the sequence is confirmed, the plasmid harboring the correct construct is cotransformed into *E. coli* BL21(DE3) (Stratagene) with vectors expressing the *E. coli suf ABCDSE* operon (pPH151) genes, which encodes proteins involved in Fe–S cluster repair and assembly (Nachin, Loiseau, Expert, & Barras, 2003). The vectors can also be cotransformed in *E. coli* Rosetta 2 (DE3) (Novagen), which contain rare codons that are seldom used in *E. coli* (Saichana et al., 2016). The pET vector contains a kanamycin (Kan) resistance gene, while the pPH151 is resistant to chloramphenicol (Cam); thus, the transformed strain is Kan and Cam resistant and can be grown in selective media with both antibiotics.

2.2 PqqE Overexpression

Depending on the bacterial source of PqqE, the protein expression must be done anaerobically (e.g., *K. pneumoniae*) or it can be done in aerobic conditions (e.g., *Mex*). A summary of PqqE expression techniques is shown in Table 1.

2.2.1 Growth and Expression of K. pneumoniae *PqqE in Anaerobic Conditions (Wecksler et al., 2009)*

1. Select a single isolated colony of *E. coli* BL21 (DE3) on agar plates of Luria–Bertolli (LB) medium containing 50 μg/mL Kan and Cam antibiotics (LBKan50Cam50).
2. Use this colony to inoculate a flask with 100 mL of LBKan50Cam50 media and grow overnight at 37°C, 250 rpm.
3. Use 20 mL of the overnight culture to inoculate 2 L of the same medium in Erlenmeyer flasks with a rubber septum and purge the head space with argon.

Table 1 Methods for Expression of PqqE in Anaerobic (*K. pneumoniae* or *Mex*) or Aerobic (*Mex*) Conditions

	Anaerobic (Latham et al., 2015; Wecksler et al., 2009)	Aerobic (Barr et al., 2016)
Heterologous protein	His_6-PqqE or PqqE-His_6	His_6-PqqE
Vectors	pET28a or pET24b (Novagen)	pET28a (Novagen)
Vector hosts	*E. coli* BL21(DE3) (with pPH151) (Stratagene)	*E. coli* BL21 (DE3) (with pPH51) (Stratagene)
Starter cultures/ inoculum	Cells grown aerobically overnight at 37°C	Cells grown aerobically overnight at 37°C
Growth medium	LBKan50Cam50 (*Klbp*) TBKan50Cam50 + 50 m*M* fumarate + 50 μ*M* Fe(III) citrate (*Mex*)	TBKan50Cam50 + 100 μ*M* Fe(III) citrate
Induction	OD_{600} = 0.1 with 100 μ*M* IPTG (*Klbp*) + 5 mg/L $Fe(NH_4)_2(SO_4)_2{\cdot}6(H_2O)$ OD_{600} = 0.6 with 400 μ*M* IPTG (*Mex*)	OD_{600} = 0.1 with 100 μ*M* IPTG + 50 μ*M* cysteine
Expression conditions	18°C, overnight (*Klbp*) 19°C, 12 h (*Mex*)	20°C, 18 h

4. Grow at 30°C, 220 rpm until reaching an OD600 of 0.1 and induce the expression by adding isopropyl β-D-1-thiogalactopyranoside (IPTG) to a final concentration of 100 μ*M*. Simultaneously, add Fe(NH4)2(SO4)2·6 (H2O) to a final concentration of 5 mg/L, for the formation of Fe–S clusters.
5. Purge the head space with argon and allow the cultures to grow anaerobically overnight at 18°C, 220 rpm.
6. Harvest by centrifugation, freeze the cell paste in liquid nitrogen, and store at −80°C until use.

2.2.2 *Growth in Aerobic Conditions and Expression of* Mex *PqqE in Anaerobic Conditions (Latham et al., 2015)*

1. Select a single isolated colony of *E. coli* BL21 (DE3) on LBKan50Cam50 plates.
2. Prepare a starter culture by inoculating a flask with 100 mL of terrific broth (TB) media with 50 μg/mL Kan and 50 μg/mL Cam (TBKan50Cam50) and growing overnight at 37°C, 250 rpm.

3. Use 20 mL of the overnight culture to inoculate Erlenmeyer flasks containing 2 L of the same medium and grow in the same conditions until OD_{600} reaches 0.6.
4. Add 50 m*M* fumarate, 50 μ*M* Fe(III) citrate, stopper the flasks, and purge the head space with argon to transition from aerobic to anaerobic conditions.
5. Allow cells to grow for 30 min in anaerobic conditions before inducing protein expression with 400 μ*M* IPTG.
6. Incubate cells anaerobically at 19°C for 12 h.
7. Harvest the cells by centrifugation and freeze as described in Section 2.2.1.

2.2.3 Growth and Expression of Mex PqqE in Aerobic Conditions (Barr et al., 2016)

1. Select a single isolated colony of *E. coli* BL21 (DE3) on LBKan50Cam50 plates and prepare a starter culture like described in Section 2.2.2.
2. Use 20 mL of the overnight culture to inoculate Erlenmeyer flasks containing 2 L of TBKan50Cam50 with 4 g/L glycerol as carbon source, supplemented with 100 μ*M* Fe(III) citrate or ammonium Fe(III) citrate.
3. Allow the cultures to grow at 31°C, 220 rpm for 4 h and then decrease the temperature to 20°C.
4. One hour later induce protein expression by adding IPTG to a final concentration of 100 μ*M*. Simultaneously, add cysteine as a source of sulfur, to a final concentration of 50 μ*M*.
5. Grow the cells overnight in the same conditions, harvest by centrifugation, and freeze and store as described previously.

2.2.4 Expression of ^{57}Fe-Labeled PqqE to Be Used in Mössbauer Spectroscopy

Protein samples to be used in Mössbauer spectroscopy must be labeled with ^{57}Fe, and, for that purpose, the cultures are grown in the presence of labeled Fe(III) citrate (Barr et al., 2016).

1. Prepare a 50-mL starter overnight culture as described previously.
2. Prepare 20 m*M* ^{57}Fe citrate by reacting 117 mg iron oxide (Cambridge Isotopes) with 1 mL 10 *M* HCl at 80°C. Dilute the formed $^{57}FeCl_3$ 10-fold with 40 m*M* citric acid.
3. Prepare a solution of 5 g OmniPur® casamino acids (Sigma-Aldrich) in 10 mL water and chelate the free metals by incubating the solution with 2.5 g of Chelex-100 chelating ion-exchange resin (Bio-Rad) for 1 h.

4. Prepare flasks with 1 L of M9 medium and supplement with 1 mL of 1000× trace minerals solution (Sigma-Aldrich), the 10 mL casamino acids solution (5 g/L final concentration), and the 1 mL 20 m*M* ^{57}Fe citrate (20 μ*M* final concentration). Sterilize by autoclaving. Add Kan and Cam to final concentrations of 50 μg/mL.
5. Inoculate the flasks and follow the procedure described in Section 2.2.3. When inducing with IPTG, supplement with pyridoxine HCl and cysteine to final concentrations of 10 and 100 μ*M*, respectively.
6. Follow the protocol described for unlabeled PqqE (Section 2.2.3).
7. As a control, express the protein using the same protocol but with regular Fe(III) citrate as the iron source.

2.3 PqqE Purification

The expressed His_6-PqqE is purified by affinity chromatography from a nickel column. The purification is simplified by the fact that the protein has a reddish-brown color due to the presence of Fe–S clusters; thus, the PqqE-containing fractions are easily identified by their color (Barr et al., 2016; Wecksler et al., 2009).

Purification of the *K. pneumoniae* His_6 tag-PqqE is performed under strict anaerobic conditions in an inert atmosphere glovebox (Wecksler et al., 2009), while *Mex* PqqE can be purified in anaerobic (Barr et al., 2016; Latham et al., 2015) or aerobic conditions (Saichana et al., 2016). The steps used in anaerobic and aerobic purifications are identical; the only difference being that aerobic purification is done in normal lab conditions, while anaerobic purification is done in a glovebox using degassed solutions and buffers. The following protocol is for anaerobic purification. In all steps aliquots (10–50 μL) should be collected for SDS-PAGE analysis. A summary of PqqE purification procedures is shown in Table 2.

1. Prepare and degas all buffers needed for the purification: *Lysis buffer*—50 m*M* Tris buffer pH 7.9, 300 m*M* NaCl, 30 m*M* imidazole, 1 m*M* DTT or TCEP, 10% (v/v) glycerol; *Wash buffer*—50 m*M* Tris buffer pH 7.9, 300 m*M* NaCl, 40 m*M* imidazole, 1 m*M* DTT or TCEP, 10% (v/v) glycerol; *Elution buffer*—50 m*M* Tris buffer pH 7.9, 300 m*M* NaCl, 300 m*M* imidazole, 10% (v/v) glycerol; *Storage buffer*—50 m*M* Tris pH 7.9, 300 m*M* NaCl, 10% (v/v) glycerol.
2. Defrost the frozen cell pellets on ice and resuspend in five times the mass of cell paste volume of lysis buffer.

Table 2 Methods for Purification of *K. pneumoniae* or *M. extorquens* PqqE

	K. pneumoniae **(Wecksler et al., 2009)**	***M. extorquens*** **(Barr et al., 2016; Latham et al., 2015)**
Cell pellet disruption	Lysis with BugBuster (Novagen) in 50 m*M* Tris pH 7.9, 1 m*M* DTT, 300 m*M* NaCl, 10 m*M* imidazole, 5 μL benzonase nuclease; centrifugation 15,000 × *g*, 20 min, 4°C	Lysis with BugBuster (Novagen) in 50 m*M* Tris pH 7.9, 300 m*M* NaCl, 1 m*M* TCEP, 30 m*M* imidazole, 5 μL benzonase nuclease; 15,000 × *g*, 20 min, 4°C
Chromatography	(i) Ni-NTA column (12 in. long, 2.5 in wide) equilibrated with 50 m*M* Tris pH 7.9, 1 m*M* DTT, 300 m*M* NaCl, 10 m*M* imidazole. Washed with 100 mL of the same buffer, then with 100 mL of 25 m*M* imidazole in buffer, followed by 100 mL of 50 m*M* imidazole in buffer. Eluted with 200 m*M* imidazole in buffer. Fractions selected based on color, pooled and concentrated to 5 mL by ultrafiltration (Amicon Ultra 30K membrane) (ii) Gel filtration in PD-10 column equilibrated with 50 m*M* Tris pH 7.9, 1 m*M* DTT, 300 m*M* NaCl to remove imidazole. Protein collected off the column and concentrated to ~10 mg/mL by ultrafiltration (Amicon Ultra 30K membrane)	(i) His-Trap FF column (Novagen) equilibrated with 50 m*M* Tris pH 7.9, 300 m*M* NaCl, 1 m*M* TCEP, 30 m*M* imidazole buffer. Washed with the same buffer and eluted with 300 m*M* imidazole in buffer. Fractions pooled and concentrated to 2.5 mL by ultrafiltration (Amicon Ultra 30K membrane) (ii) Gel filtration in PD-10 columns equilibrated with 50 m*M* Tris pH 7.9, 300 m*M* NaCl, 10% (v/v) glycerol
Yield	0.5–1.5 mg/L LB medium	18 mg/L TB medium
Reconstitution	Not possible, the protein precipitates	Chemical reconstitution with Na_2S and ammonium Fe(III) citrate
Fe and S content	10.4 ± 0.9 mol Fe and 7.0 ± 1.0 mol S/monomer	7–10 mol Fe/monomer (as-purified) 13 mol Fe and 12.2 mol S/monomer (reconstituted)

All procedures are done in anaerobic conditions, in a glovebox.

3. Lyse by the addition of 0.2% (w/v) CHAPS or 1 × BugBuster (Novagen), 0.1 mg/mL lysozyme (final concentrations), and 5 μL benzonase nuclease, for 30 min, following the manufacturer's guidelines.
4. Transfer the lysates to anaerobic centrifuge tubes (Beckman) and centrifuge at 15,000 × *g*, 15 min, 4°C.
5. Reintroduce the centrifuge tubes into the glovebox and collect the supernatant.
6. Equilibrate a Nickel column (Ni-NTA resin or His-Trap FF column (GE Healthcare)) with cold lysis buffer. Load the supernatant into the column manually with a pipette or using a peristaltic pump.
7. Upon protein binding, wash the column with three volumes of cold wash buffer.
8. Elute the protein with cold elution buffer. Collect and pool all the reddish-brown fractions.
9. Continuing in the glove box, concentrate by centrifugal ultrafiltration in an Amicon Ultra 30K filtration device (Millipore).
10. To remove the imidazole, pass the concentrated protein sample through a PD-10 column previously equilibrated with 50 m*M* Tris pH 7.9, 300 m*M* NaCl, and 10% (v/v) glycerol.
11. If needed, concentrate the protein again using centrifugal ultrafiltration as described previously.
12. Aliquot in sealed vials, flash-freeze in liquid nitrogen, and store in a liquid nitrogen dewar.
13. Untagged PqqE needed for SPR protein interaction assays (Section 3.4.2) is obtained by cleavage of the His-tag with a thrombin kit (Novagen), according to the manufacturer's instructions.

2.4 Chemical Reconstitution of the As-Purified PqqE

K. pneumoniae PqqE purified anaerobically cannot be reconstituted with excess of iron and sulfide since the protein precipitates upon Fe^{2+}/Fe^{3+} and S^{2-} addition (Wecksler et al., 2009). As-purified *Mex* PqqE can contain incomplete Fe–S clusters. Thus, for activity assays and the majority of spectroscopic studies, the enzyme was routinely reconstituted with iron and sulfide. The protein reconstitution is done under anaerobic conditions, in a glovebox with <5 ppm O_2 (Barr et al., 2016). The ^{57}Fe-labeled PqqE (described in Section 2.2.4) is reconstituted following a similar protocol, but using solutions (except for ammonium Fe(III) sulfate) treated with Chelex-100 chelating ion-exchange resin (Bio-Rad) for 30 min, to minimize the iron contamination.

1. Prepare 100 mM fresh stock solutions of DTT, ammonium Fe(III) citrate, and Na_2S in 1 M Tris pH 7.9 buffer: 15.4 mg DTT/1 mL buffer, 24.3 mg ammonium Fe(III) citrate/1 mL buffer, and 24 mg Na_2S/1 mL buffer.
2. Add 1.5 μL ofDTT stock solution into 1 mL 100 μM PqqE sample while stirring.
3. Then add 1 μL of the ammonium Fe(III) citrate stock solution to the sample and stir slowly for 15 min at room temperature.
4. Finally add 1 μL of the Na_2S stock solution to the sample and stir slowly for 15 more min at room temperature. Repeat steps 3 and 4 four times.
5. Pass the sample through a PD-10 column preequilibrated with storage buffer (see Section 2.3) to remove excess of iron and sulfide.
6. Flash-freeze in liquid nitrogen and store in a liquid nitrogen dewar.

2.5 Cloning, Expression, and Purification of Flavodoxin Reductase and Flavodoxin A

Because the PqqE activity in vitro requires the presence of a flavodoxin A (FldA) and flavodoxin reductase (FNR)-based reduction system (other reductants lead to an uncoupled cleavage of the cosubstrate SAM) (Barr et al., 2016), these proteins are also expressed and purified in our lab. Because no FldA could be identified in the *Mex* genome, FldA was amplified from the *Azotobacter vinelandii* ATCC478 genome.

2.5.1 Mex *FNR*

All the buffers used are the ones described in Section 2.3 for PqqE purification but without DTT/TCEP.

1. Amplify FNR from *Mex* AM1 genomic DNA and clone into the pET28a plasmid.
2. Transform the plasmid into *E. coli* BL21 (DE3) competent cells and plate in LBKan50 agar plates.
3. Pick up a single, isolated colony to inoculate 50 mL of LBKan50 medium and incubate at 37°C overnight, at 250 rpm.
4. Use 10 mL of the overnight culture to inoculate 1 L of fresh LBKan50 medium, grow in the same conditions, and when OD_{600} reaches 0.5 induce the expression by adding IPTG to a final concentration of 500 μM.
5. Simultaneously add riboflavin to a final concentration of 100 μM and express the protein overnight at 31°C, 220 rpm.

6. Harvest cells by centrifugation and freeze the cell pellets at −80°C or proceed to purification.
7. The protein is purified at 4°C using a Ni-NTA column (Thermo Fisher Scientific).
8. Add 20 mL of lysis buffer and FAD to a final concentration of 10 m*M* to the cell pellet obtained from 1 L of culture.
9. Sonicate the cell suspension on ice, with cycles of 20 s *on* and 30 s *off*, for a total *on* time of 2 min.
10. Centrifuge the cell lysate at 16,000 × *g* for 30 min at 4°C.
11. Load the supernatant (cell-free extract) in the column.
12. Wash the column with three-column volumes of washing buffer.
13. Elute the protein using three-column volumes of elution buffer and collect 15 mL bright yellow fractions.
14. Pool the fractions and concentrate by centrifugal ultrafiltration using a 15-mL Amicon Ultra 3K filter device (Millipore).
15. To remove the imidazole, pass the concentrated protein sample through a PD-10 column (GE Healthcare) previously equilibrated with storage buffer and collect the bright yellow protein fractions.
16. Flash-freeze 200 μL aliquots and store at −80°C.

2.5.2 A. vinelandii *FldA*

All the buffers used are the ones described in Section 2.3 for PqqE purification.

1. Amplify FldA from the *A. vinelandii* genomic DNA using an appropriate set of primers and clone into the pGEX-6p-1 plasmid (GE Healthcare).
2. Transform the plasmid into *E. coli* BL21 (DE3) competent cells and select positive transformants in LB agar plates with 100 μg/mL Amp (LBAmp100).
3. Use a single, isolated colony to inoculate 50 mL of LBAmp100 medium and incubate overnight at 37°C, 250 rpm.
4. Use 10 mL of overnight culture to inoculate 1 L of fresh LBAmp100 medium, grow in the same conditions, and induce with IPTG to a final concentration of 500 μ*M*, when OD_{600} is 0.5.
5. Simultaneously add FMN to a final concentration of 50 μ*M* and express the protein overnight at 31°C, 220 rpm.
6. Harvest the cells by centrifugation and freeze the cell pellets at −80°C or proceed to purification.
7. Purify the protein by affinity chromatography in a glutathione Sepharose column (GE Healthcare) as follows.

8. Add 20 mL of storage buffer to the cell pellet obtained from 1 L of culture.
9. Prepare the cell-free extract as described in Section 2.5.1 for cells expressing FNR.
10. Equilibrate the 5 mL glutathione-Sepharose column with 50 m*M* Tris–HCl pH 7.9, 300 m*M* NaCl, and load the dark-greenish supernatant.
11. Wash with three-column volumes of storage buffer and discard the flow-through.
12. Elute the protein with a 30-mL of a 15-m*M* solution of glutathione (GSH) in storage buffer.
13. To remove the GST-tag add 40 μL PreScission protease (GE Healthcare) to the ~15 mL of eluted protein solution and shake on a nutator (Clay Adams brand) for 12 h at 4°C.
14. Concentrate the protein sample to 5 mL using an Amicon Ultra 10K filtration device (Millipore).
15. Run the sample in a size-exclusion column HiPrep 26/60 Sephacryl S-200 High Resolution (GE Healthcare) equilibrated with storage buffer, outfitted to an AKTA FPLC (GE Healthcare) using a flow rate of 0.5 mL/min (this purification step must be performed at 4°C).
16. Run aliquots of the fractions with protein in SDS-PAGE and identify the ones containing untagged FldA.
17. Pool the fractions with untagged FldA and concentrate the protein solution to 3 m*M* using an Amicon Ultra 3K membrane (Millipore).
18. Aliquot into 50 μL samples, flash-freeze in liquid nitrogen, and store at −80°C.

2.6 Cloning, Expression, and Purification of PqqCD and PqqD

For activity assays and interaction experiments of PqqE with PqqD and PqqA, *Mex* PqqD was also cloned, expressed, and purified in our lab. In vivo *Mex* PqqD is N-terminally fused to PqqC, and this fusion was amplified from the *Mex* genome as a His_6-PqqCD (Section 2.6.1) (Latham et al., 2015) and used for interaction assays (Section 3.4.2). Additionally PqqD was expressed and purified as a single protein, either as a His_6-PqqD (Latham et al., 2015), for interaction assays, or as untagged PqqD (Barr et al., 2016), for activity assays.

2.6.1 His_6-PqqCD

All the buffers used are the ones described in Section 2.3 for PqqE purification. PqqCD is amplified from *Mex* AM1 genomic DNA and cloned into the pET28a plasmid. The expression and purification procedures are similar to the ones described for FNR (Section 2.5.1), except that the protein was expressed at 19°C and the lysis buffer does not contain FAD.

2.6.2 His_6-PqqD

The sequence of a truncation of PqqD not containing the PqqC part (amino acids 280–372 of the PqqCD sequence) was obtained as described by Latham et al. (2015). This sequence was amplified from the *Mex* AM1 genome and cloned into pET28a. The expression and purification procedures are similar to the ones described in Section 2.6.1.

2.6.3 Untagged PqqD

1. Amplify PqqD from *Mex* AM1 genomic DNA and clone into the pGEX-6p-1 plasmid (GE Healthcare).
2. Transform into *E. coli* BL21 (DE3) competent cells and plate in LBAmp100 agar plates.
3. Prepare a starter culture by inoculating 50 mL LBAmp100 medium with a single, isolated colony from the LBAmp100 plate, and incubate overnight at 37°C, 220 rpm.
4. Use 10 mL of overnight culture to inoculate 1 L of fresh LBAmp100 medium, grow in the same conditions until OD_{600} is 0.5, and induce the expression by adding IPTG to a final concentration of 500 μ*M*.
5. Incubate the cultures overnight at 19°C, 220 rpm to allow the protein to express.
6. Harvest the cells by centrifugation, freeze, and store at −80°C until use.
7. Purify the protein in a glutathione-Sepharose column followed by size-exclusion chromatography using the same methodology described in Section 2.5.

3. PROTEIN CHARACTERIZATION

3.1 Determination of Protein Concentration and Purity Evaluation

The purity of the PqqE, FNR, FldA, FNR, PqqCD, and PqqD preparations during the purification process is monitored by SDS-PAGE using the method of Laemmli (1970).

The concentration of protein is determined by the Bradford assay (Bradford, 1976) and its comparison with the absorbance at 280 nm (A_{280}), as described by Barr et al. (2016). Correct protein quantification is very important to determine the number of Fe–S clusters per polypeptide. Thus, a conversion factor relating protein determined by the Bradford assay, which relies on a standard curve usually prepared with a different protein (e.g., BSA), with that determined by absorbance at 280 nm, must be calculated. This is done using the following procedure (Barr et al., 2016):

1. Denature the protein by incubating for 1 h with 6 *M* guanidinium HCl, and 0.1 *M* citric acid to destroy the Fe–S clusters and chelate the free iron.
2. Pass the protein through a PD-10 column to remove the citric acid.
3. Determine the protein concentration by measuring the absorbance at 280 nm and determine the molar extinction coefficient using the method of Pace, Vajdos, Fee, Grimsley, and Gray (1995). Alternatively, a theoretical molar extinction coefficient of 57,870 M^{-1} cm^{-1} can be calculated in ProtParam, assuming that all Cys residues are reduced (Gasteiger et al., 2005).
4. Determine the concentration of protein in the same sample using the Bradford assay with a BSA standard curve linearized by dividing the absorbance at 590 nm by the absorbance at 450 nm (Zor & Selinger, 1996).
5. Compare the values obtained by the two methods and determine the conversion factor.

3.2 Characterization of Fe–S Clusters

3.2.1 Iron and Sulfide Quantification

The catalysis of PqqE depends on the correct incorporation of the Fe–S clusters that are formed in the protein either during expression or by chemical reconstitution after purification. It is necessary to quantitatively determine the total content of iron and sulfide for each batch of expressed and purified PqqE. We have adopted the method of Crack, Green, Thomson, and Le Brun (2014) for both determinations.

Iron content determination

Since the as-purified PqqE generally contains roughly 10 iron molecules per monomer (Barr et al., 2016; Latham et al., 2015), the protein should be diluted 10-fold prior to the assays. It is also worth noting that the use of screwed cap vials and the cooling process are necessary in order to avoid the change of concentration due to the water loss by evaporation during the heating process.

1. Add 100 μL of PqqE sample into screwed cap tubes in triplicate—tubes #1–3.

2. Set up a control vial by adding 100 μL of storage buffer (see Section 2.3) to tube #4.
3. Add 100 μL of each concentration of iron ICP-MS standard solution (Sigma)—0, 25, 50, 100, 150, 200, and 350 μ*M*—to tubes #5–11.
4. Add 100 μL of 22% (v/v) HNO_3 to each tube.
5. Incubate the tubes in a dry bath at 90°C for 30 min.
6. Cool the samples to room temperature and spin in a microcentrifuge at 12,000 × *g* for 1 min.
7. Neutralize the samples by adding 600 μL of 7.5% (w/v) ammonium acetate to each tube and mix by vortexing.
8. Reduce the samples by adding of 100 μL of 12.7% (w/v) freshly made ascorbic acid and mix by vortexing.
9. Add 100 μL of 10 m*M* ferrozine solution and mix by vortexing.
10. Spin the samples in a microcentrifuge at 12,000 × *g* for 1 min.
11. Transfer the solutions into disposable cuvettes and read the absorbance at 562 nm on the UV–vis spectrometer (Cary 50 spectrophotometer with a spectral bandwidth 1.5 nm).
12. Prepare a standard curve using the solutions with different concentrations of standard iron (tubes #5–11) and calculate the total iron content in the protein solution (tubes #1–3).
13. Calculate the iron/monomer of PqqE based on the protein concentration of the same PqqE sample that is used in the total iron content measurement (see Section 3.1).
14. Repeat the experiment individually three times and calculate the average iron content of the protein sample.

Sulfide content determination

1. Add 200 μL of an appropriately diluted PqqE sample into tubes #1–3.
2. Prepare a control vial by adding 200 μL of storage buffer (see Section 2.3) in tube #4.
3. Prepare a 2-m*M* sulfide stock solution by dissolving 240 mg of sodium sulfide nonahydrate ($Na_2S{\cdot}9H_2O$) in 500 mL 10 m*M* NaOH.
4. Prepare sulfide standard solutions by dilution of the 2 m*M* stock to 0, 10, 50, 100, 150, 200, and 250 μ*M*. Use a 10-m*M* NaOH solution as a blank control.
5. Add 200 μL of the sulfide standard solutions prepared in step 2 to tubes #5–11.
6. Immediately add 600 μL of freshly prepared 1% (w/v) zinc acetate into each tube.
7. Add 50 μL of 12% (v/v) NaOH into each tube.

8. Cap the tubes well and mix by inversion. Incubate at room temperature for 15 min
9. Centrifuge the tubes at 16,000 × *g* for 1 min.
10. Add 150 μL of freshly prepared 0.1% (w/v) *N,N*-dimethyl-*p*-phenylenediamine dihydrochloride in 5 *M* HCl.
11. Add 150 μL of freshly prepared 10 m*M* $FeCl_3$ in 1 *M* HCl.
12. Vortex the solutions to mix and incubate at room temperature for 30 min.
13. Transfer the solutions into disposable cuvettes and read the absorbance at 670 nm on UV–vis spectrometer (Cary 50 spectrophotometer, spectral bandwidth 1.5 nm).
14. Prepare a standard curve using standard solutions of sodium sulfide prepared in step 2 (tubes #5–11) and use it to calculate the total sulfide content in the protein samples (tubes #1–3).
15. Calculate the sulfide/monomer of PqqE based on the protein concentration of the same PqqE sample used in the total sulfide content measurement (see Section 3.1).
16. Repeat the experiment individually three times and calculate the average sulfide content of the protein sample.

3.2.2 EPR Characterization

EPR spectroscopy has been successfully used in the characterization of the Fe–S clusters in structure–function studies of radical SAM enzymes (Stich, Myers, & Britt, 2014). EPR spectra can provide useful information, such as the identity of Fe–S clusters, the protein environment around the clusters, and the quantity/quality of the iron and sulfur in the protein sample. Without the addition of reducing reagent, the oxidized [3Fe–4S] cluster presented in the as-purified/reconstituted PqqE can be easily detected by EPR at 40 K, while the oxidized iron in the [4Fe–4S] clusters is EPR silent (Pandelia, Lanz, Booker, & Krebs, 2015). Therefore, the sample preparation for characterization of PqqE using EPR spectroscopy involves the use of both, the oxidized and the reduced form of enzyme.

The sodium dithionite (DTH) solution must be freshly prepared anaerobically, inside the glovebox, using an aliquot of DTH powder. The commercially available DTH is usually less than 85% pure, and the purity decreases after long-term storage, even if in the glovebox. In this case, it is necessary to determine the quality of DTH using a redox titration method (McKenna, Gutheil, & Song, 1991). If highly concentrated PqqE is needed for better spectroscopy quality, the concentration step is done during reconstitution. Protein precipitated during the reconstitution steps is removed by centrifugation at 14,000 × *g* for 10 min.

The representative procedure described here is focused on the comparative Fe–S cluster characterization of *K. pneumoniae* PqqE and the PqqE–SAM complex, with the control being as-isolated enzyme in the absence of SAM/DTH (Fig. 5). The EPR spectrometer used was a Bruker ECS106 X-band spectrometer with a TE_{102} cavity (ER4102ST) resonating at 9.5 GHz, equipped with an Oxford ESR900 liquid helium cryostat in conjunction with an ITC503 temperature controller.

1. Mix a 10-fold excess of DTH solution (freshly prepared in 50 m*M* Tris, 300 m*M* NaCl pH 7.9, under anaerobic conditions) with 100 μ*M* PqqE in 50 m*M* Tris pH 7.9, 300 m*M* NaCl with or without 20% (v/v) glycerol by stirring the reaction mixture at room temperature in the glovebox for 10 min. Alternatively, a fivefold excess of DTH can be used with an incubation time extended to 30 min.
2. For the experiments with PqqE–SAM, after the reduction reaction, add a 10-fold excess of a 10 m*M* SAM solution in water (prepared as described

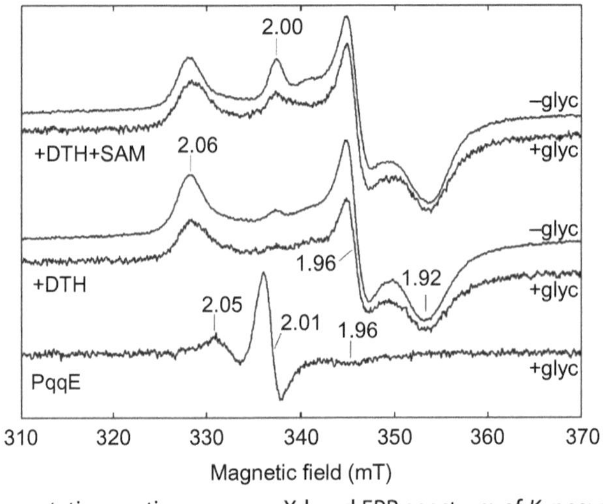

Fig. 5 A representative continuous wave X-band EPR spectrum of *K. pneumoniae* PqqE samples in various conditions. The signals are as follows, from *top* to *bottom*: dithionite (DTH)-reduced PqqE with a 10-fold excess of SAM (with and without glycerol), DTH-reduced PqqE (with and without glycerol), and as-isolated PqqE with glycerol. *Adapted from Wecksler, S. R., Stoll, S., Tran, H., Magnusson, O. T., Wu, S. P., King, D., et al. (2009). Pyrroloquinoline quinone biogenesis: demonstration that PqqE from Klebsiella pneumoniae is a radical S-adenosyl-L-methionine enzyme.* Biochemistry, *48(42), 10151–10161. Copyright © 2009 American Chemical Society.*

in Zhang & Klinman, 2015) to the reduced PqqE solution and incubate for approximately 5 min.

3. Aliquot 300 μL of the DTH-reduced PqqE sample into an X-band EPR tube.
4. Prepare the control samples by adding only PqqE or reduced PqqE (prepared in step 1) to EPR tubes.
5. Seal the tube with a rubber stopper in the glovebox and fast-freeze the sample in liquid nitrogen immediately after removal from the glovebox.
6. The spectrometer settings are set as follows: EPR conditions: temperature, 40 K; microwave power, 6.3 mW; microwave frequency, 9.480 GHz.
7. Determine the spectra of the PqqE samples at 40 K.
8. Spectra are simulated using the EasySpin MATLAB® package.

3.2.3 Mössbauer Spectroscopy

Compared to EPR, Mössbauer spectroscopy provides valuable information about the total iron present either in Fe–S clusters or adventitiously bound, without the need of prior reduction of the protein. Based on the parameters provided by the spectra, one can distinguish the different iron oxidation state and the [Fe–S] species in a mixed protein environment. This is especially useful for characterization of the clusters in PqqE, which has been found to contain a mix of [2Fe–2S] and [4Fe–4S] clusters (Barr et al., 2016; Saichana et al., 2016; Fig. 6). By analyzing proteins with different Fe–S site knockouts, one may also be able to identify the source of certain signals in the Mössbauer spectra. Here, we provide a general method for quantitatively determining the different iron species in the Fe–S clusters of as-purified WT PqqE labeled with ^{57}Fe (prepared as described in Section 2.2.4). We used a Mössbauer spectrometer (See Co., Edina, MN) connected to a Janis Research Co. cryostat (Wilmington, MA).

1. Add the purified and reconstituted ^{57}Fe-labeled PqqE solution into the Teflon sample holder (thickness 0.2 in.) in the glovebox with $O_2 < 1$ ppm.
2. Place the sample holder snuggly in the sample rod holder and wrap in Kapton® Tape prior to introduction into the spectrometer.
3. Collect spectra and analyze with the WMOSS software package (Prisecaru, 2009–2016).

3.3 Activity

3.3.1 SAM Cleavage Activity

The PqqE-catalyzed PqqA cross-linking reaction is initiated by the generation of a SAM-derived radical, concomitant with oxidation of the adjacent

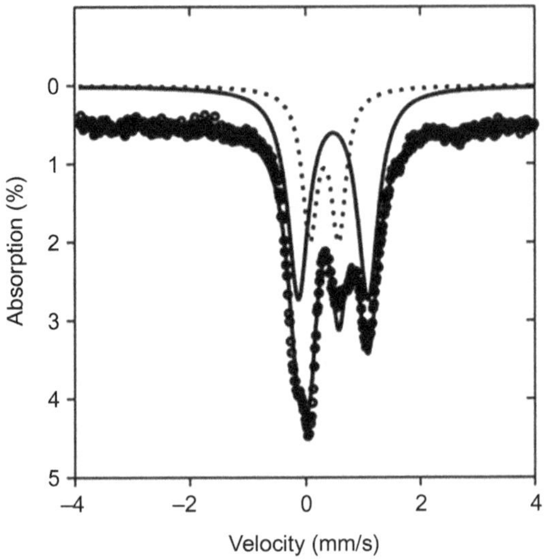

Fig. 6 Illustrative zero-field Mössbauer spectrum of as-purified *Mex* ^{57}Fe-PqqE at 4K. *Black circles* show the raw data, which have been offset from the axis by 0.5% for ease of interpretation. These data are decomposed into 4Fe–4S (*solid line*) and 2Fe–2S (*dotted line*) components. *Adapted from Barr, I., Latham, J. A., Iavarone, A. T., Chantarojsiri, T., Hwang, J. D., & Klinman, J. P. (2016). Demonstration that the radical S-adenosylmethionine (SAM) enzyme PqqE catalyzes de novo carbon-carbon cross-linking within a peptide substrate PqqA in the presence of the peptide chaperone PqqD.* Journal of Biological Chemistry, 291*(17), 8877–8884. Copyright ©2016 by American Society for Biochemistry and Molecular Biology.*

RS $[4Fe–4S]^+$ cluster to $[4Fe–4S]^{2+}$. This can lead to peptide (PqqA) cross-linking (Section 3.3.2) or uncoupled cleavage to yield an equivalent of 5′-deoxyadenosine (5′-dA). The formation of 5′-dA can be easily quantified by HPLC. In this experiment, we used a Beckman HPLC equipped with a diode array detector and operated by 32 Karat version 8.0 software. The column used was a reverse-phase C18 Jupiter® 4 μm Proteo 90 Å, 250 mm × 4.60 mm C18 column.

1. Set up reactions in the glovebox with $O_2 < 0.8$ ppm.
2. Prepare 100 m*M* DTH in 50 m*M* Tris pH 7.9, 1 m*M* DTT, 300 m*M* NaCl, 20% (v/v) glycerol.
3. Add 89 μL PqqE stock solution in the same buffer into tubes #1–8.
4. Add 1 μL 100 m*M* DTH into each sample.
5. Leave the samples in the glovebox at room temperature for 10 min to complete the reduction.
6. Add 10 μL 5 m*M* SAM into each sample.

7. Quench the reactions at various time points (1 min–16 h) by addition of 5 μL 88% (v/v) formic acid per 100 μL of the sample.
8. Take the quenched reactions out of the glovebox and centrifuge at 16,000 × *g* for 1 min.
9. Inject 100 μL of each sample into HPLC for 5′-dA determination; the conditions are: 1 mL/min flow rate, 0.05% (v/v) formic acid, 1.0% (v/v) acetonitrile in water for 30 min, followed by a linear gradient from 1% to 100% acetonitrile over 10 min. Spectra are recorded at 215 and 260 nm (Fig. 7).
10. Prepare a 5′-dA standard curve by analyzing solutions with different concentrations—0, 5, 10, 50, 100, 150, 200, and 250 μ*M*, in the same conditions.

3.3.2 PqqA Modification Assay

The SAM cleavage reaction described in Section 3.3.1 provides insight about the catalytic activity of the radical SAM $[4Fe–4S]^+$ cluster that generates the active radical in the initial steps of the PqqE reaction, but it does not reflect the catalytic efficiency of PqqE to perform the biologically active peptide modification reaction. The direct measurement of the PqqE bioactivity can be determined by assaying the production of the modified peptide by LC-MS.

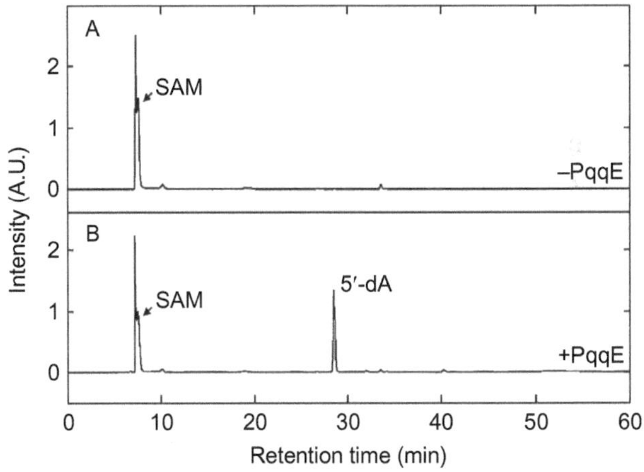

Fig. 7 LC-MS analysis of SAM cleavage reaction by *K. pneumoniae* DTH-reduced PqqE. (A) LC-MS elution profile (monitored at 260 nm) of a control sample without PqqE. (B) LC-MS elution profile of an anaerobic reaction mixture containing PqqE, SAM, and DTH. *Adapted from Wecksler, S. R., Stoll, S., Tran, H., Magnusson, O. T., Wu, S. P., King, D., et al. (2009). Pyrroloquinoline quinone biogenesis: demonstration that PqqE from Klebsiella pneumoniae is a radical S-adenosyl-L-methionine enzyme.* Biochemistry, *48(42), 10151–10161. Copyright © 2009 American Chemical Society.*

The cross-linked peptide product has a lower retention time compared to the unmodified PqqA in LC and can be identified from the m/z in the MS spectra.

The reaction mixture must contain both the peptide chaperone PqqD and an appropriate reduction system, in this instance FldA/FNR/NADPH (Barr et al., 2016). While the ratio of each component can be further optimized for better PqqA conversion, the FNR concentration should always be lower than the FldA concentration, due to the FldA-binding competition between PqqE and FNR, which is expected to reduce the yield of the cross-linked product.

In the following protocol, we used an Agilent 1200 LC with a C4 column (Restek) connected in-line with a Thermo LTQ-Orbitrap-XL mass spectrometer equipped with an electrospray ionization source and operated in the positive ion mode.

PqqA (with a nonconserved cysteine mutated to serine for increased stability) was custom synthesized and obtained from CPC Scientific.

All solutions and enzymes used in this protocol are prepared in storage buffer. Stock solutions are: 100 μ*M* PqqE, 1 m*M* PqqA, 3 m*M* PqqD, 5 m*M* SAM, 10 m*M* NADPH, 2 m*M* FNR, and 2 m*M* FldA. All concentrations mentioned in the following protocol are final concentrations in the mixtures.

1. Mix and incubate 1 m*M* NADPH, 5 μ*M* FNR, and 20 μ*M* FldA in tube #1 in the glovebox for 30 min.
2. Mix 50 μ*M* PqqE and 500 μ*M* SAM in tube #2 in the glovebox.
3. Mix 50 μ*M* PqqD and 50 μ*M* PqqA in tube #3 in the glovebox.
4. Mix the reaction mixtures from the three tubes and complete the volume to 200 μL with 50 m*M* Tris pH 7.9, 300 m*M* NaCl, 10% (v/v) glycerol.
5. Incubate at room temperature for 16 h and quench the reaction by addition of 10 μL 88% (v/v) formic acid.
6. Centrifuge the quenched reaction mixture at 14,000 × *g* for 5 min and dilute 20 μL of the supernatant to 200 μL.
7. Apply 5 μL of the sample onto LC-MS and elute the sample with a linear gradient 20%–70% water (2% formic acid)/acetonitrile (2% formic acid).
8. Analyze the sample peak containing the PqqA and PqqA cross-linked product using the Xcalibur software (version 2.0.7, Thermo). The extent of cross-linking of the PqqA peptide is determined by integrating extracted ion chromatograms for the $[M+2H]^{2+}$ ions of cross-linked and unmodified forms of PqqA (occurring at $m/z = 1536.7$ and 1537.7, respectively) (Fig. 8).

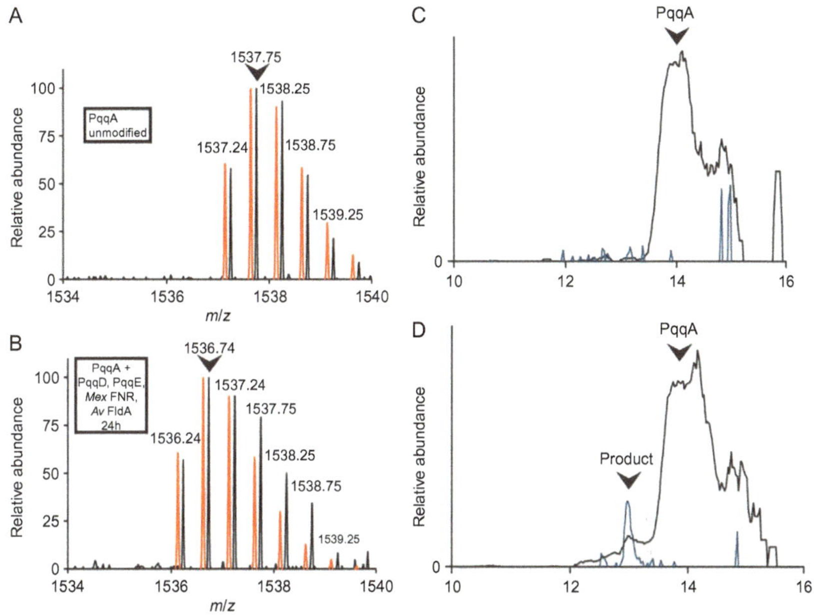

Fig. 8 LC-MS analysis of modification of PqqA by PqqE and PqqD. (A) Initial ion mass envelope of unmodified PqqA with two charge state. (B) Mass envelope of a minor peak, eluting 1 min earlier, seen following a 24-h reaction under anaerobic conditions. A noticeable shift in mass by 2 Da is observed, consistent with cross-linking of residues in PqqA. The calculated mass envelopes for PqqA (A) and modified PqqA (B) are shown in *red* and slightly offset for ease of comparison. The most abundant ions are indicated by the *arrows*, and these were used to quantify the relative amount of modified PqqA. (C) Chromatograph showing the elution profile of 1537.7 (*black*) and 1536.7 (*blue*) ions in an unreacted PqqA sample. (D) Chromatograph showing the elution profile of 1537.7 (*black*) and 1536.7 (*blue*) ions in a 24-h reaction mixture; a small peak containing cross-linked PqqA is seen to elute earlier than the unreacted PqqA. *Adapted from Barr, I., Latham, J. A., Iavarone, A. T., Chantarojsiri, T., Hwang, J. D., & Klinman, J. P. (2016). Demonstration that the radical S-adenosylmethionine (SAM) enzyme PqqE catalyzes de novo carbon-carbon cross-linking within a peptide substrate PqqA in the presence of the peptide chaperone PqqD.* Journal of Biological Chemistry, 291*(17), 8877–8884. Copyright ©2016 by American Society for Biochemistry and Molecular Biology.*

3.4 Protein Interaction Studies

3.4.1 Nanoelectrospray Ionization Mass Spectrometry of Native Proteins

In the study of the interaction between PqqE with other proteins in the PQQ biosynthesis, nanoelectrospray ionization mass spectrometry (NanoESI-MS) provides a direct measurement of the binding ratio of PqqE to other proteins in the pathway. The procedure for the determination of the binding ratio in the PqqE/D complex in solution is described here. Based on the ions of PqqE monomer, PqqD monomer, and the *m/z* of the PqqE/D complex, one can determine the number of PqqE and PqqD in the PqqE/D complex. The instrument used was a Quadrupole time-of-flight (Q-TOF) mass spectrometer equipped with a nano-*Z*-spray nano-ESI source (Q-TOF Premier, Waters, Beverly, MA).

1. Exchange the purified PqqE and PqqD (in 50 m*M* Tris, 300 m*M* NaCl pH 7.9 buffer) into 25 m*M* ammonium bicarbonate buffer pH 7.9 using preequilibrated PD-10 columns (GE Healthcare).
2. Prepare the PqqD and PqqE complex sample by mixing 10 μ*M* PqqE and 50 μ*M* PqqD in ammonium bicarbonate buffer pH 7.9.
3. Load 10 μL of the complex sample into the nano-ESI emitter using a gas-tight syringe.
4. Initiate the electrospray by increasing the DC potential. Set up the instrument parameters as previously described by Beeson, Iavarone, Hausmann, Cate, and Marletta (2011).
5. Record mass spectra as previously described (Wecksler et al., 2010).
6. Collect the MS data of the intact proteins and the complex and process using MassLynx software (version 4.1, Waters) (Fig. 9).

3.4.2 Surface Plasmon Resonance

For determination of the dissociation constants between PqqD and PqqE, and PqqA and PqqCD fusion, surface plasmon resonance (SPR) measurements and isothermal titration calorimetry (ITC) analysis have both been applied. However, the ITC measurements require highly concentrated protein samples due to the relatively high K_d of PqqD and PqqE (Latham et al., 2015). Therefore, SPR gives more accurate measurements in this system.

In the following experiment, we used a Biacore T100 SPR (Biacore, GE Healthcare). In order to reduce the buffer effect on the measurements, all the solutions should be exchanged to the same stock buffer prior to the determinations. For the PqqCD/PqqA interaction, His_6-PqqCD is immobilized on the Ni-NTA chip and the PqqA solution is injected as the analyte. In the study of PqqD/E interaction, His_6-PqqD is immobilized, and native PqqE is injected as the analyte. For both experiments, the analyte is serially diluted seven- to ninefold, depending on the number of injections in the

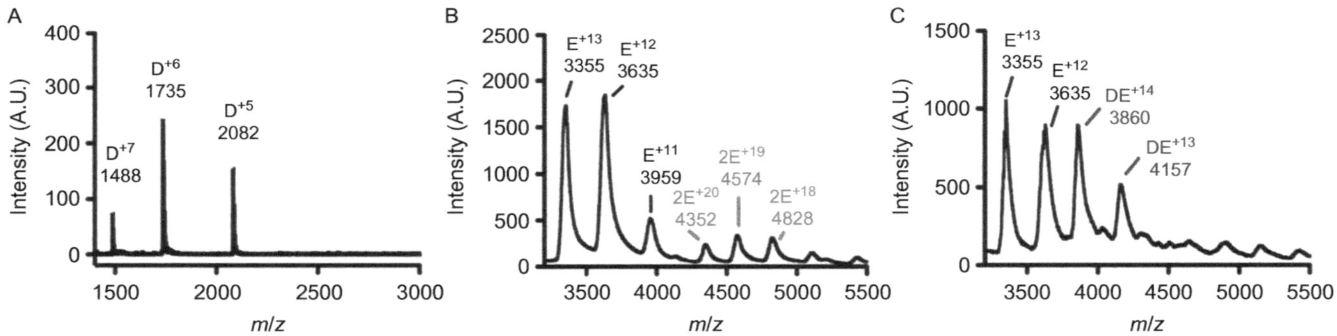

Fig. 9 Nano-Q-TOF-ESI-MS characterization of the PqqD:PqqE complex. (A) The *Mex* PqqD native mass spectrum shows a monomer population with a molecular weight of 10.4 kDa calculated from the charge-state distribution. (B) The native mass spectrum of *Mex* PqqE shows two sets of charge-state distributions with masses of 43.6 and 86.9 kDa, indicating that both the monomeric and dimeric species are present. (C) In the presence of *Mex* PqqD, a 54.0-kDa charge-state distribution is formed corresponding to a 1:1 *Mex* PqqE:PqqD complex. *Adapted from Latham, J. A., Iavarone, A. T., Barr, I., Juthani, P. V., & Klinman, J. P. (2015). PqqD is a novel peptide chaperone that forms a ternary complex with the radical S-adenosylmethionine protein PqqE in the pyrroloquinoline quinone biosynthetic pathway.* Journal of Biological Chemistry, 290*(20), 12908–12918. Copyright ©2015 by American Society for Biochemistry and Molecular Biology.*

experiment. A careful wash must be done after each cycle so that the binding capacity is kept optimal throughout the entire experiment.

Interaction of PqqCD and PqqA

1. Exchange His_6-PqqCD and PqqA (both in 50 m*M* Tris 300 m*M* NaCl pH 7.9 buffer) into HBS-P buffer (GE Healthcare) and determine the concentration.
2. In the Biacore T100 SPR leave Flow cell 1 with HBS-P buffer as the reference channel.
3. Immobilize His_6-PqqCD by injecting 10 n*M* on the NTA chip (Biacore, GE Healthcare) at room temperature.
4. Inject the serial dilutions of native PqqA solution for 30 s at 30 μL/min followed by a 60-s dissociation phase.
5. After each cycle (15 injections), regenerate the surface injecting 300 m*M* EDTA, followed by 500 μ*M* $NiCl_2$ and 3 m*M* EDTA.
6. Calculate the binding affinities from steady-state analysis of the maximum steady-state response.
7. Analyze the data by plotting the time and response units using the Prism 5 Software (Prism) (Fig. 10).

Interaction of PqqD and PqqE

The protocol is identical to the previous one, with His_6-PqqD replacing His_6-PqqCD and untagged PqqE substituting PqqA as the analyte.

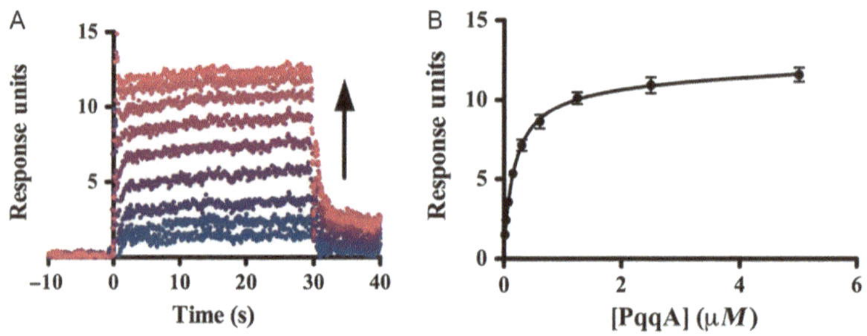

Fig. 10 Binding of PqqCD and PqqA observed by SPR. (A) Representative dataset (*Mex* His_6-PqqCD as ligand and *Mex* PqqA as analyte) showing that the SPR response increases as *Mex* PqqA concentration increases (*blue* to *violet*). (B) The steady-state fits were used to calculate the dissociation constant (error bars represent the standard deviation of three independent experiments). *Adapted from Latham, J. A., Iavarone, A. T., Barr, I., Juthani, P. V., & Klinman, J. P. (2015). PqqD is a novel peptide chaperone that forms a ternary complex with the radical S-adenosylmethionine protein PqqE in the pyrroloquinoline quinone biosynthetic pathway.* Journal of Biological Chemistry, 290*(20), 12908–12918. Copyright ©2015 by American Society for Biochemistry and Molecular Biology.*

4. SUMMARY AND CONCLUSIONS

PQQ, although only produced by bacteria, has been widely recognized as an important nutrient for growth and development in plant and animals (Mure, 2004). PQQ exhibits growth-promoting activity, anti-diabetic effects, antioxidant, and neuroprotection properties (Akagawa, Nakano, & Ikemoto, 2015). Despite the significant importance of PQQ to human health, the details of the biosynthetic pathway of this molecule have been, until recently, largely unknown. Recent studies by our group revealed that a key step in the PQQ biosynthesis involves the formation of a C–C bond between two amino acids in the peptide substrate PqqA, catalyzed by the radical SAM enzyme PqqE in complex with the chaperone PqqD (Barr et al., 2016; Latham, Barr, & Klinman, 2017; Latham et al., 2015). This reaction is both complicated and fascinating not only because it is an O_2-sensitive reaction occurring in an aerobic bacteria (*Mex*) but also due to the fact that it requires precise redox control and interaction among multiprotein complexes. The chaperone protein PqqD is most generally encoded by an individual gene *pqqD* in the gene cluster (e.g., *K. pneumoniae* operon), whereas in *Mex* it is fused to the C-terminus of the *pqqC* gene. In the latter case, PqqD that is fused to the C-terminus of PqqC may only bind PqqE transiently during the cross-linking of PqqA, dissociating prior to the downstream catalytic steps required for PQQ production. This contrasts with the other RiPP proteins such as CteB where the peptide-binding domain is fused to the N-terminus of the radical SAM protein (Grove et al., 2017). For the activity studies of PqqA cross-linking described herein, PqqD was cloned from the *Mex* PqqCD gene and expressed as a separate protein.

Even though the O_2 tolerance is relatively high for *Mex* PqqE, it is necessary to purify the enzyme under strict anaerobic conditions, to preserve the Fe–S clusters essential for the enzyme activity. With regard to the activity assays, DTH can be used for examination of the SAM cleavage activity. However, a biological reducing system, such as FldA/FNR/NADPH, is essential for successful modification of PqqA (Barr et al., 2016). While a *Mex* FldA/FNR reducing pair would have been ideal for assaying the *Mex* PqqE reaction, *A. vinelandii* FldA was used in our studies due to the inability of obtaining the FldA from *Mex*. The requirement to use chemically reconstituted PqqE to catalyze the PqqA modification is almost certainly related to the fact that PqqE requires three intact Fe–S centers of varying structures and stabilities for substrate turnover. The exact nature of the two auxiliary clusters is an ongoing and exciting direction to research on PqqE structure–function (Barr et al., 2016; Saichana et al., 2017).

REFERENCES

Akagawa, M., Nakano, M., & Ikemoto, K. (2015). Recent progress in studies on the health benefits of pyrroloquinoline quinone. *Bioscience, Biotechnology, and Biochemistry*, *80*(1), 13–22.

Banci, L., Ciofi-Baffoni, S., Mikolajczyk, M., Winkelmann, J., Bill, E., & Pandelia, M. E. (2013). Human anamorsin binds [2Fe-2S] clusters with unique electronic properties. *Journal of Biological Inorganic Chemistry*, *18*(8), 883–893.

Barr, I., Latham, J. A., Iavarone, A. T., Chantarojsiri, T., Hwang, J. D., & Klinman, J. P. (2016). Demonstration that the radical S-adenosylmethionine (SAM) enzyme PqqE catalyzes de novo carbon-carbon cross-linking within a peptide substrate PqqA in the presence of the peptide chaperone PqqD. *Journal of Biological Chemistry*, *291*(17), 8877–8884.

Beeson, W. T., Iavarone, A. T., Hausmann, C. D., Cate, J. H., & Marletta, M. A. (2011). Extracellular aldonolactonase from Myceliophthora thermophila. *Applied and Environmental Microbiology*, 77(2), 650–656.

Bonnot, F., Iavarone, A. T., & Klinman, J. P. (2013). Multistep, eight-electron oxidation catalyzed by the cofactorless oxidase, PqqC: Identification of chemical intermediates and their dependence on molecular oxygen. *Biochemistry*, *52*(27), 4667–4675.

Bradford, M. M. (1976). A rapid and sensitive method for the quantitation of microgram quantities of protein utilizing the principle of protein-dye binding. *Analytical Biochemistry*, *72*, 248–254.

Crack, J. C., Green, J., Thomson, A. J., & Le Brun, N. E. (2014). Techniques for the production, isolation, and analysis of iron-sulfur proteins. *Methods in Molecular Biology*, *1122*, 33–48.

Duine, J. A. (1999). The PQQ story. *Journal of Bioscience and Bioengineering*, *88*(3), 231–236.

Gasteiger, E., Hoogland, C., Gattiker, A., Duvaud, S., Wilkins, M. R., Appel, R. D., et al. (2005). Protein identification and analysis tools on the ExPASy server. In J. M. Walker (Ed.), *The proteomics protocols handbook* (pp. 571–607). New York: Humana Press.

Goldman, P. J., Grove, T. L., Sites, L. A., McLaughlin, M. I., Booker, S. J., & Drennan, C. L. (2013). X-ray structure of an AdoMet radical activase reveals an anaerobic solution for formylglycine posttranslational modification. *Proceedings of the National Academy of Sciences of the United States of America*, *110*(21), 8519–8524.

Grell, T. A., Goldman, P. J., & Drennan, C. L. (2015). SPASM and twitch domains in S-adenosylmethionine (SAM) radical enzymes. *Journal of Biological Chemistry*, *290*(7), 3964–3971.

Grove, T. L., Himes, P. M., Hwang, S., Yumerefendi, H., Bonanno, J. B., Kuhlman, B., et al. (2017). Structural insights into thioether bond formation in the biosynthesis of sactipeptides. *Journal of the American Chemical Society*, *139*(34), 11734–11744.

Hauge, J. G. (1964). Glucose dehydrogenase of Bacterium anitratum: An enzyme with a novel prosthetic group. *Journal of Biological Chemistry*, *239*, 3630–3639.

Klinman, J. P., & Bonnot, F. (2014). Intrigues and intricacies of the biosynthetic pathways for the enzymatic quinocofactors: PQQ, TTQ, CTQ, TPQ, and LTQ. *Chemical Reviews*, *114*(8), 4343–4365.

Laemmli, U. K. (1970). Cleavage of structural proteins during the assembly of the head of bacteriophage T4. *Nature*, *227*(5259), 680–685.

Latham, J. A., Barr, I., & Klinman, J. P. (2017). At the confluence of ribosomally synthesized peptide modification and radical S-adenosylmethionine (SAM) enzymology. *Journal of Biological Chemistry*, *292*(40), 16397–16405.

Latham, J. A., Iavarone, A. T., Barr, I., Juthani, P. V., & Klinman, J. P. (2015). PqqD is a novel peptide chaperone that forms a ternary complex with the radical S-adenosylmethionine protein PqqE in the pyrroloquinoline quinone biosynthetic pathway. *Journal of Biological Chemistry*, *290*(20), 12908–12918.

Magnusson, O. T., Toyama, H., Saeki, M., Rojas, A., Reed, J. C., Liddington, R. C., et al. (2004). Quinone biogenesis: Structure and mechanism of PqqC, the final catalyst in the production of pyrroloquinoline quinone. *Proceedings of the National Academy of Sciences of the United States of America*, *101*(21), 7913–7918.

McKenna, C. E., Gutheil, W. G., & Song, W. (1991). A method for preparing analytically pure sodium dithionite. Dithionite quality and observed nitrogenase-specific activities. *Biochimica et Biophysica Acta*, *1075*(1), 109–117.

Mure, M. (2004). Tyrosine-derived quinone cofactors. *Accounts of Chemical Research*, *37*(2), 131–139.

Nachin, L., Loiseau, L., Expert, D., & Barras, F. (2003). SufC: An unorthodox cytoplasmic ABC/ATPase required for [Fe-S] biogenesis under oxidative stress. *The EMBO Journal*, *22*(3), 427–437.

Pace, C. N., Vajdos, F., Fee, L., Grimsley, G., & Gray, T. (1995). How to measure and predict the molar absorption coefficient of a protein. *Protein Science*, *4*(11), 2411–2423.

Pandelia, M. E., Lanz, N. D., Booker, S. J., & Krebs, C. (2015). Mössbauer spectroscopy of Fe/S proteins. *Biochimica et Biophysica Acta*, *1853*(6), 1395–1405.

Prisecaru, I. (2009–2016). *WMOSS4 Mössbauer spectral analysis software*. Retrieved from www.wmoss.org.

Saichana, N., Tanizawa, K., Pechousek, J., Novak, P., Yakushi, T., Toyama, H., et al. (2016). PqqE from Methylobacterium extorquens AM1: A radical S-adenosyl-l-methionine enzyme with an unusual tolerance to oxygen. *Journal of Biochemistry*, *159*(1), 87–99.

Saichana, N., Tanizawa, K., Ueno, H., Pechoušek, J., Novák, P., & Frébortová, J. (2017). Characterization of auxiliary iron–sulfur clusters in a radical S-adenosylmethionine enzyme PqqE from Methylobacterium extorquens AM1. *FEBS Open Bio*, 7, 1864–1879. Retrieved from https://doi.org/10.1002/2211-5463.12314.

Shen, Y. Q., Bonnot, F., Imsand, E. M., RoseFigura, J. M., Sjolander, K., & Klinman, J. P. (2012). Distribution and properties of the genes encoding the biosynthesis of the bacterial cofactor, pyrroloquinoline quinone. *Biochemistry*, *51*(11), 2265–2275.

Stich, T. A., Myers, W. K., & Britt, R. D. (2014). Paramagnetic intermediates generated by radical S-adenosylmethionine (SAM) enzymes. *Accounts of Chemical Research*, *47*(8), 2235–2243.

Velterop, J. S., Sellink, E., Meulenberg, J. J., David, S., Bulder, I., & Postma, P. W. (1995). Synthesis of pyrroloquinoline quinone in vivo and in vitro and detection of an intermediate in the biosynthetic pathway. *Journal of Bacteriology*, *177*(17), 5088–5098.

Wagner, T., Koch, J., Ermler, U., & Shima, S. (2017). Methanogenic heterodisulfide reductase (HdrABC-MvhAGD) uses two noncubane [4Fe-4S] clusters for reduction. *Science*, *357*(6352), 699–703.

Wecksler, S. R., Stoll, S., Iavarone, A. T., Imsand, E. M., Tran, H., Britt, R. D., et al. (2010). Interaction of PqqE and PqqD in the pyrroloquinoline quinone (PQQ) biosynthetic pathway links PqqD to the radical SAM superfamily. *Chemical communications (Cambridge)*, *46*(37), 7031–7033.

Weckslar, S. R., Stoll, S., Tran, H., Magnusson, O. T., Wu, S. P., King, D., et al. (2009). Pyrroloquinoline quinone biogenesis: Demonstration that PqqE from Klebsiella pneumoniae is a radical S-adenosyl-L-methionine enzyme. *Biochemistry, 48*(42), 10151–10161.
Zhang, J., & Klinman, J. P. (2015). High-performance liquid chromatography separation of the (S,S)- and (R,S)-forms of S-adenosyl-L-methionine. *Analytical Biochemistry, 476*, 81–83.
Zor, T., & Selinger, Z. (1996). Linearization of the Bradford protein assay increases its sensitivity: Theoretical and experimental studies. *Analytical Biochemistry, 236*(2), 302–308.

CHAPTER FIFTEEN

Methods for Studying the Radical SAM Enzymes in Diphthamide Biosynthesis

Min Dong*, Yugang Zhang*, Hening Lin†,1

*Department of Chemistry and Chemical Biology, Cornell University, Ithaca, NY, United States
†Howard Hughes Medical Institute, Cornell University, Ithaca, NY, United States
[1]Corresponding author: e-mail address: hl379@cornell.edu

Contents

Abstract

Diphthamide is a unique posttranslational modification on translation elongation factor 2 (EF2) in archaea and eukaryotes. Biosynthesis of diphthamide was proposed to involve four steps. The first step is a C—C bond forming reaction catalyzed by unique radical *S*-adenosylmethionine (SAM) enzymes. Classical radical SAM enzymes use SAM and [4Fe–4S] clusters to generate a 5′-deoxyadenynal radical and catalyze numerous reactions. Radical SAM enzymes in diphthamide biosynthesis cleave a different C—S bond in SAM to generate a 3-amino-3-carboxypropyl radical and modify a histidine residue of substrate protein EF2. Here, we describe our investigations on these unique radical SAM enzymes, including the preparation, characterization, and activity assays we have developed.

Methods in Enzymology, Volume 606
ISSN 0076-6879
https://doi.org/10.1016/bs.mie.2018.04.001

ABBREVIATION

EF2 elongation factor 2
EPR electron paramagnetic resonance
IPTG isopropyl-β-D-thiogalactopyranoside
LB lysogeny broth
NAD nicotinamide adenine dinucleotide
OD optical density
SAM *S*-adenosylmethionine
SEC-MALS size exclusion chromatography-coupled multiangle light scattering
TFA trifluoroacetic acid
UV–vis ultraviolet–visible

1. INTRODUCTION

Diphthamide is a unique posttranslational modification on translation elongation factor 2 (EF2) in archaea and eukaryotes (Robinson, Henriksen, & Maxwell, 1974; Van Ness, Howard, & Bodley, 1980a, 1980b). Diphtheria toxin and Pseudomonas exotoxin A recognize diphthamide and catalyze the transfer of an adenosine diphosphate ribose (ADP-ribose) from nicotinamide adenine dinucleotide (NAD) to one of the nitrogen atoms in the imidazole ring of the histidine residue of EF2 (Fig. 1) (Collier, 2001). The ADP-ribosylation inhibits the function of EF2 and thus protein synthesis, leading to cell death (Honjo, Nishizuka, Hayaishi, & Kato, 1968).

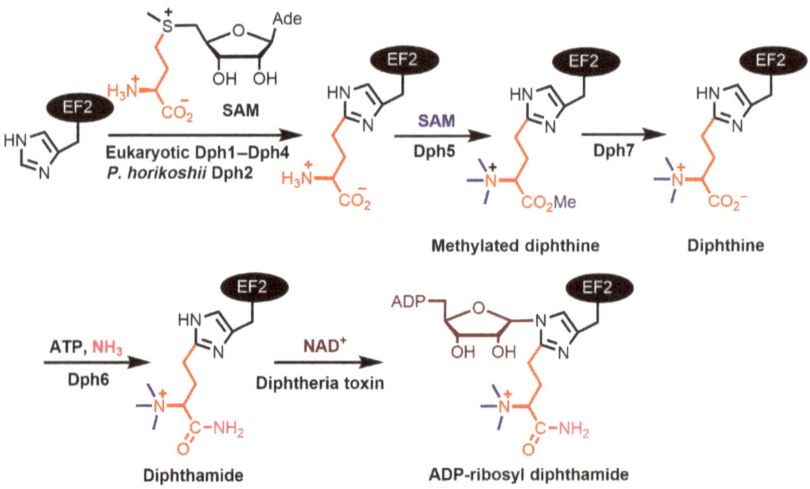

Fig. 1 Diphthamide biosynthesis pathway.

The biosynthesis of diphthamide was proposed to involve at least four steps and seven proteins (Schaffrath, Abdel-Fattah, Klassen, & Stark, 2014). The first step is the transfer of the 3-amino-3-carboxypropyl group from *S*-adenosyl-L-methionine (SAM) to the histidine residue of EF2, forming a C—C bond. The enzymes that perform this reaction are Dph2 (diphthamide biosynthesis protein 2) homodimer in archaea, like *Pyrococcus horikoshii* (*Ph*Dph2) (Zhang et al., 2010) or a Dph1–Dph2 heterodimer in eukaryotes (Dong et al., 2014). These enzymes form a unique family of radical SAM (RS) enzymes. Classical RS enzymes, a family of enzymes with more than 113,000 members found in nature (Akiva et al., 2014), all contain a [4Fe–4S] cluster and bind SAM. In these enzymes, the reduced state of the [4Fe–4S] cluster provides one electron to cleave the $C_{5',Ade}$-S bond of SAM, generating a 5′-deoxyadenosyl (5′-dA) radical that then initiates downstream reactions (Broderick, Duffus, Duschene, & Shepard, 2014). In contrast, the *Ph*Dph2 homodimer or Dph1–Dph2 heterodimer binds [4Fe–4S] clusters, cleaves the $C_{\gamma,Met}$-S bond of SAM, and generates a 3-amino-3-carboxypropyl (ACP) radical. The formation of the ACP radical is supported by the *Ph*Dph2-catalyzed generation of 2-aminobutyric acid and homocysteine sulfinic acid in the absence of the substrate protein, *Ph*EF2 (Zhang et al., 2010). Further support for a radical mechanism was provided by the reaction with an SAM analogue, SAM_{CA} (Dong et al., 2016).

In this chapter, we detail the preparation, characterization, and activity assays of the RS enzymes involved in diphthamide biosynthesis, *Ph*Dph2 and yeast Dph1–Dph2. We also describe the preparation of the related proteins in the first step of diphthamide biosynthesis, such as substrate protein *Ph*EF2 and yeast EF2, electron carrier Dph3, and the reductase Cbr1 in yeast.

2. PROTEIN EXPRESSION, PURIFICATION, AND CHARACTERIZATION

2.1 *Ph*Dph2 and *Ph*EF2

The first step of diphthamide biosynthesis is the cleavage of the ACP group from SAM and formation of a new C—C bond (Lin, 2011). From genetic studies, this step requires four proteins in eukaryotes, Dph1, Dph2, Dph3, and Dph4 (Liu, Milne, Kuremsky, Fink, & Leppla, 2004). Eukaryotic Dph1 and Dph2 share about 20% sequence identity. Iterative BLAST searches (Altschul, Gish, Miller, Myers, & Lipman, 1990) starting with *Saccharomyces cerevisiae* Dph1 or Dph2 generate both proteins from other eukaryotic species.

In contrast, BLAST searches identify only one protein, Dph2, in archaeal species (Zhang et al., 2010). Eukaryotic Dph3 and Dph4 have no orthologues in archaea. To better understand the first step of diphthamide biosynthesis, we initially studied the relatively simpler archaeal system *P. horikoshii*, which is a thermophile and its proteins typically have better stability.

Procedure for *Ph*Dph2 expression

1. Amplify *Phdph2* gene from *P. horikoshii* genomic DNA by PCR and insert into pENTR/TEV/D-TOPO entry vector (Invitrogen), followed by recombination with pDESTF1 destination vector to create expression clones with an N-terminal His_6 tag. This vector also contains an ampicillin (Amp) resistance marker for in vivo selection.
2. Transform the generated plasmid into the *E. coli* expression strain BL21 (DE3) with pRARE2, which encodes rarely used tRNAs in *E. coli* and contains a chloramphenicol (Cam) resistance gene. The transformed cells were selected on lysogeny broth (LB) agar plates containing 0.1 g/L Amp and 0.02 g/L Cam.
3. Pick a single colony to inoculate 20 mL of LB containing 0.1 g/L Amp and 0.02 g/L Cam at 37°C and 200 rpm overnight.
4. Inoculate the baffled shaker flasks containing 2 L of LB supplemented with 0.02 g/L Cam and 0.1 g/L Amp with the 20 mL starter culture.
5. Grow cells at 37°C and 200 rpm. Cool down the cells to 18°C by ice water when the cultures reach an optical density (OD) of 0.6 measured at 600 nm. Add $FeCl_3$, $Fe(NH_4)_2(SO_4)_2$, and L-cysteine to final concentrations of 50, 50, and 400 μ*M*, respectively.
6. Induce the expression with 0.1 m*M* isopropyl-β-D-thiogalactopyranoside (IPTG), at which point seal the culture flasks to limit the amount of oxygen in the system. Incubate the cells in a shaker at 18°C and 200 rpm for 20 h.
7. Harvest the cells by centrifugation at 6000 × *g* (Beckman Coulter Avanti J-E) for 10 min and flash freeze the cell paste in liquid nitrogen before storage at −80°C.

Procedure for *Ph*Dph2 anaerobic purification

As in classical RS enzymes, the [4Fe–4S] clusters in *Ph*Dph2 and Dph1–Dph2 are also oxygen-labile. Therefore, purification of these proteins, as well as the related assays for the activity test needs to be performed in an anaerobic environment. A vinyl anaerobic chamber (Coy LAB PRODUCTS) with an auto vacuum airlock is used. Details about the anaerobic chamber and usage were well described in a previous chapter by Booker's Lab (Lanz et al., 2012). A slightly different method is used here for preparing anaerobic buffers.

Autoclave the lysis buffer (20 m*M* Tris–HCl buffer pH 8.0, 500 m*M* NaCl, 10 m*M* $MgCl_2$, 5 m*M* imidazole) and desalting buffer (200 m*M* Tris–HCl at pH 7.4 containing 150 m*M* NaCl, and 5% glycerol) at 125°C for 25 min. Cap the buffer bottles while cooling down. After cooled down to room temperature, transfer the buffers to round flasks and further degas by Schlenk line for 2 h with magnetic stirring. Protect the buffers with N_2 and transfer them into the anaerobic chamber.

1. Transfer the frozen cell paste from the equivalent of 4 L of bacterial culture into a Coy anaerobic chamber and resuspend in 80 mL of degassed lysis buffer. Add 1 m*M* dithiothreitol (DTT), 1 g/L lysozyme (VWR), and 500 units benzonase (Sigma-Aldrich) in the buffer and incubate at 26°C for 30 min.
2. Split the solution into two 50-mL tubes. Seal the tubes in the chamber with black tape, transfer them outside of the chamber, and freeze them in liquid nitrogen. Move the tubes back into the chamber and thaw. Repeat this freeze–thaw step again to more efficiently lysis the cells.
3. Transfer the lysate to 50-mL centrifuge tubes. Seal the tubes, transfer out of the chamber, centrifuge for 30 min at 20,000 × *g*, and then transfer back into the chamber.
4. Preequilibrate 3 mL of Ni-NTA Agarose resin (Qiagen) with the lysis buffer inside the chamber. Incubate the supernatant from centrifuge tubes with the resin for 30 min. Load the resin onto a polypropylene column and wash with 20 mL of lysis buffer.
5. Wash the resin again with 20 mL of 30 m*M* imidazole in lysis buffer. Elute *Ph*Dph2 (dark brown color) from the column with elution buffers (200 m*M* imidazole in the lysis buffer).
6. Take small aliquots out of the chamber and analyze by SDS-PAGE. Combine the fractions inside the chamber based on the SDS-PAGE analysis. Buffer-exchange combined *Ph*Dph2 to the desalting buffer with 1 m*M* DTT using a Bio-Rad 10-DG desalting column inside the chamber.
7. Concentrate the purified *Ph*Dph2 protein using Amicon Ultra 0.5 mL centrifugal filter devices (Millipore) inside the chamber. Aliquot protein into 0.5-mL microcentrifuge tubes and put these tubes into a 50-mL tube. Seal the tube for storage at −80°C. Determine protein concentration by Bradford assay.

Procedure for *Ph*EF2 expression and purification

1. Clone *Phef2* gene following the same protocol as that of *Phdph2*. Transform the plasmid into *E. coli* expression strain BL21 (DE3) with a pRARE2 plasmid.

2. Grow the cells in 2 L of LB medium supplemented with 0.035 g/L Cam and 0.1 g/L Amp at 37°C. Induce protein expression with 0.1 m*M* IPTG when the cultures reach an OD of 1.0 measured at 600 nm. Grow the cells for 3 h at 37°C before harvesting by centrifugation at 6000 × *g* for 10 min.
3. Resuspend the cell pellet in 50 mL of buffer containing 20 m*M* Tris–HCl (pH 8.0), 500 m*M* NaCl, and 0.5 m*M* phenylmethylsulfonyl fluoride.
4. Lyse cells by an EmulsiFlex-C3 cell disruptor (Avestin, Inc. Canada). Remove cell debris by centrifugation at 12,000 × *g* for 30 min.
5. Purify the protein from the supernatant on a BioLogic DuoFlow 10 System (Bio-Rad, Hercules, CA) with a HisTrap HP column (GE Healthcare, Piscataway, NJ) with a linear gradient from 30 m*M* imidazole to 500 m*M* imidazole in 30 min.
6. Collect the fractions and dialyze against 200 m*M* Tris–HCl at pH 7.4 containing 150 m*M* NaCl, 1 m*M* DTT, and 5% glycerol. Concentrate the purified protein and determine concentration by Bradford assay.

2.2 *S. cerevisiae* Dph1–Dph2

Yeast Dph1 or Dph2 cannot be singly expressed in *E. coli* as soluble proteins. They need to be coexpressed to obtain the soluble Dph1–Dph2 heterodimer. *Dph1* and *dph2* genes were initially put in the coexpression vector pETduet-1. The expression level was not very good. Interestingly, when the whole *dph1–dph2* gene piece was cut from pETduet-1 and put into pET28b vector, we obtained higher Dph1–Dph2 expression. To increase the Fe–S cluster loading, Dph1–Dph2 was coexpressed with the ISC (iron–sulfur cluster) proteins, a set of bacterial proteins that assemble and transfer Fe–S clusters to client proteins (Frazzon & Dean, 2003). Consistent with *Ph*Dph2, anaerobically purified yeast Dph1–Dph2 complex displayed a dark brown color and showed a typical $[4Fe–4S]^{2+}$ cluster absorption band at 410 nm. The 410 nm absorption decreased when dithionite was added. Quantification based on the 410 nm absorption, as well as iron and sulfur content assays suggested the presence of 0.6∼1 $[4Fe–4S]^{2+}$ clusters per Dph1–Dph2 complex. The electron paramagnetic resonance (EPR) spectrum of the reduced Dph1–Dph2 complex showed *g* values of 2.04 and 1.92, consistent with a $[4Fe–4S]^{+}$ cluster (Dong et al., 2014). Size exclusion chromatography-multiangle light scattering (SEC-MALS) experiment of the Dph1–Dph2 complex showed that the molecular weight is ∼110 kDa, suggesting that these two proteins form a heterodimer.

Procedure for yeast Dph1–Dph2 expression

1. Extract yeast genomic DNA from BY4741 strain using Pierce Yeast DNA Extraction Kit. Amplify yeast *dph1* and *dph2* genes from yeast genomic DNA by PCR.
2. First insert *dph2* gene with an N-terminal His tag into multiple cloning sites (MCS) 2 of the pETduet-1 vector using *Nde*I and *Xoh*I restriction sites, then insert *dph1* gene into MCS1 using *Eco*RI and *Sal*I restriction sites.
3. Cut the whole *dph1–dph2* gene piece by *Eco*RI and *Xoh*I, insert into a pET28b vector digested by *Eco*RI and *Xoh*I.
4. Cotransform the resulting plasmid containing *dph1–dph2* gene with pDB1282 (Frazzon & Dean, 2003), which contains the *Azotobacter vinelandii* ISC operon, into BL21 (DE3) pRARE2 strain.
5. Grow the cells in 2L LB medium with 0.1 g/L Amp, 0.02 g/L Cam, and 0.05 g/L kanamycin (Kan) at 37°C and 200 rpm.
6. At an OD_{600nm} of 0.3, add solid arabinose to each flask to a final concentration of 0.1% (w/v) to induce ISC proteins expression.
7. At an OD_{600nm} of 0.6, cool down the cultures with an ice-water bath and supplement the cultures with $FeCl_3$, $Fe(NH_4)_2(SO_4)_2$, and L-cysteine to final concentrations of 50, 50, and 400 μ*M*, respectively.
8. Induce Dph1 and Dph2 protein expression by adding 0.1 m*M* IPTG. Seal the culture flasks with balloons to limit the amount of oxygen in the system. Incubate cells in a shaker at 18°C and 200 rpm for 20 h before harvest.

Procedure for anaerobic purification of yeast Dph1–Dph2 complex

1. The purification is carried out in a Coy anaerobic chamber. Suspend cell pellet from 2L culture in 40 mL lysis buffer (500 m*M* NaCl, 10 m*M* $MgCl_2$, 5 m*M* imidazole, 1 m*M* DTT, and 200 m*M* Tris–HCl at pH 7.4).
2. Add lysozyme (200 mg) and nuclease (4 μL, 25 U/mL of lysis buffer, Thermo) to the cell suspension and incubate at 25°C for 0.5 h. Seal the tubes, transfer out of the chamber, and freeze them in liquid nitrogen. Move the tubes back into the chamber and thaw at 25°C. Repeat this freeze–thaw step to obtain better lysis.
3. Remove the cell debris by centrifugation at 20,000 × *g* for 30 min outside of the anaerobic chamber. Transfer the tubes back into the chamber. Incubate the supernatant for 0.5 h with 1.2 mL of Ni-NTA resin preequilibrated with the lysis buffer. Load the Ni-NTA resin onto a polypropylene column and wash with 20 mL of lysis buffer, followed by 20 mL of 30 m*M* imidazole in the lysis buffer.

4. Elute Dph1–Dph2 from the column with elution buffers (100, 150, and 200 m*M* imidazole in the lysis buffer, 1.5 mL each). Buffer-exchange the brown-colored elution fractions to 150 m*M* NaCl, 200 m*M* Tris–HCl at pH 7.4 and 5% glycerol using a Bio-Rad 10-DG desalting column.
5. Concentrate the purified proteins using Amicon Ultra 0.5-mL centrifugal filter devices inside the chamber. Determine protein concentration by Bradford assay.

Procedure for the preparation of ^{57}Fe-enriched Dph1–Dph2

To prepare ^{57}Fe enriched Dph1–Dph2, grow the transformed cells in M9 minimal medium supplemented with 0.2% (w/v) glucose, 2 m*M* $MgSO_4$, and 0.1 m*M* $CaCl_2$. Dissolve 50 mg of ^{57}Fe powder (Isoflex, USA) in 2 *M* HCl (1.7 mL) and adjust the solution pH to 5–6 as a stock solution. Add the ^{57}Fe stock solution and L-cysteine to M9 medium to final concentrations of 100 and 400 μ*M*, respectively. Express and purify ^{57}Fe enriched Dph1–Dph2 following the same protocols as those for Dph1–Dph2 described above.

2.3 Dph1–Dph2 Characterization

2.3.1 SEC-MALS for Dph1–Dph2 Heterodimer

Apply purified Dph1–Dph2 protein (~10 μg/μL) to SEC using a Yarra SEC 4000 column (Phenomenex Inc.) equilibrated in MALS buffer (200 m*M* Tris–HCl pH 7.4, 150 m*M* NaCl). The SEC was coupled to a static 18-angle light scattering detector (DAWN HELEOS-II) and a refractive index detector (Optilab T-rex) (Wyatt Technology). Collect data every second at a flow rate of 1 mL/min. Analyze the data using the program ASTRA, giving the molar mass and mass distribution (polydispersity) of the sample. Monomeric BSA (Sigma) was used for normalization of the light scattering detectors and data quality control.

2.3.2 UV–Vis Spectroscopy of Dph1–Dph2

1. Prepare 100 μL Dph1–Dph2 complex (25 μ*M*) anaerobically in 150 m*M* NaCl and 200 m*M* Tris–HCl at pH 7.4 and 5% glycerol.
2. Transfer the sample to a quartz cell and seal it before taking out of the anaerobic chamber. Detect UV–vis spectra on a Cary 50 Bio UV–vis spectrophotometer (Varian), scanning from 200 to 800 nm.
3. Correct the baseline with the buffer used to prepare the samples before measurement.
4. Put the quartz cell back into the chamber and treat the sample with dithionite for 10 min at a final concentration of 0.5 m*M*. Seal the sample and detect by UV–vis spectrophotometer again.

2.3.3 EPR Spectroscopy of Dph1–Dph2

1. Add dithionite to the purified Dph1–Dph2 complex (350 μ*M*) to a final concentration of 1 m*M* in an anaerobic chamber. Incubate the mixture for 5 min and transfer to the bottom of an EPR tube (715-PQ-250M, Wilmad–LabGlass) using long glass pipette.
2. Freeze the sample by dipping the EPR tube in liquid N_2 promptly several times to prevent the tube from cracking. Cap the EPR tube and transfer out of the chamber.
3. Record EPR spectra on a Bruker ElexSys E500 EPR spectrometer at a frequency of 9.34 GHz at 12 K using an ESR910 liquid-helium cryostat (Oxford Instruments) with the settings as follows: modulation frequency, 100 kHz; modulation amplitude, 8 G; microwave power, 0.63 mW. Calibrate the field sweeps with a BRUKER ER 035 Gauss meter. The microwave frequency can be monitored with a frequency counter. Xepr software can be used for data acquisition and manipulation.

2.4 Preparation of *S. cerevisiae* EF2, Dph3, and Cbr1

2.4.1 Expression and Purification of S. cerevisiae *EF2*

Overexpression of *S. cerevisiae* EF2 in *E. coli* strains was not successful. We then prepared EF2 with a C-terminal His_6 tag in yeast cells. After testing different promoter systems, better protein quality was obtained with the vectors having a MET25 promoter. Yeast cells can be grown either in culture flask or fermenter at 30°C, the latter gives much higher cells density and more proteins.

1. PCR amplify the *S. cerevisiae ef2* coding sequence from yeast genomic DNA with an added C-terminal His_6 tag. Clone into p423MET25 vector (ATCC, Manassas, VA), which has a His3 selection marker.
2. Transform the p423MET25-EF2 plasmid into *S. cerevisiae* BY4741 strain with deletion of *dph2* gene (OpenBiosystems, Huntsville, AL) using a Frozen-EZ Yeast Transformation kit (ZYMO RESEARCH CORP).
3. To express the EF2 protein, grow the transformed cells in 6 L synthetic complete media lacking histidine for 48 h in a fermenter (BioFlo 110, New Brunswick Scientific) at 30°C with agitation.
4. Harvest cells by centrifugation until the OD_{600nm} is 4–6. Resuspend the collected cells in 200 mL of lysis buffer, which contains 20 m*M* Tris–HCl (pH 8.0), 400 m*M* $(NH_4)_2SO_4$, 10 m*M* $MgCl_2$, and 1 m*M* protease inhibitor phenylmethylsulfonyl fluoride.
5. Lyse the cells using glass beads (400 μm, OPS DIAGNOSTICS) with a bead beater (Biospec). Fill the beater chamber with 200 mL glass beads and 200 mL lysis buffer containing 50–80 g of cells. Install the ice-water

jacket and fill it with ice. As a cycle, run the bead beater for 1 min and let it sit for 2 min to cool down. Repeat this cycle four times.

6. Remove cell debris by centrifugation at 12,000 × *g* for 30 min. Purify the protein from the supernatant on a BioLogic DuoFlow 10 System with a HisTrap HP column with a linear gradient from 30 m*M* imidazole to 500 m*M* imidazole in 30 min.
7. Collect the fractions and dialyze against 200 m*M* Tris–HCl buffer (pH 7.4) containing 500 m*M* NaCl and 5% glycerol. Concentrate the purified proteins and determine protein concentration by Bradford assay.

2.4.2 Cloning, Expression, and Purification of Dph3

Yeast *dph3* was found to be identical to *Kluyveromyces lactis* killer toxin zymocin insensitive 11 (*kti11*) gene from *S. cerevisiae* (Fichtner & Schaffrath, 2002). Dph3 is a small protein less than 10 kDa that contains a CSL zinc finger domain. The CSL zinc finger uses four conserved Cys residues to coordinate Zn^{2+} (Sun et al., 2005), and the last Cys residue is followed by a Ser and a Leu residue. The first solution NMR structure of yeast Dph3 (Sun et al., 2005) shows a single Zn^{2+} ion bound by four conserved cysteine residues of Dph3. However, the Dph3 protein was purified from *E. coli* grown on minimal medium M9 supplemented with $ZnSO_4$. We (Dong et al., 2014) and others (Proudfoot et al., 2008) demonstrated that yeast Dph3 binds both zinc and iron when prepared from *E. coli* cells grown on rich medium (Proudfoot et al., 2008). Crystal structures of Dph3 bind only iron (Kolaj-Robin, McEwen, Cavarelli, & Seraphin, 2015) and both zinc and iron (Glatt et al., 2015) were reported. Our in vitro reconstitution experiment demonstrated that the iron-bound form of Dph3 helps to reduce Dph1–Dph2.

1. Amplify yeast *dph3* gene from yeast genomic DNA, which was extracted from BY4741 using Pierce Yeast DNA Extraction Kit.
2. Clone *dph3* gene following the same protocol as that of *Phdph2*. Transform this plasmid to BL21(DE3) pRARE2 strain.
3. Grow cells in 2 L LB medium supplemented with 0.035 g/L Cam and 0.1 g/L Amp at 37°C and 200 rpm. At an OD_{600nm} of 0.6, cool down the cultures to 18°C and supplement with (or without) $Fe(NH_4)_2(SO_4)_2$ to a final concentration of 70 μ*M*.
4. Induce protein expression by adding IPTG to 0.1 m*M*. Harvest cells after incubation at 18°C for 20 h. Resuspend the cell pellet in 30 mL of buffer containing 20 m*M* Tris–HCl (pH 7.4), 500 m*M* NaCl, and 0.5 m*M* phenylmethylsulfonyl fluoride.

5. Lyse cells by an EmulsiFlex-C3 cell disruptor. Purify Dph3 through Ni-NTA affinity chromatography following the same protocol as that used for the purification of *Ph*EF2.
6. Collect the eluted imidazole fractions and dialyze against 20 m*M* Tris–HCl buffer (pH 7.4) containing 500 m*M* NaCl and 5% glycerol. Concentrate the purified protein using Amicon Ultra-4 centrifugal filter devices.

2.4.3 Dph3 Characterization by UV–Vis Spectroscopy

Dph3 can be characterized by UV–vis spectroscopy. The cysteine-coordinated ferric iron exhibits 350, 490, and 570 nm absorptions. Prepare 100 μL of Dph3 (80 μ*M*) in 20 m*M* Tris–HCl buffer (pH 7.4) containing 500 m*M* NaCl and 5% glycerol in a quartz cell. Correct the baseline using the same buffer. Obtain the UV–vis spectrum of Dph3 by scanning from 200 to 800 nm. Then add dithionite to the sample to the final concentration of 0.5 m*M* and obtain the UV–vis spectrum again.

2.4.4 Cloning, Expression, and Purification of Cbr1

To reduce Dph1–Dph2 and Dph3 using electrons from NADH, a flavin-dependent enzyme is needed. We discovered that cytochrome *b*5 reductase 1 (Cbr1) is the physiological reductase of Dph3 (Lin, Dong, Zhang, Lee, & Lin, 2016). Cbr1 is a flavin-containing protein located on the outer membrane of endoplasmic reticulum and mitochondria. The first 27 amino acids are a transmembrane helix. The soluble form of Cbr1 with truncation of the first 27 amino acids is cloned from the DNA of yeast BY4741 strain.

1. PCR-amplify the truncated *cbr1* gene from yeast genomic DNA with primers introducing *Eco*RI and *Xho*I restriction sites.
2. Double digest 1.5 μg of pET28a vector and the PCR product in 50 μL solution containing 5 μL Cutsmart buffer, 2.5 μL *Eco*RI-HF, and 2.5 μL *Xho*I in 37°C for 3 h.
3. Insert the truncated *cbr1* gene into pET28a vector.
4. Transform the resulting plasmid into BL21R2 *E. coli* strain. Pick single colony to inoculate overnight starter culture with 5 mL of LB media containing 50 μg/mL Kan and 25 μg/mL Cam in a 37°C shaker.
5. Inoculate 2 L of LB media containing 50 μg/mL Kan and 25 μg/mL Cam with overnight starter culture in a 37°C shaker until OD_{600nm} reaches 0.6.
6. Cool down the culture with ice-water bath for 20 min. Induce the expression of truncated Cbr1 with 1.5 m*M* of IPTG and 50 μM riboflavin for 20 h at 16°C.

7. Collect the pellet, resuspend in 30 mL of lysis buffer (25 m*M* Tris–HCl pH 8.0, 150 mL NaCl, and 10% glycerol), and lyse with EmulsiFlex-C3 cell disruptor.
8. Centrifuge the lysate at 20,000 × *g* for 30 min. Incubate the supernatant with 0.5 mL of Ni-NTA resin preequilibrated with the lysis buffer for 30 min at 4°C.
9. Load the incubated Ni-NTA resin onto a polypropylene column at 4°C. Wash with 10 mL lysis buffer, followed by 40 mL of 30 m*M* imidazole in lysis buffer.
10. Elute protein with elution buffers (50, 75, 100, and 200 m*M* imidazole in the lysis buffer, 5 mL each). Collect the yellow fractions to run a 12% SDS-PAGE to check protein purity.
11. Dialyze the pure fraction against 25 m*M* Tris–HCl pH 8.0 with 150 mL NaCl.

3. IN VITRO ANALYSIS OF SAM CLEAVAGE BY Dph1–Dph2 WITH HIGH-PERFORMANCE LIQUID CHROMATOGRAPHY

1. Set up reactions under anaerobic condition with 100 μ*M* Dph1–Dph2, 2 m*M* SAM, 10 m*M* dithionite, 150 m*M* NaCl, 1 m*M* DTT, and 200 m*M* Tris–HCl at pH 7.4 in a final volume of 30 μL. Also carry out negative control reactions without dithionite (Fig. 2).
2. Incubate the mixture at 30°C for 1 h.
3. Quench reactions with 30 μL of 10% trifluoroacetic acid (TFA), followed by centrifugation to separate the precipitated proteins and the supernatant.
4. Analyze the supernatant by high-performance liquid chromatography (Shimadzu) on a C18 column (Kinetex) monitored at 254 nm, using a linear gradient from 0% to 40% solvent B over 15 min at a flow rate of 0.5 mL/min (solvent A: 0.1% aqueous TFA and solvent B: 0.1% TFA in acetonitrile).

4. IN VITRO RECONSTITUTION OF THE FIRST STEP OF DIPHTHAMIDE BIOSYNTHESIS

4.1 Anaerobic Reconstitution of *Ph*Dph2 Activity

1. Set up a reaction mixture containing 20 μ*M* *Ph*Dph2, 7 μ*M* *Ph*EF2, and 5 m*M* dithionite in 150 m*M* NaCl, 1 m*M* DTT, and 200 m*M* Tris–HCl at pH 7.4 in an anaerobic chamber. Set up a same reaction but without

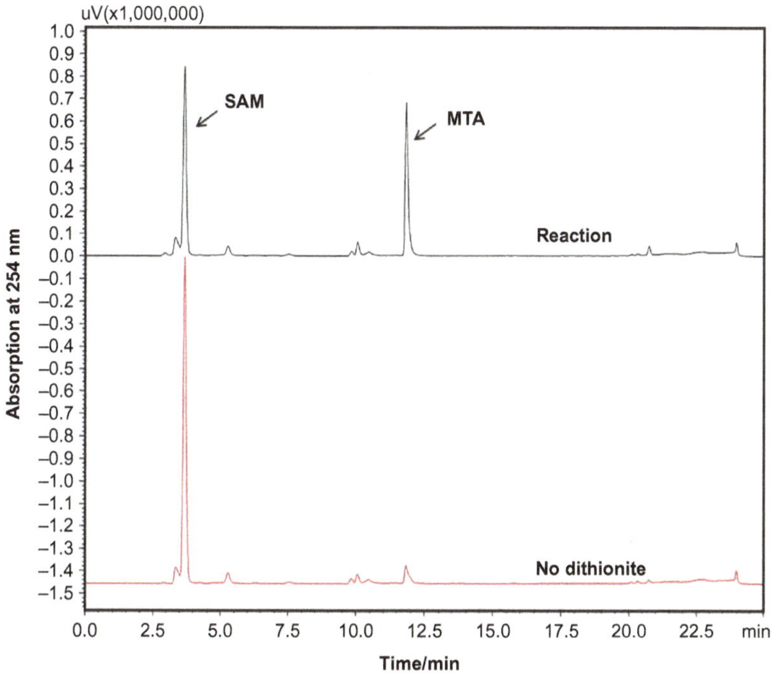

Fig. 2 HPLC assay for Dph1–Dph2 activity. Reduced Dph1–Dph2 is active in the SAM cleavage reaction, generating MTA. SAM is not cleaved without dithionite.

dithionite added as a control reaction. Seal the reaction vials with rubber stoppers (Sigma-Aldrich Z564680) and then transfer the vials out of the anaerobic chamber.

2. Inject carboxyl-^{14}C-SAM (18 μ*M*; 55 mCi/mmol, American Radiolabeled Chemicals Inc.) into the reaction vials to start the reaction. Vortex the reaction mixture briefly and incubate at 65°C for 10 min.
3. Stop the reactions by adding protein loading dye and heating the vial at 95°C for 5 min.
4. Analyze the reaction mixture with 12% SDS-PAGE gel electrophoresis (Fig. 3A).
5. Expose the dried gel to a PhosphorImaging screen (GE Healthcare) and detect the radioactivity using a *Typhoon FLA 7000* (GE Healthcare).

4.2 Anaerobic Reconstitution of the First Step of Diphthamide Biosynthesis Using Yeast Dph1–3

Truncated Cbr1 uses NADH to reduce Dph3 and Dph1–Dph2 in vitro (Lin et al., 2016). We can use Cbr1 with NADH and Dph3 to reconstitute

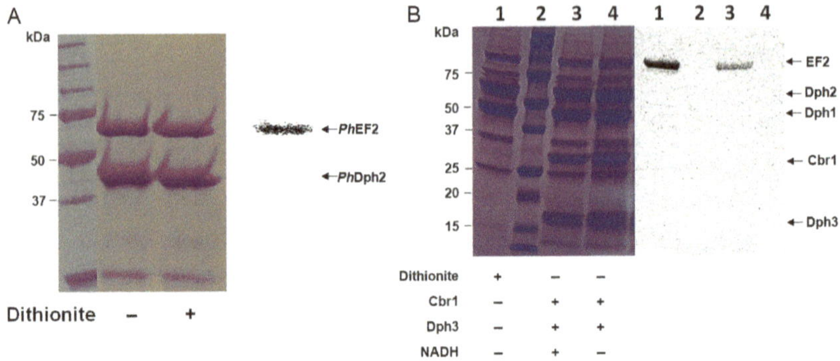

Fig. 3 In vitro reconstitution of the first step of diphthamide biosynthesis with ^{14}C-SAM. (A) In vitro reconstitution of *Ph*Dph2 activity using dithionite. (B) In vitro reconstitution of yeast Dph1–Dph3 activity. Lane 2 is protein ladder. All the reactions in lanes 1, 3, and 4 contained EF2 and Dph1–Dph2. The presence of other reagents is indicated below each lane. In both (A) and (B), the *left panels* show the Coomassie blue-stained gel; the *right panels* show the autoradiography.

the first step of diphthamide biosynthesis. By using carboxyl-^{14}C-SAM, we can label yeast EF2 with ^{14}C and the reaction can be monitored with autoradiography using a Typhoon scanner (Fig. 3B).

1. Transfer at least 100 μL of Dph3, Cbr1, and EF2 separately on ice into three 10 mL round-bottom flasks equipped with magnetic spin bars. Degas these proteins using a Schlenk line before use.
2. Add proteins and reagents in the following sequence in an anaerobic chamber: 6 μL of desalting buffer (150 m*M* NaCl, 200 m*M* Tris–HCl at pH 8.0, 5% glycerol), Dph3 (10 μ*M*, final concentration), Cbr1 (5 μ*M*), NADH (200 μ*M*), Dph1–Dph2 (5 μ*M*), and EF2 (2 μ*M*). Add water to make the final volume of the reaction 23 μL. Seal the Eppendorf tubes with rubber stoppers. Transfer all the sealed Eppendorf tubes out of the glove box.
3. Prepare diluted carboxyl-^{14}C-SAM by adding 4 μL of carboxyl-^{14}C-SAM with a microliter syringe into a tube with 16 μL of desalting buffer (prepared ahead of time in the glove box). Vertex the tube and spin down briefly.
4. Add 2 μL of diluted carboxyl-^{14}C-SAM into each reaction with a microliter syringe. Vertex and spin down briefly.
5. Incubate all the reactions at 30°C for 1 h.
6. Add 5 μL loading dye to quench the reactions. Spin down all the samples briefly after quenching and boil at 95°C for 5 min.

7. Resolve the reaction mixture using 12% SDS-PAGE. Stain the gel with Coomassie staining solution and then destain the gel.
8. Dry the destained gel with cellophane (GelAir™, Bio-Rad) support in the gel dryer for 3 h. Expose the dried gel with phosphor screen for 24 h and scan the phosphor screen with a Typhoon scanner.

Note

1. After quenching, the samples can be stored in −20°C freezer.
2. The destaining step is essential to get rid of the free carboxyl-^{14}C-SAM. Free carboxyl-^{14}C-SAM in the gel will result in higher background signal.
3. Allow the gel to dry fully to prevent the contamination of the phosphor screen and typhoon scanner.
4. A positive control with dithionite as reductant and a negative control without reductant should be carried out at the same time.

4.3 In Vitro ADP-Ribosylation of Diphthamide Using Rhodamine NAD (Rh-NAD)

To facilitate the detection of diphthamide formation, we also developed a fluorescent assay to monitor the reaction (Du, Jiang, & Lin, 2009). This assay can be used on both purified EF2 protein and cell lysate. The assay takes advantage of the selective recognition of both diphthamide and diphthine (the intermediate before the amidation step, see Fig. 1) by DT. We synthesized a rhodamine-conjugated NAD molecule, Rh-NAD. If diphthamide or diphthine is formed in the reaction mixture, then incubating the reaction mixture with Rh-NAD and DT should allow fluorescent labeling of EF2. The synthesis of Rh-NAD has been reported (Du et al., 2009).

Incubate the purified yeast EF2 and Rh-NAD (50 μ*M*) with DT (100 n*M*) at 30°C in 25 m*M* Tris–HCl (pH 8.0), 50 m*M* NaCl, 30 m*M* DTT, and 2 m*M* ethylenediaminetetraacetic acid (EDTA) for 1 h. Resolve the reaction mixture by SDS-PAGE (Fig. 4A). Visualize rhodamine fluorescence signal from protein gel on a Fisher Scientific Ultraviolet Transilluminators or quantify the signal using a Typhoon scanner.

The labeling method can be further used on cell lysate from wild type (wt) yeast strain and Δ*dph1* strain (Fig. 4B). The EF2 from wt strain has diphthamide modification and thus can be labeled by diphtheria toxin, while the EF2 from Δ*dph1* cells lacks diphthamide modification and thus cannot be labeled.

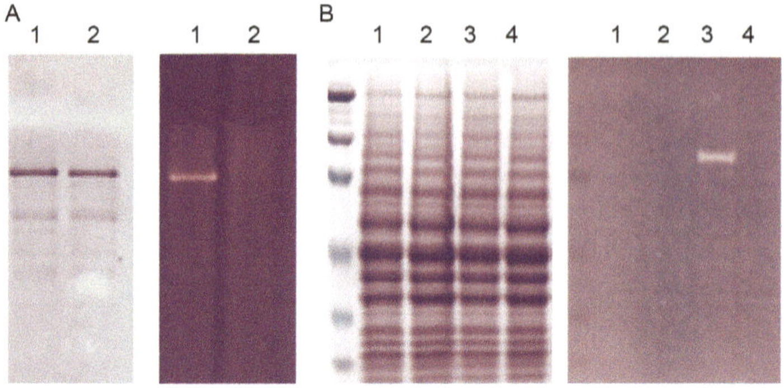

Fig. 4 Rh-NAD labeling diphthamide using diphtheria toxin (DT)-catalyzed ADP-ribosylation of EF-2. (A) Fluorescent labeling of purified eEF-2 with diphthamide modification. Lane 1, with DT; lane 2, without DT. (B) Fluorescent labeling of EF-2 in cell lysate. Lane 1, wild type cell lysate without DT; lane 2, Δ*dph1* cell lysate without DT; lanes 3, wild type cell lysate with DT; lanes 4, Δ*dph1* cell lysate with DT. *Panel (B) is adapted with permission from Du, J., Jiang, H., & Lin, H. (2009). Investigating the ADP-ribosyltransferase activity of sirtuins with NAD analogues and 32P-NAD.* Biochemistry, *48(13), 2878–2890. doi:10.1021/bi802093g. Copyright (2009) American Chemical Society.*

5. CONCLUSION

The first step of diphthamide biosynthesis is the transfer of an ACP group from SAM to substrate EF2 by unique radical SAM enzymes. In this chapter, we have discussed the preparation and characterization of these enzymes, *Ph*Dph2 homodimer and yeast Dph1–Dph2 heterodimer, as well as the preparation of related proteins in the first step of diphthamide biosynthesis, including substrate protein *Ph*EF2 and yeast EF2, yeast Dph3, and Cbr1. We have highlighted three activity assays for the analysis of diphthamide radical SAM enzymes: HPLC assay to detect the cleavage product MTA, radioactive labeling assay using carboxyl-^{14}C-SAM, and a fluorescent assay using Rh-NAD.

ACKNOWLEDGMENTS

This work was supported by grants from the National Institute of Health, National Institute of General Medical Sciences (GM 088276).

REFERENCES

Akiva, E., Brown, S., Almonacid, D. E., Barber, A. E., Custer, A. F., Hicks, M. A., et al. (2014). The structure–function linkage database. *Nucleic Acids Research*, *42*(D1), D521–D530. https://doi.org/10.1093/nar/gkt1130.

Altschul, S. F., Gish, W., Miller, W., Myers, E. W., & Lipman, D. J. (1990). Basic local alignment search tool. *Journal of Molecular Biology*, *215*(3), 403–410. https://doi.org/10.1016/S0022-2836(05)80360-2.

Broderick, J. B., Duffus, B. R., Duschene, K. S., & Shepard, E. M. (2014). Radical S-adenosylmethionine enzymes. *Chemical Reviews*, *114*(8), 4229–4317. https://doi.org/10.1021/cr4004709.

Collier, R. J. (2001). Understanding the mode of action of diphtheria toxin: A perspective on progress during the 20th century. *Toxicon*, *39*(11), 1793–1803.

Dong, M., Horitani, M., Dzikovski, B., Pandelia, M. E., Krebs, C., Freed, J. H., et al. (2016). Organometallic complex formed by an unconventional radical S-Adenosylmethionine enzyme. *Journal of the American Chemical Society*, *138*(31), 9755–9758. https://doi.org/10.1021/jacs.6b04155.

Dong, M., Su, X., Dzikovski, B., Dando, E. E., Zhu, X., Du, J., et al. (2014). Dph3 is an electron donor for Dph1-Dph2 in the first step of eukaryotic diphthamide biosynthesis. *Journal of the American Chemical Society*, *136*(5), 1754–1757. https://doi.org/10.1021/ja4118957.

Du, J., Jiang, H., & Lin, H. (2009). Investigating the ADP-ribosyltransferase activity of sirtuins with NAD analogues and 32P-NAD. *Biochemistry*, *48*(13), 2878–2890. https://doi.org/10.1021/bi802093g.

Fichtner, L., & Schaffrath, R. (2002). *KTI11* and *KTI13*, *Saccharomyces cerevisiae* genes controlling sensitivity to G1 arrest induced by *Kluyveromyces lactis* zymocin. *Molecular Microbiology*, *44*(3), 865–875.

Frazzon, J., & Dean, D. R. (2003). Formation of iron-sulfur clusters in bacteria: An emerging field in bioinorganic chemistry. *Current Opinion in Chemical Biology*, 7(2), 166–173.

Glatt, S., Zabel, R., Vonkova, I., Kumar, A., Netz, D. J., Pierik, A. J., et al. (2015). Structure of the Kti11/Kti13 heterodimer and its double role in modifications of tRNA and eukaryotic elongation factor 2. *Structure*, *23*(1), 149–160. https://doi.org/10.1016/j.str.2014.11.008.

Honjo, T., Nishizuka, Y., Hayaishi, O., & Kato, I. (1968). Diphtheria toxin-dependent adenosine diphosphate ribosylation of aminoacyl transferase II and inhibition of protein synthesis. *The Journal of Biological Chemistry*, *243*(12), 3553–3555.

Kolaj-Robin, O., McEwen, A. G., Cavarelli, J., & Seraphin, B. (2015). Structure of the Elongator cofactor complex Kti11/Kti13 provides insight into the role of Kti13 in Elongator-dependent tRNA modification. *The FEBS Journal*, *282*(5), 819–833. https://doi.org/10.1111/febs.13199.

Lanz, N. D., Grove, T. L., Gogonea, C. B., Lee, K. H., Krebs, C., & Booker, S. J. (2012). RlmN and AtsB as models for the overproduction and characterization of radical SAM proteins. *Methods in Enzymology*, *516*, 125–152. https://doi.org/10.1016/B978-0-12-394291-3.00030-7.

Lin, H. (2011). S-Adenosylmethionine-dependent alkylation reactions: When are radical reactions used? *Bioorganic Chemistry*, *39*(5–6), 161–170. https://doi.org/10.1016/j.bioorg.2011.06.001.

Lin, Z., Dong, M., Zhang, Y., Lee, E. A., & Lin, H. (2016). Cbr1 is a Dph3 reductase required for the tRNA wobble uridine modification. *Nature Chemical Biology*, *12*(12), 995–997. https://doi.org/10.1038/nchembio.2190.

Liu, S., Milne, G. T., Kuremsky, J. G., Fink, G. R., & Leppla, S. H. (2004). Identification of the proteins required for biosynthesis of diphthamide, the target of bacterial ADP-ribosylating toxins on translation elongation factor 2. *Molecular and Cellular Biology*, *24*, 9487–9497.

Proudfoot, M., Sanders, S. A., Singer, A., Zhang, R., Brown, G., Binkowski, A., et al. (2008). Biochemical and structural characterization of a novel family of cystathionine beta-synthase domain proteins fused to a Zn ribbon-like domain. *Journal of Molecular Biology*, *375*(1), 301–315. https://doi.org/10.1016/j.jmb.2007.10.060.

Robinson, E. A., Henriksen, O., & Maxwell, E. S. (1974). Elongation factor 2. Amino acid sequence at the site of adenosine diphosphate ribosylation. *The Journal of Biological Chemistry*, *249*(16), 5088–5093.

Schaffrath, R., Abdel-Fattah, W., Klassen, R., & Stark, M. J. (2014). The diphthamide modification pathway from *Saccharomyces cerevisiae*—Revisited. *Molecular Microbiology*, *94*(6), 1213–1226. https://doi.org/10.1111/mmi.12845.

Sun, J., Zhang, J., Wu, F., Xu, C., Li, S., Zhao, W., et al. (2005). Solution structure of Kti11p from *Saccharomyces cerevisiae* reveals a novel zinc-binding module. *Biochemistry*, *44*, 8801–8809.

Van Ness, B. G., Howard, J. B., & Bodley, J. W. (1980a). ADP-ribosylation of elongation factor 2 by diphtheria toxin. Isolation and properties of the novel ribosyl-amino acid and its hydrolysis products. *The Journal of Biological Chemistry*, *255*(22), 10717–10720.

Van Ness, B. G., Howard, J. B., & Bodley, J. W. (1980b). ADP-ribosylation of elongation factor 2 by diphtheria toxin. NMR spectra and proposed structures of ribosyl-diphthamide and its hydrolysis products. *The Journal of Biological Chemistry*, *255*(22), 10710–10716.

Zhang, Y., Zhu, X., Torelli, A. T., Lee, M., Dzikovski, B., Koralewski, R. M., et al. (2010). Diphthamide biosynthesis requires an organic radical generated by an iron-sulphur enzyme. *Nature*, *465*(7300), 891–896. https://doi.org/10.1038/nature09138.

CHAPTER SIXTEEN

Guidelines for Determining the Structures of Radical SAM Enzyme-Catalyzed Modifications in the Biosynthesis of RiPP Natural Products

Leah B. Bushin*, Mohammad R. Seyedsayamdost*,†,1
*Department of Chemistry, Princeton University, Princeton, NJ, United States
†Department of Molecular Biology, Princeton University, Princeton, NJ, United States
[1]Corresponding author: e-mail address: mrseyed@princeton.edu

Contents

Abstract

Radical *S*-adenosylmethionine (RaS) enzymes catalyze some of the most fascinating transformations in Nature. With only ~100 of the >300,000 members studied to date, it is safe to assume that a plethora of new reactions and reaction mechanisms remain to be elucidated. It is by now relatively easy to spot RaS enzymes in microbial genomes. However, to determine the reactions that they carry out, detailed structural characterization of the product(s) is necessary, a process that still represents a significant roadblock in the study of RaS enzymes. We have recently combined natural products

Methods in Enzymology, Volume 606
ISSN 0076-6879
https://doi.org/10.1016/bs.mie.2018.04.016

structural elucidation along with RaS enzymology to provide a proof of concept for how the confluence of these approaches can lead to the discovery of new natural products and RaS enzyme-mediated transformations. Herein, we provide guidelines for expressing, purifying, and reconstituting a subclass of RaS enzymes that contain a so-called SPASM domain, as well as characterizing the reactions that they catalyze using a combination of HR/MS^n and NMR investigations. Application of these approaches will aid in expanding the chemical and biosynthetic repertoire of RaS enzymes in the future.

1. INTRODUCTION

Radical *S*-adenosylmethionine (RaS) enzymes catalyze some of the most unusual transformations in Nature (Broderick, Duffus, Duschene, & Shepard, 2014; Frey, 2001; Landgraf, McCarthy, & Booker, 2016; Young & Bandarian, 2013). They are involved in numerous physiological processes, including deoxyribonucleotide synthesis, cofactor biogenesis, catabolic pathways, and secondary metabolism. Of the >300,000 members within the superfamily that can be spotted bioinformatically, less than 0.1% have been characterized to date, indicating that RaS enzymes remain a vast and largely untapped store for new enzymatic chemistries (Akiva et al., 2014). However, one of the more challenging aspects in studying a given new RaS enzyme with little homology to previous ones is identifying the substrate and physiological pathway in which it is involved. These aspects are simplified in a subclass of natural products known as RiPPs (ribosomally synthesized and posttranslationally modified peptides), as the substrate is genetically encoded by a "substrate gene" and the mature product is involved in secondary metabolism (Arnison et al., 2013; McIntosh, Donia, & Schmidt, 2009).

RiPP natural products are generated by an elegant biosynthetic logic: a substrate peptide is expressed as a multipartite precursor, containing usually an N-terminal leader motif followed by a core sequence, and in some cases, a C-terminal follower sequence (Arnison et al., 2013; McIntosh et al., 2009; Oman & van der Donk, 2010). The leader facilitates proper interaction with the tailoring enzyme(s) and thereby provides specificity. After ribosomal synthesis, the substrate peptide is modified at the core by one or more tailoring enzymes. Subsequently, the leader (and follower) sequences are removed proteolytically to deliver the mature natural product. The genomic synteny of a typical RaS enzyme-modified RiPP, which we refer to as RaS RiPP, is shown using the streptide (*str*) gene cluster as an example (Fig. 1A; Ibrahim et al., 2007; Schramma, Bushin, & Seyedsayamdost, 2015). The substrate is encoded by

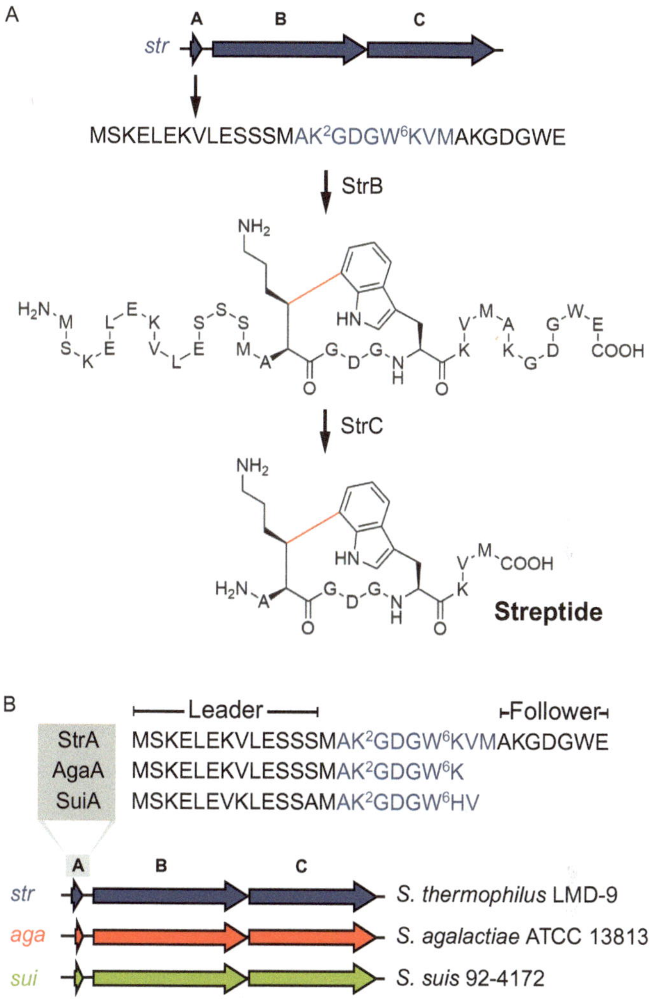

Fig. 1 Streptide biosynthesis and the *str* gene cluster. (A) *StrA* encodes a precursor peptide, the sequence of which is shown. The internal nine residues that form the mature product are colored *blue*; residues to be cross-linked are labeled. *StrB* encodes a RaS enzyme that installs the unusual Lys-Trp cross-link. *StrC* codes for a peptidase/transporter, which removes the leader and follower sequences, possibly in conjunction with another protease, to complete streptide biosynthesis. (B) Comparison of the homologous *str*, *aga*, and *sui* gene clusters from three Streptococci. The sequence of each precursor peptide (StrA, AgaA, and SuiA) is shown, with the residues in the Lys-Trp cross-link labeled.

strA, a 30-mer precursor peptide with the sequence shown. The RaS enzyme StrB installs an unusual posttranslational modification (PTM) consisting of a Lys-Trp carbon–carbon cross-link with the shown regio- and stereochemistry (Schramma et al., 2015). The leader and follower sequences are then removed, by StrC and possibly a second protease, to deliver mature streptide. The inclusion of the substrate gene within the biosynthetic gene cluster significantly simplifies the search for the substrate of the adjacent RaS enzyme.

The remaining challenge, then, in studying novel RaS RiPPs, is elucidating the reaction carried out by the tailoring enzyme. In cases where the structure of the final natural product is known, the modification installed by the RaS enzyme can be intuited. We have previously combined natural product discovery and RaS enzymology to provide an example of this process (Schramma et al., 2015). When the mature product is not known, however, uncovering the reaction carried out by the RaS enzyme can be much more difficult. As the substrates are typically large, usually >25 amino acids in length, determining the structure of the resulting product presents a formidable challenge that is not easily addressed by a single method. Herein, we discuss several strategies for tackling this problem as well as the considerations that can be used in choosing among them. Focusing on in vitro approaches, we provide detailed procedures for characterizing the products of new RaS enzymes to enable downstream mechanistic investigations.

2. STRATEGIES FOR CHARACTERIZING PTMs IN RaS RiPP BIOSYNTHESIS

The simplest case in determining the PTM installed by a RaS enzyme is one where the structure of the mature natural product is known. When this is not the case, two main strategies can be employed to solve the structure of the resulting product.

The simplest and arguably most powerful approach is one where the RaS enzyme is expressed recombinantly, the substrate is generated by solid-phase peptide synthesis (SPPS), and the two are mixed together, in the presence of the appropriate cofactors and other protein domains, to assess product formation. Detailed procedures for this workflow are provided below. The advantages of this general strategy are that any probes, structural or mechanistic, can be incorporated into the substrate. Moreover, steady-state and presteady-state kinetics can be conducted, thus providing insights into the reaction mechanism. Lastly, studies based on stoichiometry, which is especially important for RaS enzymes, are also easily addressed. While the approach aids in mechanistic aspects, it does require purification of the RaS enzyme, which must be performed anaerobically and therefore can be challenging. Moreover, structural characterization of the product is not straightforward, in large part because in vitro reactions, even if thermodynamically favored, rarely go to completion. As such, the product and substrate must be separated for detailed studies by nuclear magnetic resonance (NMR) spectroscopy and mass spectrometry (MS). Even if the product is resolved chromatographically from the substrate, NMR analysis is by no means trivial as the modification only affects a small number of the single- and multidimensional NMR correlations in a sea of cross-peaks, which need to be identified with high fidelity. Nonetheless, depending on the purpose of the study, the mechanistic advantages may outweigh any possible disadvantages.

A variant of the approach above is to use available proteases to cleave the N-terminal leader sequence and therefore reduce the number of NMR cross-peaks, facilitating the identification of those that are relevant to the PTM. Proteases, such as trypsin, chymotrypsin, AspN, GluC, and LysC to name a few, are available for this purpose, thus minimizing the sequence dependence of the approach.

A second general strategy that has been successful as well involves carrying out the modification reaction in vivo (Morinaka et al., 2018; Shi, Yang, Garg, & van der Donk, 2011). The precursor peptide and RaS enzyme are expressed recombinantly from the same vector, with the precursor carrying, usually, an N-terminal hexa-His tag. Upon expression of both genes, the enzyme installs the PTM intracellularly, thus obviating the need for challenging anaerobic purification of the RaS enzyme. The cells are then lysed, and the product purified using metal ion affinity chromatography. The chief advantages of this strategy are good conversion of substrate to product and ease of purification, both of which aid in structural elucidation, again by MS and NMR, especially when combined with a proteolytic step to remove most or all of the leader peptide (including the hexa-His tag). The primary disadvantage is that mechanistic or structural probes can only be incorporated within the confines of natural amino acids. Incorporation of isotopomeric amino acids is possible, as long as expression of the peptide/protein combination is carried out in minimal media, which may limit the yield of the product. Because it relies on ribosomal synthesis, this process would globally, rather than site-specifically, introduce the isotopolog. Lastly, kinetic or stoichiometric studies are not possible, as the reaction takes place in vivo. Nonetheless, this process typically allows for more than sufficient quantities of purified product for structural elucidation by spectroscopic analysis and may be favored, if mechanistic studies are not a chief goal of the study.

Especially important to both strategies above are incorporation of isotopically labeled amino acids for both structural and mechanistic studies. ^{13}C-labeled amino acids "isolate" the NMR contributions by the modified residue(s), thus aiding in the identification of the relevant cross-peaks in a pool of other correlations. Likewise, ^{2}H-labels can either simplify complex regions of NMR spectra or serve as probes for H-atom abstraction in mechanistic investigations.

With these considerations in mind, we will next describe detailed procedures that we have used for determining the cross-link installed by SuiB, which is encoded by the *sui* gene cluster from the pathogen *Streptococcus suis* (Fig. 1B; Schramma, Forneris, Caruso, & Seyedsayamdost, 2018; Schramma & Seyedsayamdost, 2017). Much like the *str* cluster, the *sui* cluster encodes a precursor peptide, SuiA, with homology to StrA, as well as a RaS enzyme (SuiB) and a protease/transporter (SuiC) with high sequence similarity to StrB and StrC, respectively. The PTM installed by SuiB was not known, though the high homology between StrB and SuiB suggested that both enzymes would catalyze similar transformations. While in this case our efforts were guided by the results from StrB, the approaches outlined herein can be used to determine the structures of new PTMs installed by other RaS RiPP enzymes.

3. EXPRESSION, PURIFICATION, AND RECONSTITUTION OF RaS ENZYMES

The first step in applying the in vitro approach described above involves preparation of all the necessary components. The gene encoding the RaS enzyme of interest is cloned into a pET vector, usually as an N-terminally His-tagged construct. If the producing strain is available and can be cultured under standard laboratory conditions, genomic DNA can be isolated and used as template for PCR amplification. Otherwise, the gene can be synthesized. In either case, a codon-optimized gene may prove necessary for ample protein expression. In the case of SuiB, genomic DNA from *S. suis* was used for amplification, and *suiB* was subsequently cloned into pET-28b with an N-terminal hexa-His-tag. The exact sequence of the protein so encoded contained the N-terminal appendage M-G-S-S-H-H-H-H-H-H-S-S-G-L-V-P-R-G-S-H upstream of the first Met residue of SuiB (Schramma & Seyedsayamdost, 2017).

The pET construct and pDB1282, which encodes the *isc* operon required for assembly of Fe–S clusters (Johnson, Unciuleac, & Dean, 2006; Lanz et al., 2012), are then cotransformed into a suitable expression host, such as *Escherichia coli* BL21(DE3), and grown as described below. When targeting an uncharacterized RaS enzyme, it is best to carry out test expressions on a small scale to determine the optimal IPTG concentration, temperature, and duration of induction. Once optimal conditions have been identified, expression cultures can be scaled up accordingly. The protocol that we developed for expression of SuiB follows.

3.1 Protocol for Expression of SuiB

1. Inoculate a 14-mL sterile culture tube containing 5 mL LB (+50 μg/mL kanamycin (Kan) and 100 μg/mL ampicillin (Amp)) with a single colony of *E. coli* Bl21(DE3) cells carrying pET-28b-*suiB* + pDB1282. Incubate at 37°C/200 rpm overnight.
2. The next morning, inoculate a 250-mL Erlenmeyer flask containing 100 mL LB (+50 μg/mL Kan and 100 μg/mL Amp) with 1 mL overnight culture (1% dilution). Grow at 37°C/200 rpm for 4–6 h.
3. Prepare 4 × 4 L Erlenmeyer flasks each containing 2.5 L of 1 × M9 salts supplemented with final concentrations of 2 m*M* $MgSO_4$, 0.1 m*M* $CaCl_2$, 20 m*M* glucose, 25 μ*M* $(NH_4)_2Fe(SO_4)$, 50 μg/mL Kan, and 50 μg/mL Amp.
4. Inoculate each 4-L flask with 25 mL intermediate culture (1% dilution). Grow at 37°C/180 rpm to an optical density at 600 nm (OD_{600nm}) of 0.3.
5. At OD_{600nm} = 0.3, supplement cultures with final concentrations of 2 m*M* L-Cys, 25 μ*M* $(NH_4)_2Fe(SO_4)$, and 0.2% (w/v) L-(+)-arabinose. Grow at 37°C/180 rpm to OD_{600nm} = 0.6.
6. At OD_{600nm} = 0.6, remove flasks from shaker and cool in an ice-water bath for ~10 min. Remove 0.5 mL aseptically for SDS-PAGE analysis later. Then add IPTG to a final concentration of 0.25 m*M*. Grow statically for 18 h at 18°C.
7. After 18 h, remove 0.5 mL for SDS-PAGE analysis. Then harvest cells by centrifugation (15,000 × *g*, 30 min, 4°C). Flash-freeze cells in liquid N_2 and store the cell pellet at −80°C.
8. The pre- and post-induction samples should be analyzed by SDS-PAGE prior to purification to ensure expression of the desired protein (Fig. 2A).

3.2 Protocol for Purification of SuiB

Our purification protocol is based on that previously reported by Lanz et al. (2012). Purification of RaS enzymes must be carried out in an inert atmosphere. All necessary materials should be brought into the glovebox 1–2 days prior to the day of purification. Buffers and other solutions should be degassed on a Schlenk line prior to transport into the anaerobic chamber

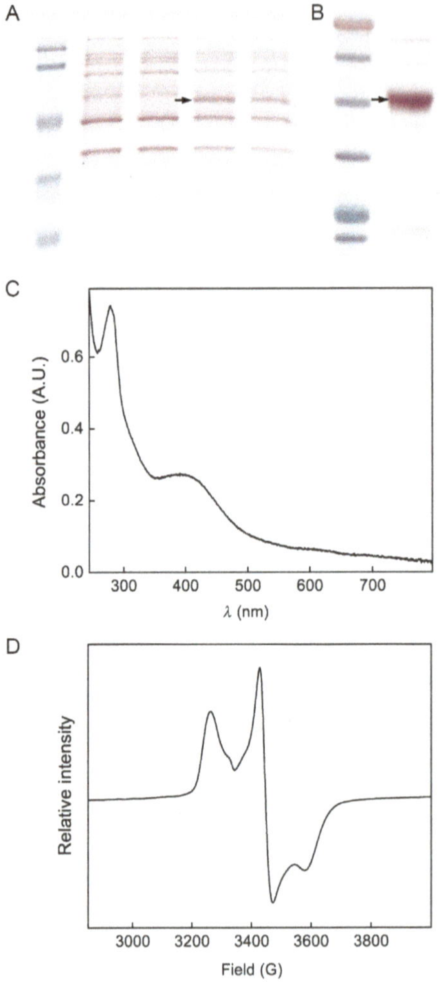

Fig. 2 Expression, purification, and characterization of SuiB. (A, B) SDS-PAGE analysis of SuiB expression under the conditions described (A) and of purified SuiB (B). The *arrow* points to the SuiB band. (C, D) UV–Vis absorption (C) and X-band EPR (D) spectra of purified, reconstituted SuiB.

at least a day before the purification. Chemicals should be brought in as powders and prepared as solutions inside the box, whenever possible:

1. Transport the frozen cell pellet into the glovebox.
2. Thaw the cell pellet and resuspend in lysis buffer (5 mL buffer/g of cell pellet), which consists of 50 m*M* HEPES, 300 m*M* KCl, 10% glycerol, 0.05% Triton X-100, 2 m*M* imidazole, pH 7.5. Supplement with final concentrations of 5 m*M* β-mercaptoethanol (βME) and 0.1 m*M* of phenylmethanesulfonyl fluoride (PMSF). PMSF can be prepared as a 10 m*M* solution in isopropanol.

3. Transfer the mixture to a 150-mL beaker and stir on ice until the mixture is homogenous. Then add 1 mg solid, lyophilized lysozyme per mL of suspension. Stir for an additional 30 min.
4. Sonicate inside the glovebox for 2 min in 15 s on/15 s off cycles at 30% power (~150 W). Allow the mixture to cool for 1 min on ice. Carry out another 2 min sonication cycle. Repeat this entire process a second time. Ensure that the crude extract remains on ice throughout the sonication protocol and that the probe is removed between cycles to allow the mixture to cool. At the end of this process, remove ~50 μL for analysis (see step 14).
5. Transfer the cell lysate to 40-mL centrifuge tubes and seal with parafilm. Bring the tubes out of the glovebox and remove cell debris by centrifugation (18,000 × *g*, 65 min, 4°C).
6. Immediately following centrifugation, bring the tubes back into the glovebox, being careful not to disrupt the pellet. Transfer the cleared crude extract to a clean 150-mL beaker and remove ~50 μL for analysis (see step 14).
7. Prepare 0.2 volumes (relative to the volume of crude extract) of 6% (w/v) streptomycin sulfate in lysis buffer and add dropwise to the crude extract over 15 min while gently stirring on ice. Stir the mixture for an additional 15 min.
8. Transfer the mixture to 40-mL centrifuge tubes, seal with parafilm, and take the tubes out of the glovebox. Remove precipitated DNA by centrifugation (35,000 × *g*, 40 min, 4°C).
9. Immediately following centrifugation, bring the tubes back into the glovebox and carefully transfer the supernatant to a clean receptacle, making sure not to disturb the pellet. Isolate ~50 μL for analysis (see step 14).
10. Load the supernatant onto a HisPur Co column (Thermo-Fisher, 5 mL) equilibrated with 10 column volumes (CV) of lysis buffer (+5 m*M* βME). Keep the supernatant on ice throughout the loading process. Upon loading the cleared crude extract, wash with 6 CV lysis buffer (+5 m*M* βME), followed by 6 CV of wash buffer consisting of 50 m*M* HEPES, 300 m*M* KCl, 10% glycerol, 30 m*M* imidazole, pH 7.5 (+5 m*M* βME).
11. Finally, elute bound protein with 4 CV of elution buffer, consisting of 50 m*M* HEPES, 300 m*M* KCl, 15% glycerol, 300 m*M* imidazole, pH 75 (+5 m*M* βME). Begin collecting when the eluant turns a brownish hue and stop as soon as the eluant color returns to clear. This process typically delivers ~5–6 mL of RaS enzyme at a concentration of 3–15 mg/mL. Be sure to remove ~50 μL of the flowthrough, the wash, and the eluant for subsequent analysis (see step 14).

12. Load the eluant onto a Sephadex G-25 column equilibrated with reconstitution buffer, which consists of 100 m*M* HEPES, 300 m*M* KCl, 10% glycerol, pH 7.5, 5 m*M* DTT. Once again, the RaS enzyme can be tracked by its brown color. All protein should elute off the column within 1 CV. Remove ~50 µL for analysis (see step 14).
13. Aliquot the purified protein into 1.5-mL Eppendorf tubes (~0.5 mL each). Store either at –80°C outside of the glovebox or, if available, in a –40°C freezer inside the glovebox.
14. Analyze all the aliquots collected throughout the purification by SDS-PAGE. This will allow the desired protein to be tracked throughout the process; it is especially useful when low yields or poor purity is obtained, and the protocol needs troubleshooting. The final as-isolated enzyme should also be analyzed by UV–Vis spectroscopy, EPR spectroscopy, and Fe/labile S^{2-} quantification, as previously described in detail (Beinert, 1983; Broderick et al., 2014; Cutsail, Telser, & Hoffman, 2015; Fish, 1988; Frey, 2001; Landgraf et al., 2016; Pandelia, Lanz, Booker, & Krebs, 2015; Young & Bandarian, 2013). An SDS-PAGE gel of purified SuiB is shown in Fig. 2B.

3.3 Protocol for Reconstitution of SuiB

In our hands, as-isolated RaS enzymes have been inactive, and they have therefore necessitated chemical reconstitution. When working with a new, uncharacterized RaS enzyme, the nature ([4Fe–4S] vs [2Fe–2S] vs other) and stoichiometry (1, 2, 3, or more per protomer) of Fe–S clusters are unknown. Methods to determine these aspects by using spectroscopic techniques have been reported previously and will not be reiterated here (Cutsail et al., 2015; Lanz et al., 2012). If this type of analysis is not feasible, sequence comparisons with characterized RaS enzymes can be undertaken to estimate the number of Fe–S clusters per protomer, which then would inform the stoichiometry of Fe and S^{2-} to be added during chemical reconstitution. In general, a 10-fold excess of Fe and S^{2-} relative to the protein is a good starting point for SPASM domain-containing RaS enzymes.

Prior to reconstitution, the protein concentration must be determined. This can be accomplished by denaturation followed by UV–Vis spectroscopy (using the calculated extinction coefficient), as detailed by the Klinman lab (Barr et al., 2016), or using the Bradford assay followed by determining the appropriate correction factor, as detailed by the Booker group

(Lanz et al., 2012). With the concentration of the purified protein in hand along with the number and type of anticipated Fe–S cluster(s), chemical reconstitution can ensue. The scale of the reconstitution reaction is flexible. We typically carry it out on a ~50-mg scale. The procedure we have developed for SuiB is as follows:

1. Thaw the RaS enzyme inside the glovebox. If precipitation is visible, remove the precipitate by centrifugation (on a microcentrifuge inside the box). Transfer the RaS enzyme to a 25-mL pear-shaped flask equipped with a stir bar.
2. Gently stir the protein solution on ice and supplement with a final concentration of 2 m*M* sodium dithionite (DT), which can be prepared as a 200 m*M* stock solution in H_2O inside the glovebox.
3. Prepare a ~20-m*M* stock of $(NH_4)_2Fe(SO_4)$ in H_2O. Add 4 equiv. of Fe^{II} per protomer dropwise to the flask. Stir on ice for 45 min.
4. Add another 4 equiv. of Fe^{II} dropwise. Stir on ice for an additional 45 min.
5. Prepare a ~50-m*M* Na_2S solution in H_2O. Dropwise, add 4 equiv. S^{2-} per protomer. Stir on ice for 45 min.
6. Repeat step 5.
7. Add 2 equiv. Fe^{II} and stir on ice for 30 min.
8. Add 2 equiv. S^{2-} and DT to a final concentration of 2 m*M*. Stir overnight and allow to warm to room temperature.
9. The next morning, remove any precipitant by centrifugation (inside the box) and load the reconstituted RaS enzyme onto a Sephadex G-25 column (50 mL) that has been equilibrated with reconstitution buffer supplemented with 2 m*M* DTT. Elution can be monitored by the dark brown color of the RaS enzyme.
10. Determine the concentration of reconstituted RaS enzyme. If necessary dilute or concentrate to ~100–300 μ*M*. Aliquot into 1.5-mL Eppendorf tubes and freeze as earlier (step 13 of Section 3.2). The protein can be concentrated inside the glovebox using Amicon Ultracel centrifugal filter concentrators and a small benchtop centrifuge.
11. As in step 14 of Section 3.2, the purified, reconstituted protein needs to be characterized thoroughly by UV–Vis and EPR spectroscopy as well as labile Fe and sulfide quantification. Where possible, Mossbauer spectroscopy should be employed as well to further characterize the nature and stoichiometry of Fe–S clusters. This completes the preparation of the RaS enzyme. UV–Vis and EPR spectra of reconstituted, reduced SuiB are shown in Fig. 2C.

4. SOLID-PHASE SYNTHESIS OF PEPTIDE SUBSTRATES

We prepare the substrates for the RaS enzyme of interest by SPPS, which, as mentioned earlier, allows site-specific incorporation of various desired probes. An automated microwave peptide synthesizer is highly recommended, such as the CEM Liberty Blue (Pedersen, Tofteng, Malik, & Jensen, 2012; Vanier, 2013), which we have used routinely. Manual synthesis by SPPS is also possible, although slow and low yielding for the typically long peptide substrates found in RaS RiPP biosynthetic gene clusters.

Due to the length of precursor peptides, a low-loading resin is highly preferred. These can be purchased in preloaded form, that is, with the C-terminal amino acid attached to the resin (note that SPPS proceeds in the direction of C- to N-terminus). N_α-Fmoc- and side chain-protected amino acids are also commercially available. The scale of peptide syntheses is flexible; we typically perform them on a 50–100 μmol scale using 5–6 equiv. of the next N_α-Fmoc amino acid in the sequence and 5–6 equiv. of coupling reagent. In cases, where expensive, synthetic isotopomers are incorporated, the number of equiv. can be reduced to ~threefold along with an extended coupling reaction time. Beyond the 20th residue in a peptide sequence, we typically double-couple subsequent amino acids. Because SPPS is fairly routine, detailed methods are not enumerated here. The reader is referred to several thorough texts on this topic (Behrendt, White, & Offer, 2016; El-Faham & Albericio, 2011; Fields, 1997; Pennington & Dunn, 1994).

4.1 Protocol for Cleavage of Peptide from Resin

Upon completion of the synthesis, the peptide must be cleaved from the resin. To do so, the resin is transferred from the reaction vessel to an Econo-Pac column, washed several times with DCM, and dried thoroughly under vacuum. Cleavage from the resin follows:

1. Add freshly prepared cleavage cocktail (5 mL per 100 mg resin) to the column. Cleavage cocktail consists of TFA, water, triisopropylsilane, and 3,6-dioxa-1,8-octanedithiol in a 92.5:2.5:2.5:2.5 mixture. Stir for 1–3 h at room temperature.
2. Drain the solution from the column and rinse the remaining resin with TFA several times.
3. Concentrate the combined filtrate and TFA rinses using a stream of $N_{2(g)}$ until the peptide begins to precipitate. At this point, add 10 volumes of ice cold diethyl ether.

4. Isolate the precipitated peptide by centrifugation (4000 × *g*, 10 min, 4°C). Pour off the diethyl ether and allow the precipitated peptide to dry overnight in a chemical fume hood.
5. Confirm the identity of the peptide by HPLC-MS and HR-MS^n.

4.2 Purification of Substrate Peptides

Purification of the crude peptide should be carried out by preparative HPLC. Depending on the peptide sequence, especially considering charge and hydrophobicity, the optimal column, solvent, and gradient will vary. These parameters need to be determined empirically. As a first pass, we typically utilize a Phenomenex Luna C18 column with H_2O/MeCN (+0.1% FA) as solvent and a gradient of 5%–95% MeCN over 40 min. Depending on the elution profile, the parameters can then be further optimized. An HPLC profile of pure SuiA is shown in Fig. 3.

Once pure substrate peptide and RaS enzyme have been obtained, the enzyme can be assessed for activity (see below). Note that certain RiPP gene clusters encode discrete RiPP recognition elements (RREs), proteins that are involved in substrate recognition (Burkhart, Hudson, Dunbar, & Mitchell, 2015). In StrB, AgaB, and SuiB, the RRE is encoded as a domain within the RaS enzyme, but that is not always the case. If a discrete RRE is encoded, it must also be cloned, expressed, and purified for use as a component in the assay. The RRE does not contain air-sensitive cofactors and can be purified as a tagged construct under aerobic conditions, though the ideal tag needs to be determined experimentally.

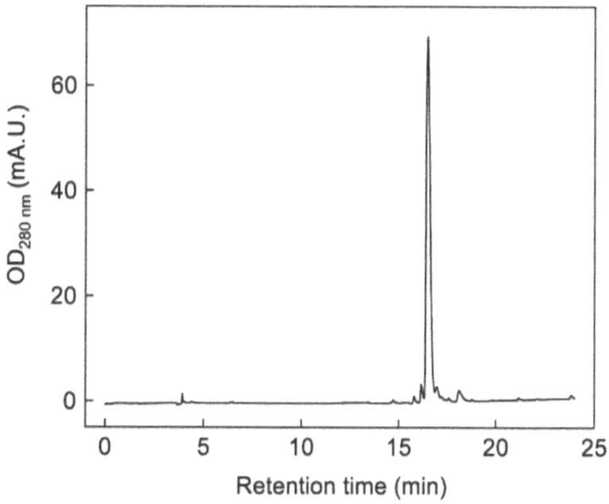

Fig. 3 HPLC analysis of purified SuiA. The main peak observed corresponds to SuiA (>90% pure). Minor impurity peaks are truncated SuiA peptides that likely form as a result of incomplete coupling or deprotection reactions during SPPS.

5. ENZYMATIC ACTIVITY ASSAYS

Enzymatic assays are performed in an inert atmosphere. The peptide substrate, DT, DTT, and SAM are typically transported into the glovebox as powders, where stock solutions can be prepared. We typically prepare stock solutions of DT and DTT in water (~0.1–0.5 *M* each), and stocks of SAM and peptide in reconstitution buffer, which we also use as the reaction buffer (see step 12, Section 3.2). The concentrations of stocks of SAM and peptide are verified spectrophotometrically after they have been prepared. The extinction coefficient for SAM is 15,400 M^{-1} cm^{-1} at 260 nm. Peptide extinction coefficients can be calculated using online tools such as ExPASy.

Since the product of the transformation is unknown and finding it is the objective of the assay, it is imperative to include several controls (-peptide, -RaS enzyme, and -SAM). Therefore, initial assays should consist of, at a minimum, four reactions (three controls and one full reaction). If desired, additional reactions containing various ratios of enzyme to substrate may be included as well. We recommend running the reactions on a 30-μL scale, which should provide ample material for HPLC-MS and MS/MS analysis. The first step is to identify the product (Section 5.1), followed by structural elucidation (Section 5.2).

5.1 Protocol for Enzymatic Assay and Analysis by HPLC-MS

1. Set up four reactions side by side in 0.5-mL or 1.5-mL Eppendorf tubes, consisting of (a) the full reaction as well as no-substrate (b), no-RaS enzyme (c), and no-SAM (d) controls. Start by clearly labeling each tube and adding the components in Table 1 in the order listed. SAM is added last (except in the no-SAM control) to initiate the

Table 1 Stock and Final Concentrations of Components Used Routinely in Enzymatic Assays

Component	Stock Concentration (m*M*)	Final Concentration (m*M*)
DTT	100	5
DT	100	5
Substrate	5	0.5–1
RaS enzyme	~0.1–0.3	0.01–0.03
SAM	30	1

reaction. Reconstitution buffer (same as assay buffer) can be used to adjust the final reaction volume to 30 μL.

2. Upon initiation with SAM, the reactions are incubated at room temperature inside the glovebox for 1–24 h. The optimal incubation time is of course enzyme dependent, but a range of incubation times can be tested initially by setting up reaction replicates.
3. At the end of the incubation period, transfer the reaction tubes out of the glovebox and immediately quench with an equal volume of 100 m*M* H_2SO_4. Invert each tube several times to mix.
4. Remove precipitated protein by centrifugation (16,000 × *g*, 2 min, RT).
5. Isolate the supernatant taking care not to transfer any precipitated components. Transfer the supernatant to labeled HPLC-MS vials for analysis.
6. Inject 10–15 μL of the supernatant directly onto an HPLC-MS. We have routinely used an Agilent HPLC-Qtof-MS (equipped with an automated sampler, a 1260 Infinity series HPLC, a photodiode detector, a JetStream ESI source, and the 6540 Qtof), or an Agilent single-quadrupole HPLC-MS (equipped with a liquid autosampler, 1260 Infinity series HPLC, a photodiode array detector, and a 6120 series ESI mass spectrometer). This analysis can also be carried out on similar instruments by different manufacturers (e.g., Thermo, Waters, AB-Sciex, etc.). Results of HPLC-MS analysis of the reaction of SuiB with SuiA are shown in Fig. 4.

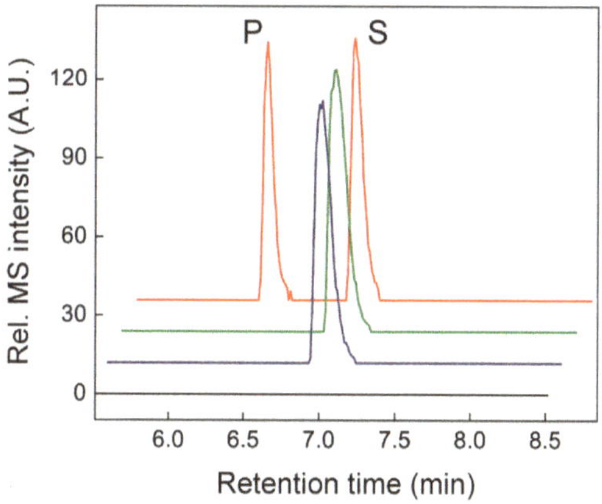

Fig. 4 SuiB enzymatic assay monitored by HPLC-Qtof-MS. Shown are extracted ion chromatograms for the substrate (S) and product (P) peptides in reactions containing all components (*red trace*), or lacking SuiA (*black*), SuiB (*blue*), and SAM (*green*). The traces are offset in both axes for clarity.

7. Because the samples are not desalted, a significant dead time must be programmed into the elution profile, that is, where the first ~5–10 min of the eluant is not injected onto the mass spectrometer (depending on flow rate and column size). If the peptide of interest is particularly polar or the detection of 5′-deoxyadenosine (5′-dA) by MS proves necessary, solid-phase extraction can be performed to desalt the samples.
8. Examine the traces for new peaks, which differ in m/z, retention time, and/or UV–Vis spectrum from the substrate. The candidate product peak should not be observed in the various controls and only be detectable in the full reaction (Fig. 4). It should also accumulate as a function of time. If a candidate product peak is identified, further analysis can be performed as outlined in step 12 and the ensuing sections.
9. If no product peak is identified, two general avenues should be explored: (a) lack of product detection due to polarity, precipitation, sensitivity, etc., and (b) lack of product formation. If (a) is the case, typically a decrease in the substrate peak will be observed without a concomitant appearance of product. Moreover, this scenario typically leads to formation 5′-dA, the production of which can be assessed independently by HPLC-MS. If the peptide product is not detected, different workup procedures should be explored, for example, a heat-quench method rather than acid-quench, use of HPLC resins with differential retention, solid-phase extraction prior to HPLC-MS analysis, or analysis by MALDI-MS, to name a few.
10. If product is not formed in scenario (b), the ability of the RaS enzyme to generate 5′-dA in the absence of substrate should be verified to ensure that the enzyme is functional in reductive cleavage of SAM. Futile cleavage of SAM in the absence of substrate is a characteristic feature of RaS enzymes. If the enzyme is functional, additional assays may be carried out by varying enzyme/substrate ratios, reaction times, reaction conditions (pH, temperature), and supplementation with other suspected and necessary components. If the product is still not formed, the reconstitution protocol should be optimized, by increasing/decreasing the equivalent of Fe and S^{2-}, and by monitoring the resulting clusters spectroscopically. Note that DT may not be a strong enough reductant for some RaS enzymes. Other reductants, such as Ti^{III}, photoreduced deazaflavin, or an endogenous flavodoxin, may be considered, followed by assessment of the Fe–S cluster(s) spectroscopically as well as formation of 5′-dA in the absence of substrate (Barr et al., 2016; Bruender, Young, & Bandarian, 2016; Henshaw,

Cheek, & Broderick, 2000; Walsby et al., 2002). Orthologous enzymes that are better behaved may be examined as well.

11. Once a candidate product has been identified, it should be examined further by HR-MS/MS. Parent ions of various charge states can be selected and a range of collision energies utilized to maximize fragmentation. The resulting fragmentation pattern should be analyzed in detail in comparison to that of the substrate. By compiling which b and y ions are observed in the substrate and product MS/MS profiles, the analysis should point to the exact residues that are modified by the RaS enzyme. This information is critical and informs downstream structure elucidation efforts by NMR. In the case of the SuiB product, the lack of fragmentation within the K-GDG-W region indicated that a Lys-Trp cross-link is installed (Fig. 5A).

5.2 Incorporation of Site-Specific NMR Probes

Structural elucidation of the modification installed, especially if it is new, will require analytical methods in addition to MS. The sheer length of the peptides in RiPP biosynthetic pathways complicates NMR analysis. By selectively labeling the residues that are modified, one can home in on the relevant signals. This "highlighting" or "amplification" approach enables more rapid structural elucidation, while at the same time reducing the amount of material required for NMR experiments involving ^{13}C or ^{15}N nuclei. ^{2}H-labeled amino acids can be equally useful in ^{2}H NMR applications and in simplification of spectral regions in cases where many overlapping signals cannot be unequivocally assigned. Isotopomers containing ^{13}C, ^{15}N, or ^{2}H nuclei are not always commercially available as Fmoc- and side chain-protected derivatives. And when they are, they may be prohibitively expensive. As a consequence, we often purchase labeled free amino acid isotopologs and install the necessary protecting groups. The reader is referred to detailed protocols for addition of N_{α}-Fmoc and side chain-protection groups (Behrendt et al., 2016; El-Faham & Albericio, 2011; Fields, 1997; Isidro-Llobet, Álvarez, & Albericio, 2009). These can then be used in SPPS, followed by purification, and verification of the site of incorporation by HR-MS and HR-MS/MS (see Section 4).

5.3 Protocol for Large-Scale Reactions and Analysis by NMR

Once the product has been identified and preliminarily characterized, NMR analysis will be necessary to fully assign its structure. With the appropriate

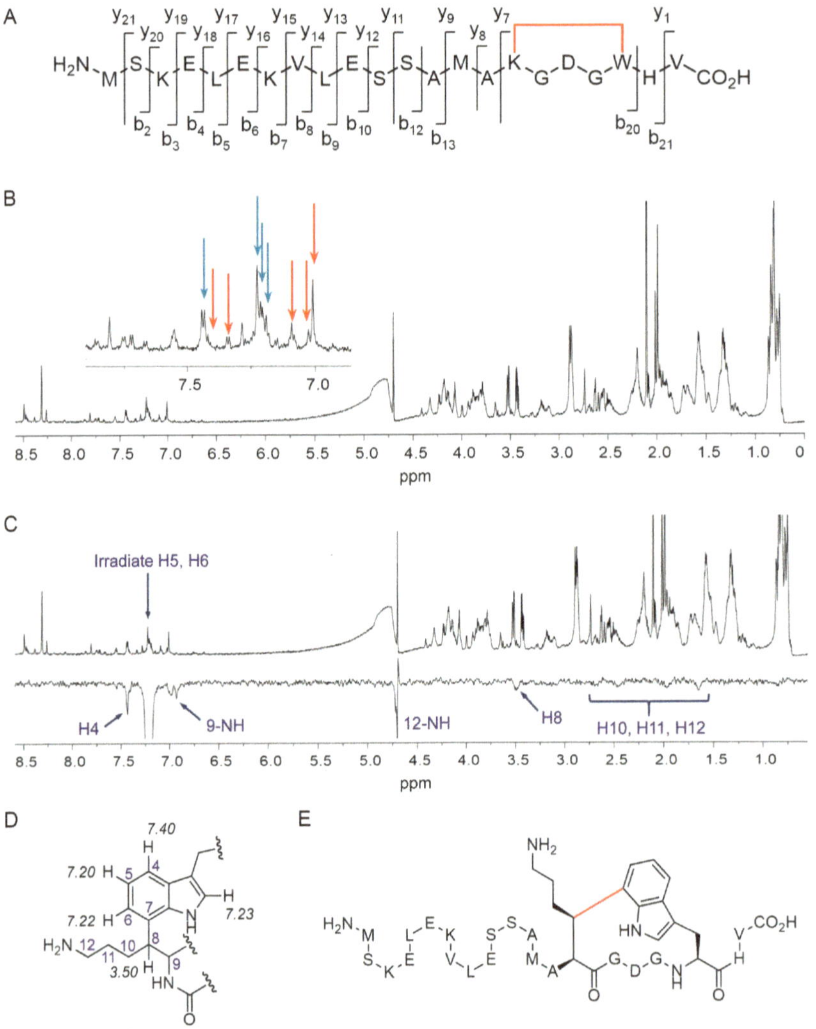

Fig. 5 Structural elucidation of the PTM installed by SuiB. (A) HR-MS/MS analysis of SuiB product. All b and y ions are observed except for those with the K-GDG-W motif, symptomatic for a Lys-Trp cross-link. (B) ^1H NMR analysis of purified SuiB product. The *inset* shows a magnified view of the downfield region of the spectrum. *Blue arrows* point to the modified Trp-^1Hs. *Red arrows* mark Trp-^1Hs of contaminating substrate. (C) 1D selective NOESY spectrum, in which the indole-H5/H6 signals of the product were irradiated (*top spectrum*). Observed signal peaks are shown (*bottom spectrum*). (D) Numbering scheme used to annotate peaks in (C). Some chemical shift values are shown (in ppm). (E) Structure of the SuiB product as elucidated from HR-MS/MS and 1D/2D NMR analysis.

building blocks in hand (Section 5.2), SPPS can be carried out as described in Section 4 on a ~50 μmol scale. Following purification by HPLC, the substrate containing the isotopomer can be used in a large-scale enzymatic reaction:

1. Maintaining the same concentrations as used in the small-scale activity assays, scale up the reaction 100–500-fold, depending on the amount of substrate/enzyme available. Typically, from a 50-μmol peptide synthesis, sufficient material should be obtained to set up a ~10-mL large-scale reaction, which can yield 5–15 mg of purified product.
2. Incubate the reaction under the optimal conditions determined in Section 5.1.
3. Depending on the isotope incorporated and conversion to product, the product may need to be purified from substrate. The advantage of ^{13}C-enriched amino acids is that product purification is not required. We do recommend it in order to simplify the NMR analysis.
4. Purification of product from substrate may call for multiple steps. We typically use preparative HPLC as a first step, followed by finer resolution on an analytical HPLC/column. For both steps, especially the latter, the appropriate resin and gradient need to be selected that maximize separation of product from substrate. It is best to develop a purification scheme with unlabeled material first.
5. Once pure product is obtained, it should be dried completely in vacuo.
6. The peptide is then dissolved in a minimal amount of solvent for maximal concentration. Two key aspects need to be considered: (a) choice of solvent, and (b) concentration. The choice of solvent depends on solubility and information to be gleaned. DMSO-d_6 and H_2O/D_2O (9:1) are good choices as superior peptide solvents and because they allow exchangeable amide-NHs to be observed. If exchangeable ^{1}Hs are not involved in the modification and amide-NHs do not need to be identified, D_2O should be used as it will significantly simplify 2D NMR spectral analysis. As for concentration, we usually use 5/2.5 mm NMR tubes, which only require ~100 μL sample. Capillary tubes, which hold as little as ~30 μL volumes, may be used as well, depending on the solubility of the peptide.
7. Typically, the following spectra are recorded and analyzed on both the substrate and product peptide: ^{1}H, ^{13}C, COSY, TOCSY, HSQC, HMBC, and NOESY. ^{13}C COSY spectra, or the INADEQUATE experiment, are useful as well with ^{13}C-enriched samples. 1D selective NOESY or HMBC experiments, specifically irradiating certain protons

in the product, are also useful in delineating correlations emanating from a selected position (Fig. 5B; Schramma & Seyedsayamdost, 2017).

8. These spectra are then analyzed in detail to map the structure of the modification. Many good references are available for solving structures of organic compounds using 1D/2D NMR analysis (Crews, Rodriguez, & Jaspars, 2009).
9. The proposed structure, based on MS and NMR analysis, can be further verified using other synthetic probes or modifications to the substrate, followed by reaction with the RaS enzyme. Ultimately, by combining insights gleaned from HR-MS^n and 1D/2D NMR analysis of both natural abundance and isotopomeric substrates, as well as analysis by other spectroscopic methods where appropriate (such as UV–Vis or IR spectroscopy), the structure of the product can be inferred. HR-MS/MS and some NMR analysis of the product of SuiB are shown in Fig. 5.

6. CONCLUSIONS

Focusing on a subclass of RaS enzymes involved in the biosynthesis of RiPP natural products, we have enumerated detailed procedures for expressing, purifying, and reconstituting RaS enzymes in vitro, synthesizing the peptide substrate, and characterizing the structure of the ensuing product. Due to length and space considerations, some aspects in this process were not discussed in great depth and readers are referred to other texts and the primary literature for other salient features. Spotting RaS enzymes in microbial and macrobial genomes is common place. But with less than 0.1% of the bioinformatically predicted enzymes characterized to date, RaS enzymes represent a vast frontier of unexplored chemistries. Targeting those that act on peptide substrates in RiPP biosynthetic pathways will likely reveal a host of novel structural motifs. The approaches described above will aid in the discovery of new RaS enzyme-mediated transformations and their underlying mechanisms. These in turn will inform the discovery of new, diverse RiPP natural products.

ACKNOWLEDGMENTS

We thank the Pew Biomedical Scholars Program and the Burroughs Wellcome Fund PATH Investigator Award (to M.R.S.) for financial support, as well as Dr. Kelsey R. Schramma, Alessio Caruso, and Kenzie A. Clark for their contributions toward the methods described. L.B.B. was supported by a National Science Foundation Graduate Research Fellowship.

REFERENCES

Akiva, E., Brown, S., Almonacid, D. E., Barber, A. E., Custer, A. F., Hicks, M. A., et al. (2014). The structure-function linkage database. *Nucleic Acids Research*, *42*, D521–D530.

Arnison, P. G., Bibb, M. J., Bierbaum, G., Bowers, A. A., Bugni, T. S., Bulaj, G., et al. (2013). Ribosomally synthesized and post-translationally modified peptide natural products: Overview and recommendations for a universal nomenclature. *Natural Product Reports*, *30*, 108–160.

Barr, I., Latham, J. A., lavarone, A. T., Chantarojsiri, T., Hwang, J. D., & Klinman, J. P. (2016). Demonstration that the radical S-adenosylmethionine (SAM) enzyme PqqE catalyzes de novo carbon-carbon cross-linking within a peptide substrate PqqA in the presence of the peptide chaperone PqqD. *The Journal of Biological Chemistry*, *291*, 8877–8884.

Behrendt, R., White, P., & Offer, J. (2016). Advances in Fmoc solid-phase peptide synthesis. *Journal of Peptide Science*, *22*, 4–27.

Beinert, H. (1983). Semi-micro methods for analysis of labile sulfide and of labile sulfide plus sulfane sulfur in unusually stable iron-sulfur proteins. *Analytical Biochemistry*, *131*, 373–378.

Broderick, J. B., Duffus, B. R., Duschene, K. S., & Shepard, E. M. (2014). Radical S-adenosylmethionine enzymes. *Chemical Reviews*, *114*, 4229–4317.

Bruender, N. A., Young, A. P., & Bandarian, V. (2016). Chemical and biological reduction of the radical SAM enzyme 7-carboxy-7-deazaguanine synthase. *Biochemistry*, *55*, 1939.

Burkhart, B. J., Hudson, G. A., Dunbar, K. L., & Mitchell, D. A. (2015). A prevalent peptide-binding domain guides ribosomal natural product biosynthesis. *Nature Chemical Biology*, *11*, 564–570.

Crews, P., Rodriguez, J., & Jaspars, M. (2009). *Organic structure analysis*. Oxford University Press.

Cutsail, G. E., Telser, J., & Hoffman, B. M. (2015). Advanced paramagnetic resonance spectroscopies of iron–sulfur proteins: Electron nuclear double resonance (ENDOR) and electron spin echo envelope modulation (ESEEM). *Biochimica et Biophysica Acta*, *1853*, 1370–1394.

El-Faham, A., & Albericio, F. (2011). Peptide coupling reagents, more than a letter soup. *Chemical Reviews*, *111*, 6557–6602.

Fields, G. B. (1997). *Solid-phase peptide synthesis*. San Diego: Academic Press.

Fish, W. W. (1988). Rapid colorimetric micromethod for the quantitation of complexed iron in biological samples. *Methods in Enzymology*, *158*, 357–364.

Frey, P. A. (2001). Radical mechanisms of enzymatic catalysis. *Annual Review of Biochemistry*, *70*, 121–148.

Henshaw, T. F., Cheek, J., & Broderick, J. B. (2000). The $[4Fe\text{-}4S]^{1+}$ cluster of pyruvate formate-lyase activating enzyme generates the glycyl radical on pyruvate formate-lyase: EPR-detected single turnover. *Journal of the American Chemical Society*, *122*, 8331–8332.

Ibrahim, M., Guillot, A., Wessner, F., Algaron, F., Besset, C., Courtin, P., et al. (2007). Control of the transcription of a short gene encoding a cyclic peptide in *Streptococcus thermophilus*: A new quorum-sensing system? *Journal of Bacteriology*, *189*, 8844–8854.

Isidro-Llobet, A., Álvarez, M., & Albericio, F. (2009). Amino acid-protecting groups. *Chemical Reviews*, *109*, 2455–2504.

Johnson, D. C., Unciuleac, M.-C., & Dean, D. R. (2006). Controlled expression and functional analysis of iron-sulfur cluster biosynthetic components within *Azotobacter vinelandii*. *Journal of Bacteriology*, *188*, 7551–7561.

Landgraf, B. J., McCarthy, E. L., & Booker, S. J. (2016). Radical S-adenosylmethionine enzymes in human health and disease. *Annual Review of Biochemistry*, *85*, 485–514.

Lanz, N. D., Grove, T. L., Gogonea, C. B., Lee, K.-H., Krebs, C., & Booker, S. J. (2012). RlmN and AtsB as models for the overproduction and characterization of radical SAM proteins. *Methods in Enzymology*, *516*, 125–152.

McIntosh, J. A., Donia, M. S., & Schmidt, E. W. (2009). Ribosomal peptide natural products: Bridging the ribosomal and nonribosomal worlds. *Natural Product Reports*, *26*, 537–559.

Morinaka, B. I., Lakis, E., Verest, M., Helf, M. J., Scalvenzi, T., Vagstad, A. L., et al. (2018). Natural noncanonical protein splicing yields products with diverse β-amino acid residues. *Science*, *359*, 779–782.

Oman, T. J., & van der Donk, W. A. (2010). Follow the leader: The use of leader peptides to guide natural product biosynthesis. *Nature Chemical Biology*, *6*, 9–18.

Pandelia, M.-E., Lanz, N. D., Booker, S. J., & Krebs, C. (2015). Mössbauer spectroscopy of Fe/S proteins. *Biochimica et Biophysica Acta*, *1853*, 1395–1405.

Pedersen, S. L., Tofteng, A. P., Malik, L., & Jensen, K. J. (2012). Microwave heating in solid-phase peptide synthesis. *Chemical Society Reviews*, *41*, 1826–1844.

Pennington, M. W., & Dunn, B. M. (1994). *Peptide synthesis protocols*. Totowa, NJ: Humana Press.

Schramma, K. R., Bushin, L. B., & Seyedsayamdost, M. R. (2015). Structure and biosynthesis of a macrocyclic peptide containing an unprecedented lysine-to-tryptophan crosslink. *Nature Chemistry*, *7*, 431–437.

Schramma, K. R., Forneris, C. F., Caruso, A., & Seyedsayamdost, M. R. (2018). Mechanistic investigations of lysine-tryptophan crosslink formation catalyzed by streptococcal radical SAM enzymes. *Biochemistry*, *57*, 461–468.

Schramma, K. R., & Seyedsayamdost, M. R. (2017). Lysine-tryptophan-crosslinked peptides produced by radical SAM enzymes in pathogenic streptococci. *ACS Chemical Biology*, *12*, 922–927.

Shi, Y., Yang, X., Garg, N., & van der Donk, W. A. (2011). Production of lantipeptides in *Escherichia coli*. *Journal of the American Chemical Society*, *133*, 2338–2341.

Vanier, G. S. (2013). Microwave-assisted solid-phase peptide synthesis based on the fmoc protecting group strategy (CEM). *Methods in Molecular Biology*, *1047*, 235–249.

Walsby, C. J., Hong, W., Broderick, W. E., Cheek, J., Ortillo, D., Broderick, J. B., et al. (2002). Electron-nuclear double resonance spectroscopic evidence that S-adenosylmethionine binds in contact with the catalytically active $[4Fe-4S]^+$ cluster of pyruvate formate-lyase activating enzyme. *Journal of the American Chemical Society*, *124*, 3143–3151.

Young, A. P., & Bandarian, V. (2013). Radical mediated ring formation in the biosynthesis of the hypermodified tRNA base wybutosine. *Current Opinion in Chemical Biology*, *17*, 613–618.

CHAPTER SEVENTEEN

Identification of the Radical SAM Enzymes Involved in the Biosynthesis of Methanopterin and Coenzyme F_{420} in Methanogens

Kylie D. Allen[*,2], **Robert H. White**[†,1]
[*]Department of Chemistry and Biochemistry, Gonzaga University, Spokane, WA, United States
[†]Virginia Polytechnic Institute and State University, Blacksburg, VA, United States
[1]Corresponding author: e-mail address: rhwhite@vt.edu

Contents

[2] Current address: Department of Biochemistry, Virginia Polytechnic Institute and State University, Blacksburg, VA, United States.

Methods in Enzymology, Volume 606
ISSN 0076-6879
https://doi.org/10.1016/bs.mie.2018.04.015

Abstract

Methanogenic archaea represent a source of unique and fascinating anaerobic biochemistry that includes the involvement of many radical *S*-adenosyl-L-methionine (SAM) enzymes, some of which have well-established functions, while the majority have currently unknown or only partially understood functions. Here, we describe our strategy for the identification of the radical SAM enzyme that catalyzes the two methylation reactions in methanopterin biosynthesis in *Methanocaldococcus jannaschii*. Additionally, we describe the similar strategy carried out for the identification of the two radical SAM enzymes required for the biosynthesis of the 7,8-didemethyl-8-hydroxy-5-deazariboflavin (F_0) moiety of coenzyme F_{420} in *M. jannaschii*. This approach can be employed for future functional identification of radical SAM enzymes with currently unknown functions.

1. INTRODUCTION

1.1 Radical SAM Enzymes in Coenzyme Biosynthesis in Methanogens

Radical *S*-adenosyl-L-methionine (SAM) enzymes utilize a [4Fe–4S] cluster and SAM to catalyze an impressive variety of complicated radical reactions in diverse biochemical pathways across all domains of life. Many of these enzymes are involved in the biosynthesis of complex secondary natural products, while others are required for the biosynthesis of coenzymes required for primary metabolism. Radical SAM enzymes are involved in the biosynthesis of several widely encountered coenzymes including biotin, lipoic acid, thiamin, heme, and molybdopterin (Jarrett, 2015; Landgraf, McCarthy, & Booker, 2016; Mehta et al., 2015), as well as some less common, but nonetheless equally as important, coenzymes such as coenzyme F_{420} and methanopterin (MPT), which will be discussed here. We propose that the involvement of radical SAM enzymes in coenzyme biosynthesis relates to the early evolution of coenzyme biosynthesis, which occurred in the anaerobic environment of ancient earth where radical SAM enzymes thrived.

Methanogenic archaea are strictly anaerobic microorganisms with a unique energy metabolism dependent on the formation of methane via the reduction of various substrates such as CO_2, small methylated compounds, and acetate (Costa & Leigh, 2014). In addition to utilizing common coenzymes found in other organisms, methanogens also contain several unique coenzymes required for methanogenesis. Interestingly, the percentage of putative radical SAM enzymes per proteome is much higher in methanogens (1.5%) than in anaerobic bacteria (0.8%), aerobic bacteria (0.44%), and eukaryotes such as yeast (0.05%). Most radical SAM enzymes encoded in the genomes of methanogens are hypothetical proteins with currently unknown functions, while the majority of methanogenic radical SAM enzymes that do have known or expected functions are involved in coenzyme biosynthesis. This chapter discusses the strategy used for the functional identification of the radical SAM methylase required for the addition of two methyl groups to a pterin substrate in MPT biosynthesis. It also covers the identification of F_0 synthase, composed of two radical SAM enzymes in methanogens that catalyze the formation of a complex heterocycle to produce the catalytic portion of coenzyme F_{420}.

1.2 Strategy for Determining Functions of Methanogenic Proteins

The overall goal of our research is to establish the functions of hypothetical enzymes present in methanogenic archaea. We have especially focused on the hyperthermophilic methanogen, *Methanocaldococcus jannaschii*, originally isolated from a white smoker deep-sea hydrothermal vent (Jones, Leigh, Mayer, Woese, & Wolfe, 1983), a harsh marine environment that mimics the conditions of early earth (Nisbet & Sleep, 2001). The *M. jannaschii* genome can be mapped to the deepest phylogenetic branches of life that include sulfur-chemolithoautotrophic (hyper)thermophiles (Liu, Beer, & Whitman, 2012), suggesting that *M. jannaschii* can be considered a model organism for the study of primitive metabolism. When the *M. jannaschii* genome was sequenced, only about one-third of the encoded proteins were initially assigned a role with high confidence (Bult et al., 1996). Some of these proteins, such as the numerous [4Fe–4S] cluster-containing enzymes, are clearly related to the prebiotic chemistry of hydrothermal vents and the origin of life (Braakman, 2013; Cody et al., 2000; Huber & Wachtershauser, 1997; Novikov & Copley, 2013).

Our general strategy is to analyze the genomes of methanogens to identify possible genes that may be involved in a given biosynthetic pathway, express and purify the encoded protein, and finally test it for the hypothesized catalytic activity. In the case of radical SAM enzymes from methanogens, two primary experimental roadblocks impede this strategy. The first is that the putative substrates to be tested are in many cases not readily available and, second, that radical SAM enzymes can often be difficult to isolate in an active form especially due to their high intolerance to oxygen (Lanz et al., 2012). Here, we show that these challenges can be minimized by taking advantage of *Escherichia coli* for both recombinant protein production and potential substrate production. As long as *E. coli* contains similar metabolites to the host organism for the radical SAM enzyme of interest, these enzymes can function in vivo to produce the desired metabolite or its analog. This final metabolite, resulting from the radical SAM enzyme reaction of interest, can then be extracted from the cells and identified to demonstrate the enzyme's function. Thus, we only need to develop analytical methods to establish if the expected reaction product was produced in the cells expressing the radical SAM enzyme of interest. This strategy removes the need for complicated and time-consuming protein purification and substrate synthesis steps early on, which can be carried out later once firm evidence for the enzyme's function has been established in *E. coli*.

2. OVEREXPRESSION OF METHANOGENIC PROTEINS IN *E. COLI*

2.1 Cloning

Fairly routine cloning procedures are carried out to overexpress the radical SAM enzyme of interest in *E. coli*. We generally clone into pT7-7 (Tabor & Richardson, 1985) for expression of the native, nontagged enzyme, or into pET-19b (Novagen) for expression of the fusion enzyme with an N-terminal histidine (10X His) tag, which can be removed after purification if necessary by treatment with enterokinase. Each of these plasmids contains a canonical T7 RNA polymerase promoter sequence and the lac operator upstream of the inserted DNA encoding for the protein of interest. DH5α *E. coli* cells are utilized for transformation of the ligated plasmid and storage of the transformed strain. For overexpression of methanogenic proteins, BL21-CodonPlus (DE3)-RIL cells (Agilent Technologies) are usually used in our laboratory, but traditional BL21 *E. coli* cells also often support successful overexpression of methanogenic proteins.

2.2 Growth Medium and General Expression Methods

For routine expression, we use Luria-Bertani (LB, 200 mL) growth medium supplemented with 100 μg/mL ampicillin in a 1-L Erlenmeyer flask inoculated with an overnight culture (2 mL) of expression cells containing the plasmid with the cloned gene of interest. The culture is incubated at 37°C with shaking at ~300 rpm until it reaches an OD_{600} of ~0.8. Protein expression is induced using either lactose (28 m*M*) or isopropyl β-D-1-thiogalactopyranoside (IPTG, 0.5 m*M*) and the culture is grown at 37°C with shaking for an additional 3–4 h before harvesting and storing the resulting cell pellet at −20°C.

In many cases, it is useful to add stable isotopically labeled compounds to the *E. coli* growth medium to determine if these compounds are precursors to the metabolite of interest. Since our products are identified by mass spectrometry, this procedure allows one to establish the extent of label incorporation based on the increases in the intensities of the isotope-containing ions. To maximize the incorporation of these potential precursors and to make the analysis of the incorporation less complicated, we often perform the overexpression in M9 minimal medium, usually with glucose (2%) instead of LB medium. When M9 medium is used, the cells are grown for an additional 18 h after induction.

To analyze and confirm expression of the protein of interest, an induction study is performed where the whole cells at various time points after induction are analyzed on an SDS-PAGE gel. A small piece of the gel band corresponding to the expected molecular weight of the overexpressed protein of interest is cut out, digested with trypsin, and analyzed by matrix-assisted laser desorption/ionization (MALDI) mass spectrometry to confirm the identity of the protein.

3. IDENTIFICATION OF THE RADICAL SAM ENZYME INVOLVED IN THE BIOSYNTHESIS OF METHANOPTERIN

3.1 Introduction to Methanopterin

MPT (Fig. 1A) is a coenzyme found in methanogens that is functionally and structurally similar to folate (Fig. 1B; Escalante-Semerena, Rinehart, & Wolfe, 1984). Like folate, MPT is biologically active in its 5,6,7,8-tetrahydro (H_4) form (Fig. 1D) and functions to carry and transfer C_1 groups, bound via N^5 or N^{10} or both, of varying oxidation states including formyl, methylene, and methyl groups. Methanogens use MPT in place of folate for canonical folate-dependent reactions and use MPT in methanogenesis. Some methanogens contain sarcinopterin instead of MPT, which has a glutamate residue linked to the hydroxyglutarate

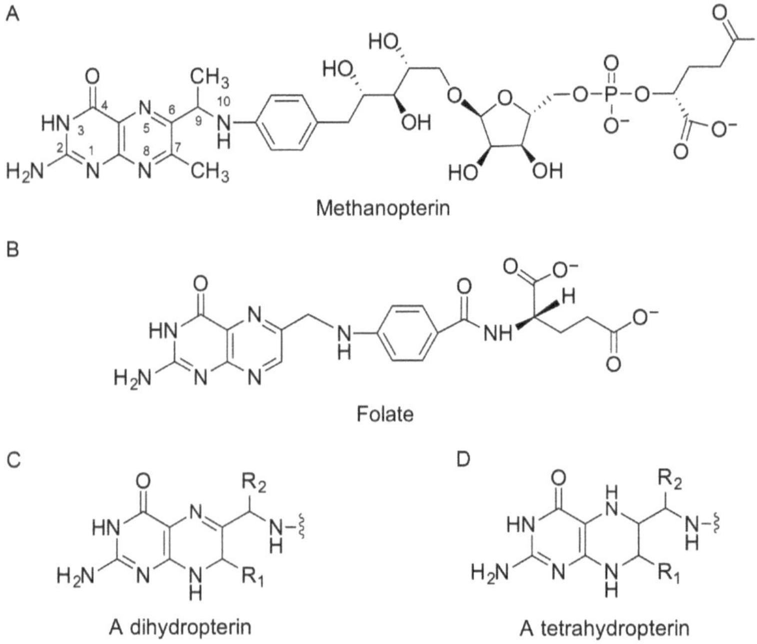

Fig. 1 Structures of methanopterin (A) compared to folate (B) in their fully oxidized forms. During the biosynthesis of both methanopterin and folate, the pterin exists in the dihydro form (C), while the final active coenzyme is in the tetrahydro form (D).

residue in its side chain (van Beelen et al., 1984). Additionally, methylotrophic bacteria contain a truncated form of MPT called dephospho-MPT, where the phosphate and α-hydroxyglutarate in the side chain are absent (Chistoserdova, Vorholt, Thauer, & Lidstrom, 1998).

One major difference in the structure of MPT compared to folate is the electron-donating methylene group conjugated to N^{10} through the aromatic ring; folate has an electron-withdrawing carbonyl group in this position. The other major difference is the presence of methyl groups at the C-7 and C-9 positions of the pterin ring of MPT (Fig. 1A), which distinguish MPT from all other pterin-containing natural products. Originally, it was presumed that these methyl groups were added via traditional SAM-dependent nucleophilic substitution chemistry. However, a potential canonical SAM-dependent methyltransferase associated with MPT biosynthesis in the genomes of methanogens or methylotrophs was never identified. Given the increasing examples of radical SAM enzymes catalyzing methylation reactions (Bauerle, Schwalm, & Booker, 2015; Zhang, van der Donk, & Liu, 2012), we expanded our genomic analysis to consider the possibility that the enzyme catalyzing the methylation reaction in MPT biosynthesis could be a radical SAM enzyme. This revealed a radical SAM-encoding gene in some methanogens found near the gene encoding β-ribofuranosylaminobenzene 5′-phosphate synthase, a characterized enzyme involved in MPT biosynthesis (Rasche & White, 1998; Scott & Rasche, 2002).

We tested our proposal that the identified radical SAM enzyme from *M. jannaschii*, MJ0619, catalyzed the methylation reaction(s) in MPT biosynthesis by taking advantage of the fact that *E. coli* naturally contains folate, which is structurally similar to MPT. We, therefore, reasoned that MJ0619 could potentially methylate folate and/or a folate precursor when expressed in *E. coli*. Our results are consistent with MJ0619 catalyzing both methylation reactions in MPT biosynthesis (Allen, Xu, & White, 2014) and our detailed methods leading to this conclusion are described later.

3.2 Protein Expression

Traditional cloning techniques were used to clone MJ0619 (Swiss-Prot Accession No. Q58036) into pET-19b. The enzyme is overexpressed using the routine procedures described earlier, except that the medium (either LB or M9) is supplemented with 0.4 m*M* ferrous ammonium sulfate. Under our standard expression conditions, MJ0169 expresses well in *E. coli* to produce an intense band on SDS-PAGE gels corresponding to its expected molecular weight of 57,351 kDa. When the enzyme is extracted from the cells via sonication in an anaerobic chamber, it is largely insoluble, but roughly 10% is recovered in the soluble fraction. We have purified His-tagged MJ0619 in small scales using a metal affinity resin under anaerobic conditions, but will not discuss that here.

When isotopically labeled compounds are added to M9 medium, we have used L-[methyl-$^{2}H_3$]methionine at a concentration of 5 m*M* or [methyl-$^{2}H_3$] acetate at a concentration of 0.3% as a carbon/energy source instead of glucose. These are added as filter-sterilized solutions after the M9 medium has been autoclaved.

3.3 Isolation and Partial Purification of Oxidized Pterins From *E. coli* Cells

Since H_4folate is the biologically active form of folate, most of the folate in the cells will be in this fully reduced form. When isolated under typical aerobic conditions, a large percentage of H_4folate spontaneously undergoes

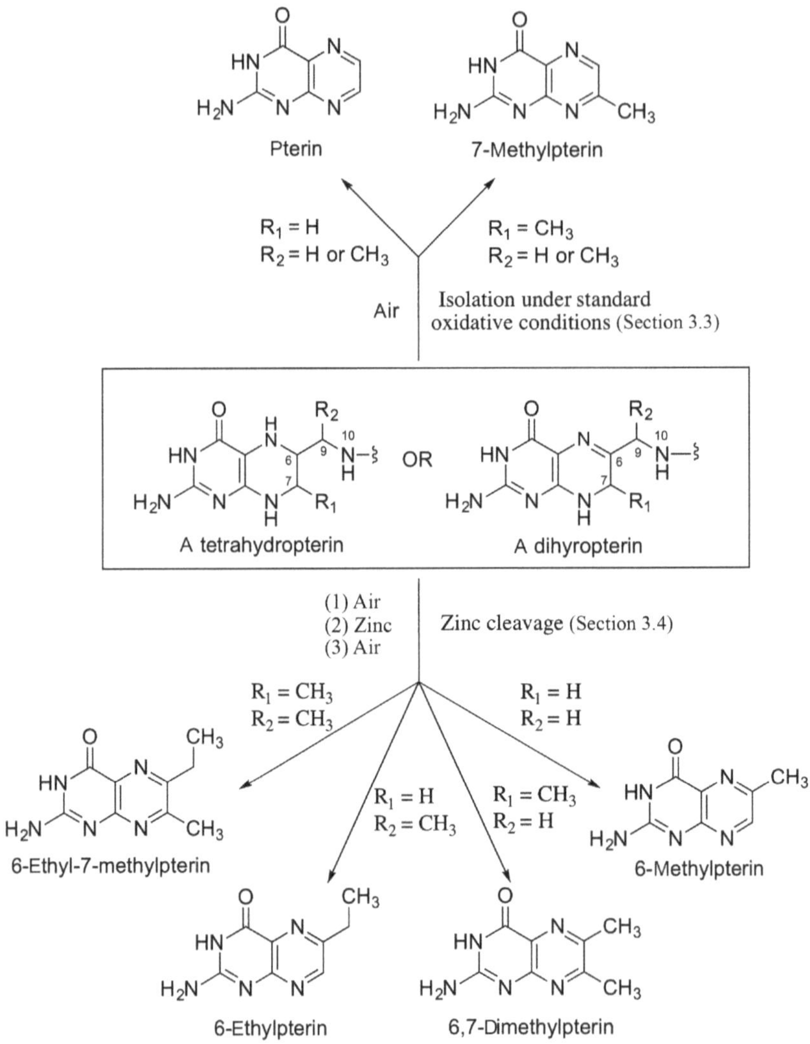

Fig. 2 Degradation products of H_4pterin-containing and H_2pterin-containing molecules after exposure to air and/or Zn/HCl reduction.

oxidative cleavage to produce pterin as one of the major products (Fig. 2; Reed & Archer, 1980). Similarly, 7-methylpterin is produced as a result of oxidative degradation of H_4MPT (Fig. 2; White, 1985). During the biosynthesis of folate and MPT, the pterin exists in a partially reduced form as 7,8-dihydropterin. The oxidative degradation of H_2pterin-containing compounds is more complicated and depends on many variables, but the oxidized pterins are also a common product similar to what is observed during H_4pterin oxidation (Fig. 2; Dantola, Vignoni, Capparelli, Lorente, & Thomas, 2008; Vignoni et al., 2010). Only the analysis of the oxidized forms of pterins will be described here.

E. coli cell pellets (typically around 0.2–0.4 g wet weight) are resuspended in 1 mL 50% methanol in a glass sample vial with a fitted cap. This suspension is incubated at 100°C for 10 min and then centrifuged (14,000 × *g*, 5 min) to pellet insoluble cell debris. The resulting soluble extract is concentrated to ~200 μL by evaporation with a stream of nitrogen gas and is then applied to a Dowex 50W-8X-H^+ column (1 × 5 mm). The column is washed with water (300 μL) and the pterins are eluted with 6 *M* aqueous ammonia (300 μL). The pterin-containing fraction is evaporated to dryness with a stream of nitrogen gas and then resuspended in 50 μL of 50% methanol for subsequent purification by preparative thin-layer chromatography (TLC) (described in Section 3.5).

3.4 Reductive Cleavage of the Side Chains of Pterin-Containing Molecules

The above procedure for isolation of the pterin portion of pterin-containing molecules, resulting from spontaneous oxidative cleavage of the entire side chain, allows one to determine whether a methyl group is present at the C-7 position of the pterin ring. In order to observe whether a methyl group is at the C-9 position, a zinc cleavage must be performed. This procedure reductively cleaves the bond between the 9 and 10 positions and the oxidized pterin ring results after further purification and manipulation under aerobic conditions (lower portion of Fig. 2). If both the 7 and 9 positions are methylated, as is analogous to MPT, 6-ethyl-7-methylpterin is observed after zinc reduction. If neither position is methylated, as is the case for folate, then 6-methylpterin is observed after zinc reduction. If only the C-7 position were methylated, then 6,7-dimethylpterin would be observed after zinc reduction.

E. coli cells are extracted with 50% methanol as described earlier. After centrifugation, the methanol is evaporated from the soluble extract and the resulting aqueous solution is made 1 *M* in HCl in a total volume of 1 mL. Zinc dust (~10 mg) is then added and the solution is incubated at room temperature with consistent agitation/shaking for at least 10 min (Foo, Cichowicz, & Shane, 1980). The excess zinc is then pelleted by centrifugation (14,000 × *g*, 5 min) and the resulting pterins are purified on Dowex resin as described earlier.

3.5 Purification of Pterins by TLC

The desired pterins, either from the direct oxidative isolation (Section 3.3) or from the reductive cleavage procedure (Section 3.4), can be further purified by TLC if desired. We use silica gel 60 TLC plates (E. Merck, Darmstadt, Germany) and the pterins are observed as fluorescent spots on the plates. The solvent system that we generally use for this is acetonitrile, water, and 88% formic acid (8:2:1). Under these chromatographic conditions, the R_f values for some select pterins are: 7-methylpterin, 0.56; pterin, ~0.3; 6-ethyl-7-methylpterin, 0.75; 6-hydroxymethylpterin, 0.36; 6-hydroxyethyl-7-methylpterin, 0.69. The area of the plate containing the pterin of interest (based upon known R_f of a standard) is removed from the plate and the silica containing the pterin of interest is placed into a pipet tip column with a cotton plug at the bottom. The pterin is then eluted with the TLC solvent into a clean vial. This solvent is evaporated off and the resulting purified pterin is resuspended in 80 μL of water for high-pressure liquid chromatography (HPLC) and liquid chromatography–mass spectrometry (LC–MS) analysis.

3.6 Isolation and Partial Purification of Folates From *E. coli* Cells

Intact folates from the cells must be analyzed to gain insight into the possible substrate for methylation when MJ0619 is expressed in *E. coli*. For isolation of folates, the *E. coli* cell pellet is extracted with 50% methanol as described earlier. After centrifugation, the methanol is removed from the soluble extract by evaporation. Folates in *E. coli* contain several γ-linked glutamate residues that must be removed to make the subsequent analysis more straightforward. Approximately 500 μL of the aqueous cell extract is added to 200 μL of Sprague Dawley rat plasma (Innovative Research, Inc.). The rat plasma contains γ-glutamyl hydrolase, which will deglutamylate folates to the monoglutamate level (Pribat et al., 2010). The cell extract and rat plasma mixture is incubated overnight at 37°C and then centrifuged (14,000 rpm, 5 min) to remove any precipitate. To partially purify folates from the cells, the supernatant is applied to a DEAE-Sephadex A-25 column (2 × 10 mm) equilibrated with water. The column is washed with water (1 mL) and 0.5 *M* ammonium bicarbonate (1 mL), followed by elution with 2 *M* ammonium bicarbonate (1 mL) to give the folate-containing fraction. This is evaporated with a stream of nitrogen gas to remove the ammonium bicarbonate and dissolved in a total volume of 80 μL water for LC–MS analysis.

3.7 Analysis of Pterins by HPLC With Fluorescence Detection

Oxidized pterins are highly fluorescent, which is very useful for their specific detection in complex mixtures at low concentrations. We use a Shimadzu HPLC system equipped with a photodiode array detector and a fluorescence detector with a Pursuit XRs 5 C18 column (Agilent, 250 × 4.6 mm, 2.6 μm particle size). The standard elution profile consists of 5 min at 95% A (25 m*M* sodium acetate pH 6.0) and 5% B (methanol), followed by a linear gradient from 5% to 50% buffer B over 25 min at 1 mL/min. The pterins are detected by excitation at 356 nm and emission at 450 nm. Under these LC conditions, pterin elutes at 10.3 min and 7-methylpterin elutes at 15.1 min.

3.8 Analysis of Pterins and Folates by LC–MS

LC–MS is utilized for confirmation of pterin identity and for identification of folate and, potentially, methylated folates. We use an AB Sciex 3200 QTRAP mass spectrometer system attached to an Agilent 1200 Series liquid chromatograph with a Zorbax Eclipse XDB-C18 column (Agilent, 4.6 × 50 mm, 1.8 μm particle size). For pterins, the elution profile consists of a 10-min linear gradient from 95% solvent A (25 m*M* ammonium acetate) and 5% solvent B (methanol) to 35% solvent A and 65% solvent B at 0.5 mL/min. For folate and methylated folates, the elution profile consists of a 10-min linear gradient from 95% solvent A (0.1% formic acid in water) and 5% solvent B (0.1% formic acid in methanol) to 90% solvent A and 10% solvent B at 0.5 mL/min. MS data are acquired in positive mode for pterins and negative mode for folate. For pterins, electrospray ionization (ESI) is set at 4500 V and a temperature of 400°C. For folate and the methylated folates, ESI is set at −4500 V and 500°C. The curtain gas is 35 au, GS1 is 60 au, and GS2 is 50 au. Using various chemically synthesized standards, we have developed sensitive multiple reaction monitoring (MRM) methods for detection of specific pterins and folates (Table 1). Analyst software (Applied Biosystems/MDS SCIEX) is used for system operation and data processing.

3.9 Analysis of Amino Acids From *E. coli*

To compare and interpret the incorporation of isotopically labeled precursors such as L-[methyl-$^2H_3$]methionine or [methyl-$^2H_3$]acetate into the pterins of interest, it is necessary to analyze the amino acids in the cells. The *E. coli* cells are extracted with 50% methanol as described earlier. The supernatant from

Table 1 Parameters for Multiple Reaction Monitoring (MRM) LC-ESI-MS Detection of Pterins and Folates in *E. coli* Cell Extracts

Compound	Precursor ion (*m/z*)	Product ion (*m/z*)	Dwell Time (ms)	DP (V)	CE (V)
7-Methylpterin	178.1	133.1	100	20	33
	178.1	106.1	100	20	33
	178.1	92.1	75	20	33
6-Ethyl-7-methylpterin	206.1	179.1	100	20	30
	206.1	163.1	100	20	30
6-Hydroxyethyl-7-methylpterin	222.1	204.2	100	20	30
	222.1	161.1	100	20	30
	222.1	132.1	75	20	30
Folate	440.3	311.1	100	−55	−30
Folate + 14[a]	454.3	325.3	100	−55	−30
Folate + 28[b]	468.3	339.3	100	−55	−30

[a]This corresponds to folate with one methyl group, either at C-7 or at C-9.
[b]This corresponds to folate with two methyl groups or a single formyl group. The product ions cannot distinguish between the two possible compounds.
Pterins are observed best in positive mode and ions reported are $(M+H)^+$ ions. Folates are observed best in negative mode and ions reported are $(M-H)^-$ ions. These parameters were generated from synthetic methylated folates, but we have never observed singly or doubly methylated folates in *E. coli* extracts with or without MJ0619.

the extraction is removed and the pellet is resuspended in 1 mL of 6 *M* HCl. This mixture is heated at 100°C overnight and then the 6 *M* HCl is removed by evaporation using a stream of nitrogen gas. The sample is resuspended in water, filtered, and the amino acids purified on a Dowex 50W-8X-H^+ column (1 × 5 mm). The amino acids are converted to their *N*-trifluoroacetyl methyl ester derivatives and analyzed by gas chromatography–mass spectrometry (White, 1988).

3.10 Source of Methylated Pterin and Folate Standards

Pterin can be purchased from Sigma-Aldrich. 6-Methylpterin and 6,7-dimethylpterin can be purchased from Schircks Laboratories. The remainder of the methylated pterin and folate standards must be synthesized. 7-Methylpterin is prepared by the condensation of 2,4,5-triamino-6-hydroxypyrimidine with 2-oxo-propanal (pyruvaldehyde) (White, 1985).

6,7-Dimethylpterin, 6-ethylpterin, 6-ethyl-7-methylpterin, and 6-ethyl-7-methylpterin are prepared by the condensation of 2,4,5-triamino-6-hydroxypyrimidine with 2,3-butanedione, 1,2-butanedione, 2,3-pentanedione, 1,2-butanedione, or 2,3-pentanedione, respectively (Mowat, Boothe, et al., 1948). 7-Methylfolate and 7,9-dimethylfolate are prepared as described (Boothe et al., 1952). 6-Hydroxyethyl-7-methylpterin is prepared by reduction of 6-acetyl-7-methylpterin with $NaBH_4$ (White, 1990).

3.11 Summary of Past Results

When MJ0619 is expressed in *E. coli*, 7-methylpterin is observed in the cell extracts (Fig. 3). Since *E. coli* does not normally produce methylated pterins,

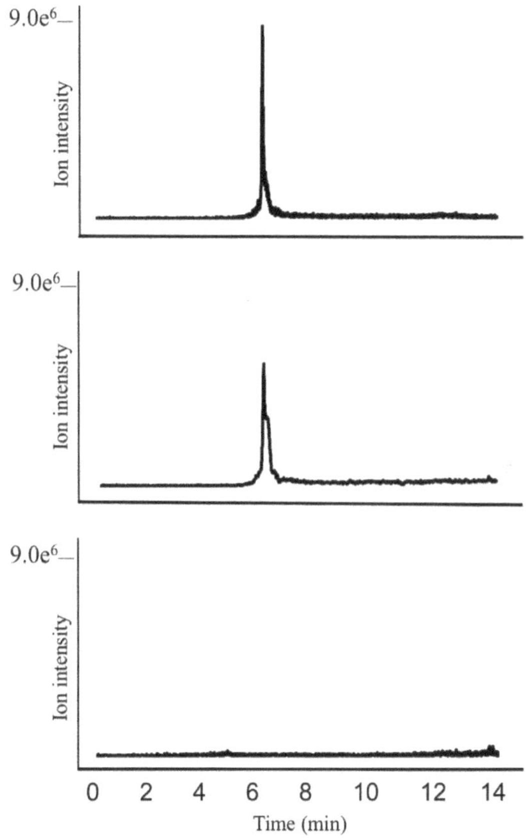

Fig. 3 Representative LC–MS MRM trace for 7-methylpterin, which elutes at 6.1 min under the conditions described in the text. *Top* is authentic standard, *middle* is the partially purified pterin extract from *E. coli* cells expressing MJ0619, and *bottom* is the control *E. coli* cells lacking MJ0619. The mass spectrometer parameters are listed in Table 1.

this indicates that MJ0619 methylates the C-7 position of folate or a folate precursor (Fig. 2). Analysis of the zinc-treated extracts reveals the presence of 6-ethyl-7-methylpterin, indicating that MJ0619 also methylates the C-9 position of folate or a folate precursor. Monomethylated pterins are not observed in the zinc-treated extracts, indicating that MJ0619 methylates both positions consecutively. Methylated folates are not observed in *E. coli* cells expressing MJ0619, but 6-hydroxyethyl-7-methylpterin is observed in these cells. This indicates that the substrate for methylation catalyzed by MJ0619 when expressed in *E. coli* could be 6-hyroxymethyl-H_2pterin or, more likely, 6-hydroxymethyl-H_2pterin-PP (Fig. 4), a common intermediate in both folate and MPT biosynthesis. Since methylated folates are not observed, the methylated pterin precursor is not a substrate for the subsequent enzyme(s) in folate biosynthesis in *E. coli*.

When *E. coli* expressing MJ0619 is grown in M9 medium in the presence of L-[methyl-$^2H_3$]methionine (C^2H_3-methionine), the 7-methylpterin is completely unlabeled even though the C^2H_3-methionine is incorporated into the *E. coli* proteins with retention of all three deuteriums. The well-characterized radical SAM methyltransferases, RlmN and Cfr, methylate an adenosine residue in 23S rRNA, where the final methylated product retains two hydrogens from the original SAM/methionine-derived methyl group (Grove et al., 2011; Grove, Radle, Krebs, & Booker, 2011). Since the methyl group added by MJ0619 does not have any deuterium incorporation from C^2H_3-methionine, MJ0619 likely does not utilize a mechanism similar to RlmN and Cfr.

When *E. coli* expressing MJ0619 is grown in M9 medium in the presence of C^2H_3-acetate as the sole carbon source, the methyl of 7-methylpterin contains the following distribution of deuterium: $^2H_0=21\%$, $^2H_1=12\%$, $^2H_2=28\%$, $^2H_3=39\%$. The methyl group of methionine in these same cells is mostly unlabeled with the following distribution of deuterium: $^2H_0=46\%$, $^2H_1=35\%$, $^2H_2=14\%$, $^2H_3=5\%$. The large difference in

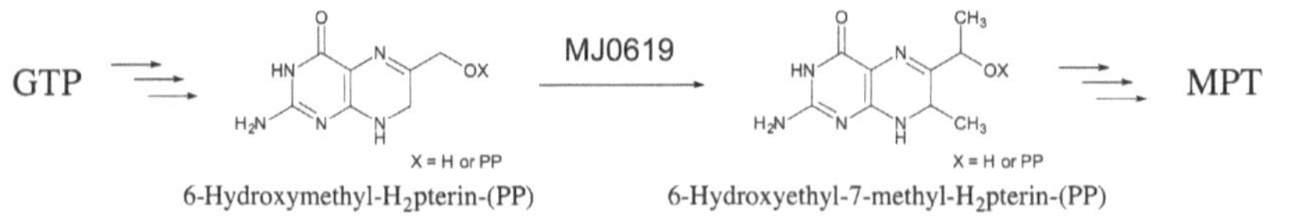

Fig. 4 Proposed reaction catalyzed by MJ0619 that is consistent with the current results obtained in our lab. The methylated pterin product is a precursor for MPT biosynthesis in methanogens, but is not a substrate for subsequent folate biosynthesis enzymes in *E. coli*.

deuterium distribution for the methyl group of 7-methylpterin compared to the methyl group of methionine further indicates that the methyl group added by MJ0619 is not derived from methionine. This result also indicates that the methyl group is not derived from N^5-methyl-H_4folate (CH_3H_4folate) since methionine is synthesized from CH_3H_4folate.

Interestingly, the distribution of deuterium labeling on the methyl group of the 7-methylpterin is similar to the labeling observed for the methyl group of thymine when *E. coli* is grown with deuterated serine as a sole carbon source (White, 1983). Thymine is biosynthesized from 2′-deoxyuridine-5′-monophosphate (dUMP) to give 2′-deoxythymidine-5′-monophosphate (dTMP) in a reaction catalyzed by thymidylate synthase. In this reaction, N^5,N^{10}-methylene-H_4folate (CH_2H_4folate) functions as a methylene group donor as well as a hydride ion donor to generate the final methyl group on dTMP. The dUMP substrate is activated for nucleophilic attack through formation of a covalent cysteine adduct. We hypothesize that MJ0619 catalyzes a related methylation reaction using CH_2H_4folate as a methyl group source when expressed in *E. coli*. Since the H_2pterin substrate in the MJ0619-catalyzed reaction cannot be activated by a covalent cysteine intermediate, activation is likely achieved instead via substrate radical formation (Allen et al., 2014). Given that *M. jannaschii* uses MPT instead of folate for all folate-requiring chemistry, this implies that MPT, in the CH_2H_4MPT form, may be a coenzyme in its own methylation.

Early work on MPT methylation in two methanogens, *Methanobrevibacter ruminantium* and *Methanococcus voltae*, grown with C^2H_3-methionine, indicated that the methyl groups on MPT are derived from methionine since 7-methylpterin molecules with 0, 1, 2, and 3 deuteriums were observed (White, 1986). One explanation for this is that different methanogens use different mechanisms, methyl group donors, and genes/enzymes for the methylation reactions in MPT biosynthesis. Alternatively, enzymatic activity may be present in methanogens, but not in *E. coli*, that can efficiently generate labeled CH_2H_4MPT from the provided C^2H_3-methionine. This could explain why early experiments in methanogens show C^2H_3-methionine incorporated into the methyl groups of MPT, while the experiments with MJ0619 expressed in *E. coli* show that the methyl groups of MPT are not derived from methionine. Further studies are required to reveal the unusual mechanism for methylation catalyzed by MJ0619 and to identify the genes/enzymes catalyzing the MPT methylation reactions in other organisms since homologues to MJ0619 are not present in the genomes of many organisms that contain MPT.

4. IDENTIFICATION OF THE RADICAL SAM ENZYMES INVOLVED IN THE BIOSYNTHESIS OF F_0

4.1 Introduction to Coenzyme F_{420}

Coenzyme F_{420} (Fig. 5) is a deazaflavin hydride carrier coenzyme. The oxidized coenzyme is intensely fluorescent upon excitation at 420 nm and was first discovered in mycobacteria (Cousins, 1960). Later, F_{420} was purified and structurally characterized from a methanogen, *Methanobacterium* sp. strain M.o.H. (Cheeseman, Toms-Wood, & Wolfe, 1972; Eirich, Vogels, & Wolfe, 1978), and has since been found in numerous actinomycetes and in all methanogens. The redox active core ring system of F_{420}, 7,8-didemethyl-8-hydroxy-5-deazariboflavin, with the 5′ ribityl group, is known as F_0 (Fig. 5), while the complete F_{420} contains an added phosphate as well as a bridging lactyl moiety followed by a γ-linked polyglutamate tail (Fig. 5). F_0 is structurally similar to the more common flavin found in riboflavin, flavin adenine dinucleotide (FAD), and flavin mononucleotide (FMN); however, F_0 contains a carbon instead of nitrogen at the fifth position in the tricyclic ring (Fig. 5). This resulting *deaza*isoalloxazine ring has a lowered reduction potential and can only participate in two-electron chemistry, as opposed to the characteristic one- or two-electron chemistry catalyzed by common flavins (Walsh, 1986). The bacteria containing F_{420} use the coenzyme in various redox reactions catalyzed by enzymes that often share homology to FMN- or FAD-utilizing enzymes (Selengut & Haft, 2010). Methanogens employ F_{420} for two reduction steps in methanogenesis going from methenyltetrahydromethanopterin to methyltetrahydromethanopterin as well as

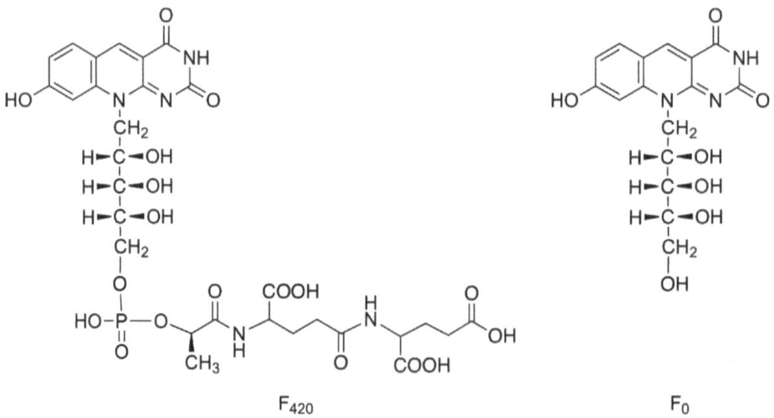

Fig. 5 Structures of F_{420} and F_0.

for other hydride transfer reactions. Additionally, most cyanobacteria and halophilic archaea contain a DNA photolyase enzyme that uses just the F_0 portion of F_{420} as a coenzyme (Dewit & Eker, 1987; Eker, Kooiman, Hessels, & Yasui, 1990; Lin & White, 1986).

Transposon mutagenesis studies in the F_{420}-producing bacterium *Mycobacterium bovis* identified a single gene called *fbiC* that was required for F_0 synthesis (Choi, Kendrick, & Daniels, 2002). Analysis of the *M. jannaschii* genome revealed two genes, MJ0446 (*cofG*) and MJ1431 (*cofH*), that are homologous to the amino-terminal and carboxy-terminal halves of the *M. bovis fbiC* gene, respectively. These two genes both contain the conserved CX_3CX_2C radical SAM sequence motif. Early isotopic labeling studies indicated that the precursors to F_0 were 4-hydroxyphenylpyruvate (HPP) and 5-amino-6-ribitylamino-2,4(1*H*,3*H*)-pyrimidinedione, an intermediate in riboflavin biosynthesis (Jaenchen, Schonheit, & Thauer, 1984; Reuke, Korn, Eisenreich, & Bacher, 1992). Since both molecules are present in *E. coli*, we used heterologous expression and analysis of *E. coli* medium for F_0 to verify the genes required for F_0 biosynthesis (Graham, Xu, & White, 2003). The methods for this approach are described in the following sections.

4.2 Protein Expression in *E. coli*

M. jannaschii cofG (locus MJ0446, Swiss-Prot Accession No. Q57888), *M. jannaschii cofH* (locus MJ1431, Swiss-Prot Accession No. Q58826), and *Mycobacterium smegmatis fbiC* (http://www.tigr.org) were cloned into pT7-7 using standard techniques to generate the recombinant plasmids pMjCofG, pMjCofH, and pMsFbiC. Both MjCofG and MjCofH express well in *E. coli* and migrate on an SDS-PAGE gel close to their expected masses of 41,985 and 40,806 Da, respectively. While both proteins are strongly expressed, only MjCofG is substantially soluble in cell-free extract. MjCofH was also cloned in pET19b for expression as a fusion protein with an N-terminal polyhistidine tag. MjCofH-His is also strongly expressed in *E. coli* and is slightly more soluble than the untagged protein. MsFbiC is also highly expressed, but the majority is insoluble under routine aerobic expression and isolation conditions. Heterologously expressed FbiC from *M. bovis* is also insoluble in cell-free extracts (Choi et al., 2002). For the coexpression of *M. jannschii* CofG and CofH, *cofH* containing the 10X-his tag was PCR amplified with restriction sites for *Bgl*II introduced at the 5′ and 3′ ends of the PCR product. This was then ligated into a *Bgl*II site in the plasmid pMjCofG, resulting in pMjCofGH.

For partial purification of the desired proteins, the *E. coli* cells are sonicated in 20 m*M* bis-Tris–HCl (pH 6.5) and the insoluble material is removed by centrifugation. We have partially purified the recombinant FbiC from the resulting supernatant by salting out under anoxic conditions with ammonium sulfate followed by desalting on a Sephadex G-25 column. FbiC precipitates out between 25% and 40% ammonium sulfate. Since proteins from *M. jannaschii* generally have the advantage of being heat stable, soluble extracts containing MjCofG or MjCofH are heated at 70°C for 10 min before centrifugation to remove nonheat-stable *E. coli* proteins. Both proteins are stable to heating under these conditions.

4.3 Isolation and Partial Purification of F_0 From *E. coli* Cells

Different methods are used for F_0 analysis depending on the source and amount of F_0 in the sample. For the analysis of low concentrations of F_0 in growth medium, a 6 mL sample is concentrated down to about 500 μL by evaporation under a stream of nitrogen gas. The concentrate is then applied to a C18 silica column (0.5 × 5 cm) equilibrated in water. The column is washed with 4 mL water followed by 2 mL 10% methanol in water and the F_0 is eluted from the column with 2–3 mL of 30% methanol until the yellow fluorescent riboflavin elutes, monitored using a 366-nm hand-held illuminator (UVP, San Gabriel, CA, United States). This fraction containing F_0 is concentrated with a stream of nitrogen gas and the F_0 further purified by preparative TLC on silica gel 60 TLC plates (E. Merck, Darmstadt, Germany) with a solvent system of acetonitrile/water/formic acid (88%), 40:10:5. F_0 migrates with an R_f of 0.53 and is identified by its blue fluorescence upon illumination with long-wavelength ultraviolet light. The area of the TLC plate that contains F_0 is removed and the F_0 is eluted with 50% methanol. After concentration by evaporation with a stream of nitrogen gas, F_0 in the samples is analyzed by HPLC. For the analysis of growth medium containing higher amounts of F_0, direct quantification is possible by measuring the fluorescence intensity of the samples (excitation wavelength, 420 nm; emission wavelength, 480 nm) compared to standards with known F_0 concentrations. A synthetic standard of F_0 is prepared using described methods (Ashton, Brown, Jacobson, & Walsh, 1979).

F_0 present in cell pellets can be isolated by suspending the cells (0.2–0.5 g) in 50% methanol (3 mL) and incubating at 70°C for 5 min followed by centrifugation. The resulting pellets are then reextracted in the same manner. The soluble extracts are combined and the methanol removed by evaporation with a stream of nitrogen gas. F_0 in the samples is then purified by preparative TLC as explained above and assayed by HPLC. For F_0 present in cell-free incubations, proteins are precipitated from the incubation mixture with three volumes of methanol and insoluble material is removed by centrifugation ($14,000 \times g$, 10 min). F_0 contained in the soluble extract is then purified by preparative TLC and analyzed by HPLC.

4.4 Analysis of F_0 by HPLC With Fluorescence Detection

F_0 can be identified at very low levels by HPLC due to its intense fluorescence. Our analyses have previously been done on a Shimadzu SCL-6B HPLC using a C18 column (AXXI Chrom octadecylsilane, 5 μm, 4.6 mm × 25 cm) preceded by a guard column (RP-18 NEW-GUARD, 7 μm, 1.5 cm). Sodium acetate buffer (25 m*M*, pH 6.0, 0.02% NaN_3) containing 35% methanol is used for isocratic elution at a flow rate of 0.5 mL/min. The eluent is monitored by fluorescence (excitation wavelength, 420 nm, emission wavelength, 480 nm) using an RF-551 fluorescence HPLC monitor (Shimadzu), and by absorbance at 280 nm using an SPD-6A V UV–Vis spectrophotometric detector (Shimadzu). Under these conditions F_0 elutes at 9.6 min and separates from all other known F_0 or riboflavin-containing compounds (Graupner & White, 2003; Li, Graupner, Xu, & White, 2003).

4.5 Summary of Past Results

A summary of the amounts of F_0 present in *E. coli* medium under various conditions is listed in Table 2. A small but consistently detectable amount of F_0 is observed in medium from control experiments with *E. coli* lacking any *cof* genes (Experiment 1, Table 2). This low level of synthesis could arise from the chemical condensation of the precursors for its synthesis that are naturally present in *E. coli*; this spontaneous reaction has been observed in vitro (R.H.W., unpublished results). Alternatively, the extracts may contain another fluorescent molecule that cochromatographs with F_0. *E. coli* cells containing either MjCofG or MjCofH expressed alone produce only slightly more than this background level of F_0 (Experiments 2 and 3, Table 2). However, when *cofG* and polyhistidine-tagged *cofH* are coexpressed, the cells release nearly 50-fold more F_0 into the medium compared to control cells (Experiment 4, Table 2). Additionally, cells expressing the fused bifunctional enzyme from *M. smegmatis*, MsFbiC, release 84-fold more F_0 into the minimal growth medium than cells

Table 2 Amounts of F_0 Observed in the Growth Medium of *E. coli* Control Cells (Experiment #1) Compared to CofG (Experiment #2), CofH (Experiment #3), CofG and CofH (Experiment #4), or FbiC From *M. smegmatis* (Experiments #5 and 6)

Experiment	Strain	Growth Medium	$[F_0]$ (n*M*)
#1	*E. coli* control cells (BL21-CodonPlus (DE3)-RIL)	M9 + 0.5% peptone	0.02
#2	*E. coli* + pMjCofG	M9 + 0.5% peptone	0.04
#3	*E. coli* + pMjCofH	M9 + 0.5% peptone	0.05
#4	*E. coli* + pMjCofGH	M9 + 0.5% peptone	1.9
#5	*E. coli* + pMsFbiC	M9	160
#6	*E. coli* + pMsFbiC	M9 + 6 m*M* tyrosine	640

Quantitation is based on comparison to a standard solution of F_0.

coexpressing MjCofG and MjCofH (Experiment 5, Table 1). The same cells grown in minimal medium saturated with tyrosine release fourfold more F_0 (Experiment 6, Table 2), suggesting that tyrosine is converted into the HPP substrate or that tyrosine itself is the substrate and, thus, stimulates activity. Subsequent work on CofG and CofH from *M. jannaschii* as well as a bifunctional enzyme from *Thermobifida fusca* has demonstrated that F_0 synthase uses 5-amino-6-ribitylamino-2,4(1*H*,3*H*)-pyrimidinedione and tyrosine (instead of HPP as previously suggested) as substrates and generates two 5′-deoxyadenosyl radicals during catalysis (Decamps et al., 2012; Philmus, Decamps, Berteau, & Begley, 2015).

REFERENCES

Allen, K. D., Xu, H., & White, R. H. (2014). Identification of a unique radical S-adenosylmethionine methylase likely involved in methanopterin biosynthesis in *Methanocaldococcus jannaschii*. *Journal of Bacteriology*, *196*(18), 3315–3323.

Ashton, W. T., Brown, R. D., Jacobson, F., & Walsh, C. (1979). Synthesis of 7,8-didemethyl-8-hydroxy-5-deazariboflavin and confirmation of its identity with the deazaisoalloxazine chromophore of Methanobacterium redox coenzyme F420. *Journal of the American Chemical Society*, *101*(15), 4419–4420.

Bauerle, M. R., Schwalm, E. L., & Booker, S. J. (2015). Mechanistic diversity of radical S-adenosylmethionine (SAM)-dependent methylation. *Journal of Biological Chemistry*, *290*(7), 3995–4002.

Boothe, J. H., Mowat, J. H., Waller, C. W., Angier, R. B., Semb, J., & Gazzola, A. L. (1952). 7-Methylpteroylglutamic acid and some related compounds. *Journal of the American Chemical Society*, *74*(21), 5407–5409.

Braakman, R. (2013). Mapping metabolism onto the prebiotic organic chemistry of hydrothermal vents. *Proceedings of the National Academy of Sciences of the United States of America*, *110*(33), 13236–13237.

Bult, C. J., White, O., Olsen, G. J., Zhou, L., Fleischmann, R. D., Sutton, G. G., et al. (1996). Complete genome sequence of the methanogenic archaeon, *Methanococcus jannaschii*. *Science*, *273*(5278), 1058–1073.

Cheeseman, P., Toms-Wood, A., & Wolfe, R. S. (1972). Isolation and properties of a fluorescent compound, factor 420, from Methanobacterium strain M.o.H. *Journal of Bacteriology*, *112*(1), 527–531.

Chistoserdova, L., Vorholt, J. A., Thauer, R. K., & Lidstrom, M. E. (1998). C1 transfer enzymes and coenzymes linking methylotrophic bacteria and methanogenic Archaea. *Science*, *281*(5373), 99–102.

Choi, K. P., Kendrick, N., & Daniels, L. (2002). Demonstration that fbiC is required by *Mycobacterium bovis* BCG for coenzyme F(420) and FO biosynthesis. *Journal of Bacteriology*, *184*(9), 2420–2428.

Cody, G. D., Boctor, N. Z., Filley, T. R., Hazen, R. M., Scott, J. H., Sharma, A., et al. (2000). Primordial carbonylated iron-sulfur compounds and the synthesis of pyruvate. *Science*, *289*(5483), 1337–1340.

Costa, K. C., & Leigh, J. A. (2014). Metabolic versatility in methanogens. *Current Opinion in Biotechnology*, *29*, 70–75.

Cousins, F. B. (1960). The prosthetic group of a chromoprotin from mycobacteria. *Biochimica et Biophysica Acta*, *40*, 532–534.

Dantola, M. L., Vignoni, M., Capparelli, A. L., Lorente, C., & Thomas, A. H. (2008). Stability of 7,8-dihydropterins in air-equilibrated aqueous solutions. *Helvetica Chimica Acta*, *91*(3), 411–425.

Decamps, L., Philmus, B., Benjdia, A., White, R., Begley, T. P., & Berteau, O. (2012). Biosynthesis of F_o, precursor of the F_{420} cofactor, requires a unique two radical-SAM domain enzyme and tyrosine as substrate. *Journal of the American Chemical Society*, *134*(44), 18173–18176.

Dewit, L. E. A., & Eker, A. P. M. (1987). 8-Hydroxy-5-deazaflavin-dependent electron-transfer in the extreme halophile Halobacterium cutirubrum. *FEMS Microbiology Letters*, *48*(1–2), 121–125.

Eirich, L. D., Vogels, G. D., & Wolfe, R. S. (1978). Proposed structure for coenzyme F420 from Methanobacterium. *Biochemistry*, *17*(22), 4583–4593.

Eker, A. P. M., Kooiman, P., Hessels, J. K. C., & Yasui, A. (1990). DNA photoreactivating enzyme from the cyanobacterium Anacystis nidulans. *Journal of Biological Chemistry*, *265*(14), 8009–8015.

Escalante-Semerena, J. C., Rinehart, K. L., Jr., & Wolfe, R. S. (1984). Tetrahydromethanopterin, a carbon carrier in methanogenesis. *Journal of Biological Chemistry*, *259*(15), 9447–9455.

Foo, S. K., Cichowicz, D. J., & Shane, B. (1980). Cleavage of naturally occurring folates to unsubstituted p-aminobenzoylpoly-gamma-glutamates. *Analytical Biochemistry*, *107*(1), 109–115.

Graham, D. E., Xu, H., & White, R. H. (2003). Identification of the 7,8-didemethyl-8-hydroxy-5-deazariboflavin synthase required for coenzyme F(420) biosynthesis. *Archives of Microbiology*, *180*(6), 455–464.

Graupner, M., & White, R. H. (2003). *Methanococcus jannaschii* coenzyme F420 analogs contain a terminal alpha-linked glutamate. *Journal of Bacteriology*, *185*(15), 4662–4665.

Grove, T. L., Benner, J. S., Radle, M. I., Ahlum, J. H., Landgraf, B. J., Krebs, C., et al. (2011). A radically different mechanism for S-adenosylmethionine-dependent methyltransferases. *Science*, *332*(6029), 604–607.

Grove, T. L., Radle, M. I., Krebs, C., & Booker, S. J. (2011). Cfr and RlmN contain a single [4Fe-4S] cluster, which directs two distinct reactivities for S-adenosylmethionine: Methyl transfer by SN2 displacement and radical generation. *Journal of the American Chemical Society*, *133*(49), 19586–19589.

Huber, C., & Wachtershauser, G. (1997). Activated acetic acid by carbon fixation on (Fe,Ni) S under primordial conditions. *Science, 276*(5310), 245–247.

Jaenchen, R., Schonheit, P., & Thauer, R. K. (1984). Studies on the biosynthesis of coenzyme-F420 in methanogenic bacteria. *Archives of Microbiology, 137*(4), 362–365.

Jarrett, J. T. (2015). The biosynthesis of thiol- and thioether-containing cofactors and secondary metabolites catalyzed by radical S-adenosylmethionine enzymes. *Journal of Biological Chemistry, 290*(7), 3972–3979.

Jones, J. B., Leigh, J. A., Mayer, F., Woese, C. R., & Wolfe, R. S. (1983). *Methanococcus jannaschii* sp. nov., an extremely thermophilic methanogen from a submarine hydrothermal vent. *Archives of Microbiology, 136*, 254–261.

Landgraf, B. J., McCarthy, E. L., & Booker, S. J. (2016). Radical S-adenosylmethionine enzymes in human health and disease. *Annual Review of Biochemistry, 85*, 485–514.

Lanz, N. D., Grove, T. L., Gogonea, C. B., Lee, K. H., Krebs, C., & Booker, S. J. (2012). RlmN and AtsB as models for the overproduction and characterization of radical SAM proteins. *Methods in Enzymology, 516*, 125–152.

Li, H., Graupner, M., Xu, H., & White, R. H. (2003). CofE catalyzes the addition of two glutamates to F_{420}-0 in F_{420} coenzyme biosynthesis in *Methanococcus jannaschii. Biochemistry, 42*(32), 9771–9778.

Lin, X. L., & White, R. H. (1986). Occurrence of coenzyme F420 and its gamma-monoglutamyl derivative in nonmethanogenic archaebacteria. *Journal of Bacteriology, 168*(1), 444–448.

Liu, Y., Beer, L. L., & Whitman, W. B. (2012). Sulfur metabolism in archaea reveals novel processes. *Environmental Microbiology, 14*(10), 2632–2644.

Mehta, A. P., Abdelwahed, S. H., Mahanta, N., Fedoseyenko, D., Philmus, B., Cooper, L. E., et al. (2015). Radical S-adenosylmethionine (SAM) enzymes in cofactor biosynthesis: A treasure trove of complex organic radical rearrangement reactions. *Journal of Biological Chemistry, 290*(7), 3980–3986.

Mowat, J. H., Boothe, J. H., et al. (1948). The structure of the liver L. casei factor. *Journal of the American Chemical Society, 70*(1), 14–18.

Nisbet, E. G., & Sleep, N. H. (2001). The habitat and nature of early life. *Nature, 409*(6823), 1083–1091.

Novikov, Y., & Copley, S. D. (2013). Reactivity landscape of pyruvate under simulated hydrothermal vent conditions. *Proceedings of the National Academy of Sciences of the United States of America, 110*(33), 13283–13288.

Philmus, B., Decamps, L., Berteau, O., & Begley, T. P. (2015). Biosynthetic versatility and coordinated action of 5′-deoxyadenosyl radicals in deazaflavin biosynthesis. *Journal of the American Chemical Society, 137*(16), 5406–5413.

Pribat, A., Blaby, I. K., Lara-Nunez, A., Gregory, J. F., 3rd, de Crecy-Lagard, V., & Hanson, A. D. (2010). FolX and FolM are essential for tetrahydromonapterin synthesis in Escherichia coli and Pseudomonas aeruginosa. *Journal of Bacteriology, 192*(2), 475–482.

Rasche, M. E., & White, R. H. (1998). Mechanism for the enzymatic formation of 4-(beta-D-ribofuranosyl)aminobenzene 5′-phosphate during the biosynthesis of methanopterin. *Biochemistry, 37*(32), 11343–11351.

Reed, L. S., & Archer, M. C. (1980). Oxidation of tetrahydrofolic acid by air. *Journal of Agricultural and Food Chemistry, 28*, 801–805.

Reuke, B., Korn, S., Eisenreich, W., & Bacher, A. (1992). Biosynthetic precursors of deazaflavins. *Journal of Bacteriology, 174*(12), 4042–4049.

Scott, J. W., & Rasche, M. E. (2002). Purification, overproduction, and partial characterization of beta-RFAP synthase, a key enzyme in the methanopterin biosynthesis pathway. *Journal of Bacteriology, 184*(16), 4442–4448.

Selengut, J. D., & Haft, D. H. (2010). Unexpected abundance of coenzyme F(420)-dependent enzymes in *Mycobacterium tuberculosis* and other actinobacteria. *Journal of Bacteriology, 192*(21), 5788–5798.

Tabor, S., & Richardson, C. C. (1985). A bacteriophage T7 RNA polymerase/promoter system for controlled exclusive expression of specific genes. *Proceedings of the National Academy of Sciences of the United States of America*, *82*(4), 1074–1078.
van Beelen, P., Labro, J. F., Keltjens, J. T., Geerts, W. J., Vogels, G. D., Laarhoven, W. H., et al. (1984). Derivatives of methanopterin, a coenzyme involved in methanogenesis. *European Journal of Biochemistry*, *139*(2), 359–365.
Vignoni, M., Cabrerizo, F. M., Lorente, C., Claparols, C., Oliveros, E., & Thomas, A. H. (2010). Photochemistry of dihydrobiopterin in aqueous solution. *Organic & Biomolecular Chemistry*, *8*(4), 800–810.
Walsh, C. (1986). Naturally-occurring 5-deazaflavin coenzymes—Biological redox roles. *Accounts of Chemical Research*, *19*(7), 216–221.
White, R. H. (1983). Proton exchange on carbons 2 and 3 of serine during their conversion into methyl groups of methionine and thymine in *Escherichia coli*. *Biochemistry*, *22*(8), 1883–1888.
White, R. H. (1985). 7-Methylpterin and 7-methyllumizine: Oxidative degradation products of 7-methyl-substituted pteridines in methanogenic bacteria. *Journal of Bacteriology*, *162*(2), 516–520.
White, R. H. (1986). Biosynthesis of the 7-methylated pterin of methanopterin. *Journal of Bacteriology*, *165*(1), 215–218.
White, R. H. (1988). Biosynthesis of the 2-(aminomethyl)-4-(hydroxymethyl)furan subunit of methanofuran. *Biochemistry*, *27*(12), 4415–4420.
White, R. H. (1990). Biosynthesis of methanopterin. *Biochemistry*, *29*(22), 5397–5404.
Zhang, Q., van der Donk, W. A., & Liu, W. (2012). Radical-mediated enzymatic methylation: A tale of two SAMS. *Accounts of Chemical Research*, *45*(4), 555–564.

FURTHER READING

Fischer, R., Gartner, P., Yeliseev, A., & Thauer, R. K. (1992). N5-methyl-tetrahydromethanopterin: Coenzyme M methyltransferase in methanogenic archaebacteria is a membrane protein. *Archives of Microbiology*, *158*(3), 208–217.

CHAPTER EIGHTEEN

Lessons From the Studies of a C—C Bond Forming Radical SAM Enzyme in Molybdenum Cofactor Biosynthesis

Haoran Pang, Kenichi Yokoyama[1]

Department of Biochemistry, Duke University Medical Center, Durham, NC, United States

[1]Corresponding author: e-mail address: ken.yoko@duke.edu

Contents

Abstract

MoaA is one of the founding members of the radical *S*-adenosyl-$_L$-methionine (SAM) superfamily, and together with the second enzyme, MoaC, catalyzes the construction of the pyranopterin backbone structure of the molybdenum cofactor (Moco). However, the exact functions of both MoaA and MoaC had remained ambiguous for more than 2 decades. Recently, their functions were finally elucidated through successful characterization of the MoaA product as 3′,8-cyclo-7,8-dihydro-GTP (3′,8-cH_2GTP),

Methods in Enzymology, Volume 606
ISSN 0076-6879
https://doi.org/10.1016/bs.mie.2018.04.014

which was shown to be converted to cyclic pyranopterin monophosphate (cPMP) by MoaC. 3′,8-cH_2GTP was produced in a small quantity and was highly oxygen sensitive, which explains why this compound had previously eluded characterization. This chapter describes the methodologies for the characterization of MoaA, MoaC, and 3′,8-cH_2GTP, which together significantly altered the view of the mechanism of the pyranopterin backbone construction during the Moco biosynthesis. Through this chapter, we hope to share not only the protocols to study the first step of Moco biosynthesis but also the lessons we learned from the characterization of the chemically labile biosynthetic intermediate, which would be informative for the study of many other metabolic pathways and enzymes.

ABBREVIATIONS

3′,8-cH_2GTP	3′,8-cyclo-7,8-dihydro-GTP
5′-dA	5′-deoxyadenosine
5′-dA$^•$	5′-deoxyadenosyl radical
Amp	ampicillin
BSA	bovine serum albumin
Cam	chloramphenicol
COSY	homonuclear correlation spectroscopy
cPMP	cyclic pyranopterin monophosphate
DMPT	dimethylpterin
DTT	dithiothreitol
EDTA	ethylenediaminetetraacetic acid
ENDOR	electron nuclear double resonance
ESI-TOF-MS	electrospray ionization time-of-flight mass spectrometry
GTP	guanosine 5′-triphosphate
HMBC	heteronuclear multiple bond correlation
HMQC	heteronuclear multiple quantum coherence
HPAEC	high-performance anion exchange chromatography
HPLC	high-performance liquid chromatography
IPTG	isopropyl β-D-1-thiogalactopyranoside
Kan	kanamycin
LB	Lysogeny broth
LC–MS	liquid chromatography–mass spectrometry
MoCD	molybdenum cofactor deficiency
Moco	molybdenum cofactor
MPT	molybdopterin
MS	mass spectrometry
NMR	nuclear magnetic resonance
NOESY	nuclear Overhauser effect spectroscopy
SAM	*S*-adenosyl-L-methionine
SDS-PAGE	sodium dodecyl sulfate polyacrylamide gel electrophoresis
TCA	trichloroacetic acid
TFA	trifluoroacetic acid
βME	β-mercaptoethanol

1. INTRODUCTION

Molybdenum cofactor (Moco) is an essential organometallic cofactor found in nearly all organisms (Mendel & Schwarz, 2011). Moco-dependent enzymes take advantage of the redox ability of Moco and catalyze biologically important reactions. Unlike many other organic cofactors, Moco cannot be directly taken up as a nutrient, thus requiring de novo biosynthesis. Therefore, Moco biosynthesis is essential for the survival of many organisms (Leimkuhler, Wuebbens, & Rajagopalan, 2011). In human, Moco deficiency (MoCD) caused by genetic mutations in one of the Moco biosynthesis enzymes can lead to activity loss of all Moco-dependent enzymes, resulting in fatal neurodegeneration and early childhood death (Johnson et al., 1980). Moco biosynthesis is also found essential for virulence of several clinically important bacterial pathogens, such as *Mycobacterium tuberculosis* (Boshoff & Barry, 2005) and *Pseudomonas aeruginosa* (Folkesson et al., 2012). These organisms rely on several Moco-dependent enzymes in the anaerobic respiration to survive the hypoxic conditions of the host body (Hurdle, O'Neill, Chopra, & Lee, 2011). More recently, Moco biosynthesis in enterobacteria in gut microbiome was shown to be essential for these organisms to cause inflammation, and small molecule inhibitors of Moco biosynthesis was shown effective to prevent such inflammation (Zhu et al., 2018).

Moco biosynthetic pathway is conserved among most organisms and can be divided into three major steps (Fig. 1) (Leimkuhler et al., 2011; Mendel & Schwarz, 2011). The biosynthesis is initiated by the rearrangement of guanosine 5′-triphosphate (GTP) into cyclic pyranopterin monophosphate (cPMP), followed by insertion of two sulfur atoms to form molybdopterin (MPT). Finally, molybdate, the only absorptive form of molybdenum, is ligated to the dithiolene moiety of MPT to form Moco. The conversion of GTP into cPMP is of particular mechanistic interests, as it establishes the pyranopterin backbone of Moco from GTP through highly complex and unique rearrangement reaction. This transformation has been affected in more than 60% of the human MoCD patients (Reiss, 2000). Early isotope-labeling experiments had demonstrated that the transformation involves a unique ring rearrangement where guanine C-8 is inserted between C-2′ and C-3′ of ribose (Fig. 1; see symbols on GTP and cPMP) (Rieder et al., 1998; Wuebbens & Rajagopalan, 1995). This transformation is distinct from other pterin biosynthetic pathways, in which C-8 is hydrolyzed

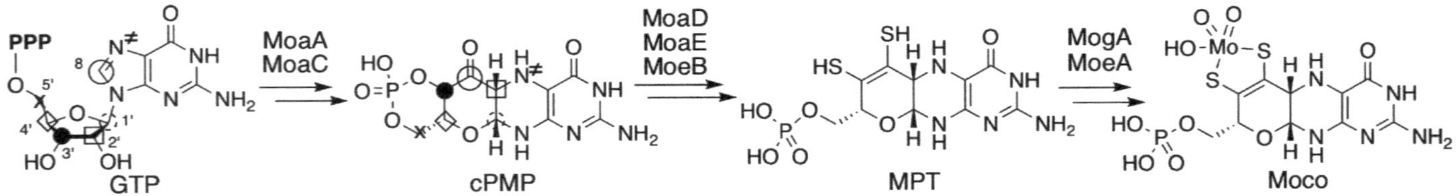

Fig. 1 Moco biosynthetic pathway in bacteria. The *symbols* on GTP and cPMP show the fates of carbon and nitrogen atoms during the construction of the pyranopterin backbone as determined by isotope-labeling studies (Rieder et al., 1998; Wuebbens & Rajagopalan, 1995).

as formate and does not retain in the pterin product (Burg & Brown, 1968; Shiota, Palumbo, & Tsai, 1967). Thus, the mechanism of pyranopterin ring formation during the Moco biosynthesis was considered as a critical mystery in cofactor biosynthesis and enzymology fields.

In bacteria, the transformation from GTP to cPMP is catalyzed by two enzymes, MoaA and MoaC. MoaA belongs to the radical *S*-adenosyl-L-methionine (SAM) superfamily (Sofia, Chen, Hetzler, Reyes-Spindola, & Miller, 2001), while MoaC shows no obvious sequence or structural similarities to other characterized enzymes (Wuebbens, Liu, Rajagopalan, & Schindelin, 2000). Because of the general awareness of radical SAM enzymes catalyzing novel chemical reactions, MoaA was thought to catalyze the complex reactions to rearrange GTP to cPMP (Hänzelmann & Schindelin, 2004; Leimkuhler et al., 2011). MoaC was thought to be either a regulatory protein (Hänzelmann & Schindelin, 2004) or an enzyme responsible only for the cyclic phosphate formation (Leimkuhler et al., 2011). In 2006, Hänzelmann and Schindelin mentioned in their publication that when MoaA was assayed with GTP and SAM in the absence of MoaC, they observed a compound that can be chemically derivatized to dimethylpterin (DMPT) (Hänzelmann & Schindelin, 2006). Although no data were published, the statement implied the presence of a potential intermediate during pyranopterin ring formation. By the early 2010s, the predominant view in the field was that MoaA catalyzes the transformation of GTP into pyranopterin triphosphate, while MoaC catalyzes the cyclic phosphate formation (Fig. 2A) (Kanaujia et al., 2010; Leimkuhler et al., 2011). However, pyranopterin triphosphate would not be converted to DMPT

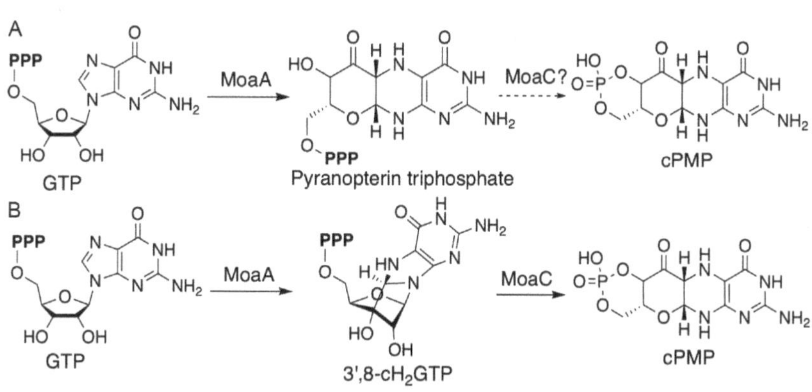

Fig. 2 Proposed models for the pyranopterin ring formation catalyzed by MoaA and MoaC during the Moco biosynthesis. (A) Pyranopterin triphosphate as the MoaA product and (B) 3′,8-cH_2GTP as the MoaA product.

(Wuebbens & Rajagopalan, 1993). Thus, such model is inconsistent with the statement by Hänzelmann and Schindelin (2006), and the function of MoaA remained ambiguous.

When we initiated the project in 2011, no experimental data were available to support even the presence of the putative product of MoaA. Therefore, we initially focused on detection of the putative MoaA product using a stepwise assay (Hover, Loksztejn, Ribeiro, & Yokoyama, 2013). In this assay, MoaA was first incubated with GTP and SAM in the absence of MoaC, followed by separation of MoaA and its small molecule products by ultrafiltration. Incubation of the small molecule fraction with MoaC resulted in the formation of cPMP. The putative MoaA product was also derivatized to DMPT (Hover et al., 2013), consistent with the statement by Hänzelmann and Schindelin (2006). The quantitative agreement between cPMP and DMPT suggested that the MoaA product was quantitatively converted into cPMP (Hover et al., 2013). These observations were the first quantitative evidence for the presence of a small molecule that is diffusible from MoaA and converted to cPMP by MoaC.

The isolation and structural characterization of this small molecule, however, were challenging due to the substoichiometric turnover of MoaA and the limited chemical stability of the MoaA product. In particular, the MoaA product was quickly decomposed in the presence of oxygen ($t_{1/2}$ ~70 min) or at acidic pH ($t_{1/2} < 10$ min) (Hover et al., 2013). These challenges were overcome by the development of large-scale preparation of MoaA and the purification of the MoaA product under strictly anaerobic conditions at mild basic pH and low temperature. The isolated MoaA product was characterized using comprehensive approaches including chemical and enzymological derivatizations, UV–vis absorption spectroscopy, mass spectrometry (MS), and nuclear magnetic resonance (NMR) spectroscopy, based on which the structure was proposed as 3′,8-cyclo-7,8-dihydro-GTP (3′,8-cH_2GTP, Fig. 2B) (Hover et al., 2013). 3′,8-cH_2GTP has an acid-labile 6-hydroxy-2,4,5-triaminopyrimidine partial structure that can be converted to DMPT, which is consistent with the Hänzelmann and Schindelin's statement (Hänzelmann & Schindelin, 2006). In parallel to our efforts, Mehta et al. reported an liquid chromatography–mass spectrometry (LC–MS) analysis of a MoaA product that absorbs light at ~320 nm and shows a low-resolution mass signal at $m/z = 524$ $[M+H]^+$ (Mehta et al., 2013). While the authors assigned this compound to pyranopterin triphosphate, the compound was never isolated. Since 3′,8-cH_2GTP absorbs light at 322 nm and has a molecular weight of 523, the compound observed by Mehta et al.

was most likely 3′,8-cH_2GTP as well. Thus, 3′,8-cH_2GTP was observed by multiple groups, but its structure was assigned incorrectly due to the lack of its isolation.

The identification of 3′,8-cH_2GTP as the MoaA product revealed that MoaA is responsible for the C—C bond formation between guanine C-8 and ribose C-3′, while MoaC, in contrast to the previous consensus, catalyzes the complicated rearrangement reaction during the conversion of 3′,8-cH_2GTP into cPMP (Fig. 2B). Therefore, the successful characterization of the MoaA product revealed novel enzyme functions and mechanisms during the construction of the pyranopterin ring of Moco. This success was made possible through quantitative and comprehensive characterization of MoaA, MoaC, and 3′,8-cH_2GTP. Therefore, in this chapter, we will describe the protocols for preparation and characterization of MoaA and MoaC, and for the isolation and characterization of 3′,8-cH_2GTP. These protocols can be used to study MoaA and MoaC from other organisms, including humans, to investigate the generality of their functions, and to elucidate detailed mechanisms of these enzymes.

Also, the characterization of 3′,8-cH_2GTP highlights important lessons for characterization of enzymes and biosynthetic pathways in general. As will be detailed in this chapter, the study emphasizes the importance of quantitative characterization of enzymes and their products for accurate determination of enzyme functions. The study also demonstrated the importance of understanding the chemical properties of small molecules for their isolation and structural characterization. We believe these are important lessons applicable to many other biosynthetic enzymes and pathways, especially those with unstable intermediates.

2. PREPARATION AND CHARACTERIZATION OF MoaA

MoaA catalyzes the transformation of GTP into 3′,8-cH_2GTP. Fig. 3A shows the current model for the catalytic mechanism (Hover et al., 2013). The reaction is initiated by reductive cleavage of SAM to generate 5′-deoxyadenosyl radical (5′-dA$^{\bullet}$) and L-methionine. The resulting 5′-dA$^{\bullet}$ abstracts the H-atom at the 3′-position of GTP and produces C3′-radical intermediate (step 1), which then attacks C-8 of guanine base (step 2). The resulting aminyl radical is reduced by transfers of a proton and an electron to produce the final product 3′,8-cH_2GTP (step 3). The C-terminal cluster was proposed to be the electron donor (Hover et al., 2013; Mehta et al., 2013), but no evidence is currently available.

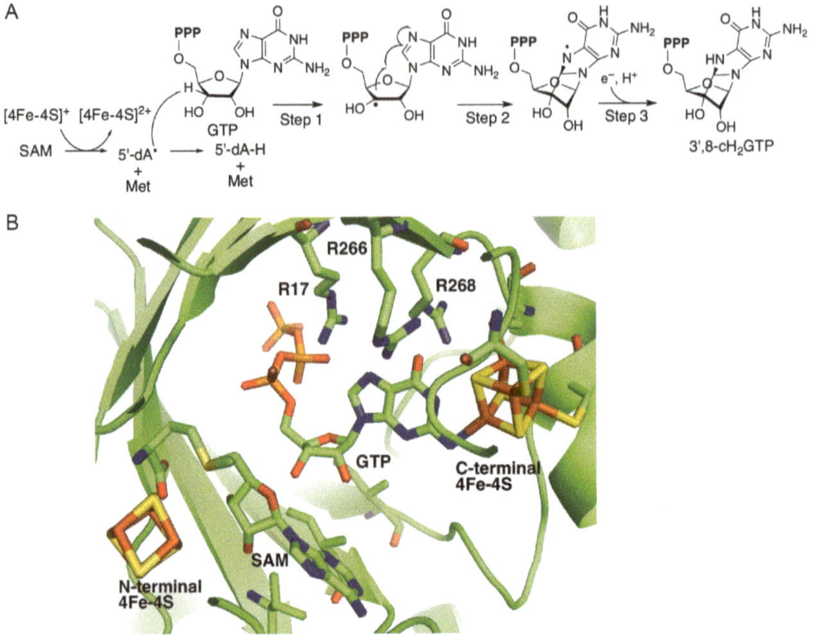

Fig. 3 Proposed mechanism and crystal structure of MoaA. (A) Current model for the catalytic mechanism of MoaA and (B) model of the MoaA active site. The position of SAM is modeled into the crystal structure of MoaA in complex with GTP (PDB ID: 2FB3) (Hänzelmann & Schindelin, 2006) by structural alignment with the crystal structure of MoaA in complex with SAM (PDB ID: 1TV8) (Hänzelmann & Schindelin, 2004). *Reprinted with permission from Hover, B. M., Loksztejn, A., Ribeiro, A. A., & Yokoyama, K. (2013). Identification of a cyclic nucleotide as a cryptic intermediate in molybdenum cofactor biosynthesis.* Journal of the American Chemical Society, 135*(18), 7019–7032. doi:10.1021/ja401781t. Copyright 2013 American Chemical Society.*

MoaA is a representative member of the twitch subfamily (Goldman, Grove, Booker, & Drennan, 2013), one of the subclasses in the radical SAM superfamily. Members of this subfamily are characterized by the presence of two [4Fe–4S] clusters; one is the canonical radical SAM cluster and the other is an auxiliary [4Fe–4S] cluster in an extended C-terminal domain (Goldman et al., 2013). The crystal structure of *Staphylococcus aureus* MoaA in complex with SAM (Fig. 3B) showed that the canonical radical SAM cluster was bound at the N-terminus of MoaA, and ligated by three Cys residues while the fourth Fe ligated by SAM (Hänzelmann & Schindelin, 2004). The auxiliary cluster at the C-terminus of MoaA is also ligated by three Cys. The crystal structure of *S. aureus* MoaA in complex with GTP (Hänzelmann & Schindelin, 2006), together with an electron nuclear double resonance (ENDOR) spectroscopy analysis (Lees et al., 2009), revealed that the fourth Fe interacts with N−1 of the guanine base of GTP. This interaction was proposed to change the tautomerization of guanine base to generally unfavorable enol tautomer (Lees et al., 2009), although the role of tautomerization on the catalytic mechanism of MoaA is not known. The C-terminal cluster is also critical for the stability of MoaA, and mutagenesis of any of the Cys ligands results in expression of insoluble proteins (Hänzelmann & Schindelin, 2006). Both [4Fe–4S] clusters are essential for the catalytic function of MoaA and are both oxygen sensitive. Therefore, the preparation and characterization of MoaA require anaerobic methods. In this section, we will describe our protocols for expression, purification, and characterization of MoaA using the enzyme from *S. aureus* as a model.

2.1 Expression and Purification of MoaA

In general, purification of radical SAM enzymes is performed under strictly anaerobic conditions because of the oxygen-sensitive nature of the [4Fe–4S] clusters. The previous protocol for MoaA purification described by Hänzelmann and Schindelin also used anaerobic conditions (Hänzelmann & Schindelin, 2004). However, anaerobic purification limits the amount of protein that can be purified in a single experiment. In our current protocol, ~1.6 g of MoaA is required for isolation of 3–5 mg of the MoaA product (see Section 4.2). To facilitate such isolation effort, we developed a protocol for MoaA purification under aerobic conditions. In this procedure, coexpression of the *suf* operon encoding Fe–S cluster assembly enzymes (SufA, SufB, SufC, SufD, SufS, and SufE) (Hänzelmann et al., 2004) was critical. Also, to maintain the reduced form of all the Cys residues, thiol-reducing reagent, either β-mercaptoethanol (βME) or dithiothreitol (DTT), was always included in the buffer throughout the purification.

With these modifications, MoaA can be overexpressed and purified under aerobic conditions. Our typical purification of MoaA is performed from *Escherichia coli* harvested from over 50 L of culture (usually around 200 g of cell paste) using Ni-NTA agarose resin in the presence of βME as a Cys protection reagent, followed by Sephadex G-25 size exclusion chromatography for exchange into buffer containing DTT. The aerobically purified MoaA harbors only 1.5–2 equiv. of iron and sulfide bound to each monomer, suggesting that even the structurally essential C-terminal cluster is only partially loaded. Therefore, to avoid inactivation of MoaA in the absence of the fully loaded C-terminal cluster, we usually reconstitute the cluster immediately after the purification. The cluster reconstitution and all the subsequent characterizations are carried out under strict anaerobic conditions (<0.1 ppm O_2).

1. Subclone *S. aureus moaA* gene into the *Nde*I and *BamHI* sites of the pET15b vector to give pET15b-HisMoaA (Hover et al., 2013).
2. Transform *E. coli* BL21(DE3) cells with the pET15b-HisMoaA and SUF plasmids (Hänzelmann et al., 2004) and culture the transformant overnight on an Lysogeny broth (LB) agar plate with 100 mg/L ampicillin (Amp) and 35 mg/L chloramphenicol (Cam) at 37°C.
3. Pick up a single colony and culture overnight in 5 mL of LB medium (100 mg/L Amp and 35 mg/L Cam) at 37°C.
4. Inoculate 250 mL of LB medium (100 mg/L Amp and 35 mg/L Cam) with the 5 mL overnight culture and incubate at 37°C and 200 rpm until saturation (~18 h).
5. Inoculate each of 6 × 1.5 L of LB medium (100 mg/L Amp and 35 mg/L Cam) with 40 mL of saturated overnight culture and incubate at 37°C, 200 rpm.
6. When OD_{600nm} reaches 0.9–1.1, cool down the culture to 30°C and induce the overexpression by adding isopropyl β-D-1-thiogalactopyranoside (IPTG) to the final concentration of 0.5 m*M*. Continue the culture at 30°C and 200 rpm for 4–5 h.
7. Harvest the cells by centrifugation at 5000 × *g*, 20 min, 4°C. The 9 L culture typically yields ~30 g of cell paste. The cell paste can be stored at −20°C.
8. Repeat steps 2–7 to accumulate ~200 g cell paste (usually requires over 50 L of LB culture).
9. Resuspend the cell paste (~200 g) with 500 mL of buffer A (50 m*M* Tris–HCl, pH 7.6, 0.3 *M* NaCl, 10% glycerol) supplemented with 40 m*M* imidazole, 3 m*M* βME, and 10 μL benzonase nuclease (EMD Millipore) and homogenize the cell suspension.
10. Lyse the cells by two passages through French Press at 14,000 psi.
11. Remove the cell debris by centrifugation at 25,000 × *g* for 30 min at 4°C. The supernatant should be dark brown.
12. Load the supernatant onto a Ni-NTA column (100–150 mL resin).
13. Wash the column with buffer A supplemented with 40 m*M* imidazole and 3 m*M* βME until no protein is eluted based on Bradford assay (typically need ~1 L buffer).
14. Elute MoaA with buffer A supplemented with 400 m*M* imidazole and 3 m*M* βME until no protein is eluted based on Bradford assay (typically need ~200 mL buffer).

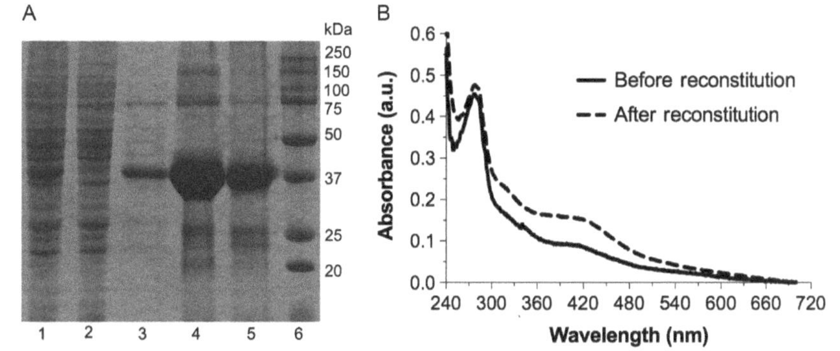

Fig. 4 Purification and reconstitution of *S. aureus* MoaA. (A) SDS-PAGE of MoaA at different purification steps. Lanes: 1, crude cell lysate; 2, Ni-NTA flow through; 3, Ni-NTA wash with 40 m*M* imidazole; 4, Ni-NTA elution with 400 m*M* imidazole; 5, after G-25; and 6, molecular weight marker. (B) UV–vis absorption spectra of MoaA (11.4 μ*M*) before (*solid line*) and after reconstitution (*dashed line*).

15. Concentrate the resultant solution to ~100 mL by ultrafiltration (30 kDa MWCO).
16. Load the eluted protein onto 1 L Sephadex G-25 column equilibrated in buffer A supplemented with 5 m*M* DTT.
17. Elute the protein with buffer A supplemented with 5 m*M* DTT. Check the fractions by UV–vis absorption spectra and pool the fractions containing MoaA.
18. Analyze each step of purification by sodium dodecyl sulfate polyacrylamide gel electrophoresis (SDS-PAGE) to check the purity and recovery (Fig. 4A).
19. Determine UV–vis spectra of as-isolated MoaA from 700 to 200 nm (Hitachi U-3900 UV–vis spectrometer). There should be absorption band from protein at 280 nm and that from the [4Fe–4S] clusters at 300–500 nm with a shoulder at ~420 nm (Fig. 4B, solid line).

2.2 MoaA Concentration Determination

Determination of protein concentrations for radical SAM enzymes is not trivial. The residual Fe–S cluster in the as-isolated MoaA significantly contributes to the absorption at 280 nm, and thus the concentration of MoaA cannot be determined directly by the UV absorbance of proteins at 280 nm. Instead, these enzymes are the best quantified by colorimetric protein assays, such as Bradford assay, compatible with Fe and other reducing agents. Since bovine serum albumin (BSA) standards give variable errors depending on the protein of interests (for MoaA, BSA gave >10% errors), it is necessary to prepare MoaA standards with known concentrations. The MoaA standards can be prepared by first removing all the copurified iron and sulfur under a denaturing condition, followed by refolding by exchanging the protein back into the native condition. The concentration of the apo-MoaA is determined using the UV absorption at 280 nm (Gill & Von Hippel, 1989). The refolded apo-MoaA is then used as a concentration standard for the Bradford assay of as-isolated and reconstituted MoaA.

1. Mix 0.5 mL of as-isolated MoaA with 1.5 mL of 8 *M* guanidine HCl, 13.3 m*M* ethylenediaminetetraacetic acid (EDTA), pH 8. Incubate at 4°C for 1 h.
2. Load onto a G-25 column (20 mL) equilibrated in buffer A supplemented with 5 m*M* DTT and elute with the same buffer.
3. Collect fractions every 0.5 mL of elution. Identify the fractions containing MoaA judged by Bradford assay and combine.
4. Centrifuge at 16,000 × *g* for 5 min to remove precipitants.
5. Dilute one part of the resulting apo-MoaA in three parts of 8 *M* guanidine HCl. The final concentration of guanidine HCl should be 6 *M*.
6. Determine the absorption at 280 nm by UV–vis spectrometer and calculate the protein concentration using $\varepsilon_{280\text{nm}} = 26{,}200\,M^{-1}\,\text{cm}^{-1}$. The extinction coefficient of the protein in 6 *M* guanidine HCl is calculated based on the following equation (Pace, Vajdos, Fee, Grimsley, & Gray, 1995): $\varepsilon_{280\text{nm}}(M^{-1}\,\text{cm}^{-1}) = (N_{\text{Trp}})(5500) + (N_{\text{Tyr}})(1490) + (N_{\text{Cys}})(125)$.
7. Use the apo-MoaA to make concentration standards. Determine the concentration of as-isolated MoaA by the Bradford assay. Typically, each purification yields 6–7 mg of as-isolated MoaA per gram of wet cell paste.

2.3 Reconstitution of the [4Fe–4S] Clusters

The two [4Fe–4S] clusters of MoaA can be chemically reconstituted. This reconstitution and all the following steps are performed under strictly anaerobic conditions in a glove box (M. Braun) that is maintained at <0.1 ppm O_2 and 10°C. While there have been several different ways to reconstitute the [4Fe–4S] clusters of radical SAM enzymes, such as using cysteine desulfurase as a sulfide source (Mahanta, Zhang, Hudson, van der Donk, & Mitchell, 2017), we found that chemical reconstitution using $(NH_4)_2Fe(SO_4)_2$ and Na_2S as iron and sulfide source, respectively, resulted in reproducible reconstitution.

For the chemical reconstitution, it is critical to minimize the formation of insoluble ferrous sulfide precipitant. This can be achieved by slow and careful titration of the solutions of $(NH_4)_2Fe(SO_4)_2$ and Na_2S to give enough time for self-assembly of the [4Fe–4S] clusters in the active site of MoaA. Also, to avoid an unnecessary excess of Fe^{2+} and S^{2-}, the amount of Fe–S cluster copurified in the as-isolated MoaA is quantified to adjust the amount of $(NH_4)_2Fe(SO_4)_2$ and Na_2S added during the reconstitution. The successful reconstitution typically gives slightly yellowish dark brown color while ferrous sulfide forms black precipitation or colloid. The quality of the reconstituted [4Fe–4S] clusters can also be assessed by UV–vis absorption spectra, which should show strong absorption of the [4Fe–4S] clusters in the 300–500 nm region with a characteristic shoulder at ~420 nm (Fig. 4B). Contamination of ferrous sulfide colloid results in very broad and featureless absorption from 200 to 800 nm.

1. Degas 1 L of buffer A and 1 L of Sephadex G-25 resin equilibrated and suspended in ~2 L of buffer A by 15 min evacuation followed by 10 min Ar refill.
2. Evacuate the flasks containing the degassed buffer and resin after the last cycle in step 1, and bring them into the glove box. Equilibrate the buffer and resin in the glove box atmosphere for at least 15 h.
3. Bring two portions of preweighted DTT powder (5 and 10 mmol) into the glove box and add 5 mmol to the degassed buffer A and 10 mmol to the degassed Sephadex G-25 resin immediately before use. Pack the Sephadex G-25 resin to a column (5×50 cm^2).
4. Remove oxygen from the as-isolated MoaA (~0.2 m*M*) solution by exchanging the atmosphere to Ar using a Schlenk line by brief evacuation (<30 s) followed by Ar refill and equilibration for 5 min. Repeat this cycle four times. During the evacuation, it is critical to minimize formation of bubbles to avoid unfolding of MoaA.
5. Evacuate the flask after the last cycle in step 4 and bring the sample into the glove box.
6. Determine the protein concentration by the Bradford assay as described in Section 2.2.
7. Quantitate the amount of $Fe^{2+/3+}$ and S^{2-} by following the reported protocols for the ferrozine (Wayne, 1988) and sulfide assays (Beinert, 1983), respectively. Typically, there is around 1.5–2 equiv. of $Fe^{2+/3+}$ and S^{2-} per monomer of MoaA.
8. Based on the concentration of MoaA and the amounts of $Fe^{2+/3+}$ and S^{2-} determined in steps 6 and 7, respectively, determine the amounts

of $(NH_4)_2Fe(SO_4)_2$ and Na_2S needed for the reconstitution. The sum of the copurified $Fe^{2+/3+}$ and S^{2-} and those added should be 9 equiv. per monomer of MoaA.

9. Introduce the preweighted $(NH_4)_2Fe(SO_4)_2$ and Na_2S crystals into the glove box and dissolve with degassed H_2O to make the final concentration 25 m*M*. Calculate the volume needed for the reconstitution. Typically 9–10 mL each of this solution is required.
10. Dropwise and gently add 1 mL of the 25 m*M* $(NH_4)_2Fe(SO_4)_2$ solution followed by 1 mL of the 25 m*M* Na_2S solution and repeat this cycle until the required volumes are added. Perform the addition over the course of 1 h and incubate the mixture for another 45 min.
11. Load the protein solution onto the 1 L Sephadex G-25 column prepared in step 3. Elute the protein using the same buffer.
12. Identify and combine the fractions containing the reconstituted MoaA based on UV–vis absorption spectra.
13. Aliquot the reconstituted MoaA solution in appropriate volumes, flash–freeze in liquid nitrogen and store at −80°C. Store under aerobic conditions at this temperature does not influence the activity. Do not thaw under aerobic conditions.
14. Determine the UV spectra of reconstituted MoaA using a septum-sealed screw-cap quartz cuvette (Starna Cells Inc.). As shown in Fig. 4B (dashed line), the absorption intensity at 300–500 nm derived from the [4Fe–4S] clusters should be higher for the successfully reconstituted MoaA compared to the as-isolated MoaA.
15. Determine the concentration of reconstituted MoaA by Bradford assay as described in Section 2.2.
16. Determine the amount of $Fe^{2+/3+}$ and S^{2-} bound to each monomer by ferrozine (Wayne, 1988) and sulfide assays (Beinert, 1983), respectively. Typically the yield is around 4–4.5 mg of reconstituted MoaA per gram of wet cell paste, with 6–7 equiv. of $Fe^{2+/3+}$ and S^{2-} per monomer of MoaA.

2.4 Activity Assays of MoaA

To characterize the MoaA catalytic activity, we established three activity assay methods, all of which are important to verify the quality of the reconstituted MoaA. In the first method, the MoaA product is chemically derivatized to DMPT through acid hydrolysis and incubation with 2,3-butanedione (Fig. 5A). DMPT is fluorescent and thus can be quantified

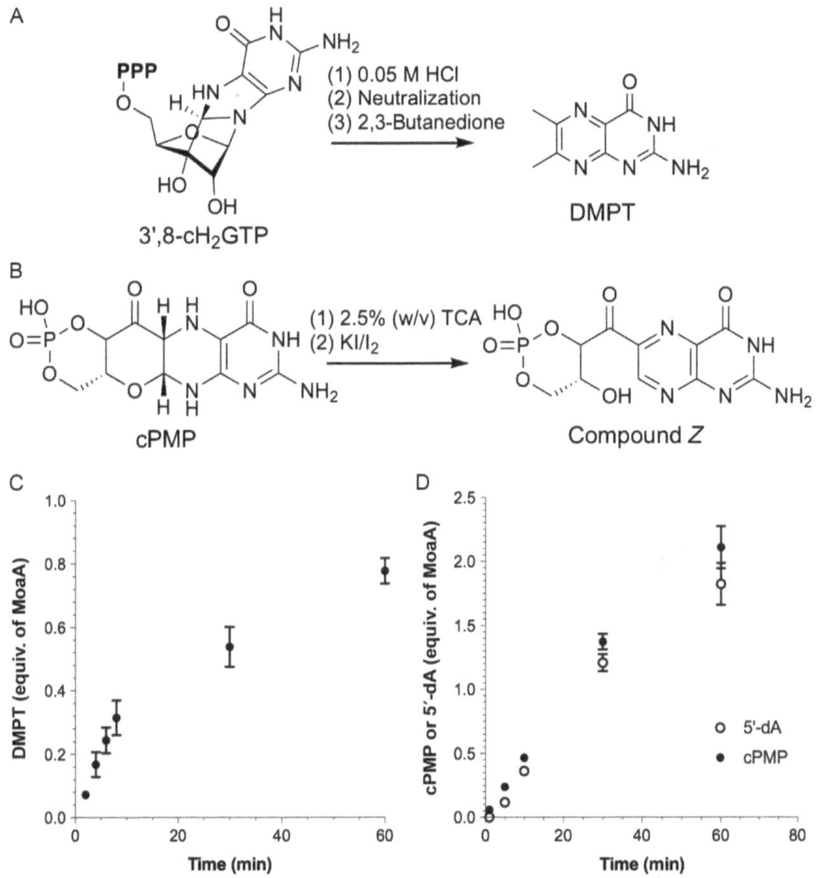

Fig. 5 Activity assays of MoaA. (A and B) Chemical derivatizations of 3′,8-cH_2GTP to DMPT (A) and cPMP to compound *Z* (B). (C) Time course of the MoaA reaction in the absence of MoaC. The amount of 3′,8-cH_2GTP formation was quantified by HPLC after its chemical derivatization to DMPT. (D) Time course of the formation of 5′-dA and cPMP during the catalysis in the coupled assay. cPMP was quantified by HPLC after its conversion to compound *Z*. Panels (C) and (D) are prepared using the previously published data (Hover et al., 2013).

by high-performance liquid chromatography (HPLC). This assay quantifies 3′,8-cH_2GTP and therefore can be used to characterize the function of MoaA independent of MoaC. However, many compounds can be converted to DMPT under the same conditions (Bracher et al., 1999; Hover, Tonthat, Schumacher, & Yokoyama, 2015). Therefore, it is advisable to confirm the observation by the coupled assay, in which the MoaA assays are performed in the presence of MoaC. In this assay, 3′,8-cH_2GTP is converted to cPMP in situ and the resulting cPMP is subsequently derivatized to compound *Z* (Fig. 5B) and quantified by HPLC. Compound *Z* authentic standard can be isolated from the *ΔmoeB E. coli* strain by following the published protocol (Wuebbens & Rajagopalan, 1993). This assay allows characterization of only the MoaA activity relevant to the cPMP formation. The third assay is the quantitation of 5′-deoxyadenosine (5′-dA) produced during the MoaA reaction, which can be combined with the coupled assay to determine the reaction stoichiometry. This stoichiometry is important for evaluating the quality of enzymes. MoaA with good quality exhibits excellent coupling of SAM cleavage to 3′,8-cH_2GTP formation. No more than 5% of abortive SAM cleavage should be observed (Hover & Yokoyama, 2015).

2.4.1 Quantitative Detection of the MoaA Product by Chemical Derivatization

3′,8-cH_2GTP has a 6-hydroxy-2,4,5-triaminopyrimidine partial structure connected to the other parts of the molecule through acid-labile bonds (aminal and glycosidic bonds; see Fig. 5A). Thus, 3′,8-cH_2GTP can be derivatized to DMPT by acid hydrolysis that releases 6-hydroxy-2,4,5-triaminopyrimidine from 3′,8-cH_2GTP, followed by neutralization and incubation with 2,3-butanedione, which derivatizes 6-hydroxy-2,4,5-triaminopyrimidine into DMPT. The resulting DMPT is fluorescent with excitation wavelength at 365 nm, and emission wavelength at 445 nm, and thus can be quantified by HPLC with excellent sensitivity. The lower limit of detection in typical MoaA assays is ~1 pmol.

1. Preincubate MoaA (10 μ*M*) and sodium dithionite (SDT, 2 m*M*) in buffer A supplemented with 5 m*M* DTT at 25°C in the glove box for 5 min.
2. Preincubate SAM (2 m*M*), GTP (2 m*M*) in buffer A supplemented with 5 m*M* DTT at 25°C in the glove box for 5 min.
3. Initiate the reaction by mixing equal volumes of solutions from steps 1 and 2, and incubate the reaction at 25°C.
4. Quench the reaction by mixing 40 μL of the reaction mixture with 5 μL of 0.5 *M* HCl at different time points (e.g., 2, 4, 6, 8, 10, 30, and 60 min). Usually, the product formation is in linear correlation with time over the first 10 min, after which it slows down significantly (Fig. 5C).
5. Transfer the quenched samples out of the glove box and incubate the samples at 95°C for 5 min.
6. Cool down and neutralize the sample by adding 3 μL of 1 *M* NaOH.

7. Add 12.5 μL of 0.66% (v/v) 2,3-butanedione in 0.9 *M* Tris–HCl (pH 8.5) and incubate at 95°C for 45 min.
8. Centrifuge at 16,000 × *g* for 10 min to remove precipitants.
9. Analyze the supernatant by HPLC (Hitachi L-2139 pump, L-2455 diode array detector, L-2485 fluorescence detector, L-2200 auto-sampler, and L-2300 column oven) equipped with an ODS hypersil column ($250 \times 4.6\,mm^2$, 3 μm) at 40°C. Chromatography is performed by isocratic elution with 92.5% 30 m*M* ammonium formate, pH 4.5, and 7.5% methanol at a flow rate of 1.5 mL/min and monitored by fluorescence (excitation 365 nm and emission 445 nm). Under this condition, DMPT elutes out at ~5.5 min.
10. To determine a standard curve for DMPT quantitation, inject different amounts of DMPT standards (e.g., 5, 10, 15, 20, and 25 pmol) to the HPLC system and analyze under the conditions described in step 9. From the resulting HPLC traces, determine the DMPT peak areas and plot against the amount of injection.
11. Determine the DMPT peak areas in the HPLC traces of the samples and quantify the amount of DMPT using the standard curve determined in step 10. As shown in Fig. 5C, the amount of DMPT increases with prolonged incubation time with a sign of product inhibition after ~10 min.

2.4.2 Coupled Assay of MoaA with MoaC

The DMPT assay quantifies the amounts of compounds with 6-hydroxy-2,4,5-triaminopyrimidine partial structure, which is present in not only 3′,8-cH_2GTP but many other compounds (Bracher et al., 1999; Hover et al., 2015). Also, as will be detailed in Section 4, 3′,8-cH_2GTP is not chemically stable. Therefore, it is preferable to assay MoaA in the presence of MoaC and detect cPMP. This assay is more relevant to the actual Moco biosynthesis inside the cells because MoaA and MoaC are always present in healthy cells. MoaC has n*M*-order K_m values toward 3′,8-cH_2GTP with the k_{cat} value 4–13 folds higher than that of MoaA (see Section 3, for details), and does not exhibit apparent product inhibition (compare Fig. 5C and D). Thus, the inclusion of MoaC in the MoaA assay allows selective and quantitative conversion of 3′,8-cH_2GTP to cPMP. The amount of cPMP can be quantified after its oxidative derivatization to compound *Z*, which fluoresces with the excitation and the emission wavelengths of 367 and 450 nm, respectively. In the protocol described below, the typical lower limit of detection for compound *Z* is ~1 pmol.

1. Preincubate MoaA (2 μ*M*) and SDT (2 m*M*) in buffer A supplemented with 5 m*M* DTT at 25°C in a glove box for 5 min.
2. Preincubate SAM (2 m*M*), GTP (2 m*M*), MoaC (20 μ*M*), and $MgCl_2$ (2 m*M*) in buffer A supplemented with 5 m*M* DTT at 25°C in a glove box for 5 min.
3. Initiate the reaction by mixing equal amounts of solutions from steps 1 to 2, and incubate the reaction at 25°C.
4. Quench the reaction by mixing 45 μL of the reaction mixture with 5 μL of 25% (w/v) trichloroacetic acid (TCA) at different time points (e.g., 2, 4, 6, 8, 10, 30, and 60 min).
5. Transfer the samples out of the glove box and add 5 μL of a solution containing 2% (w/v) KI and 1% (w/v) I_2 to each sample. Incubate at room temperature for 20 min.
6. Centrifuge at 16,000 × *g* for 10 min to remove precipitants.
7. Analyze the supernatant by the HPLC system equipped with an ODS hypersil column (250 × 4.6 mm^2, 3 μm). Perform chromatography by isocratic elution with 92.5% of 0.1% trifluoroacetic acid (TFA)/H_2O buffer, and 7.5% methanol with a flow rate of 1 mL/min and monitor the elution by fluorescence (excitation 367 nm and emission 450 nm). Under this condition, compound *Z* elutes at ~2.5 min.
8. To determine a standard curve for compound *Z* quantitation, inject different amounts of compound *Z* standards (e.g., 5, 10, 15, 20, and 25 pmol) to the HPLC system and analyze under the conditions described in step 7. From the resulting HPLC traces, determine the compound *Z* peak areas and plot against the amount of injection.
9. Determine the compound *Z* peak areas on the HPLC traces of the samples and quantify the amount of compound *Z* using the standard curve determined in step 8.

2.4.3 Determination of the Reaction Stoichiometry

MoaA initiates the radical reaction by reductive cleavage of SAM and uses a stoichiometric amount of SAM to produce 3′,8-cH_2GTP. In general, many radical SAM enzymes catalyze the abortive cleavage of SAM. In some cases, the abortive cleavage is caused by the inappropriate position of affinity tag (Grove, Ahlum, Sharma, Krebs, & Booker, 2010), or the use of non-physiological substrate (Yang et al., 2012), or reductant (Barr et al., 2016). Therefore, the extent of the abortive cleavage can be used for quality control of radical SAM enzyme preparation. For MoaA, under our assay conditions, we observed less than 5% abortive cleavage, suggesting an excellent coupling of SAM cleavage to 3′,8-cH_2GTP formation (Hover & Yokoyama, 2015). Therefore, any increase in the abortive cleavage from this level is a sign for inappropriate enzyme preparation or assay conditions. The reaction stoichiometry can be determined by quantifying the amount of 5′-dA produced during the catalysis, and its comparison with the amount of 3′,8-cH_2GTP. The amount of 5′-dA can be quantified by HPLC while the amount of 3′,8-cH_2GTP can be determined by the coupled assay as described in Section 2.4.2. The lower limit of 5′-dA detection is ~50 pmol.

1. Preincubate MoaA (40 μ*M*) and SDT (2 m*M*) in buffer A supplemented with 5 m*M* DTT at 25°C in the glove box for 5 min.
2. Preincubate SAM (2 m*M*), GTP (2 m*M*), MoaC (80 μ*M*), and $MgCl_2$ (2 m*M*) in buffer A supplemented with 5 m*M* DTT at 25°C in the glove box for 5 min.
3. Initiate the reaction by mixing equal volumes of solutions from steps 1 to 2 and incubate the reaction at 25°C.
4. Quench the reaction by mixing 90 μL reaction system with 10 μL 25% (w/v) TCA at different time points during the course of 60 min. The typical time points are 2, 4, 6, 8, 10, 30, and 60 min.
5. For quantification of cPMP, use 50 μL of the quenched samples and follow steps 4–8 in Section 2.4.2.
6. For quantification of 5′-dA, centrifuge the remaining quenched samples at 16,000 × *g* for 10 min to remove precipitants.
7. Analyze the supernatant from step 6 by the HPLC system equipped with an ODS hypersil column, 250 × 4.6 mm^2, 3 μm. Chromatography is performed by a linear gradient of 0%–15% methanol in 27 m*M* KH_2PO_4, pH 4.5 in 25 min with a flow rate of 2 mL/min and monitored by UV absorption (256 nm). Under this condition, 5′-dA elutes out at ~3.8 min.
8. To determine a standard curve for 5′-dA quantitation, inject different amounts of 5′-dA standards (e.g., 50, 100, 150, 200, and 250 pmol) to the HPLC system and analyze under the conditions described in step 7. From the resulting HPLC traces, determine the 5′-dA peak areas and plot against the amount of injection.
9. Determine the amounts of 5′-dA and cPMP formed at each time point using the standard curves. Stoichiometric formation of 5′-dA and 3′,8-cH_2GTP is expected (Fig. 5D).

3. PREPARATION AND CHARACTERIZATION OF MoaC

MoaC catalyzes the complicated rearrangement of 3′,8-cH_2GTP into cPMP (see Fig. 2B) (Hover et al., 2013). MoaC from two bacteria *E. coli* (Hover et al., 2015) and *S. aureus* (Hover et al., 2013), and MOCS1B, a human homolog of MoaC (Hover et al., 2013), have been functionally characterized. As shown in Table 1, the k_{cat} of His-tagged *S. aureus* MoaC with purified 3′,8-cH_2GTP is ~4 folds greater than that for *S. aureus* MoaA (Hover et al., 2013). The k_{cat} of His-tagged *E. coli* MoaC is even greater, although the rate for *E. coli* MoaA is unknown because *E. coli* MoaA has not been successfully expressed as a soluble protein (Hover et al., 2015). In addition, the K_m values toward 3′,8-cH_2GTP is <0.060, 0.21–0.25, and 0.79 μ*M* for His-tagged *S. aureus*, *E. coli* MoaC, and human MOCS1B, respectively (Hover et al., 2013), suggesting the recognition of 3′,8-cH_2GTP at n*M* concentrations regardless of the organismal origins. These observations suggest that the rate-determining step of cPMP formation is likely the MoaA catalysis and explain the absence of accumulation of 3′,8-cH_2GTP when both enzymes are present in vitro or in vivo.

The crystal structures have been reported for MoaC from *E. coli* (Hover et al., 2015; Wuebbens et al., 2000), *M. tuberculosis* (Srivastava, Srivastava, Arora, & Pratap, 2012), and an archaea *Sulfolobus tokodaii* (Yoshida, Yamada, Kuramitsu, & Kamitori, 2008). Based on these crystal structures, all MoaC adopts a hexameric structure, organized as a trimer of homodimers. The active site is located at the interface of two monomers in a dimer unit and consisted of six strictly conserved and catalytically essential amino acid residues (Hover et al., 2015).

Table 1 Steady-State Kinetic Parameters of MoaA and MoaC From Different Organisms (Hover et al., 2013, 2015)

Enzyme	Organism	Tag	Substrate	K_m (μ*M*)	k_{cat} (min^{-1})
MoaA	*S. aureus*	His-tagged	GTP	1.4 ± 0.2	0.045 ± 0.003
			SAM	4.1 ± 1.3	0.043 ± 0.004
MoaC	*S. aureus*	His-tagged	3′,8-cH_2GTP	<0.060	0.17 ± 0.026
	E. coli	His-tagged	3′,8-cH_2GTP	0.21 ± 0.058	0.56 ± 0.027
		Nontagged	3′,8-cH_2GTP	0.25 ± 0.040	0.53 ± 0.061
MOCS1B	*Homo sapiens*	His-tagged	3′,8-cH_2GTP	0.79 ± 0.24	0.092 ± 0.020

Since MoaC is not sensitive to oxygen, the enzyme can be purified under aerobic conditions. The activity assays, on the other hand, should be performed under anaerobic conditions because 3′,8-cH_2GTP is oxygen sensitive. Therefore, the purified MoaC is degassed and assayed in the anaerobic glove box. The ability to prepare and assay MoaC is critical for elucidation of cPMP biosynthesis, as this enzyme is responsible for the complicated rearrangement reaction of 3′,8-cH_2GTP into cPMP. Also, MoaC is used for quantitative detection of 3′,8-cH_2GTP (Section 4.1.2), as well as the coupled assay of MoaA (Section 2.4.2). Therefore, in this section, we will describe the preparation and characterization of *E. coli* MoaC as a model.

3.1 Expression and Purification of MoaC

The recombinant *E. coli* MoaC can be homologously expressed in *E. coli* with robust expression level and solubility. While *E. coli* MoaC was initially purified without an affinity tag by taking advantage of its uniquely hydrophobic nature (Wuebbens et al., 2000), we have demonstrated that N-terminal His-tag provides even shorter purification without significantly affecting the activity or crystallization (Hover et al., 2015). The His-tagged and nontagged *E. coli* MoaC exhibited essentially indistinguishable kinetic parameters (Table 1) (Hover et al., 2015). Therefore, here we describe His-tag affinity purification of *E. coli* MoaC.

1. Subclone *E. coli moaC* gene into the *Nco*I and *BamHI* sites of the pET30b vector to give pET30b-HisMoaC (Hover et al., 2015).
2. Transform *E. coli* BL21(DE3) with pET30b-HisMoaC plasmid and culture the transformant overnight on LB agar plate with 50 mg/L kanamycin (Kan) at 37°C.
3. Pick up a single colony and culture overnight in 5 mL of LB medium (50 mg/L Kan) at 37°C.
4. Inoculate 250 mL of LB medium (50 mg/L Kan) with the 5 mL of overnight culture and incubate at 37°C, 200 rpm until saturation (~18 h).
5. Inoculate each of 6 × 1.5 L of LB medium (50 mg/L Kan) with 40 mL of saturated overnight culture and incubate at 37°C, 200 rpm.
6. When OD_{600nm} reaches 0.9–1.1, induce the overexpression by adding IPTG to the final concentration of 0.5 m*M*. Continue the culture at 37°C, 200 rpm for 4–5 h.
7. Harvest the cells by centrifugation at 5000 × *g*, 20 min, 4°C. The 9 L culture typically yields ~30 g of cell paste. The cell paste can be stored at −20°C.

8. Resuspend the cell pellet with 75 mL of buffer A supplemented with 40 m*M* imidazole and 3 m*M* βME.
9. Lyse the cells by two passages through French Press at 14,000 psi.
10. Remove the cell debris by centrifugation at 25,000 × *g* for 30 min at 4°C.
11. To remove DNA, add 0.2 volumes of 6% (w/v) streptomycin sulfate in buffer A dropwise over 15 min at 4°C and stir the solution at 4°C for another 15 min.
12. Remove the precipitation by centrifugation at 25,000 × *g* for 20 min at 4°C.
13. Load the supernatant onto a Ni-NTA column (~15 mL resin).
14. Wash the column with buffer A supplemented with 40 m*M* imidazole and 3 m*M* βME until no protein is eluted based on Bradford assay (typically need ~200 mL buffer).
15. Elute MoaC with buffer A supplemented with 400 m*M* imidazole and 3 m*M* βME until no protein is eluted based on Bradford assay (typically need ~30 mL buffer).
16. Concentrate the purified MoaC to less than 15 mL by ultrafiltration (3 kDa MWCO).
17. To remove imidazole, load the protein solution onto a ~150 mL Sephadex G-25 column equilibrated in buffer A and elute MoaC with buffer A. Check the fractions by Bradford assay and pool the fractions containing MoaC.
18. Analyze each step of purification by SDS-PAGE to make sure the purity and recovery (Fig. 6A).

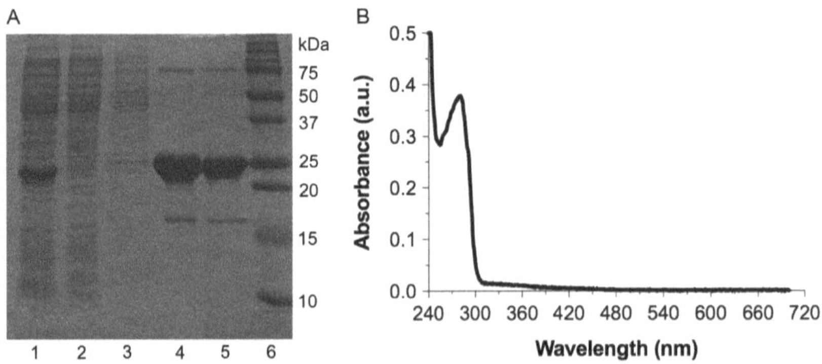

Fig. 6 Purification of *E. coli* MoaC. (A) SDS-PAGE of MoaC at different purification steps. Lanes: 1, crude cell lysate; 2, Ni-NTA flow through; 3, Ni-NTA wash with 40 m*M* imidazole; 4, Ni-NTA elution with 400 m*M* imidazole; 5, after G-25; and 6, molecular weight marker. (B) UV–vis absorption spectrum of MoaC.

19. Determine the UV–vis absorption spectra of purified MoaC by scanning from 700 to 200 nm (Fig. 6B). Pure MoaC should not absorb above 300 nm.
20. Concentrate MoaC to 0.8–1 m*M* by ultrafiltration (3 kDa MWCO). MoaC is stable for storage at these concentrations.
21. Remove oxygen from the MoaC solution by exchanging the atmosphere to Ar using a Schlenk line by brief evacuation (<30 s) followed by equilibration for 5 min. Repeat this cycle four times. During the evacuation, it is critical to minimize formation of bubbles to avoid unfolding of MoaC. Repeat this cycle for four times.
22. Evacuate the flask after the last cycle of step 21 and bring the sample into the glove box.
23. Determine MoaC concentration by the Edelhoch's method ($\varepsilon_{280\text{nm}} = 7370\, M^{-1}\ \text{cm}^{-1}$) (Gill & Von Hippel, 1989; Pace et al., 1995). Typically, 9–10 mg of MoaC is purified from each gram of wet cell paste.
24. Aliquot into appropriate volume, flash-freeze in liquid nitrogen and store at −80°C.

3.2 Activity Assay of MoaC Using a Crude MoaA Product

As described earlier, the k_{cat} value for MoaC is much greater than that for MoaA. As a result, in the presence of both enzymes, the rate of MoaC catalysis is masked by the rate-determining MoaA reaction. Therefore, the coupled assay described in Section 2.4.2 cannot be used to determine the activity of MoaC. Thus, we designed a stepwise assay which uses a crude MoaA product as MoaC substrate. In this assay, MoaA reaction is performed in the absence of MoaC and the resulting solution is passed through ultrafiltration. The resulting flow through fraction contains 3′,8-cH_2GTP. While this fraction also contains SAM, 5′-dA, L-methionine, GTP, and SDT, these molecules do not affect MoaC at the concentrations used for the MoaA assay. Therefore, this crude 3′,8-cH_2GTP can be used for MoaC characterization in the absence of purified 3′,8-cH_2GTP. The concentration of 3′,8-cH_2GTP can be determined by the DMPT derivatization, and the MoaC product, cPMP, can be quantified by HPLC after its chemical derivatization to compound *Z*.

1. Perform 5 mL MoaA reaction using 200 μ*M* MoaA, 200 μ*M* GTP, 1 m*M* SAM, and 1 m*M* SDT in buffer A with 5 m*M* DTT. Incubate the reaction at 25°C in a glove box for 1 h.
2. Remove MoaA by ultrafiltration (3 kDa MWCO).

3. Collect the small molecule fraction (typically around 4 mL). This crude MoaA product can be aliquoted, frozen in liquid nitrogen, and stored at −80°C.
4. Take out 40 μL of the small molecule fraction and mix with 5 μL 0.5 *M* HCl. Quantify 3′,8-cH_2GTP by DMPT assay as described in Section 2.4.1 (typically contains 60–100 μ*M* 3′,8-cH_2GTP).
5. Initiate MoaC reaction using 0.1 μ*M* MoaC, 10 μ*M* crude MoaA product in buffer A with 5 m*M* DTT and incubate the reaction at 25°C in a glove box.
6. Quench the reaction by mixing 45 μL of reaction mixture with 5 μL of 25% (w/v) TCA at different time points (e.g., 0.5, 1, 1.5, 2, 3, 5, 8, and 10 min).
7. Quantify the amount of cPMP by chemical derivatization as described in Section 2.4.2 (steps 4–8).

4. ISOLATION AND CHARACTERIZATION OF 3′,8-cH_2GTP

Although the presence of the MoaA product was implied in 2006 (Hänzelmann & Schindelin, 2006), the compound was not isolated and structurally characterized until our report in 2013 (Hover et al., 2013). In our study, it quickly became apparent that the MoaA product is not chemically stable. To identify the conditions in which the MoaA product can be handled, it was critical to understand its chemical properties. Therefore, we performed a stability test (Fig. 7A) (Hover et al., 2013), in which the MoaA product was incubated under various conditions. Such investigation revealed that the compound was particularly sensitive to the presence of oxygen or at acidic pH. Temperature also played some role because the degradation was significantly slower at lower temperatures. In contrast, no degradation was detectable for at least 160 min under an anaerobic condition at pH 9 and 10°C. These initial experiments formed the basis for the conditions to handle the MoaA product.

For the successful purification, facile approaches were required to quantitatively detect the MoaA product to identify the fractions after each chromatography step. Thus, we developed three approaches: UV–vis absorption spectra, MoaC assay, and anaerobic high-performance anion exchange chromatography (HPAEC). As detailed later, these assays have complementary benefits, and therefore, a combination of these assays allowed us to identify fractions containing the MoaA product after each chromatography.

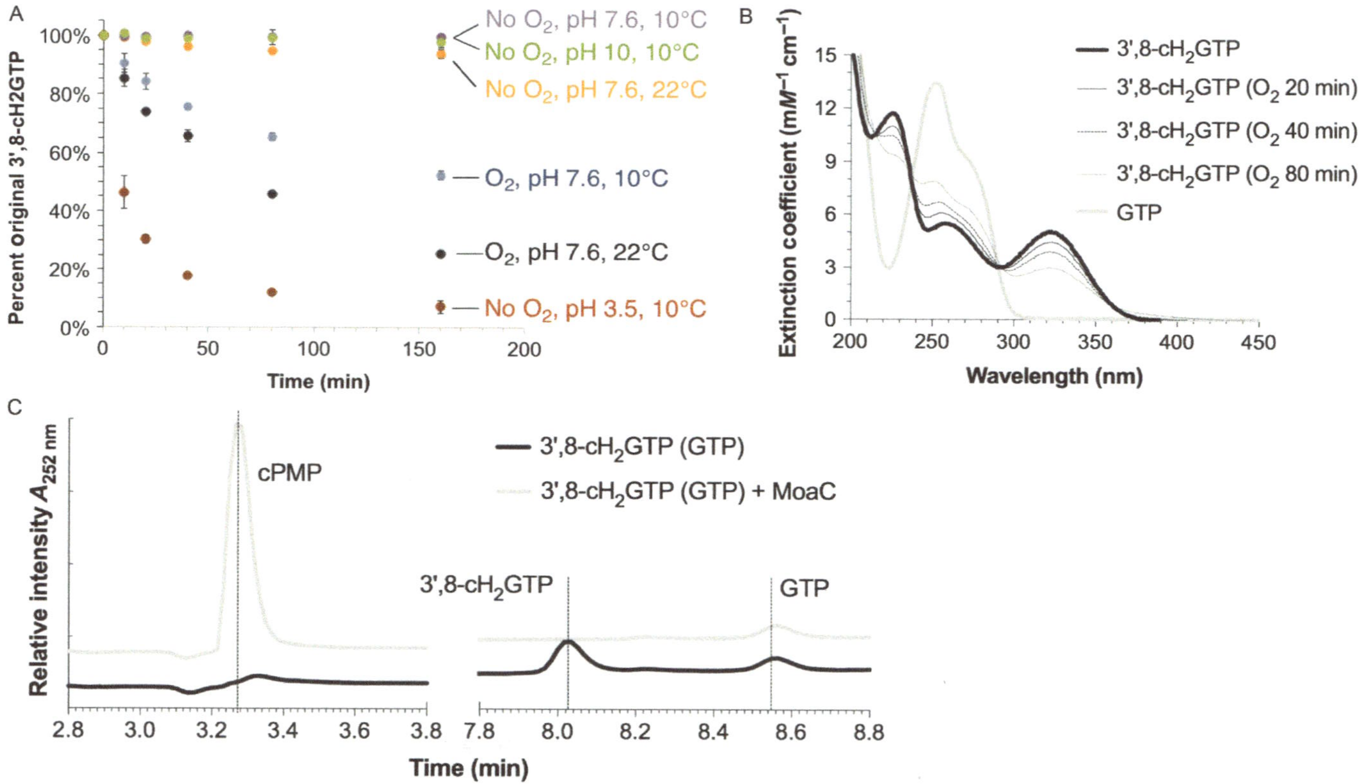

Fig. 7 Detection and quantitation of 3′,8-cH$_2$GTP. (A) Stability test of 3′,8-cH$_2$GTP. Crude MoaA product was incubated under different conditions and quantified after its conversion to cPMP by MoaC. (B) UV–vis absorption spectra of 3′,8-cH$_2$GTP (*solid black line*) and GTP (*solid gray line*). The thin lines represent the spectra of 3′,8-cH$_2$GTP after exposure to O_2 for the specified period. (C) Anaerobic HPAEC analysis of 3′,8-cH$_2$GTP, cPMP, and GTP monitored at 252 nm. *Panel (A) was reprinted with permission from Hover, B. M., Loksztejn, A., Ribeiro, A. A., & Yokoyama, K. (2013). Identification of a cyclic nucleotide as a cryptic intermediate in molybdenum cofactor biosynthesis.* Journal of the American Chemical Society, 135*(18), 7019–7032. doi:10.1021/ja401781t. Copyright 2013 American Chemical Society.*

Even after the above-described developments, when we attempted isolation of 3′,8-cH_2GTP, we encountered two major problems. First, the rate of MoaA product formation was slow and did not exceed the amount of MoaA even after 60 min of incubation (see Fig. 5C). Prolonged incubation resulted in the competition between the formation of 3′,8-cH_2GTP and its decomposition. Second, the separation of the MoaA product with some of the components in the regular assay conditions, such as DTT, glycerol, and GTP, was challenging. As detailed in Section 4.2, these challenges were overcome by optimizing the MoaA reaction conditions.

In this section, we will describe protocols for the quantitative detection of 3′,8-cH_2GTP, the optimized protocol for the 3′,8-cH_2GTP isolation and the structural characterization of 3′,8-cH_2GTP. We also describe our newly developed in situ NMR characterization, which allows structural characterization of the MoaA product without its isolation. These protocols will allow comprehensive characterization of the long-sought MoaA product, 3′,8-cH_2GTP, and can be used to characterize MoaA and MoaC homologs from other organisms.

4.1 Detection and Quantitation of 3′,8-cH2GTP

We established three methods for the detection and quantitation informative during the isolation of 3′,8-cH_2GTP. The first method uses UV–vis spectroscopy. 3′,8-cH_2GTP absorbs light at 322 nm, which is well outside the envelope of GTP, and thus detectable even in samples dominated by GTP. The second method incubates 3′,8-cH_2GTP with MoaC and detects cPMP after its conversion to compound *Z*. The third approach is an anaerobic HPAEC for quantitative detection of the intact 3′,8-cH_2GTP without derivatization.

4.1.1 Detection by UV–Vis Absorption Spectroscopy

Based on the unique UV absorption feature of 3′,8-cH_2GTP at 322 nm, UV–vis spectroscopy is an easy way to detect the compound. Fig. 7B shows the UV–vis absorption spectrum of 3′,8-cH_2GTP, together with the UV spectra of GTP which does not have UV absorbance above 300 nm. In the presence of oxygen, 3′,8-cH_2GTP has a relatively short half-life (~70 min at 10°C, Fig. 7A), and thus the detection has to be either quick or under anaerobic conditions. We found Nanodrop 1000 (Thermo Fisher Scientific) to be a convenient instrument to test multiple fractions quickly and it consumes only 1 μL of solution for each measurement. 3′,8-cH_2GTP can be quantified using the reported extinction coefficient ($5.1 \pm 1.1\ mM^{-1}\ cm^{-1}$)

at its λ_{max} (322 nm) (Hover et al., 2013). For a more accurate determination, we recommend the use of wavelength scanning UV–vis absorption spectrometer (U-3900H, Hitachi High Technologies America Inc., TX) and a septum-sealed screw-cap quartz cuvette (Starna Cells Inc., CA) to protect the anaerobic samples from the atmosphere.

The UV–vis absorption spectra can also be used to roughly estimate the purity of the compound. The isolated MoaA product has two major absorption bands at 252 and 322 nm with the extinction coefficients of 5.8 and 5.1 mM^{-1} cm^{-1}, respectively. Since GTP absorbs at 252 nm but not at 322 nm, the contamination of GTP can be measured based on the ratio of A_{252nm}/A_{322nm}. An increase of this ratio from 1.0 ± 0.1 indicates a GTP contamination.

It is important to note that the 322 nm absorption feature is derived from the base moiety, and thus this analysis does not distinguish di- and monophosphates (3′,8-cH_2GDP and 3′,8-cH_2GMP, respectively), which we frequently detected in our purification (see Section 4.2, for details). Therefore, complementary assays are required to distinguish different phosphorylation states.

4.1.2 Detection by MoaC Assay

More accurate amount of 3′,8-cH_2GTP can be determined based on MoaC assays. In a typical analysis, 10 μL of a sample is incubated with 80 μL of 20 μ*M* MoaC in buffer A at 25°C for 30 min to make sure complete conversion to cPMP. The reaction is then quenched by TCA and chemically derivatized using the method described in Section 2.4.2 (steps 4–8). Since the HPLC analysis is time consuming (15–20 min per sample), this assay is usually performed during lyophilization of the pooled fractions (see Section 4.2). 3′,8-cH_2GDP and 3′,8-cH_2GMP are not the substrate of MoaC, and thus this MoaC assay distinguishes these decomposed compounds from 3′,8-cH_2GTP and allows accurate quantitation of 3′,8-cH_2GTP in each fraction.

4.1.3 Detection by Anaerobic Anion Exchange HPLC (HPAEC)

HPLC under anaerobic conditions can be used to characterize the intact 3′,8-cH_2GTP. While we previously reported an ion-pairing HPLC under anaerobic conditions to detect 3′,8-cH_2GTP without derivatization (Hover et al., 2015), this approach gives a significant fluctuation in the retention times as well as relatively broad peak shapes, which limits the sensitivity. Therefore, we recently developed an anaerobic anion exchange HPLC method. In particular, we used Dionex ICS-5000+ HPAEC system equipped with DNAPac PA-100 column ($4 \times 250\,mm^2$, 2.7 μm), a column specialized for separation of nucleotides. This system features the ability to place the solvents under the inert gas atmosphere, allowing chromatography under anaerobic conditions.

Fig. 7C shows our representative analysis of 3′,8-cH_2GTP, GTP, and cPMP. Using GTP as a concentration standard, the amount of 3′,8-cH_2GTP could also be estimated even in the absence of purified 3′,8-cH_2GTP (note that accurate quantitation still requires purified 3′,8-cH_2GTP as the standard). Based on such quantitation, the purity of 3′,8-cH_2GTP can be determined. Thus, this HPAEC approach is complementary to the two methods described earlier.

1. Inside the glove box, place aliquots (~30 μL) of purification fractions into septum-sealed screw-cap HPLC vials with inserts for small volume samples.
2. Take out the sample vials and place into the autosampler rack maintained at 4°C.
3. Using an autosampler, inject an aliquot (5–25 μL) of the sample into an HPAEC system (Dionex ICS-5000+) equipped with a DNAPac PA100 column. The system and the column should be equilibrated in anaerobic H_2O for at least 30 min at 1.5 mL/min.
4. Perform chromatography by a gradient of 0–50% anaerobic 1 *M* NH_4OAc, pH 6 at 1 mL/min over 11 min, and monitor the elution by an ICS-3000 PDA diode array. Under these conditions, 3′,8-cH_2GTP elutes at ~8.0 min, GTP at ~8.6 min, and cPMP at ~3.2 min (Fig. 7C).
5. Using the diode array data, identify peaks with UV–vis absorption spectra that agree with 3′,8-cH_2GTP ($\lambda_{max} = 322$ nm, Fig. 7B).
6. Using the chromatograms at 252 nm, determine the areas for the 3′,8-cH_2GTP peaks.
7. Analyze GTP standards with a series of injection amounts (e.g., 10, 30, 50, 70, and 90 pmol) by HPAEC under the conditions identical to step 1.
8. Using the chromatograms at 252 nm, determine the areas for the GTP peaks and compute the standard curve.
9. Using the extinction coefficient of GTP ($13.6\,mM^{-1}\,cm^{-1}$) and 3′,8-cH_2GTP ($5.3 \pm 1.1\,mM^{-1}\,cm^{-1}$) at 252 nm (Hover et al., 2013), and the GTP standard curve determined in step 8, quantify the peak areas for 3′,8-cH_2GTP.

4.2 Isolation of 3′,8-cH2GTP

As described earlier, we met two major challenges during the isolation of 3′,8-cH_2GTP, the low turnover number of MoaA and the difficulty in the separation of 3′,8-cH_2GTP from other components in the assays. These problems were solved based on following optimizations. First, the concentration of GTP for the MoaA reaction was optimized to minimize the residual GTP after the reaction. Based on our investigation, the use of a stoichiometric amount of GTP and MoaA (200 μ*M*) gave the maximum (~50%) conversion of GTP to 3′,8-cH_2GTP. Second, the assay buffer was modified by eliminating DTT and glycerol that were difficult to separate from 3′,8-cH_2GTP. Glycerol, in particular, binds to anion exchange resin, elutes broadly, and contaminates the sample. We found that MoaA remains active at least for ~1 h without the stabilizing reagents (note that longer-term storage of MoaA requires DTT and glycerol). Therefore, we exchanged MoaA into 50 m*M* Tris–HCl, pH 7.6 immediately before the MoaA reaction. When MoaA reaction was performed in this buffer, no molecules other than GTP migrated close to 3′,8-cH_2GTP during the QAE Sephadex A25 anion exchange chromatography (Fig. 8A, left). The separation of GTP and 3′,8-cH_2GTP is still close, but they are separable on DEAE Sepharose FF column when GTP consumption is maximized by limiting the initial concentration of GTP (Fig. 8A, right) (Hover et al., 2013).

Fig. 8B shows the overview of the purification procedure, which typically provides >90% pure 3′,8-cH_2GTP in a 7.5–12.5% overall yield from GTP. The current major sources of loss are (1) the hydrolysis of 3′,8-cH_2GTP to 3′,8-cH_2GDP and 3′,8-cH_2GMP, (2) ultrafiltration, in which some of 3′,8-cH_2GTP binds MoaA and does not come through the membrane, and (3) the DEAE chromatography, where 3′,8-cH_2GTP elution partially overlaps with GTP. The hydrolysis is particularly problematic, and its cause is still unknown. Our current hypothesis is that it is caused by a minor contamination of *E. coli* endogenous phosphatase activity because the extent of hydrolysis varies by the MoaA preparation. The use of purer MoaA in the future may solve the problem. The protocol described below is our current optimized conditions, in which all the problems discussed earlier have been taken into consideration.

1. Prepare the following degassed buffer: 1.5 L of 1 *M* NH_4HCO_3 (pH 9.0), 3 L of H_2O, 1 L of 50 m*M* Tris–HCl buffer (pH 7.6).
2. Prepare the following degassed resin: Sephadex G-25 resin (~1 L, equilibrated in 50 m*M* Tris–HCl buffer, pH 7.6), QAE Sephadex

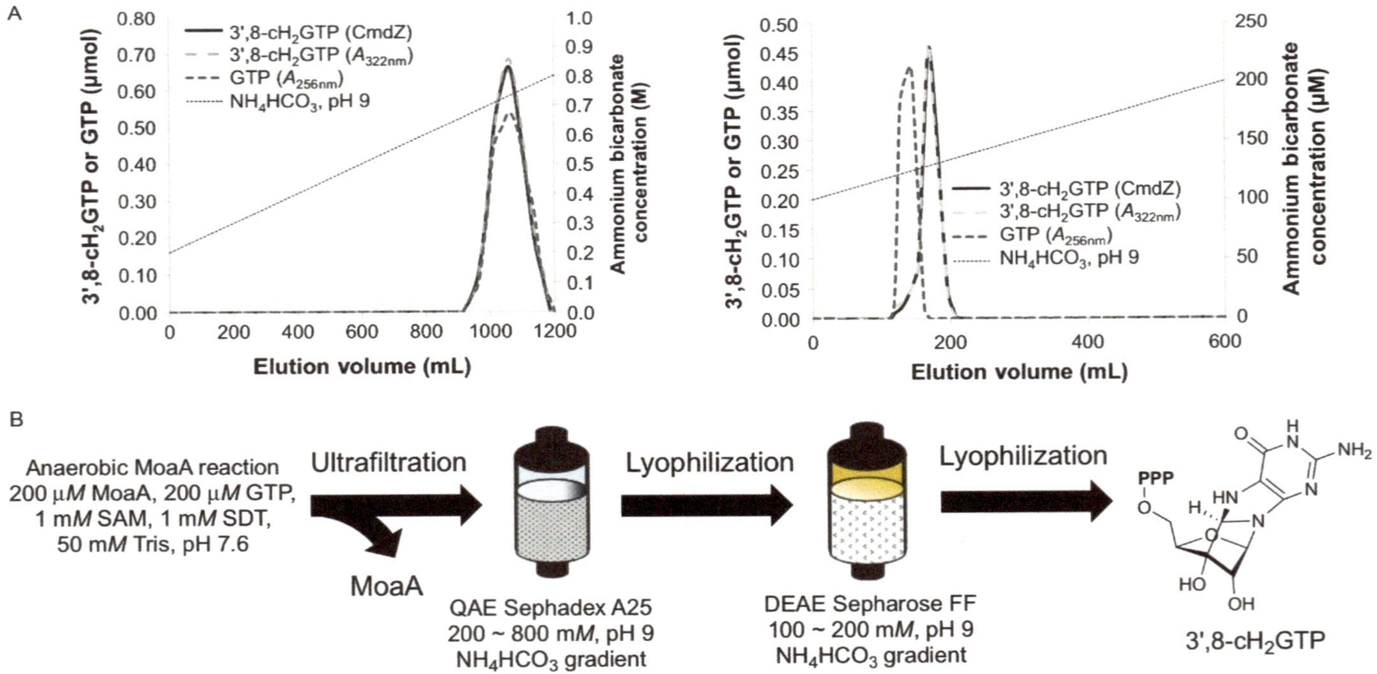

Fig. 8 Isolation of 3′,8-cH$_2$GTP. (A) Chromatograms of 3′,8-cH$_2$GTP purification on QAE Sephadex A25 (*left*) and DEAE Sepharose FF (*right*) columns. The gradient of NH_4HCO_3 (pH 9) buffer for each chromatography is indicated by *dotted lines*. The amount of 3′,8-cH$_2$GTP is quantified by either UV absorbance (*gray dashed lines*) at 322 nm or MoaC assay (*black solid lines*). The amount of GTP is quantified by UV absorbance at 256 nm after subtraction of the contribution from 3′,8-cH$_2$GTP based on its extinction coefficient at 256 nm (*black dashed lines*). (B) Overall scheme for the 3′,8-cH$_2$GTP isolation. *Panel (A) is reprinted with permission from Hover, B. M., Loksztejn, A., Ribeiro, A. A., & Yokoyama, K. (2013). Identification of a cyclic nucleotide as a cryptic intermediate in molybdenum cofactor biosynthesis.* Journal of the American Chemical Society, 135*(18), 7019–7032. doi:10.1021/ja401781t. Copyright 2013 American Chemical Society.*

A25 resin (~300 mL, equilibrated in 200 m*M* NH_4HCO_3, pH 9.0), and DEAE Sepharose FF resin (~20 mL, equilibrated in 100 m*M* NH_4HCO_3, pH 9.0).

3. Exchange MoaA into 50 m*M* Tris–HCl buffer (pH 7.6) by the degassed Sephadex G-25 column.
4. Perform a large-scale MoaA reaction in the glove box. 200 μ*M* MoaA (~1.6 g) is incubated with 200 μ*M* GTP, 1 m*M* SAM, and 1 m*M* SDT in 200 mL of 50 m*M* Tris–HCl buffer (pH 7.6) at 25°C in a metal beads bath for 60 min. (This scale of reaction requires MoaA from two large-scale preparation of MoaA with 200 g of *E. coli* cell paste as described earlier.)
5. Pass the reaction solution through an Amicon 30 kDa MWCO filtration membrane (Millipore) and collect the flow through.
6. Wash with 30 mL of ddH_2O for four times and combine the flow through.
7. Save an aliquot (10 μL) from steps 4–6 to perform MoaC assay.
8. Apply the combined flow through fractions onto the QAE Sephadex A25 column (300 mL) and wash with the column 750 mL of H_2O.
9. Elute the bound nucleotides by a 1200 mL linear gradient of 200–800 m*M* NH_4HCO_3 buffer and fractionate the elution every 14 mL using a fraction collector (AKTA frac-920).
10. Identify fractions containing 3′,8-cH_2GTP by Nanodrop. 3′,8-cH_2GTP usually comigrates with GTP. Save an aliquot (10 μL) from each fraction to perform MoaC assay.
11. Combine the fractions containing 3′,8-cH_2GTP and freeze in liquid nitrogen.
12. Take out from the glove box and lyophilize.
13. While lyophilizing, perform MoaC assays for the 10 μL aliquots saved in steps 7 and 10 by following the protocol described in Section 4.1.2 to make sure the detected 322 nm absorption in these fractions indeed derives from 3′,8-cH_2GTP and quantify its amount.
14. Bring the lyophilized fractions back to the glove box and dissolve the powder in 3–5 mL of 100 m*M* NH_4HCO_3 buffer in the glove box.
15. Apply the solution to a 15 mL DEAE Sepharose FF column and wash the column with 50 mL of 100 m*M* NH_4HCO_3 buffer.
16. Elute 3′,8-cH_2GTP with a 600 mL linear gradient of 100–200 m*M* NH_4HCO_3 buffer and fractionate the elution every 9 mL using a fraction collector (AKTA Frac-920).

17. Check the UV–vis absorption spectra for each fraction. Fractions with pure 3′,8-cH_2GTP should exhibit the A_{252nm}/A_{322nm} ratio of 1.0 ± 0.1. GTP contamination increases this ratio. Save an aliquot (10 μL) from each fraction to perform MoaC assay.
18. Freeze individual fractions containing pure 3′,8-cH_2GTP in liquid nitrogen.
19. Take out from the glove box and lyophilize.
20. While lyophilizing, perform MoaC assays for the 10 μL aliquots saved in step 17 by following the protocol described in Section 4.1.2 to quantify the amount of 3′,8-cH_2GTP. Alternatively, this step can be replaced by the anaerobic HPAEC analysis (Section 4.1.3), for which 40 μL aliquots should be saved in step 17.
21. Dissolve the lyophilized fractions with pure 3′,8-cH_2GTP in H_2O.
22. Measure the final concentration by the MoaC assay and UV absorption at 322 nm ($\varepsilon_{322nm} = 5.1 \pm 1.1\, mM^{-1}\, cm^{-1}$).
23. Determine the purity based on the anaerobic HPAEC analysis (Section 4.1.3). Typically, this protocol provides 3–5 μmol of 3′,8-cH_2GTP (7.5%–12.5% yield) with a purity >90%.
24. Aliquot, flash–freeze in liquid nitrogen, and store the purified 3′,8-cH_2GTP at −80°C.

4.3 Structural Characterization of 3′,8-cH2GTP

The structural characterization of 3′,8-cH_2GTP can be performed using MS and NMR spectroscopy. For NMR, it is recommended to use uniformly ^{13}C-labeled GTP ([U-$^{13}C_{10}$]GTP), whereas the MS analysis should be performed with natural isotope abundance to avoid any complication from partial labeling.

4.3.1 Mass Spectroscopy

The molecular weight of purified 3′,8-cH_2GTP can be determined by a high-resolution electrospray ionization time-of-flight mass spectrometry (ESI-TOF-MS), which revealed $m/z = 521.984$ $[M - H]^-$ (Fig. 9A, left). This mass signal is not only consistent with the molecular weight of 3′,8-cH_2GTP but also with that of GTP which migrates very close to 3′,8-cH_2GTP during the purification. To rule out the possibility that the signal is derived from contaminated GTP, incubate the sample with MoaC, remove MoaC by passing through 3 kDa MWCO membrane, and then analyze by MS. The original peak at 521.984 should disappear, with a new peak

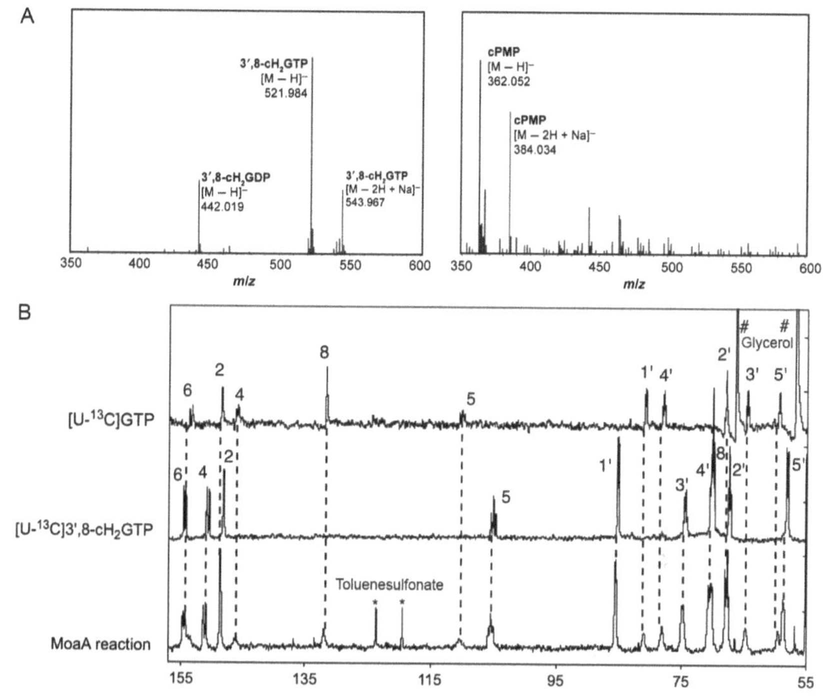

Fig. 9 Characterization of the MoaA product, 3′,8-cH_2GTP. (A) ESI-TOF-MS spectra of isolated 3′,8-cH_2GTP (*left*) and the same sample incubated with MoaC (*right*). (B) In situ ^{13}C NMR spectra. Shown are the spectra for the MoaA reaction with [U-^{13}C]GTP (*bottom*), [U-^{13}C]3′,8-cH_2GTP (*middle*), and [U-^{13}C]GTP (*top*) standards. * and # are signals derived from toluenesulfonate and glycerol, respectively. *Panel (A) is reprinted with permission from Hover, B. M., Loksztejn, A., Ribeiro, A. A., & Yokoyama, K. (2013). Identification of a cyclic nucleotide as a cryptic intermediate in molybdenum cofactor biosynthesis.* Journal of the American Chemical Society, 135*(18), 7019–7032. doi:10.1021/ja401781t. Copyright 2013 American Chemical Society. Panel (B) was adapted with permission from Hover, B. M., Tonthat, N. K., Schumacher, M. A., & Yokoyama, K. (2015). Mechanism of pyranopterin ring formation in molybdenum cofactor biosynthesis.* Proceedings of the National Academy of Sciences of the United States of America, 112*(20), 6347–6352. doi:10.1073/pnas.1500697112.*

forming at $m/z = 362.052$ $[M - H]^-$ (Fig. 9A, right), consistent with cPMP. These analyses will confirm the identity of the original mass signal indeed comes from 3′,8-cH_2GTP.

4.3.2 NMR

Detailed structural characterization is mainly accomplished using NMR. To facilitate the analysis, uniformly ^{13}C-labeled 3′,8-cH_2GTP ([U-^{13}C]3′,8-cH_2GTP) is prepared from [U-^{13}C]GTP (Sigma-Aldrich) by following the protocol identical to that for the nonlabeled compound (Section 4.2). In our analysis, ^{13}C NMR was a surprise as it demonstrated the presence of a tertiary carbon (C-3′) that was absent in any of the previously proposed MoaA product, including pyranopterin triphosphate. The most informative experiment was ^{13}C—^{13}C homonuclear correlation spectroscopy (COSY), which demonstrated the covalent bridge between C-3′ and C-8, and allowed annotation of all carbon signals. Together with ^{1}H NMR, heteronuclear multiple quantum coherence (HMQC), heteronuclear multiple bond correlation (HMBC), and nuclear Overhauser effect spectroscopy (NOESY), these experiments unambiguously determined the structure of 3′,8-cH_2GTP. The details of the NMR data and analysis can be found in our previous publication (Hover et al., 2013).

4.4 In Situ NMR Characterization

The isolation and NMR characterization of the MoaA product are powerful ways to determine its structure. However, the current protocol requires a lot of materials and efforts, which may not be affordable to many researchers. Although the chemical derivatization and UV–vis spectroscopy are simple and easy to perform, they give limited information about the structure of the molecule. Therefore, we developed the in situ ^{13}C NMR method that allows ^{13}C NMR detection of the MoaA product without isolation or derivatization (Hover et al., 2015). In this approach, the MoaA reaction is performed with [U-^{13}C]GTP (1 m*M*) as the substrate and ^{13}C NMR spectrum of the reaction mixture is determined.

1. Incubate MoaA (0.4 m*M*) with [U-^{13}C]GTP (1 m*M*), SAM (1 m*M*), and SDT (5 m*M*) in buffer A supplemented with 1 m*M* $MgCl_2$ and 2 m*M* DTT at 25°C in a glove box for 120 min.
2. Add 5% (v/v) anaerobic D_2O, transfer the solution to an NMR tube, and seal the tube with a rubber septum.
3. Perform a control reaction without MoaA and add 5% (v/v) anaerobic D_2O.
4. Collect ^{13}C NMR data for the two samples at 6°C for overnight (~16 h).

Fig. 9B shows our representative in situ ^{13}C NMR data. By comparing the signals from the MoaA reaction mixture (bottom) with the spectra of authentic [U-^{13}C]GTP (top) and [U-^{13}C]3′,8-cH_2GTP (middle), all the observed signals were assigned to either GTP or 3′,8-cH_2GTP except for some known impurities. In practice, [U-^{13}C]3′,8-cH_2GTP standard is not required because under these conditions, the chemical shifts of [U-^{13}C]

3′,8-cH_2GTP are highly reproducible. A comparison with the published spectra (Hover et al., 2013, 2015) is sufficient to assign all the signals derived from [U-^{13}C]3′,8-cH_2GTP. This in situ NMR method provides structural information about the MoaA product without any workup, and therefore avoids potential artifacts during isolation or chemical derivatization. Owing to the simplicity of the methodology, it can be used to test the generality of 3′,8-cH_2GTP as the product of MoaA from various organisms.

5. SUMMARY AND CONCLUSION

The mechanism of biosynthesis of the pterin ring of Moco was the long-standing mystery in the enzymology and cofactor biosynthesis fields. This transformation requires two enzymes, MoaA and MoaC, but their functions had remained ambiguous since the 1990s. Recently, successful isolation and structural characterization of the long-sought MoaA product, 3′,8-cH_2GTP, has finally elucidated the functions of the two enzymes and provided breakthrough insights into the early steps in the Moco biosynthesis. In this chapter, we described protocols for characterization of MoaA and MoaC, and the isolation of 3′,8-cH_2GTP.

The discovery of 3′,8-cH_2GTP was made possible based on the quantitative analyses. The quantitative activity assay of MoaA revealed its substoichiometric turnover, and the necessity to perform large-scale preparation of MoaA to isolate 3′,8-cH_2GTP. The quantitative assays of MoaC allowed identification of 3′,8-cH_2GTP during its purification. The quantitative analysis of 3′,8-cH_2GTP was critical to identify the conditions to handle this highly oxygen-sensitive compound. Thus, the quantitative characterization protocols described in this chapter all contributed to the successful identification of 3′,8-cH_2GTP.

Importantly, many of the protocols described in this chapter are based on the rich history of the biochemistry of Moco and other pterin cofactors. For example, the MoaA/MoaC coupled assay was based on the report that cPMP can be selectively and oxidatively derivatized to compound *Z* (Hänzelmann et al., 2004; Wuebbens & Rajagopalan, 1993). The chemical derivatization of MoaA product to DMPT was first mentioned in 2006 (Hänzelmann & Schindelin, 2006), but has its origin in the characterization of an intermediate in the GTP cyclohydrolase I catalysis (Bracher et al., 1999). Therefore, the discovery of 3′,8-cH_2GTP is a result of decades of studies on Moco and other pterin cofactor biosynthesis by numbers of labs in the field. We hope that the protocols described in this chapter will be useful for the future research in the Moco biosynthesis or other related fields.

The characterization of 3′,8-cH_2GTP emphasizes the importance of structural characterization of unstable intermediates existing in biosynthetic pathways. To this end, it is critical to be able to characterize the biosynthetic enzymes and the intermediates quantitatively. Therefore, while the protocols described here are specific to Moco biosynthesis, the fundamental concepts of quantitative biochemistry apply to other biosynthetic pathways and enzymes.

REFERENCES

Barr, I., Latham, J. A., Iavarone, A. T., Chantarojsiri, T., Hwang, J. D., & Klinman, J. P. (2016). Demonstration that the radical S-adenosylmethionine (SAM) enzyme PqqE catalyzes de novo carbon–carbon cross-linking within a peptide substrate PqqA in the presence of the peptide chaperone PqqD. *Journal of Biological Chemistry, 291*(17), 8877–8884. https://doi.org/10.1074/jbc.C115.699918.

Beinert, H. (1983). Semi-micro methods for analysis of labile sulfide and of labile sulfide plus sulfane sulfur in unusually stable iron–sulfur proteins. *Analytical Biochemistry, 131*(2), 373–378.

Boshoff, H. I., & Barry, C. E., 3rd. (2005). Tuberculosis—Metabolism and respiration in the absence of growth. *Nature Reviews. Microbiology, 3*(1), 70–80. https://doi.org/10.1038/nrmicro1065.

Bracher, A., Fischer, M., Eisenreich, W., Ritz, H., Schramek, N., Boyle, P., et al. (1999). Histidine 179 mutants of GTP cyclohydrolase I catalyze the formation of 2-amino-5-formylamino-6-ribofuranosylamino-4 (3H)-pyrimidinone triphosphate. *Journal of Biological Chemistry, 274*(24), 16727–16735.

Burg, A. W., & Brown, G. M. (1968). The biosynthesis of folic acid VIII. Purification and properties of the enzyme that catalyzes the production of formate from carbon atom 8 of guanosine triphosphate. *Journal of Biological Chemistry, 243*(9), 2349–2358.

Folkesson, A., Jelsbak, L., Yang, L., Johansen, H. K., Ciofu, O., Hoiby, N., et al. (2012). Adaptation of Pseudomonas aeruginosa to the cystic fibrosis airway: An evolutionary perspective. *Nature Reviews. Microbiology, 10*(12), 841–851. https://doi.org/10.1038/nrmicro2907.

Gill, S. C., & Von Hippel, P. H. (1989). Calculation of protein extinction coefficients from amino acid sequence data. *Analytical Biochemistry, 182*(2), 319–326.

Goldman, P. J., Grove, T. L., Booker, S. J., & Drennan, C. L. (2013). X-ray analysis of butirosin biosynthetic enzyme BtrN redefines structural motifs for AdoMet radical chemistry. *Proceedings of the National Academy of Sciences of the United States of America, 110*(40), 15949–15954. https://doi.org/10.1073/Pnas.1312228110.

Grove, T. L., Ahlum, J. H., Sharma, P., Krebs, C., & Booker, S. J. (2010). A consensus mechanism for radical SAM-dependent dehydrogenation? BtrN contains two [4Fe-4S] clusters. *Biochemistry, 49*(18), 3783–3785. https://doi.org/10.1021/bi9022126.

Hänzelmann, P., Hernandez, H. L., Menzel, C., Garcia-Serres, R., Huynh, B. H., Johnson, M. K., et al. (2004). Characterization of MOCS1A, an oxygen-sensitive iron–sulfur protein involved in human molybdenum cofactor biosynthesis. *Journal of Biological Chemistry, 279*(33), 34721–34732. https://doi.org/10.1074/Jbc.M313398200.

Hänzelmann, P., & Schindelin, H. (2004). Crystal structure of the S-adenosylmethionine-dependent enzyme MoaA and its implications for molybdenum cofactor deficiency in humans. *Proceedings of the National Academy of Sciences of the United States of America, 101*(35), 12870–12875. https://doi.org/10.1073/pnas.0404624101.

Hänzelmann, P., & Schindelin, H. (2006). Binding of 5′-GTP to the C-terminal FeS cluster of the radical S-adenosylmethionine enzyme MoaA provides insights into its mechanism. *Proceedings of the National Academy of Sciences of the United States of America, 103*(18), 6829–6834. https://doi.org/10.1073/pnas.0510711103.

Hover, B. M., Loksztejn, A., Ribeiro, A. A., & Yokoyama, K. (2013). Identification of a cyclic nucleotide as a cryptic intermediate in molybdenum cofactor biosynthesis. *Journal of the American Chemical Society, 135*(18), 7019–7032. https://doi.org/10.1021/ja401781t.

Hover, B. M., Tonthat, N. K., Schumacher, M. A., & Yokoyama, K. (2015). Mechanism of pyranopterin ring formation in molybdenum cofactor biosynthesis. *Proceedings of the National Academy of Sciences of the United States of America, 112*(20), 6347–6352. https://doi.org/10.1073/pnas.1500697112.

Hover, B. M., & Yokoyama, K. (2015). C-terminal glycine-gated radical initiation by GTP 3′,8-cyclase in the molybdenum cofactor biosynthesis. *Journal of the American Chemical Society, 137*(9), 3352–3359. https://doi.org/10.1021/ja512997j.

Hurdle, J. G., O'Neill, A. J., Chopra, I., & Lee, R. E. (2011). Targeting bacterial membrane function: An underexploited mechanism for treating persistent infections. *Nature Reviews. Microbiology, 9*(1), 62–75. https://doi.org/10.1038/nrmicro2474.

Johnson, J. L., Waud, W. R., Rajagopalan, K. V., Duran, M., Beemer, F. A., & Wadman, S. K. (1980). Inborn errors of molybdenum metabolism: Combined deficiencies of sulfite oxidase and xanthine dehydrogenase in a patient lacking the molybdenum cofactor. *Proceedings of the National Academy of Sciences of the United States of America, 77*(6), 3715–3719.

Kanaujia, S. P., Jeyakanthan, J., Nakagawa, N., Balasubramaniam, S., Shinkai, A., Kuramitsu, S., et al. (2010). Structures of apo and GTP-bound molybdenum cofactor biosynthesis protein MoaC from Thermus thermophilus HB8. *Acta Crystallographica. Section D, Biological Crystallography, 66*(Pt 7), 821–833. https://doi.org/10.1107/S0907444910019074.

Lees, N. S., Hänzelmann, P., Hernandez, H. L., Subramanian, S., Schindelin, H., Johnson, M. K., et al. (2009). ENDOR spectroscopy shows that guanine N1 binds to [4Fe-4S] cluster II of the S-adenosylmethionine-dependent enzyme MoaA: Mechanistic implications. *Journal of the American Chemical Society, 131*(26), 9184–9185. https://doi.org/10.1021/ja903978u.

Leimkuhler, S., Wuebbens, M. M., & Rajagopalan, K. V. (2011). The history of the discovery of the molybdenum cofactor and novel aspects of its biosynthesis in bacteria. *Coordination Chemistry Reviews, 255*(9–10), 1129–1144. https://doi.org/10.1016/j.ccr.2010.12.003.

Mahanta, N., Zhang, Z., Hudson, G. A., van der Donk, W. A., & Mitchell, D. A. (2017). Reconstitution and substrate specificity of the radical S-adenosyl-methionine thiazole C-methyltransferase in thiomuracin biosynthesis. *Journal of the American Chemical Society, 139*(12), 4310–4313. https://doi.org/10.1021/jacs.7b00693.

Mehta, A. P., Hanes, J. W., Abdelwahed, S. H., Hilmey, D. G., Hänzelmann, P., & Begley, T. P. (2013). Catalysis of a new ribose carbon-insertion reaction by the molybdenum cofactor biosynthetic enzyme MoaA. *Biochemistry, 52*(7), 1134–1136. https://doi.org/10.1021/bi3016026.

Mendel, R. R., & Schwarz, G. (2011). Molybdenum cofactor biosynthesis in plants and humans. *Coordination Chemistry Reviews, 255*, 1145–1158.

Pace, C. N., Vajdos, F., Fee, L., Grimsley, G., & Gray, T. (1995). How to measure and predict the molar absorption coefficient of a protein. *Protein Science*, *4*(11), 2411–2423.

Reiss, J. (2000). Genetics of molybdenum cofactor deficiency. *Human Genetics*, *106*(2), 157–163. https://doi.org/10.1007/s004399900223.

Rieder, C., Eisenreich, W., O'Brien, J., Richter, G., Gotze, E., Boyle, P., et al. (1998). Rearrangement reactions in the biosynthesis of molybdopterin—An NMR study with multiply 13C/15N labelled precursors. *European Journal of Biochemistry*, *255*(1), 24–36.

Shiota, T., Palumbo, M. P., & Tsai, L. (1967). A chemically prepared formamidopyrimidine derivative of guanosine triphosphate as a possible intermediate in pteridine biosynthesis. *Journal of Biological Chemistry*, *242*(8), 1961–1969.

Sofia, H. J., Chen, G., Hetzler, B. G., Reyes-Spindola, J. F., & Miller, N. E. (2001). Radical SAM, a novel protein superfamily linking unresolved steps in familiar biosynthetic pathways with radical mechanisms- functional characterization using new analysis and information visualization methods. *Nucleic Acids Research*, *29*(5), 1097–1106.

Srivastava, S., Srivastava, V. K., Arora, A., & Pratap, J. V. (2012). Overexpression, purification, crystallization and preliminary X-ray analysis of putative molybdenum cofactor biosynthesis protein C (MoaC2) from Mycobacterium tuberculosis H37Rv. *Acta Crystallographica. Section F: Structural Biology and Crystallization Communications*, *68*(Pt 6), 687–691. https://doi.org/10.1107/S174430911201665X.

Wayne, W. (1988). Rapid colorimetric micromethod for the quantitation of complexed iron in biological samples. *Methods in Enzymology*, *158*, 357–364.

Wuebbens, M. M., Liu, M. T., Rajagopalan, K., & Schindelin, H. (2000). Insights into molybdenum cofactor deficiency provided by the crystal structure of the molybdenum cofactor biosynthesis protein MoaC. *Structure*, *8*(7), 709–718.

Wuebbens, M. M., & Rajagopalan, K. (1993). Structural characterization of a molybdopterin precursor. *Journal of Biological Chemistry*, *268*(18), 13493–13498.

Wuebbens, M. M., & Rajagopalan, K. V. (1995). Investigation of the early steps of molybdopterin biosynthesis in Escherichia coli through the use of in vivo labeling studies. *The Journal of Biological Chemistry*, *270*(3), 1082–1087.

Yang, L., Lin, G., Nelson, R. S., Jian, Y., Telser, J., & Li, L. (2012). Mechanistic studies of the spore photoproduct lyase via a single cysteine mutation. *Biochemistry*, *51*(36), 7173–7188. https://doi.org/10.1021/bi3010945.

Yoshida, H., Yamada, M., Kuramitsu, S., & Kamitori, S. (2008). Structure of a putative molybdenum-cofactor biosynthesis protein C (MoaC) from Sulfolobus tokodaii (ST0472). *Acta Crystallographica. Section F: Structural Biology and Crystallization Communications*, *64*(Pt 7), 589–592. https://doi.org/10.1107/S174430910801590X.

Zhu, W., Winter, M. G., Byndloss, M. X., Spiga, L., Duerkop, B. A., Hughes, E. R., et al. (2018). Precision editing of the gut microbiota ameliorates colitis. *Nature*, *553*(7687), 208–211. https://doi.org/10.1038/nature25172.

Kolibri 360 GmbH
Ferdinand-Jühlke-Straße 7
99095 Erfurt, Deutschland
produktsicherheit@kolibri360.de